ADVANCES IN PHOTODIODES

Edited by **Gian-Franco Dalla Betta**

Advances in Photodiodes
Edited by Gian-Franco Dalla Betta

Published by InTech

Janeza Trdine 9, 51000 Rijeka, Croatia

Publishing Process Manager Katarina Lovrecic

Technical Editor Teodora Smiljanic

Cover Designer InTech Design Team

Additional hard copies can be obtained from orders@intechopen.com

Advances in Photodiodes, Edited by Gian-Franco Dalla Betta
p. cm.
ISBN 978-953-307-163-3

For EU product safety concerns:
IN TECH d.o.o., Prolaz Marije Krucifikse Kozulić 3, 51000 Rijeka, Croatia,
info@intechopen.com or visit our website at intechopen.com.

Meet the editor

Gian-Franco Dalla Betta received his M.S. degree in Electronics Engineering from the University of Bologna, Italy, in 1992, and the Ph.D. degree in Microelectronics from the University of Trento, Italy, in 1997. Since 1997 to 2002, he was with the Centre for Scientific and Technological Research (ITC-irst) of Trento, Italy. Since 2002, he has been an Associate Professor of Electronics at the University of Trento, Italy. Since 1994, he has been mainly involved in the development of silicon radiation detectors and optical sensors for high-energy physics experiments and medical imaging applications. On these subjects, he has coauthored more than 230 papers published in international journals and conference proceedings.

Prof. Dalla Betta is a Senior Member of the IEEE. In 2004, he has received a "Certificate for outstanding contributions in the field of nuclear radiation measurements" from the Radiation Instrumentation Steering Committee of the IEEE Nuclear & Plasma Science Society. Since 2008, he has served as an Associate Editor of the IEEE Transactions on Nuclear Science.

Contents

Preface XI

Part 1 Theoretical Aspects and Simulations 1

Chapter 1 **Spectral Properties of Semiconductor Photodiodes 3**
Terubumi Saito

Chapter 2 **Noise in Electronic and Photonic Devices 25**
K. K. Ghosh

Chapter 3 **Design of Thin-Film Lateral SOI PIN Photodiodes with up to Tens of GHz Bandwidth 43**
Aryan Afzalian and Denis Flandre

Chapter 4 **Modeling and Optimization of Three-Dimensional Interdigitated Lateral p-i-n Photodiodes Based on $In_{0.53}Ga_{0.47}As$ Absorbers for Optical Communications 69**
P Susthitha Menon, Abang Annuar Ehsan and Sahbudin Shaari

Chapter 5 **Simulation of Small-pitch High-density Photovoltaic Infrared Focal Plane Arrays 95**
Mikhail Nikitin, Albina Drugova, Viacheslav Kholodnov and Galina Chekanova

Part 2 Silicon Devices 121

Chapter 6 **Methodology for Design, Measurements and Characterization of Optical Devices on Integrated Circuits 123**
G. Castillo-Cabrera, J. García-Lamont and M. A. Reyes-Barranca

Chapter 7 **Performance Improvement of CMOS APS Pixels using Photodiode Peripheral Utilization Method 143**
Suat U. Ay

Chapter 8 **Color-Selective CMOS Photodiodes Based on Junction Structures and Process Recipes 159**
Oscal T.-C. Chen and Wei-Jean Liu

Chapter 9 **Extrinsic Evolution of the Stacked Gradient Poly-Homojunction Photodiode Genre 181**
Paul V. Jansz and Steven Hinckley

Chapter 10 **Silicon Photodiodes for Low Penetration Depth Beams such as DUV/VUV/EUV Light and Low-Energy Electrons 205**
Lis K. Nanver

Chapter 11 **Avalanche Photodiodes in Submicron CMOS Technologies for High-Sensitivity Imaging 225**
Gian-Franco Dalla Betta, Lucio Pancheri, David Stoppa, Robert Henderson and Justin Richardson

Chapter 12 **The Use of Avalanche Photodiodes in High Energy Electromagnetic Calorimetry 249**
Paola La Rocca and Francesco Riggi

Chapter 13 **Low-Energy Photon Detection with PWO-II Scintillators and Avalanche Photodiodes in Application to High-Energy Gamma-Ray Calorimetry 275**
Dmytro Melnychuk and Boguslaw Zwieglinski

Part 3 Emerging Technologies 289

Chapter 14 **High-Power RF Uni-Traveling-Carrier Photodiodes (UTC-PDs) and Their Applications 291**
Tadao Nagatsuma and Hiroshi Ito

Chapter 15 **n-Type β-$FeSi_2$/p-type Si Near-infrared Photodiodes Prepared by Facing-targets Direct-current Sputtering 315**
Mahmoud Shaban and Tsuyoshi Yoshitake

Chapter 16 **GaN-based Photodiodes on Silicon Substrates 331**
L.S. Chuah and Z. Hassan

Chapter 17 **Gas Source MBE Grown Wavelength Extending InGaAs Photodetectors 349**
Yong-gang Zhang and Yi Gu

Chapter 18 **Use of a-SiC:H Photodiodes in Optical Communications Applications 377**
P. Louro, M. Vieira, M. A. Vieira, M. Fernandes and J. Costa

Chapter 19 **Three Transducers Embedded into One Single SiC Photodetector: LSP Direct Image Sensor, Optical Amplifier and Demux Device 403**
M. Vieira, P. Louro, M. Fernandes, M. A. Vieira, A. Fantoni and J. Costa

Chapter 20 **InAs Infrared Photodiodes 427**
Volodymyr Tetyorkin, Andriy Sukach and Andriy Tkachuk

Chapter 21 **The InAs Electron Avalanche Photodiode 447**
Andrew R. J. Marshall

Preface

Photodiodes are the simplest but most versatile semiconductor optoelectronic devices. They can be used for direct detection of light in the ultraviolet, visible and infrared spectral regions, and of soft X rays and charged particles. When coupled with scintillators or other converting materials, they are also suitable for the detection of gamma rays and neutrons. Owing to some interesting features they can offer such as small size, ruggedness, stability, linearity, speed, low noise, etc., they are appealing to a large variety of applications, spanning from vision systems to optical interconnects, from optical storage systems to photometry and particle physics to medical imaging, etc.

The book *Advances in Photodiodes* addresses the state-of-the-art, latest developments and new trends in the field, covering theoretical aspects, design and simulation issues, processing techniques, experimental results, and applications. The book is divided into three parts.

Part 1 includes five chapters dealing with theoretical aspects, device modeling and simulations. Basic concepts, advanced models useful to describe the device operation and to predict the performance, and novel design methodologies are comprehensively reviewed. *Part 2* collects eight chapters describing recent developments in silicon photodiodes, including both CMOS-compatible and full custom devices. Design and processing issues aimed at enhancing CMOS active pixel performance for special imaging applications are reported; a new technology for very shallow junction photodiodes and use of avalanche photodiodes in calorimetry applications are also reviewed. *Part 3* includes nine chapters relevant to new developments involving technologies based on materials other than silicon (e.g., GaN, InAs, InGaAs, SiC, etc.), aimed at improved performance and extended wavelength detectivity into the ultraviolet, infrared, terahertz, and millimetric waves spectral regions.

Written by internationally renowned experts from 17 countries, with contributions from universities, research institutes and industries, the book *Advances in Photodiodes* is a valuable reference tool for students, scientists, engineers, and researchers working in such different fields as optoelectronic devices, electronic engineering, telecommunications, particle physics and medical imaging, to cite but a few.

I would like to thank all the authors for presenting their work in this book. I am also grateful to the editorial staff and the reviewers for their efforts to ensure both high quality of the book and keeping up with tight schedule for the publication. I am sure the readers will appreciate this book and find it useful.

Prof. Gian-Franco Dalla Betta
University of Trento,
Italy

Part 1

Theoretical Aspects and Simulations

1

Spectral Properties of Semiconductor Photodiodes

Terubumi Saito
*National Metrology Institute of Japan,
National Institute of Advanced Industrial Science and Technology
1-1-1, Umezono, Tsukuba-shi, Ibaraki 305-8563,
Japan*

1. Introduction

Needs for quantitative optical measurements are expanding in various applications where measurement conditions are very different. For precise measurements, uncertainties caused by difference in measurement conditions should be taken into consideration. Measurement conditions for the use of photodiodes include what kind of source is used like whether it is monochromatic or continuum spectrum, collimated or divergent, polarized or unpolarized, what the beam geometry is like whether it is oblique incident or normal incident, underfilled or overfilled, what power level the detector receives and so on.

Since photodiodes are optoelectronic devices, both optical and electronic properties are important. Contrary to electronic properties of photodiodes, optical properties, especially spectral properties like polarization dependence and beam divergence dependence have seldom been reported except from the author's group (Saito, T. et al., 1989; Saito, T. et al., 1990; Saito, T. et al., 1995; Saito, T. et al., 1996a; Saito, T. et al., 1996b; Saito, T. et al., 2000). Most photodiodes can be optically modelled by a simple layered structure consisting of a sensing semiconductor substrate covered by a thin surface layer (Saito et al., 1990). For instance, a p-n junction silicon photodiode consists of a silicon dioxide film on silicon substrate and a GaAsP Schottky photodiode consists of a gold film on GaAsP substrate. Even with a single layer, optical properties of the whole system can be very different from those for a substrate without surface layer due to the interference effect and absorption by the surface layer. To understand spectral properties like spectral responsivity and polarization responsivity dependence on angle of incidence, rigorous calculation based on Fresnel equations using complex refractive indices of the composing materials as a function of wavelength is necessary.

When the incident photon beam is parallel and there is no anisotropy in the sensing surface, there is no need to consider on polarization characteristics of photodiodes. However, when incident beam has a divergence, one has to take polarization properties into account since there are components that hit detector surface at oblique incidence (Saito et al., 1996a). To measure divergent beam power precisely, detectors ideally should have cosine response. Deviation from the cosine response also can be obtained from the theoretical model (Saito et al., 2010).

Historically, while photodiodes were started to be designed and manufactured mostly for the use in the visible and infrared, so-called semiconductor detectors like Si(Li) or pure-Ge detectors were developed independently to detect ionizing radiation like γ-rays. In these days, some photodiodes can also be used in a part of the ionizing radiation region (Korde, R. et al., 1993) by overcoming the most difficult spectral region, UV and VUV where all materials exhibit the strongest absorption. In the low photon energy region near the semiconductor bandgap, intrinsic internal quantum efficiency is expected to be unity. On the other hand, in the much higher photon energy range like in the γ-ray region, intrinsic internal quantum efficiency becomes proportional to the photon energy due to impact ionization. By combining the spectral optical properties and the intrinsic internal quantum efficiency behaviour, one can estimate absolute external quantum efficiency at any photon energy when there is no carrier recombination. Probability of surface recombination is typically dominant and becomes high when absorption in the substrate becomes strong, that is, in the UV and VUV regions.
In this chapter, after introduction and explanations for fundamentals, the above-mentioned calculation model for spectral quantum efficiency is described. Experimental results on spectral responsivity, linearity, spatial uniformity, angular dependence, divergence dependence, photoemission contribution follows to understand the spectral properties of photodiodes.

2. Basis on photodiodes

Fundamental information about photodiodes on the structure, principle, characteristics etc. can be found, for instance, in (Sze, S.M., 1981).

2.1 Terms & units

Definitions of technical terms and quantities used in this paper basically follow the CIE vocabularies (CIE, 1987). Photodetectors are devices to measure so-called intensity of the incident radiation. There are two ways to express radiation intensity; one is photon flux, Φ, defined by number of incident photons per unit time, and the other is radiant power, P, defined by radiant energy of the incident radiation per unit time. The two quantities are connected by the following equation where h is Plank constant, λ the wavelength in vacuum, and c the light velocity in vacuum.

$$\Phi = \frac{\lambda P}{hc} \tag{2.1}$$

Sensitivity, the output divided by the input, of photodetectors is also expressed in two ways corresponding to the two expressions for the input. One is quantum efficiency, η, defined by the number of photo-generated carrier pairs divided by the number of photons, and the other is responsivity defined by the photodetector output divided by the radiant power. In the case where photodetector is irradiated by monochromatic radiation, an adjective, *spectral*, which means a function of wavelength and not a spectrally integrated quantity, is added in front of each term (quantum efficiency or responsivity). When the photodetecotor is irradiated by monochromatic radiation and the photodetector output is expressed by photocurrent, spectral quantum efficiency, η, and spectral responsivity, s, in A/W are related by the following equation where e is the electronic charge in C, E the photon energy in eV, λ the wavelength in nm,

$$s = \frac{e\lambda\eta}{hc} = \frac{\eta}{E} \approx \frac{\lambda\eta}{1240} \tag{2.2}$$

It should be noted that for non-monochromatic radiation input, conversion between quantum efficiency and responsivity is impossible without the knowledge on the spectral distribution of the input radiation.

For both quantities of quantum efficiency and responsivity, further two distinct definitions exist corresponding to the two definitions for the input. One is the case when the input radiation is defined by the one *incident* to the detector and the other is the case when the input radiation is defined by the one *absorbed* in the detector. To distinguish the two cases, term, *external* (sometimes omitted) and *internal* is further added in front of each term for the former and the latter, respectively. For instance, internal spectral responsivity means photocurrent generated by the detector divided by the radiant power absorbed by the detector. When we define more specifically that internal spectral responsivity is photocurrent divided by the radiant power absorbed in the *sensitive* volume, internal spectral responsivity, s_{int}, and external spectral responsivity, s_{ext}, are connected by the following equation when reflectance of the system is R, absorptance of the surface layer A, transmittance of the surface layer (into the sensitive substrate) T.

$$\begin{aligned} s_{ext} &= (1-R-A)s_{\text{int}} \\ &= Ts_{\text{int}} \end{aligned} \tag{2.3}$$

Similarly, internal spectral quantum efficiency, η_{int}, and external spectral quantum efficiency, η_{ext}, are connected by the following equation.

$$\begin{aligned} \eta_{ext} &= (1-R-A)\eta_{\text{int}} \\ &= T\eta_{\text{int}} \end{aligned} \tag{2.4}$$

2.2 Principle & structure

A photodiode is a photodetector which has one of the structures among p-n, p-i-n, or Schottky junction where photo-generated carriers are swept by the built-in electric field. For instance, a p-on-n type silicon photodiode is constructed by doping p-type impurity to an n-type silicon substrate so that the p-type dopant density is larger than the n-type dopant density. For the purposes of anti-reflection and of passivation, silicon surface is typically thermally oxidized to form a silicon dioxide layer. Once the p-n junction is formed, each type of free carriers (holes in the p-type and electrons in the n-type) starts diffusing to its lower density side. As a result, ionized acceptors and donors generate strong built-in electric field at the junction interface. Since the built-in electric filed generates forces for holes and electrons to drift in the reverse direction to the direction due to the diffusion, the electric potential is determined so that no current flows across the junction in the dark and in the thermal equilibrium. The region where the built-in electric field is formed is called depletion region (also called space charge region). The regions before and after the depletion region where there is no electric field are called neutral region.

When the photodiode is irradiated by photons, photons are transmitted through the oxide layer, reach the silicon substrate and exponentially decay in intensity in a rate determined by the wavelength while producing electron-hole pairs. Carriers photo-generated in the depletion region are swept by the built-in field and flow as a drift current.

2.3 I-V characteristics

Current-voltage characteristics of a photodiode is given by

$$I = I_L - I_s\left[\exp\left(\frac{eV}{nkT}\right) - 1\right] \tag{2.5}$$

where, I is the current that flows in an external circuit, I_L the light-generated current, V the forward voltage across the diode, I_s the saturation current, n the ideality factor, k, Boltzman constant, and T the junction temperature.

Curve A in Fig. 1 is such a I-V characteristic under a certain irradiated condition. When the radiant power incident to the photodiode is increased, the curve moves outward as shown by curve B. When one sees short-circuit current, the current output is increased as a linear function of the radiant power (operating point moves along the vertical axis). On the other hand, if one sees open-circuit voltage, the voltage output is increased as a logarithmic function of the radiant power (operating point moves along the horizontal axis).

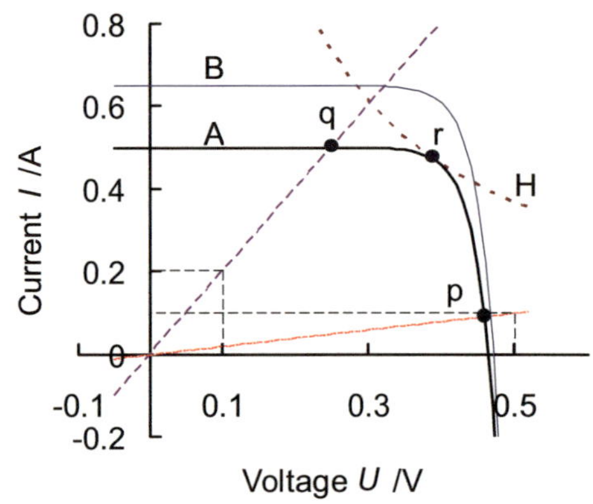

Fig. 1. Current-voltage characteristics of a virtual photodiode to illustrate its measuring conditions. See text for details.

Actual operating condition always lies between these two extremes. For instance, if the load resistance is 5 Ω in this example, the operating point is marked by point, p and the operating line is shown by the red dotted line, which exhibits non-linear response to the input radiant power. If the load resistance is changed to 0.5 Ω, the operating point and operating line become to point, q, and the purple dashed line, respectively, which results in relatively linear response below the radiant power level approximately corresponding to curve B. Therefore, it is important to have low input impedance of the current measurement circuit compared to the photodiode shunt resistance for better linear response. On the other hand, for the purpose of power generation like solar cells, it is important to match the appropriate load resistance to obtain maximum power, whose operating point is shown by point, r, which is tangent point to a hyperbolic curve, a locus to give a constant power.

3. Quantum efficiency calculation model

In this section, theoretical models to predict and affect spectral quantum efficiency (Hovel, H.J., 1975; Saito, T. et al., 1990) and considerations necessary for precise measurements (Saito, T. et al., 2000; Saito, T., 2003) are discussed.

3.1 Optical structure and model

It is known that most photodetectors like p-n junction photodiode and Schottky photodiodes are optically well-modeled as a sensitive substrate covered by a surface layer as shown in Fig. 2. Fig. 2 also illustrates beam paths when a photon beam enters a detector surface obliquely. Some of the incident photons are reflected at the detector surface. Some other photons can be lost due to absorption in a dead layer which is sometimes present in front of the photon-sensitive volume. Since reflectance for p-polarization is usually different from reflectance for s-polarization, transmittance through the surface and the dead layer into the photon-sensitive volume differs for s- and p-polarized radiation. Therefore, detectors placed obliquely to the incident radiation are usually considered to be polarization-sensitive.

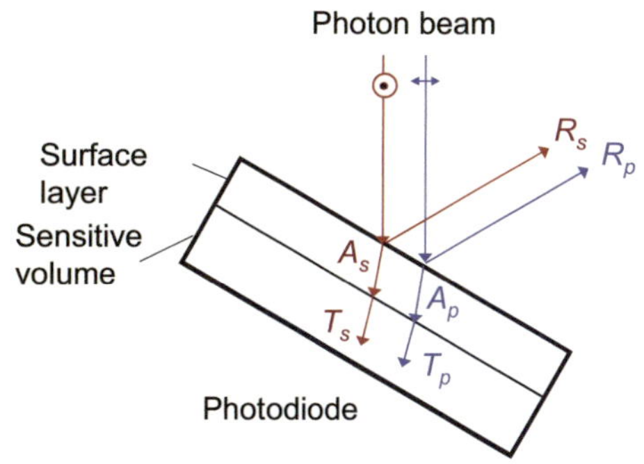

Fig. 2. Optical model for semiconductor photodiodes and possible paths of photon beams. *R*: reflectance, *A*: absorptance in the surface layer, *T*: transmittance through the surface layer to the sensitive volume. Subscript s&p: polarization components s&p.

We have already seen that external quantum efficiency is related to internal quantum efficiency by Eq. (2.4). To distinguish intrinsic and extrinsic property of internal quantum efficiency, we modify Eq. (2.4) by introducing carrier collection efficiency, *C*, and intrinsic internal spectral quantum efficiency, η'_{int}, as follows.

$$\eta_{ext} = CT\eta'_{int} \tag{3.1}$$

As we will see in the following sections, *C* and *T* can be calculated as a function of wavelength. For T, angular and polarization dependence can also be calculated. Therefore, η_{ext} can be estimated by assuming simplified η'_{int} dependence or by using independent experimental results of η'_{int}.

3.2 Intrinsic quantum efficiency

The ideal internal quantum efficiency is unity until the photon energy becomes, at least, two times the band-gap of the semiconductor used, and it begins to exceed 1 because of the impact ionization (Alig, R.C. et al., 1980). Geist et al. proved in their work on self-calibration that the spectral internal quantum efficiency is very close to unity in the wavelength range approximately from 400 nm to 800 nm (Geist, J. et al., 1979). Although the actual spectral internal quantum efficiency is, in reality, is lower than unity mainly due to surface recombination, the loss can be estimated by a simple saturation measurement by applying a retarding potential using liquid electrode to prevent minority carriers from diffusing to the interface between the silicon layer and its oxide layer.

On the other hand, in the higher photon energy region such as γ-ray region where quantum efficiency exceeds unity due to impact ionization, it is known that average energy, ε, required to create an electron-hole pair becomes almost constant to the photon energy, E_p (Ryan, R.D., 1973). In other words, the internal spectral responsivity becomes constant or the internal spectral quantum efficiency becomes proportional to the photon energy.
As a rough approximation, behaviors of internal spectral responsivity and the internal spectral quantum efficiency in the entire photon energy range are given by the following equations and are illustrated in Fig. 3 as a function of wavelength or photon energy.

$$\eta'_{\text{int}} = \begin{cases} 1 & (E \leq \varepsilon) \\ E/\varepsilon & (E \geq \varepsilon) \end{cases} \tag{3.2}$$

$$s'_{\text{int}} = \begin{cases} 1/E & (E \leq \varepsilon) \\ 1/\varepsilon & (E \geq \varepsilon) \end{cases} \tag{3.3}$$

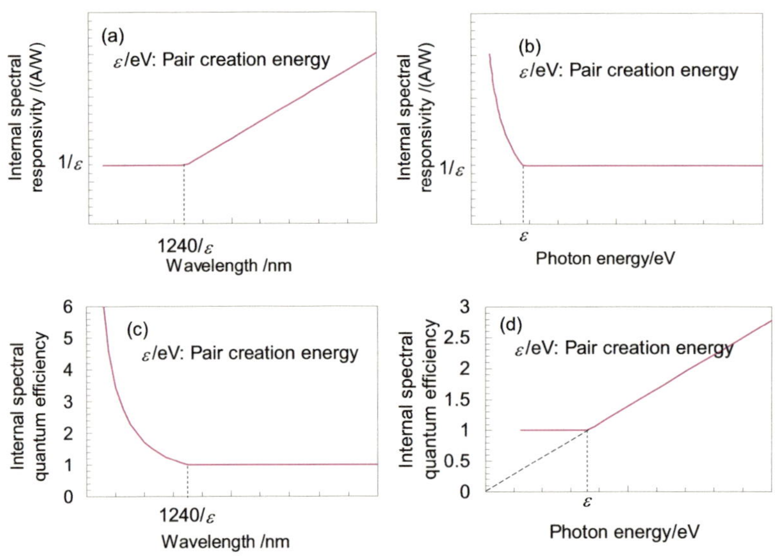

Fig. 3. Simplified spectral dependence of intrinsic internal spectral responsivity, s'_{int}, and quantum efficiency, η'_{int} expressed in all possible combinations as a function of wavelength, λ, and photon energy, E. (a): $s'_{\text{int}}(\lambda)$, (b): $s'_{\text{int}}(E)$, (c): $\eta'_{\text{int}}(\lambda)$, and (d): $\eta'_{\text{int}}(E)$.

3.3 Optical losses

The optical losses are classified, as illustrated in Fig. 2, into the loss of photons due to reflection from the surface, due to absorption in a dead layer in front of the sensitive region, and due to transmission through the sensitive region. The last case only occurs when the photon absorption coefficient is small and therefore it is negligible in the UV or VUV region because of the strong absorption. Among the optical losses, the reflection loss can be determined also by a simple reflectance measurement. However, the absorption loss cannot be determined by experiment. If the optical constants of the composing materials and the geometry are known, the optical losses can be evaluated by calculation based on the optical model.

Consider that a photodiode is placed in vacuum ($\tilde{n}_0 = 1$) and is composed of a slab of semiconductor ($\tilde{n}_2 = n_2 - ik_2$, where n_2 is a real part and k_2 is an imaginary part of the optical constant) whose thickness is large compared to the absorption length of photons considered, and a film ($\tilde{n}_1 = n_1 - ik_1$) with thickness, *d*, on the slab. When the angle of incidence on the photodiode is ϕ_0, transmittance, *T*, reflectance, *R*, and absorptance, *A*, of the film are given by the following equations:

$$T = ct^*\tilde{t}, \tag{3.4}$$

$$R = \tilde{r}^*\tilde{r}, \tag{3.5}$$

$$A = 1 - R - T, \tag{3.6}$$

$$\tilde{t} = \frac{\tilde{t}_1\tilde{t}_2 \exp(-i\tilde{\delta}_1 / 2)}{1 + \tilde{r}_1\tilde{r}_2 \exp(-i\tilde{\delta}_1)}, \tag{3.7}$$

$$\tilde{r} = \frac{\tilde{r}_1 + \tilde{r}_2 \exp(-i\tilde{\delta}_1)}{1 + \tilde{r}_1\tilde{r}_2 \exp(-i\tilde{\delta}_1)}, \tag{3.8}$$

$$\tilde{\delta}_1 = 4\pi\tilde{n}_1 d_1 \cos\tilde{\phi}_1 / \lambda, \tag{3.9}$$

where $\tilde{r}_m$ and $\tilde{t}_m$ (m=1, 2) are Fresnel's coefficients defined for s- and p-polarization as

$$\tilde{r}_{m,s} = \frac{\tilde{n}_{m-1} \cos\tilde{\phi}_{m-1} - \tilde{n}_m \cos\tilde{\phi}_m}{\tilde{n}_{m-1} \cos\tilde{\phi}_{m-1} + \tilde{n}_m \cos\tilde{\phi}}, \tag{3.10}$$

$$\tilde{r}_{m,p} = \frac{\tilde{n}_m \cos\tilde{\phi}_{m-1} - \tilde{n}_{m-1} \cos\tilde{\phi}_m}{\tilde{n}_m \cos\tilde{\phi}_{m-1} + \tilde{n}_{m-1} \cos\tilde{\phi}}, \tag{3.11}$$

$$\tilde{t}_{m,s} = \frac{2\tilde{n}_{m-1} \cos\tilde{\phi}_{m-1}}{\tilde{n}_{m-1} \cos\tilde{\phi}_{m-1} + \tilde{n}_m \cos\tilde{\phi}}, \tag{3.12}$$

$$\tilde{t}_{m,p} = \frac{2\tilde{n}_{m-1} \cos\tilde{\phi}_{m-1}}{\tilde{n}_m \cos\tilde{\phi}_{m-1} + \tilde{n}_{m-1} \cos\tilde{\phi}}, \tag{3.13}$$

Refraction angles are given by the following Snell's formula,

$$\tilde{n}_{m-1} \sin\tilde{\phi}_{m-1} = \tilde{n}_m \sin\tilde{\phi}_m. \tag{3.14}$$

The coefficient c in Eq. (3.4) is given separately for s- and p-component by

$$c = \begin{cases} \dfrac{\mathrm{Re}(\tilde{n}_2 \cos\tilde{\phi}_2)}{\mathrm{Re}(\tilde{n}_0 \cos\tilde{\phi}_0)} & \text{for s-componet} \\[2ex] \dfrac{\mathrm{Re}(\tilde{n}_2^* \cos\tilde{\phi}_2)}{\mathrm{Re}(\tilde{n}_0^* \cos\tilde{\phi}_0)} & \text{for p-componet} \end{cases} \tag{3.15}$$

An example of the calculation results for normal transmittance, T, absorptance, A, and reflectance, R, of a Si photodiode which has a 30 nm-thick SiO_2 on Si, is shown in Fig. 4.

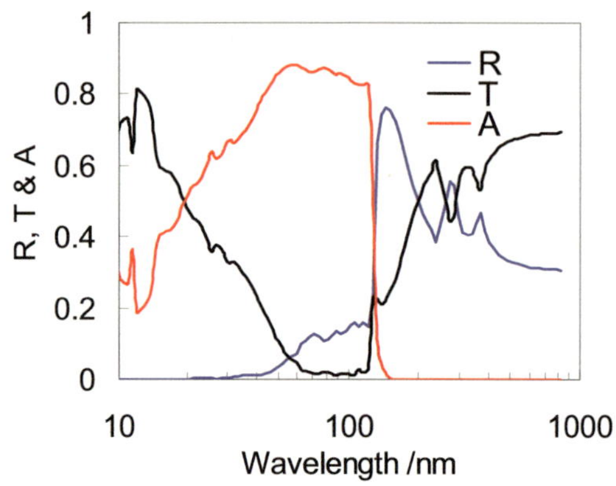

Fig. 4. Calculated spectra of transmittance (T), reflectance (R) and absorptance (A) for 30 nm-thick SiO_2 film on Si.

The detector is almost insensitive in the range from 60 to 120 nm due to the absorption by the SiO_2 layer. The major loss mechanism of photons is absorption below about 120 nm, and reflection above 120 nm. In the longer wavelength region, a change in thickness of the SiO_2, layer greatly alters the shape of the transmittance and reflectance curves due to the interference effect. For other examples including angular dependence and comparisons with experiments, see Section 4.

3.4 Carrier recombination loss

Hovel reported on carrier transport model in solar cells as a function of absorption coefficient of the incident radiation (Hovel, H.J., 1975). Carrier collection efficiency for photodiodes is given as a function of wavelength via absorption coefficients based on his model. Suppose that an n-on-p photodiode is irradiated by monochromatic radiation of wavelength λ. Carrier collection efficiency, $C(\lambda)$, which is defined by collected number of carriers divided by number of photo-generated carriers, is given by

$$C(\lambda) = C_p(\lambda) + C_{dr}(\lambda) + C_n(\lambda) \tag{3.16}$$

where $C_p(\lambda)$, $C_{dr}(\lambda)$, and $C_n(\lambda)$ are carrier collection efficiencies contributed from the front region before the depletion region by hole current, from the depletion region, and from the rear region after the depletion region by electron current, respectively. Each contribution is given by the following equations.

$$C_p(\lambda) = \frac{\alpha L_p}{\alpha^2 L_p^2 - 1}\left[\frac{\dfrac{S_p L_p}{D_p} + \alpha L_p - \exp(-\alpha x_j)\left(\dfrac{S_p L_p}{D_p}\cosh\dfrac{x_j}{L_p} + \sinh\dfrac{x_j}{L_p}\right)}{\dfrac{S_p L_p}{D_p}\sinh\dfrac{x_j}{L_p} + \cosh\dfrac{x_j}{L_p}} - \alpha L_p \exp(-\alpha x_j)\right] \tag{3.17}$$

$$C_{dr}(\lambda) = \exp(-\alpha x_j)\left[1 - \exp(-\alpha x_j)\right] \tag{3.18}$$

$$C_n(\lambda) = \frac{\alpha L_n}{\alpha^2 L_n^2 - 1}\exp\left[-\alpha(x_j + W)\right]\left[\alpha L_n - \frac{\frac{S_n L_n}{D_n}\left(\cosh\frac{H'}{L_n} - \exp(-\alpha H')\right) + \sinh\frac{H'}{L_n} + \alpha L_n \exp(-\alpha H')}{\frac{S_n L_n}{D_n}\sinh\frac{H'}{L_n} + \cosh\frac{H'}{L_n}}\right] \tag{3.19}$$

Here, notations are as follows. α: absorption coefficient of the substrate, x_j: junction depth from the substrate surface, L_p: hole diffusion length in the front region before the depletion region, D_p: hole diffusion coefficient in the front region before the depletion region, W: width of depletion region, H': width of the p-base neutral region after the depletion region. Calculation results for a p-on-n silicon photodiode are shown in Fig. 5. Parameters used are as follows; x_j=200 nm, L_n= 20 μm, D_n=2.6 cm²/s, W=9.6 μm, L_p =150 μm, D_p=12 cm²/s, S_p=10⁵ cm/s, and H'=300 μm.

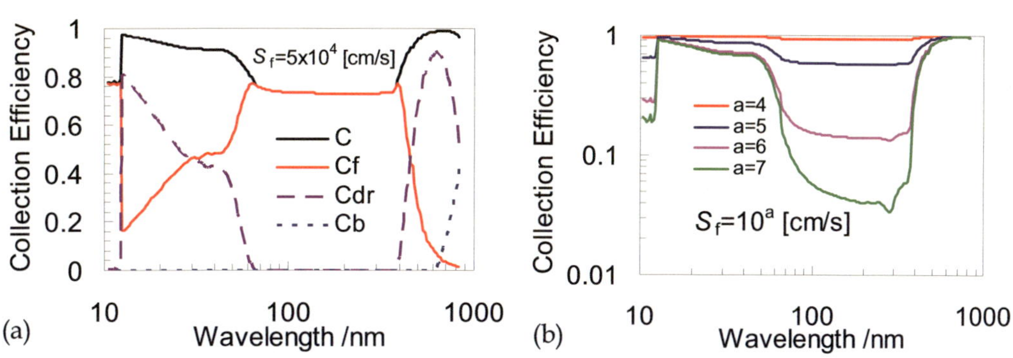

Fig. 5. Calculated carrier collection efficiency spectra for a silicon photodiode. (a): Total (C) and each contribution from the front region (C_f), deletion region (C_{dr}), back region (C_b). (b): As a function of surface recombination velocity.

As explained before, spectral dependence is brought only by the change in absorption coefficient of the semiconductor as a function of wavelength. Corresponding to the strong absorption about from 60 nm to 400 nm, collection efficiency is steeply dropped. It is known that decrease in collection efficiency becomes nearly saturated after reaching a certain level of absorption. It is clear that contribution from the front region is dominant in most of the spectral range, especially in the region mentioned above. For such a situation, one of the most important parameters to govern the efficiency is the surface recombination velocity. As Fig. 5 (b) shows, contrary to the large difference in efficiency in the UV, change in the surface recombination velocity affects little the efficiency in the visible.

3.5 Fluorescence-, Photoemission-losses etc.

There are some other possible factors that are not included in the above-mentioned theoretical model. One of the factors is an energy loss by fluorescence from the composing materials of a photodiode, which is likely to happen in the UV and VUV. Compared to the case without fluorescence, absorbed energy is decreased by emitting fluorescence and therefore the detector photocurrent may be lower than expected in a certain condition. However, if the detector is more sensitive to the longer wavelength, it is also possible that the detector photocurrent is larger than expected by receiving the fluorescence. For instance, if the covering glass emits fluorescence, the detector even becomes sensitive to the shorter wavelength radiation than the glass cut-on wavelength, where glass transmittance is zero.

Another important factor also typical in the UV&VUV is photoemission contribution (Saito, T., 2003; Saito, T. et al., 2005a). The situation is similar to the fluorescence since both cases are possible, increase and decrease in detector response depending on the measurement conditions. In a spectral region when a photon is able to cause electron photoemission, one should note the photoemission current contribution and a difference in the photodiode photocurrent depending on the polarity of the current measurement.

Fig. 6 illustrates typical photodiode measurement setups in the VUV region. Photoelectrons emitted from the front surface of the photodiode form a photoemission current circuit (denoted by i_e) in addition to the generally intended signal of the internal photocurrent, i_i, generated by the photodiode. In configuration (a), an electrometer is inserted between the ground and the front electrode of a photodiode. When the switch, SW, is kept open, the current measured by the electrometer, i, becomes $i = i_e$.

Since photoelectrons usually have non-zero kinetic energies, i_e is observable even when $U = 0$. When SW is closed the current measured by the electrometer, i_f, becomes $i_f = i_i + i_e$.

We should note that the measured current includes the photoemission current in this configuration. When the front of the photodiode is p-type, both i_e and i_i have the same sign (positive) and therefore the sum is additive. On the other hand, when the front is n-type, which happens for an n-on-p type photodiode, note that the sum of i_i and i_e becomes subtractive ($i_i < 0$ while $i_e > 0$).

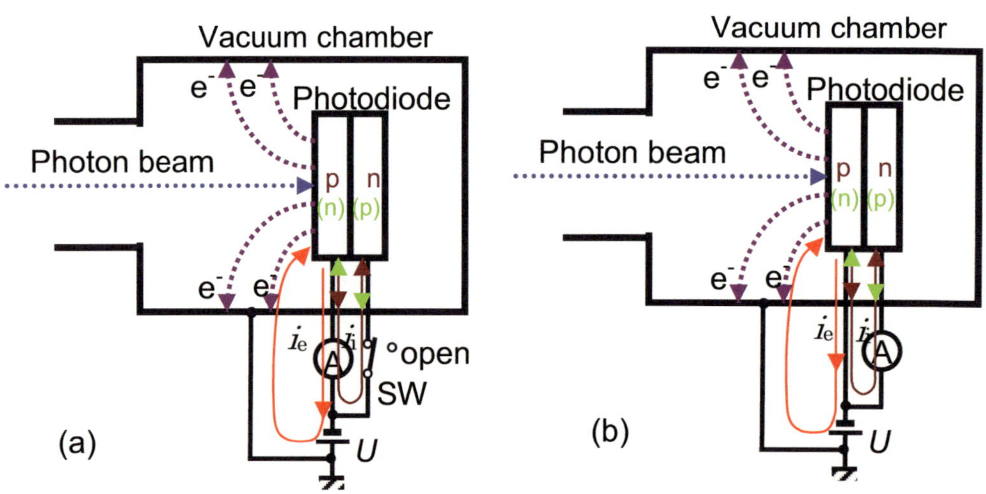

Fig. 6. Measurement circuits to illustrate difference in photocurrent due to photoemission depending on the location of the electrometer (A). (a) Rear grounding configuration: the electrometer sensing terminal is connected to the front electrode of the photodiode. For direct measurement of the photoemission current, the switch SW is set to open. (b) Front grounding configuration: the electrometer sensing terminal is connected to the rear electrode of the photodiode.

In configuration (b), an electrometer is inserted between the ground and the rear electrode of a photodiode. The current measured by the electrometer, i_r , becomes $i_r = -i_i$. Note that the photoemission current is not included and only the internal photocurrent generated by the photodiode is measured in this configuration. Also note that $i_r < 0$ for p-on-n type and $i_r > 0$ for n-on-p type.

3.6 External quantum efficiency

If we assume the simplified spectral dependence of the intrinsic quantum efficiency by Eq. (3.2), the external spectral quantum efficiency in the entire spectral range are given by applying Eq. (3.2) to Eq. (3.1) and becomes as follows.

$$\eta_{ext} = \begin{cases} CT & (E \le \varepsilon) \\ CTE / \varepsilon & (E \ge \varepsilon) \end{cases} \tag{3.20}$$

Similarly, external spectral responsivity is given by

$$s_{ext} = \begin{cases} CT / E & (E \le \varepsilon) \\ CT / \varepsilon & (E \ge \varepsilon) \end{cases} \tag{3.21}$$

Careful comparison for silicon photodiodes in the VUV range between the model calculation results with the experimental ones revealed there exists a case that large discrepancy can happen. In the spectral region where there is absorption in the silicon dioxide layer, measured quantum efficiencies of silicon photodiode are usually higher than those predicted by the above optical model. The measured data rather fit well to a calculation in which charge injection from the oxide layer to the silicon substrate is taken into account as shown below:

$$\eta_{ext} = C(T + xA)E / \varepsilon \qquad (E \ge E_a > \varepsilon) \tag{3.22}$$

where E_a is the photon energy where absorption by the oxide starts and x is an arbitrary parameter to represent the degree of charge injection and is typically 0.3, which was reported to be the best value to fit to some experimental data (Canfield, L.R. et al., 1989).

4. Spectral properties of photodiodes

We conducted a number of comparisons between the theoretical model and experiments covering most of the related characteristics like spectral dependence (Saito et al., 1989; 1990) and angular/polarization dependence (Saito et al., 1995; 1996a; 1996b). To check more precisely, we have measured spectral responsivities of silicon photodiodes for both p- and s-polarization components by using a Glan-laser prism before the detector, as a function of the angle of incidence (Saito, T. et al., 2010).

4.1 Spectral responsivity

Fig. 7 shows measured spectral dependence of various kinds of photodiodes expressed in spectral responsivity (a) and in spectral quantum efficiency (b). The detector having the highest quantum efficiency of nearly unity is a silicon trap detector (Ichino, Y. et al., 2008), which consists of three silicon photodiodes to reduce the overall reflection loss. The difference between the two Si photodiodes, A and B, mainly originates from the difference in the oxide thickness; that of A is about 30 nm, which is much thinner than that of B. The PtSi photodiode was developed to realize better stability for the UV use by forming a Schottky barrier contact of PtSi to Si (Solt, K. et al., 1996). As this spectrum shows, special care should be paid to avoid stray light contribution because the detector is much more sensitive to the radiation having longer wavelength than the one in the region of interest (UV).

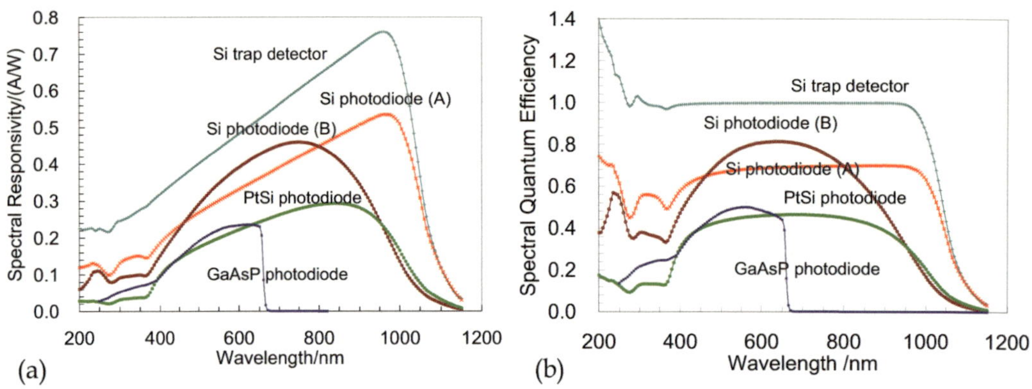

Fig. 7. (a): Measured spectral responsivities of various kinds of detectors. (b): Spectral quantum efficiency representation of the same results of (a).

To minimize the stray light contribution, it is the best to use a detector having narrower spectrum bandwidth, which can be realized by using wider bandgap semiconductors such as a GaAsP photodiode as shown here. A number of developments for better stability and solar-blindness have been reported by using various kind of wide bandgap materials like diamond (Saito, T. et al., 2005a; Saito, T. et al., 2005b; Saito, T. et al., 2006), AlN, AlGaN, InGaN etc. (Saito, T. et al., 2009a; Saito, T. et al., 2009b).

To check the validity of the calculation model described in Chapter 3, a number of comparisons have been conducted in the wide wavelength range from 10 nm to 1000 nm. One of the results for a silicon photodiode (Hamamatsu S1337 windowless type) is shown in Fig. 8 (For the intrinsic internal quantum efficiency of silicon, a separate experimental data was used below 360 nm instead of assuming the simplified spectral dependence of Eq. (3.2).). Excellent agreement was obtained especially in the visible; the calculated results agree within 0.04 % with the experiments at laser lines from 458 nm to 633 nm.

The theoretical model was also applied to a Schottky type GaAsP photodiode (Hamamatsu G2119) that consists of a 10 nm-thick Au layer on a $GaAs_{0.6}P_{0.4}$ substrate. Similarly, satisfactory agreements were obtained in the wavelength range from 100 nm to 300 nm (Saito et al., 1989).

4.2 Angular and polarization dependence

In this section, we are showing comparative results between the calculation and experiments. Experiments were carried out by measuring spectral quantum efficiency for a Si photodiode (Hamamatsu S1337-11 windowless type) in underfill condition as a function of the angle of the incidence using p- and s-polarized monochromatized beam.

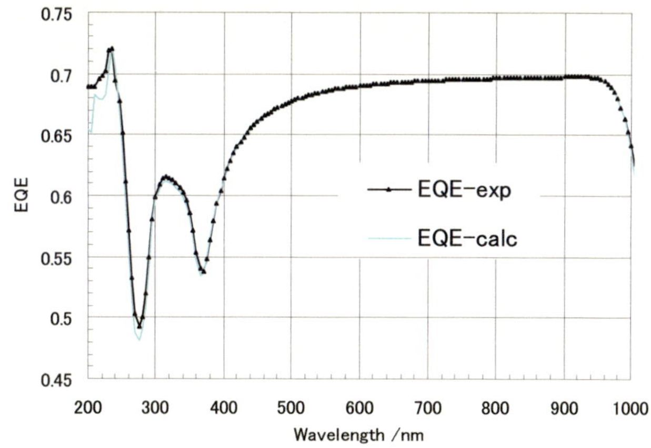

Fig. 8. Comparison between the measured external quantum efficiency of a Si photodiode and the calculated one.

Fig. 9 (a) shows such comparison results at the wavelength of 555 nm for the underfill condition. Note that excellent agreements are obtained without any adjustment in scale (no normalization was applied for the absolute values in the calculation). Slight smaller values for experiments can be explained by carrier recombination (whose correction was not applied in this case).

Since there is no absorption in the oxide layer at this wavelength, it is obvious that the increase in efficiency for p-polarization is caused by the decrease in reflectance for p-polarization and vice versa for s-polarization. If the incident radiation is unpolarized, the increase in responsivity for p-polarization is nearly cancelled out by the decrease for s-polarization except in the vicinity of $\pi/2$. As discussed in section 3.1, the same consequence is valid for any polarization state of the incident radiation if there is an axial symmetry in the angular distribution of the beam.

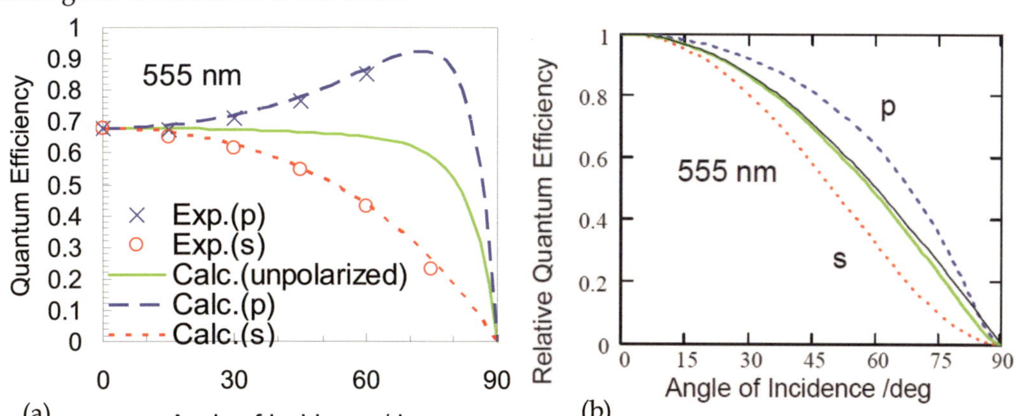

Fig. 9. Calculated quantum efficiencies at the wavelength of 555 nm of Si photodiode with a 30 nm-thick SiO_2 layer for p-, s-polarized and unpolarized radiation. (a): In underfill condition. Measured data are also shown by marks. (b): In overfill condition. Ideal cosine response is also shown by black solid curve.

Results converted from the underfill condition to overfill condition is shown in Fig. 9 (b). As expected, it results in rather good angular response close to the cosine response except in the vicinity of $\pi/2$.

The behavior of angular dependence seen above seems to be similar to those for a well-known single boundary system. However, it does not always hold for this kind of system with more than two boundaries. The situation greatly changes in the UV. Fig. 10 (a) shows the results at the wavelength of 255 nm on the same silicon photodiode as a function of the angle of incidence for the underfill condition (the relative scale for the calculation was adjusted so as to agree to the experimental one to take into account of the carrier recombination and impact ionization). Angular dependence of both for p- and s-polarization becomes almost identical. This never happens for a single boundary system but can happen for such a layered structure with a certain combination of optical indices of a surface layer and a substrate.

Results converted from the underfill condition to overfill condition are shown in Fig. 10. Since no cancellation occurs between the responses for p- and s-polarization, it results in large deviation from the cosine response for overfill condition as shown in Fig. 10 (b).

4.3 Spectral dependence of polarization property

Spectral properties of photodiodes are mainly affected by a change in transmittance to the substrate (sensitive region) through the change in complex refractive indices of the composing materials. In the visible spectral region, refractive indices of both SiO_2 and Si do not change much. In contrast, in the UV region, most materials exhibit steep change in optical indices.

Therefore, we have investigated the spectral behaviour of the polarization sensitivity, which we define as the spectral quantum efficiency at oblique incidence divided by the one at normal incidence. Both the experimental and calculated results for the Si photodiode (S1337-11) at the angles of incidence of 30° and 60° are shown in Fig. 11

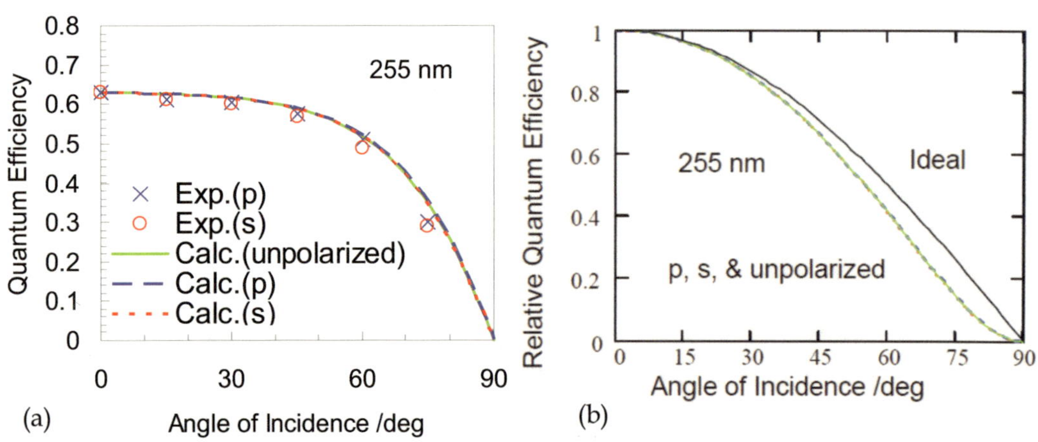

Fig. 10. Calculated quantum efficiency at the wavelength of 255 nm of Si photodiode with a 30 nm-thick SiO_2 layer for p- (in blue), s- (in red) polarized and unpolarized radiation (in green). (a): In underfill condition. Measured data are also shown by marks. (b): In overfill condition. Ideal cosine response is also shown by black solid curve.

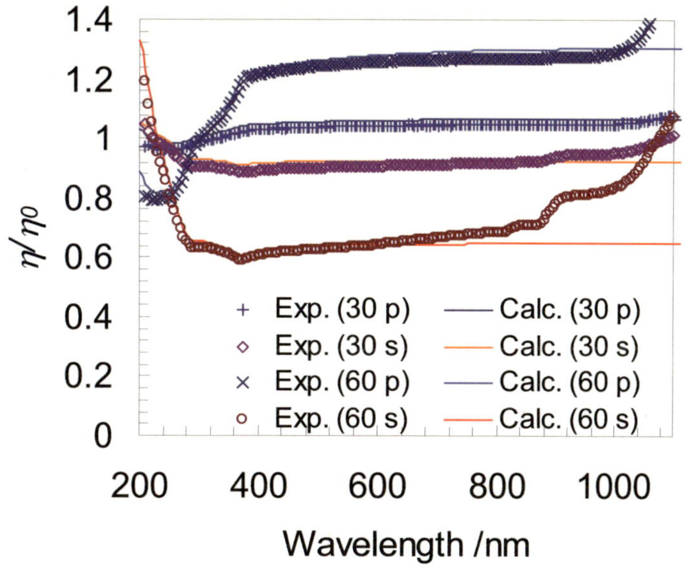

Fig. 11. Spectral dependence of polarization sensitivity at the angles of incidence of 30° and 60°. Experimental results and the calculated ones are shown by marks and lines, respectively for p- and s-polarizations. η is quantum efficiency at oblique incidence and η_0 the one at normal incidence.

Again, good agreements are obtained except for the near IR region. The deviation of the experimental data from the calculation in the near IR is highly likely to be caused by reflection (absorption becomes much weaker as the wavelength increases) by the backside of the Si substrate which was not taken into account in the model.

As expected, while there is no steep change in the visible spectra, the spectra in the UV shows complicated behaviour. At the wavelength of about 200 nm, s-polarization sensitivity is higher than p-polarization sensitivity, which is completely opposite to the behaviour in the visible and this never happens either for a single boundary system but is unique phenomenon for the layered structure.

As already shown in Fig. 10 (a), in-between the two regions, there exists a wavelength, where both polarization sensitivities become almost equal at any angle of incidence.

4.4 Divergence dependence

To experimentally measure photodiode response dependence on incident beam divergence, it is difficult to guarantee if the radiant power of the beam remains constant when the beam divergent is changed by a certain optical setup. Therefore, in this section, results of photodiode response dependence on incident beam divergence are given theoretically. The optical model used has been proved to be reliable enough in many spectral and angular/polarization measurements, as seen in the previous sections.

The calculation of photodiode response for a divergent beam (Saito et al., 2000) is made by integration of the responses for s- and p-polarization over a hemisphere (or over an apex angle of the cone-shaped incident beam) based on the following formula that gives relative response to the one for a collimated beam:

$$f(\theta) = 2\int_0^{\pi/2} [aT_s^r(\theta) + (1-a)T_p^r(\theta)]g(\theta)d\theta \tag{4.1}$$

where T_s^r (T_p^r) is the relative transmittance for s-polarization (p-polarization) normalized by the transmittance at normal incidence, $g(\theta)$ the angular distribution function, θ the angle of the incidence of a ray in the beam, and a the weighting factor for s-polarization component relative to p-component, which is set to 0.5 in this paper because of the following assumption. We assume that the incident beam has an axial symmetry around the principal axis of the beam, which is normal to the detector surface. Eq. (4.1) can also be used for the case when there is oxide charge contribution by using $T_s^r+xA_s^r$ (or $T_p^r+xA_p^r$) instead of T_s^r (T_p^r), where A_s^r (A_p^r) is the relative absorption in the oxide layer for s-polarization (p-polarization).
The calculations are conducted for two types of angular distributions: an isotropically distributed beam within a cone of rays with a certain apex angle and a beam with Gaussian angular distribution that is expressed by the following equation,

$$g(\theta) = \frac{1}{\sqrt{2\pi}\cdot\sigma}\cdot e^{-\frac{\theta^2}{2\sigma^2}} \tag{4.2}$$

where σ is the standard deviation of the angular distribution of the beam.
Fig. 12 shows the angular integrated relative response, $f(\theta)$ in Eq. (4.1), with oxide charge contribution, x=0.3, of a Si photodiode with a 27 nm-thick oxide layer for the isotropic beam and the beam with Gaussian angular distribution. The results are shown as the ratios of the responses for the beam cone with half apex angle or standard deviation of 15°, 30°, 45° and 60° to those for the collimated beam. For both beam distributions, a divergent beam usually gives a lower response than that for a collimated beam except for the spectral region approximately from 120 to 220 nm.

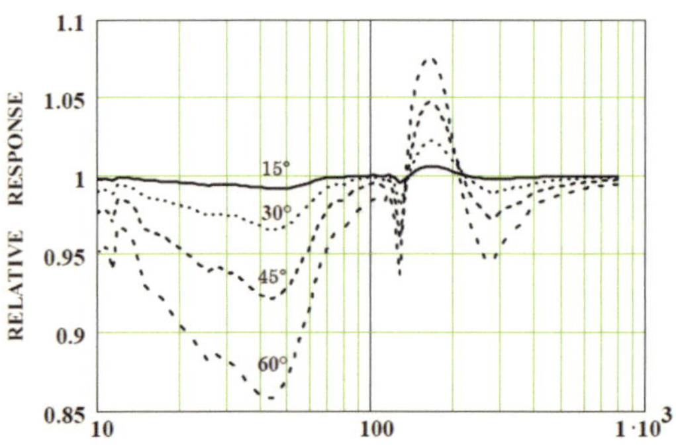

(a)

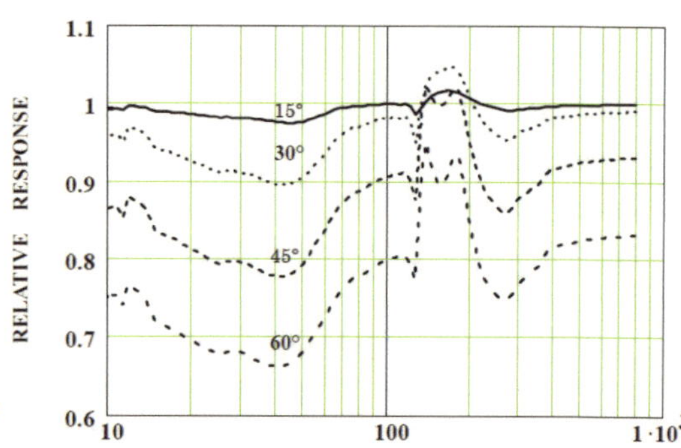

(b)

Fig. 12. Ratio spectrum of detector response for a divergent beam to the one for a parallel beam derived by angle-integration of the responses of a Si photodiode with a 27 nm-thick oxide layer. (a): For an isotropic beam with half apex angle of the beam cone of 15°, 30°, 45° and 60°. (b): For a beam with Gaussian angular distribution with standard deviation angle of 15°, 30°, 45° and 60°.

Similar results for a Si photodiode with an 8 nm-thick oxide layer, which also includes the oxide charge contribution with x =0.3 are shown in Figure 13 for the isotropic beam and the beam with Gaussian angular distribution. Compared to the case of the 27 nm oxide, the spectral region where $f(\theta)$ exceeds unity is narrower, shifts to the shorter wavelength and its peak becomes much lower.

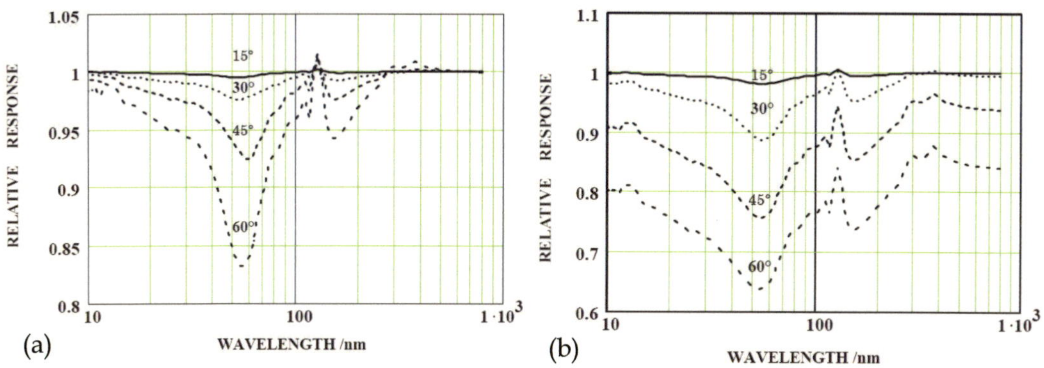

Fig. 13. Ratio spectrum of detector response for a divergent beam to the one for a parallel beam derived by angle-integration of the responses of a Si photodiode with an 8 nm-thick oxide layer. (a): For an isotropic beam with half apex angle of the beam cone of 15°, 30°, 45° and 60°. (b): For a beam with Gaussian angular distribution with standard deviation angle of 15°, 30°, 45° and 60°.

4.5 Linearity

Nonlinearity is caused partly by a detector itself and partly by its measuring instrument or operating condition; the last factor was discussed in section 2.3. The most common method to measure the nonlinearity is the superposition method (Sanders, C.L., 1962) that tests if additive law holds for the photodetector outputs corresponding to the radiant power inputs. One of the modified methods easy to use is AC-DC method (Scaefer, A.R. et al., 1983). We further modified the AC-DC method and applied to measure various kinds of photodetctors as a function of wavelength.

Fig. 14 illustrates measurement setup to test the detector linearity. A detector under test is irradiated by two beams; one is chopped (AC modulated) monochromatized radiation and the other is continuous (DC) non-monochromatized radiation. Measurement is carried out by simply changing the DC-radiation power level while the AC-radiation amplitude is kept constant. If the detector is ideally linear, AC component detected by the detector and read by the lock-in amplifier remains the same. If it changes as a function of the DC-radiation power level, the results directly shows the nonlinearity.

Measurement results for three different types of silicon photodiodes as a function of wavelength are shown in Fig. 15. Tested photodiodes are Hamamatsu S1337, UDT UV100, and UDT X-UV100. Spectral responsivity spectra of the first two correspond to the curves of Si photodiode (A) and Si photodiode (B), respectively (There is no curve for X-UV100 but its curve is close to the one of Si photodiode (B)). Surprising result is that UV-100 and X-UV100 exhibit quite large nonlinearity (more than 20 % for 10 μA) at the wavelength of 1000 nm. For the rest of data, nonlinearity was found to be mostly within 0.2 % (nearly comparable to the measurement uncertainty). Such a rising nonlinearity to the increased input radiant power is called superlinearity and is commonly found for some photodiodes. Completely opposite phenomenon called sublinearity also can happen at a certain condition, for instance, due to voltage drop by series resistance in a photodiode, or due to inappropriate high input impedance of measuring circuit compared to the photodiode shunt resistance (as discussed in 2.3). Compared to sublinearity, superlinearity may sound strange. The key to understand this phenomenon is whether there is still a space for the detector quantum efficiency to increase. Fig. 1 (b) clearly suggests that for UV-100 and X-UV100 quantum efficiency at 1000 nm is much lower than each maximum and than that of S1337. Contrary, for S1337, since the internal quantum efficiency at 1000 nm is still relatively high, near to 1, there is no space for the detector to increase in collection efficiency and thus it results in keeping good linearity even at 1000 nm. Therefore, important point to avoid nonlinear detector is to look for and use a detector whose internal quantum is nearly 100 %.

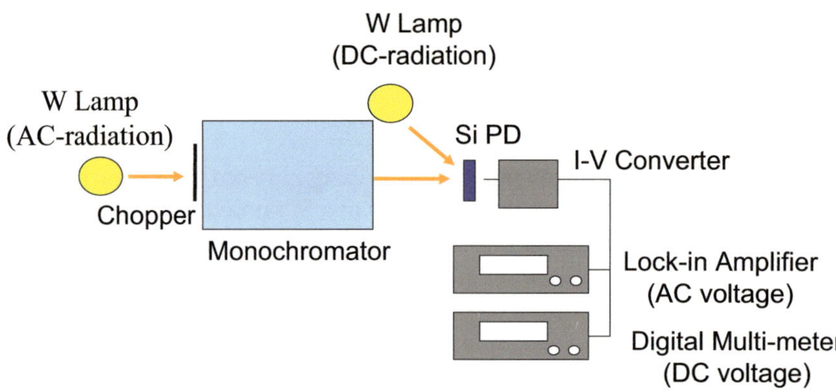

Fig. 14. Schematic diagram for linearity measurement based on AC-DC method. W Lamp: Tungsten halogen lamp, Si PD: silicon photodiode, I-V Converter: Current-to-voltage converter.

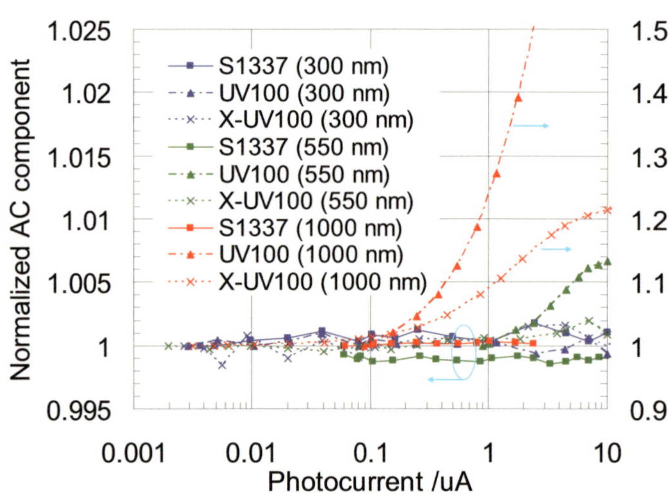

Fig. 15. Linearity measurement results for three Si photodiodes at the wavelengths of 300 nm, 550 nm and 1000 nm each. Note that two curves of UV100 and X-UV100 at 1000 nm refer to the right scale and the others refer to the left one.

4.6 Spatial uniformity

Spatial non-uniformity sometimes becomes large uncertainty component in optical measurement, especially for the use in an underfill condition. As an example, Fig. 16 shows spatial non-uniformity measurement results on a Si photodiode (Hamamatsu S1337) as a function of wavelength.

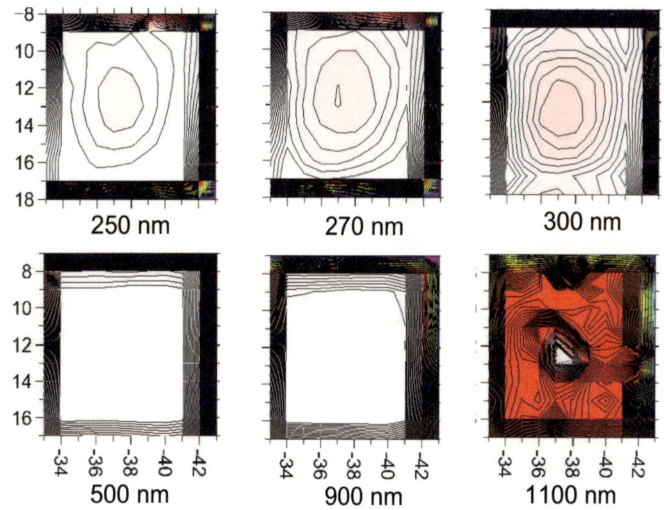

Fig. 16. Spatial uniformity measurement results for a Si photodiode as a function of wavelength. Contour spacing is 0.2 %.

It clearly shows that uniformity is also wavelength dependent as expected since absorption strongly depends on the wavelength. Except the result at the wavelength of 1000 nm, the result of 300 nm exhibits the largest non-uniformity (the central part has lower quantum efficiency). It is about 300 nm (more precisely 285 nm) that silicon has the largest absorption coefficient of 0.239 nm^{-1} (absorption length=4.18 nm) and results in large non-uniformity. It is likely the non-uniformity pattern in the UV is the pattern of surface recombination center density considering the carrier collection mechanism.

Absorption in the visible becomes moderate enough for photons to reach the depletion region and therefore, as seen in Fig. 5 (a), carrier generation from the depletion region becomes dominant. Consequently, probability to recombine at the SiO_2-Si interface becomes too low to detect its spatial distribution and result in good uniformity.

The non-uniformity at 1100 nm is exceptionally large (only the central point has sensitivity) and the pattern is different from the pattern seen in the UV.

4.7 Photoemission contribution

For quantitative measurements, it is important to know how large the photoemission current contribution is relative to its internal photocurrent. Fig. 17 (a) is an example of a spectrum of the photoemission current (i_e) divided by its internal photocurrent (i_r) for the same silicon photodiode, IRD AXUV-100G. The ratio of the photoemission current to the internal photocurrent exceeds 0.07 in the wavelength range from 100 nm to 120 nm.

Also shown in the figure are absorption coefficients of the component materials, silicon and silicon dioxide, derived from (Palik, E.D., 1985).

A similar measurement was carried out for a GaAsP Schottky photodiode, Hamamatsu G2119. The result is shown in Fig. 17 (b) together with the absorption coefficient spectra of gold (Schottky electrode) and GaAs (instead of $GaAs_{0.6}P_{0.4}$). The ratio has a larger peak of 0.26 than that for a silicon photodiode at about 100 nm.

Both results show that the photoemission contribution is significant in a wavelength region a little below the threshold where photoemission begins to occur. Therefore, it is important to specify the polarity of current measurement in this wavelength region. On the other hand, the results also imply that such enhancements are rather limited to a certain spectral range.

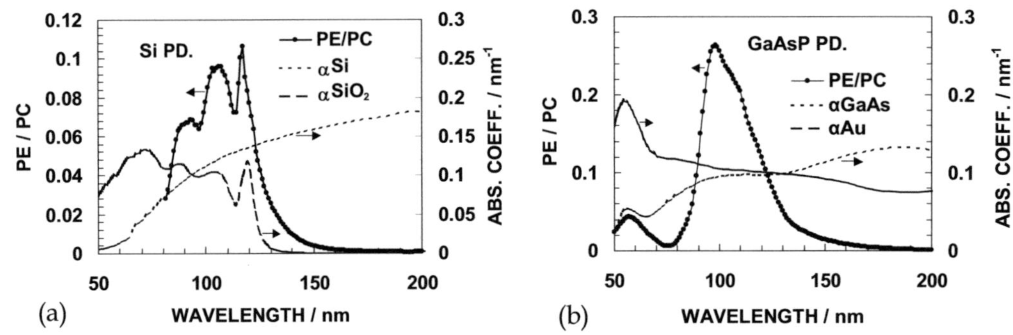

Fig. 17. Spectrum of photoemission currents (extraction voltage = 0) divided by internal photocurrents. (a): For a silicon photodiode, IRD AXUV-100G. Also, absorption coefficient spectra of silicon and silicon dioxide are shown. (b): For a GaAsP Schottky photodiode, Hamamatsu G2119, Also, absorption coefficient spectra of gold (Schottky electrode) and GaAs (instead of GaAsP) are shown.

5. Conclusion

The loss mechanisms in external quantum efficiency of semiconductor photodiodes can be classified mainly as carrier recombination loss and optical loss. The proportion of surface recombination loss for a Si photodiode shows a steep increase near the ultraviolet region and becomes constant with respect to the wavelength. The optical loss is subdivided into reflection loss and absorption loss.

The validity of the model was verified by comparison with the experiments not only for quantum efficiency at normal incidence but also for oblique incidence by taking account of polarization aspects. The experimental and theoretical results show that angular/polarization dependence does not change much as a function of wavelength in the visible but steeply changes in the UV due to the change in optical indices of the composing materials. Excellent agreements are obtained for many cases absolutely, spectrally and angularly. Therefore, it was concluded that the theoretical model is reliable enough to apply to various applications such as quantum efficiency dependence on beam divergence. The calculation results show that divergent beams usually give lower responses than those for a parallel beam except in a limited spectral region (approximately 120 nm to 220 nm for a Si photodiode with a 27 nm-thick SiO_2 layer).

For other characteristics such as spectral responsivity, linearity, spatial uniformity, and photoemission contribution, experimental results were given. The results show that all the characteristics have spectral dependence, in addition to the fore-mentioned recombination and angular properties. Therefore, it is important to characterize photodiode performances at the same wavelength as the one intended to use.

6. References

Alig, R.C.;Bloom, S.&Struck, C.W. (1980). *Phys. Rev.* B22.

Canfield, L.R.;Kerner, J.&Korde, R. (1989). Stability and Quantum Efficiency Performance of Silicon Photodiode Detectors in the Far Ultraviolet, *Applied Optics* 28(18): 3940-3943.

CIE (1987). International Lighting Vocabulary, Vienna, CIE.

Geist, J.;Zalewsky, E.F.&Schaerer, A.R. (1979). *Appl. Phys. Lett.* 35.

Hovel, H.J. (1975). Solar Cells. Semiconductors and Semimetals. R. K. a. B. Willardson, A.C. . New York, Academic. 11: 24-.

Ichino, Y.;Saito, T.&Saito, I. (2008). Optical Trap Detector with Large Acceptance Angle, *J. of Light and Visual Environment* 32: 295-301.

Korde, R.;Cable, J.S.&Canfield, L.R. (1993). IEEE Trans. Nucl. Sci. 40: 1665.

Palik, E.D., Ed. (1985). Handbook of Optical Constants of Solids. New York, Academic.

Ryan, R.D. (1973). *IEEE Trans. Nucl. Sci.* NS-20.

Saito, T. (2003). Difference in the photocurrent of semiconductor photodiodes depending on the polarity of current measurement through a contribution from the photoemission current, *Metrologia* 40(1): S159-S162.

Saito, T.&Hayashi, K. (2005a). Spectral responsivity measurements of photoconductive diamond detectors in the vacuum ultraviolet region distinguishing between internal photocurrent and photoemission current, *Applied Physics Letters* 86(12).

Saito, T.;Hayashi, K.;Ishihara, H.&Saito, I. (2005b). Characterization of temporal response, spectral responsivity and its spatial uniformity in photoconductive diamond detectors, *Diamond and Related Materials* 14(11-12): 1984-1987.

Saito, T.;Hayashi, K.;Ishihara, H.&Saito, I. (2006). Characterization of photoconductive diamond detectors as a candidate of FUV/VUV transfer standard detectors, *Metrologia* 43(2): S51-S55.

Saito, T.;Hitora, T.;Hitora, H.;Kawai, H.;Saito, I.&Yamaguchi, E. (2009a). UV/VUV Photodetectors using Group III - Nitride Semiconductors, *Phys Status Solidi* C 6: S658-S661.

Saito, T.;Hitora, T.;Ishihara, H.;Matsuoka, M.;Hitora, H.;Kawai, H.;Saito, I.&Yamaguchi, E. (2009b). Group III-nitride semiconductor Schottky barrier photodiodes for radiometric use in the UV and VUV regions, *Metrologia* 46(4): S272-S276.

Saito, T.;Hughey, L.R.;Proctor, J.E.&R., O.B.T. (1996b). Polarization characteristics of silicon photodiodes and their dependence on oxide thickness, *Rev. Sci. Instrum.* 67(9).

Saito, T.;Katori, K.;Nishi, M.&Onuki, H. (1989). Spectral Quantum Efficiencies of Semiconductor Photodiodes in the Far Ultraviolet Region, *Review of Scientific Instruments* 60(7): 2303-2306.

Saito, T.;Katori, K.&Onuki, H. (1990). Characteristics of Semiconductor Photodiodes in the Vuv Region, *Physica Scripta* 41(6): 783-787.

Saito, T.&Onuki, H. (2000). Difference in silicon photodiode response between collimated and divergent beams, *Metrologia* 37(5): 493-496.

Saito, T.; Shitomi, H. & Saito, I. (2010). Angular Dependence of Photodetector Responsivity, *Proc. Of CIE Expert Symposium on Spectral and Imaging Methods for Photometry and Radiometry*, CIE x036:2010: 141-146.

Saito, T.;Yuri, M.&Onuki, H. (1995). Application of Oblique-Incidence Detector Vacuum-Ultraviolet Polarization Analyzer, *Review of Scientific Instruments* 66(2): 1570-1572.

Saito, T.;Yuri, M.&Onuki, H. (1996a). Polarization characteristics of semiconductor photodiodes, *Metrologia* 32(6): 485-489.

Sanders, C.L. (1962). A photocell linearity tester, *Appl. Opt.* 1: 207-211

Scaefer, A.R.;Zalewski, E.F.&Geist, J. (1983). Silicon detector nonlinearity and related effects, *Appl. Opt.* 22: 1232-1236.

Solt, K.;Melchior, H.;Kroth, U.;Kuschnerus, P.;Persch, V.;Rabus, H.;Richter, M.&Ulm, G. (1996). PtSi–n–Si Schottky-barrier photodetectors with stable spectral responsivity in the 120–250 nm spectral range, *Appl. Phys. Lett.* 69(24): 3662-3664.

Sze, S.M. (1981). Physics of Semiconductor Devices. New York, Wiley.

2

Noise in Electronic and Photonic Devices

K. K. Ghosh[1], Member IEEE, Member OSA
Institute of Engineering and management, Salt Lake City, Kolkata, India

1. Introduction

Modern state-of art in the solid state technology has advanced at an almost unbelievable pace since the advent of extremely sophisticated IC fabrication technology. In the present state of microelectronic and nanoelectronic fabrication process, number of transistors embedded in a small chip area is soaring aggressively high. Any further continuance of Moore's law on the increase of transistor packing in a small chip area is now being questioned. Limitations in the increase of packing density owes as one of the reasons to the generation of electrical noise. Not only in the functioning of microchip but also in any type of electronic devices whether in discrete form or in an integrated circuit noise comes out inherently whatever be its strength. Noise is generated in circuits and devices as well. Nowadays, solid state devices include a wide variety of electronic and optoelectronic /photonic devices. All these devices are prone in some way or other to noise in one form or another, which in small signal applications appears to be a detrimental factor to limit the performance fidelity of the device. In the present chapter, attention would be paid on noise in devices with particular focus on avalanche diodes followed by a brief mathematical formality to analyze the noise. Though, tremendous amount of research work in investigating the origin of noises in devices has been made and subsequent remedial measures have been proposed to reduce it yet it is a challenging issue to the device engineers to realize a device absolutely free from any type of noise. A general theory of noise based upon the properties of random pulse trains and impulse processes is forwarded. A variety of noises arising in different devices under different physical conditions are classified under (i) thermal noise (ii) shot noise (iii) 1/f noise (iv) g-r noise (v) burst noise (vi) avalanche noise and (vii) non-equilibrium Johnson noise. In micro MOSFETs embedded in small chips the tunneling through different electrodes also give rise to noise. Sophisticated technological demands of avalanche photodiodes in optical networks has fueled the interest of the designers in the fabrication of low noise and high bandwidth in such diodes. Reduction of the avalanche noise therefore poses a great challenge to the designers. The present article will cover a short discussion on the theory of noise followed by a survey of works on noise in avalanche photodiodes.

1.1 Mathematical formalities of noise calculation

Noise is spontaneous and natural phenomena exhibited almost in every device and circuit. It is also found in the biological systems as well. However, the article in this chapter is

[1] email : kk_ghosh@rediffmail.com

limited to the device noise only. Any random variation of a physical quantity resulting in the unpredictability of its instantaneous measure in the time domain is termed as noise. Though time instant character of noisy variable is not deterministic yet an average or statistical measure may be obtained by use of probability calculation over a finite time period which agrees well with its macroscopic character. In this sense, a noise process is a stochastic process. Such a process may be stationary or non-stationary. In stationary stochastic, the statistical properties are independent of the epoch (time window) in which the noisy quantity is measured; otherwise it is non-stationary. The noise in devices, for all practical purposes, is considered to be stochastic stationary. The measure of noise of any physical quantity, say (x_T), is given by the probability density function of occurrence of the random events comprising of the noisy quantity in a finite time domain, say (T). This probability function may be first order or second order. While first order probability measure is independent of the position and width of the time-window, the second order probability measure depends. Further, the averaging procedure underlying the probability calculation may be of two types : time average and ensemble average. The time averaging is made on observations of a single event in a span of time while the ensemble averaging is made on all the individual events at fixed times throughout the observation time. In steady state situation, the time average is equivalent to the ensemble average and the system is then said to be an ergodic system. As $x_T(t)$ is a real process and vanishes at $t \rightarrow -\infty$ and $+\infty$ one may Fourier transform the time domain function into its equivalent frequency domain function X_T ($j\omega$), ω being the component frequency in the noise. Noise at a frequency component ω is measured by the average value of the spectral density of the noise signal energy per unit time and per unit frequency interval centered around ω. This is the power spectral density (PSD) of the noise signal S_x of the quantity x. The PSD of any stationary process (here it is considered to be the noise) is uniquely connected to the autocorrelation function C(t) of the process through Wiener- Khintchine theorem (Wiener,1930 & Khintchine, 1934). The theorem is stated as

$$S_x(\omega) = 4\int_0^{\infty} C(t) \cos \omega\tau \, d\omega \,, \quad \tau \text{ being the correlation time.}$$

Noise can also be conceptualized as a random pulse train consisting of a sequence of similarly shaped pulses randomly, in the microscopic scale, distributed with Poisson probability density function in time. Each pulse p (t) is originated from single and independent events which by superposition give rise to the noise signal x(t), the random pulse train. The PSD of such noises is given by the Carson theorem which is

$$S_x(\omega) = 2v\, a^2 \mid F(j\omega) \mid^2, \text{ F being the Fourier transform}$$

of the time domain noise signal x(t) and 'a^2' being the mean square value of all the component pulse amplitudes or heights. Shot noise, thermal noise and burst noise are treated in this formalism. The time averaging is more realistically connected with the noise calculations of actual physical processes.

To model noise in devices, the physical sources of the noise are to be first figured out. A detailed discussion is made by J.P. Nougier (Nougier,1981) to formulate the noise in one dimensional devices. The method was subsequently used by several workers (Shockley et.al., 1966; Mc.Gill et.al.,1974; van Vliet et.al.,1975) for calculation of noise. In a more general approach by J.P.Nougier et.al.(Nougier et.al.,1985) derived the noise formula taking into account space correlation of the different noise sources. Perhaps the two most common types of noises encountered in devices are thermal noise and shot noise.

1.1.1 Noise calculation for submicron devices

Conventional noise modeling in one dimensional devices is done by any of the three processes viz. impedance field method (IFM), Langevin method and transfer impedance method. In fact, the last two methods are, in some way or other, derived form of the IFM. The noise sources at two neighbouring points are considered to be correlated over short distances, of the order of a few mean free path lengths. Let $V_{1,2}$ be the voltage between two electrodes 1 and 2. In order to relate a local noise voltage source at a point r (say) to a noise voltage produced between two intermediate electrodes 1/ and 2/ a small ac current δI exp (jωt) is superimposed on the dc current j_0 (r) at the point r. The ac voltage produced between 1/ and 2/ is given by

$$\delta V(r - dr, f) = Z(r - dr, f).\,\delta I;$$

Z being the impedance between the point r and the electrode 2/ (the electrode 1/ is taken as reference point).

Thus, the overall voltage produced between the electrodes 1/ and 2/ is given by δI. Grad Z(r,f). dr.

Grad Z is the impedance field. With this definition of the impedance field, the noise voltage between 1/ and 2/ can be formulated as

$$S_V(f) = \int\int \text{Grad}\, Z(r, f)\, S_j(r, r/; f).\, \text{Grad}\, Z^*(r/, f)\, d^3(r)\, d^3(r/)$$

This is the three dimensional impedance formula taking into account of the space correlation of the two neighbouring sources (Nougier et.al., 1985).

2. Thermal noise

Thermal noise is present in resistive materials that are in thermal equilibrium with the surroundings. Random thermal velocity of cold carriers gives rise to thermal noise while such motion executed by hot electrons under the condition of non-equilibrium produces the Johnson noise. However, the characteristic features are not differing much and as such, in the work of noise, thermal and Johnson noises are treated equivalently under the condition of thermal equilibrium It is the noise found in all electrical conductors. Electrons in a conductor are in random thermal motion experiencing a large number of collisions with the host atoms. Macroscopically, the system of electrons and the host atoms are in a state of thermodynamic equilibrium. Departure from the thermodynamic equilibrium and relaxation back to that equilibrium state calls into play all the time during the collision processes. This is conceptualized microscopically as a statistical fluctuation of electrical charge and results in a random variation of voltage or current pulse at the terminals of a conductor (Johnson,1928). Superposition of all such pulses is the thermal noise fluctuation. In this model, the thermal noise is treated as a random pulse train. One primary reason of noise in junction diodes is the thermal fluctuation of the minority carrier flow across the junction. The underlying process is the departure from the unperturbed hole distribution in the event of the thermal motion of the minority carriers in the n-region. This leads to relaxation hole current across the junction and also within the bulk material. This tends to restore the hole distribution in its original shape. This series of departure from and restoration of the equilibrium state cause the thermal noise in junction diode. Nyquist calculated the electromotive force due to the thermal agitation of the electrons by means of principles in thermodynamics and statistical mechanics (Nyquist,1928). Application of Carson's theorem (Rice,1945) on the voltage pulse appearing at the terminals due to the mutual collisions between the electrons and the atoms leads to the expressions of power spectral densities (PSDs) of the open circuit voltage and current fluctuations as :-

$$S_V(\omega) = \frac{4\,k\,T\,R}{(1 + \omega^2\tau^2)}$$

and

$$S_I(\omega) = \frac{4\,k\,T\,/R}{(1 + \omega^2\tau^2)}$$

respectively, where k is the Boltzmann constant, T is the absolute temperature, R the resistive element, ω the Fourier frequency and τ being the dielectric relaxation time. In practice, the frequencies of interest are such that $\omega^2\,\tau^2 << 1$.

3. Shot noise

Shot noise, on the other hand, is associated with the passage of carriers crossing a potential barrier. It is, as such, very often encountered in solid state devices where junctions of various types are formed. For example, in p-n junction diodes the depletion barrier and in Schottky diodes the Schottky barrier. These are the sources of shot noises in p-n junction devices and metal-semiconductor junction devices. Shot noise results from the probabilistic nature of the barrier penetration by carriers. Thus in the event of the current contributing carriers passing through a barrier, the resulting current fluctuates randomly about a mean level. The fluctuations reflect the random and discrete nature of the carriers. A series of identically shaped decaying pulses distributed in time domain by Poisson distribution law may be a model representation of such shot noise. The spectral density of the noise power (PSD) of such Poisson distributed of the random pulse train in time domain is given by Carson's theorem (Rice, 1945)

$$\overline{S_{shot}(\omega)} = 2\,\nu a^2 = 2\,q\,I$$

assuming impulse shape function of the noise; ν and a^2 being the frequency and mean square amplitude of the pulse.
But ν = I/q and as all the pulse amplitudes are same being equal to q so

$$\overline{S_{shot}(\omega)} = 2\,\nu a^2 = 2\,q\,I$$

q and I being the electron charge and magnitude of the mean current. The spectral structure of shot noise is thus frequency independent and is a white noise.
In recent years, shot noise suppression in mesoscopic devices has drawn a lot of interest because of the potential use of these devices and because the noise contains important

information of the inherent physical processes as well. Gonzalez et.al.(Gonzalez et.al.1998) found, on the basis of the electrons' elastic scatterings, a universal shot noise suppression factor of 1/3 in non-degenerate diffusive conductors. Strong shot noise suppression has been observed in ballistic quantum point contacts, due to temporally correlated electrons, possibly a consequence of space charge effect due to Coulomb interaction (Reznikov et.al.1995). Phase coherent transport may also be a cause of shot noise suppression. Resonant tunneling of electrons through the GaAs well embedded in between two barriers of AlGaAs sets another example of suppression of shot noise (Davies et.al.,1992). Shot noise can be directly calculated from the temporal autocorrelation function of current.

4. Burst noise

Burst noise manifests itself as a bistable, step waveform of same amplitude distributed randomly in a time domain of observation. In early literatures, it is sometimes called "random telegraph signal" because of its close resemblance with telegraph signal. The burst noise appears in junction devices e.g. diodes, transistors etc., in tunnel diodes and also in carbon resistors as well. Burst noise is not much observed in devices and is seen not so common as for other types of noises. It appears that such a noise is not universally present in any devices. A typical burst noise waveform is sketched in fig.1. It consists of a random, step waveform which is superimposed with a white noise. It is believed that, the burst noise in forward junctions is due to the crystallographic defects present in the vicinity of the junction while in reversed junctions it is due to an irregular on-off switching of a surface conduction path as a result of random thermal fluctuations. Hsu and Whittier (Hsu & Whittier, 1969) dealt with an issue of determining whether the burst noise in forward junctions is a surface effect or volume effect. Extensive research has suggested that the burst noise in forward biased junctions is more a surface effect than a volume effect. Updated conclusion of the origin of the burst noise to be a surface effect has received much support. This conclusion is arrived at on the basis of noise observed as a step waveform generated by microplasmas (Champlin, 1959).

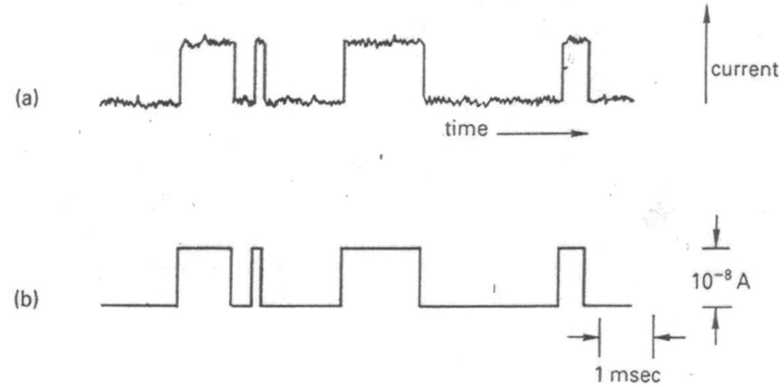

Fig. 1. Typical waveform of current burst noise (a) as observed with white noise superimposed and (b) after clipping.

The microplasmas are highly localized regions formed in the avalanche region at the reverse biased junction where the mobile charges are trapped and immobilized by flaws and crystal imperfections. The microplasma model of the burst noise gives a sequence of events that finally results in such a noise : an avalanche effect is initiated by a carrier either generated within or diffusing in the high field region. With building up of the current, the voltage drop along the high internal series resistance also increases until the voltage drop across the high field region falls below the breakdown value at which point the secondary emission of carriers stops. Some of the carriers released in the process may be trapped in the immediate vicinity of the microplasma. Subsequent to the end of the secondary emission, some of the carriers that are re-emitted from the traps trigger the action again. The process repeats by itself resulting in a series of short avalanche current bursts until by any probability there is no further re-emission of secondary carriers to trigger fresh avalanche. A number of theoretical predictions (McIntyre, 1961, 1966, 1999; Haitz, 1964;) were made to explain the noise in reverse biased diodes. The main suggestion came out of these theories was to consider the diode noise in two regimes e.g. avalanche and microplasma. Marinov et.al. (Marinov et.al., 2002) investigated the low frequency noise in rectifier diodes in its avalanche mode of working region and showed conclusively that in the breakdown region of the avalanche diode two competitive processes e.g. impact ionization and microplasma switching and conducting balance each other. The correlation of these two processes gives rise to a statistically fluctuating current wave of low frequency in the diode.

5. Low frequency noise

Electrical current through semiconductor devices are seen to exhibit low frequency fluctuations (generally below 10^5 Hz.) with 1/f spectrum. The ubiquitous 1/f fluctuations i.e. noise is still a question as to its unique origin. An enormous pool of data is there on 1/f noise and different theories as opposed to other are tried to explain this noise. The 1/f noise, also known as low frequency or Flicker noise, is an intrigue type of fluctuations seen not only in the electron devices but also found in natural phenomena like earthquakes, thunderstorms and in biological systems like heart beats, blood pressure etc. Physical origin of 1/f noise is still a debatable issue. This type of noise is the limiting factor for devices like high electron mobility transistors (HEMTs) and MOS transistors and, in fact, unlike in JFETs this is very dominant MOSFETs. A number of theoretical models on LF noise in MOS transistors are based on surface related effects. There is no universally accepted unique theory or physical model of 1/f noise. Yet, in general, it is suggested that the fluctuating mechanism is a two state physical process with a characteristic time constant τ. Each fluctuator produces a spectral density of Lorentzian spectrum with a specific characteristic time. If these characteristic times of the fluctuators vary exponentially with some parameter e.g. energy or distance, and if, in addition, there is a uniform distribution of the fluctuators in τ then a 1/f spectrum results. Further, there is some support for this noise in semiconductors to be linked with phonons, although no specific and unique mechanism has yet been proposed convincingly. The most complete model of noise caused by phonon fluctuation has been given by Jindal and van der Ziel (Jindal & van der Ziel, 1981).

The conductance depends on the product of mobility μ and carrier density n. There has been considerable discussion about which of these two quantities fluctuate? Is it the mobility fluctuation $\Delta\mu$ or carrier density fluctuation Δn or both simultaneously to fluctuate the conductance? Accordingly, there are two competing models that are invoked to figure out the reason of 1/f noise: the mobility fluctuation model devised by F.N.Hooge (Hooge,1982) and the carrier densuty fluctuation model by A.L.McWhorter (McWhorter,1955). In McWhorter model, carrier trapping resulting in immobilization and de-trapping resulting in remobilization of carriers produce the carrier number fluctuations in the current. It is believed that the number fluctuations of the carriers in the MOS channel due to tunneling between the surface states and traps in the oxide layer is the reason of LF noise in such devices. Assumption of electron-phonon scattering mechanism is also supposed to contribute to the resistance fluctuations and, in turn, to the generation of 1/f noise. A large number papers covering the works on 1/f flicker noise have been published by a number of authors. Recent interest in GaN-related compound materials have led to investigating the noise behavior in these materials. For example, there have been reported values of the Hooge parameter in GaN/AlGaN/saphire HFET devices to be higher than 10^{-2}.

6. Generation – recombination noise

This is the noise generated as a consequence of random trapping and detrapping of the carriers contributing to the current conduction through a device. These trapping centers are the Shockley-Reed- Hall (SRH) centres of single energy states found in the band gap or in depletion region or in partially ionized acceptor/ donor level in a semiconductor. The statistics of generation –recombination (g-r) through single energy level centers in the forbidden gap of the semiconductor were formulated independently by Hall (Hall, 1952) and jointly by Shockley and Reed (Shockley & Read Jr.,1952)]. The g-r noise is apparent mainly in junction devices. During a carrier diffusing from one or other of the bulk regions into the depletion region it may fall into the SRH energy trap center where it will stay for a time that is characteristic of the trap itself. This produces a recombination current pulse. Superposition of all such pulses constitutes a recombination noise current in the external circuit. Similarly, when a generation event occurs at a center, the generated carrier is swept through the depletion region by the electric field towards the bulk region. This produces a generation noise current pulse. Several authors (van der Ziel, 1950; du Pre,1950; Surdin,1951; Burgess,1955) explained the low frequency 1/f noise as a superposition of many such g-r noises and assuming the $1/\tau$ distribution in a very wide variation of relaxation times τ.

7. Noise in photonic devices

With an exception of high frequency photonic devices, important noises are 1/f noise and shot noise. A very short report on the different types of noise in different photonic devices are given here. Mainly the devices are optical fibers, light emitting diodes (LEDs), laser diodes (LDs), avalanche photodetectors (APDs) etc.

Noise in semiconductor waveguides working on the principle of total internal reflection can be studied by considering the variation of the bandgap with temperature. This is because of the fact that the bandgap itself depends upon the refractive index of the material (Herve & Vandamme, 1995) by

$$n^2 = 1 + \left[\frac{13.6}{E_g + 3.4}\right]^2$$

and for the relative temperature coefficient of refractive index it was proposed in ref. (Harve & Vandamme, 1995) as

$$\frac{1}{n}\frac{dn}{dt} = \frac{(n^2 - 1)^{3/2}}{13.6 n^2}\left[\frac{dE_g}{dT} + 2.5\,\mathrm{x}10^{-5}\right]$$

Any index difference between the core and cladding materials affects the Rayleigh scattering loss (Ohashi et.al.,1992) in the fiber. Further, variation in the index with temperature causes variation in the scattering loss. The resulting fluctuation in the fiber loss shows the character of 1/f noise (van Kemenade et.al., 1994). The 1/f fluctuations in optical systems had been studied by Kiss (Kiss, 1986).

8. Avalanche noise

At sufficiently high electric field, the accelerated free carriers (electrons and holes) by their drift motion in the semiconductor may attain so high kinetic energy as to promote electrons from the valence band to the conduction band by transfer of kinetic energy to the target electrons of the valence band by collision. In effect, this is the ionization of the atoms of the host lattice. The process of this ionization by impact is known as impact ionization. Many such individual primary impacts initiating the ionization process turn into repeated secondary impacts. These secondary impacts depend on the existing energy plus fresh gain in their kinetic energies from the electric field. Anyway, such multitude of uncontrollable and consecutive ionizing events result in the generation of a large multiplication of free carriers. This is what is known as "avalanche multiplication". A huge multiplication in the number of both types of carriers, in the form of electron-hole pairs (EHPs) takes place by the process of such avalanche multiplication. The strength of ionization of a carrier is measured by its ionization coefficient and is defined by the number of ionizing collisions the carrier suffers in unit distance of its free travel. In other words, it is the ionization rate per unit path length. The minimum energy needed to ensure an impact ionization is called the ionization threshold energy. The ionization rates (also known as ionization coefficients) of electrons and holes are, in general, different and are designated by α and β respectively. The rates are strongly dependent on the impact threshold.
There exists a probability by which the EHPs may be generated also a little bit below the threshold by highly energetic primary carriers that bombard against the valence electrons and help them to tunnel through and pass on to the conduction band. This is the tunneling-impact ionization that effectively reduces the ionization threshold (Brennan et.al.,1988). Avalanche multiplication occurs in large number of electronic devices viz. p-n junction operated in reverse breakdown voltage, JFET channel under high gate voltage, reverse biased photodiode etc. In almost a majority of devices such carrier multiplication degrades the normal operation and is the limiting factor to be cared in order to save the devices from damage. On the other hand, in case of the photo-devices e.g. photodiodes, phototransistors etc. the carrier multiplication plays the key role in operating the device. Photodiodes using the principle of avalanche multiplication of carriers are known as avalanche photodiodes (APDs). These APDs are used in optical communication systems as receivers of the weak optical signals and to convert it into a strong electrical signal by the process of carrier multiplication by avalanche impact ionization. Wide bandwidth APDs are now one of the interesting areas of research work in the field of digital communication systems, transmission of high gigabit -frequency optical signal etc. However, the ionizing collisions, the key factor in the working of such APDs, are highly stochastic by nature. This results in the creation of random number of EHPs for each photo-generated carrier undergoing random transport. Moreover, the randomness in the incoming photon flux adds to the randomness in the carrier multiplication both in temporal as well as in spatial scale. This results in what is known as multiplication or avalanche noise. In some literatures it is also

termed as excess noise. The original signal is masked by this excess noise and the signal purity is obliterated.

A detailed analysis of the multiplication noise was done by Tager (Tager, 1965) considering the two ionizing coefficients to be equal while in McIntyre's (McIntyre, 1966, 1973) work the analysis was made considering the two coefficients to be different. In the approaches of these papers continuous ionization rates were considered for both the carrier types, on the assumption that the multiplication region to be longer compared to the mean free path for an ionizing impact to occur. The noise current per unit bandwidth following McIntyre (McIntyre, 1966) is given by

$$\overline{i^2} = 2q\, I_0 M^2 F$$

where I_0 is the primary photocurrent, M is the current multiplication and F is the excess noise factor.

The validity of the continuous ionization rates for both the carriers is reasoned because of extremely large number of ionizing collisions per carrier transit. In all these conventional analyses a local field model is visualized wherein the coefficients were regarded to be the functions only of the local electric fields. It could explain the noise behavior well for long multiplication i.e. long avalanche regions.

For short regions, however, the analyses could not work and for that reason the validity of the local field effect was questionable For short avalanche region, Lukaszek et.al. (Lukaszek et.al., 1976) reported for the first time that the continuous multiplication description of avalanche process is not proper for the analysis of short region diode because here very few ionizing collisions take place per carrier transit. A very important effect, "dead space effect", may be overlooked in case of long regions but in no way for short regions. This assertion is justified if the dead space (or, "dead length") definition in relation to ionizing collision is understood. Dead space, for impact ionization, to take place is the minimum distance to be covered by an ionizing carrier from its zero or almost zero kinetic energy to attain a threshold energy to ensure an ionizing impact. Conflicting descriptions of the impact ionizations found in literatures raised confusions as to the exact nature of the dead length. It is reported through an investigation (Okuto & Crowell, 1974) that the average value of the dead space would effectively be increased for two possible reasons : one for the scattering of the carriers and consequently resulting in a longer path length to attain the threshold and secondly, because the nascent carriers at the point of just attaining the threshold are not so probabilistic (Marsland, 1987) to induce impact ionization but instead becomes more probabilistic with energy increasing non-linearly over the threshold. Based on these ideas, a parameter "p" signifying the degree of softness or hardness of the threshold is considered in subsequent works on avalanche ionization. Ideally, for no scattering the average dead length is smallest and is equal to $l_0 = \varepsilon_{th} / qE$, ε_{th} and E being considered to be a hard threshold and electric field respectively, q the charge of the carrier. As the number scatterings are increased the dead length increases and the degree of hardness of the threshold softens. Early workers used conventionally the hard threshold which resulted in some errors. Several publications (van Vliet et.al.,1979; Marsland et.al.,1992; Chandramouli & Maziar, 1993; Dunn et.al.,1997; Ong et.al., 1998) were made to investigate the nonlocal nature of impact ionization. In another approach, Ridley (Ridley,1983) for the first time introduced completely a different model based on lucky-drift mechanism for impact ionization. Subsequently, some other workers (Burt,1985; Marsland, 1987) used the model in a little modified form of the original model of Ridley (Ridley,1983) and verified with existing experimental results. In the original model or in its derivatives, the carrier motion is divided into two parts viz. the ballistic part and the lucky drift part. In the ballistic part, carriers suffer no collisions whereas in the lucky drift part carriers undergo collisions. In an attempt to thermalize the dynamic process of the carriers' motion with the crystal, energy relaxation or momentum relaxation is taken help of. Hayat et.al. (Hayat et.al., 1992) formulated a recurrence method to estimate the excess noise factor. Ong et.al.[Ong et.al.,1998) devised a very simple model to study the multiplication noise in avalanche photodiode by incorporating randomly generated ionization path lengths and the hard threshold concept. The model is shown to be in excellent agreement with the results derived by Monte Carlo model.

A more accurate analysis for avalanche effect especially for short regions was suggested by McIntyre (McIntyre,1999) considering the road map of the carriers' which includes the history of all the ionizations within the avalanche region. This reflects the fact that the impact ionization rate at a point depends simultaneously on three factors viz. (i) the local value of the electric field at that point (ii) the location of generation of the carrier and (iii) the gradient of the electric field i.e. the field profile in between the generation location and the ionizing location. Considering non-local effects and the carriers' transport history McIntyre (McIntyre,1999) presented approximate analytical expressions for the position dependent ionization coefficients. The results shown are in close agreement with those obtained from experimental measurement of noise in GaAs PIN photodiode. An exact calculation of the ionization probabilities with much more flexibilities in modeling the APDs may be achieved only with full band Monte Carlo technique (Bufler et.al., 2000). The Monte Carlo (MC) simulation method has widely been accepted to be a reliable tool of investigating successfully a great variety of transport phenomena in semiconductor devices and materials (Kosina et.al., 2000; Reggiani et.al., 1997; Kim & Hess,1986). The MC simulation offers a direct reproduction of microdynamics of the physical processes of statistical nature on the computer. The traditional drift-diffusion models rely on the assumption of equilibrium transport. They are therefore open to the question of their applicability in studying the non-equilibrium transport of hot electrons taking part in impact ionization events. Further, with the downscaling of electron devices, including the APDs as well, number of scatterings are reduced; this leads to quasiballistic and nonlocal transport; as a result the distribution function no longer remains in equilibrium.

Among the other methods, (Ridley, 1987; Herbert, 1993; Chandramouli et.al.1994) to study the impact ionization in submicron devices the MC technique of simulation has proved to be a most reliable tool as it does not suffer from any disadvantages of averaging procedures inherent in other methods. The method is recognized as the most rigorous one for carrier noise extraction as it allows the appropriate correlation functions to be calculated in a natural way from time averaging over a multi-particle history simulated during a sufficiently long time interval. At any point of time during the computer run the simulation can be stopped so that the positions of all the carriers in the real as well as in the k- space may be recorded. The frequency response of the noise is then calculated. Checked if there is sufficient accuracy, the simulation is ended; otherwise it is repeated until the desired accuracy is arrived at. Although a full band Monte Carlo (FBMC) technique (Chandramouli et.al.1994) gives a more precise and accurate result, yet the simple analytic band Monte Carlo (ABMC) method is capable of reproducing all the important high field features (Dunn et.al.1997; Di Carlo et.al., 1998). An extremely large number of ionizing collisions ($\approx 5x10^5$)

are needed to yield an adequate statistics for the simulation purpose. This makes the FBMC method an impractical one because of the requirement of huge memory and very long run time of the computer. Recently, Ghosh et.al. (Ghosh & Ghosh, 2008) used the ABMC method to study and calculate the excess noise in heterojunction APDs. The ABMC simulation is based on the hard threshold dead space effect in the displaced exponential model of distribution of random ionizing path lengths.

In the present article, the author puts forward a report of their study (Ghosh & Ghosh, 2008) of excess noise in heterojunction avalanche photodetector by Monte Carlo simulation. The MC simulation attracts much attention as it can investigate a device operation mechanism through carrier distribution dynamics and potential distribution profile. The simulation is based upon the hard threshold dead space consideration in the displaced exponential model of the distribution of ionization path lengths. As example, a material system InP / InGaAs is taken for the purpose. This heterostructure photodiode has been developed for an APD in the 1 - 1.6 μm. wavelength region for optic fiber communication system (Susa et.al. 1980; Stillman et.al.1982) and for a switching photodiode in an optoelectronic switch (Hara et.al., 1981). Noise in devices may be minimized by either of the two processes viz. tailoring the bandgap profile (Capasso et.al., 1983) and engineering the electric field profile (Hu et.al., 1996). Introduction of a heterojunction may help the less energetic carriers flowing through the large bandgap material to gain sufficient energy to ionize the low bandgap material. This results in relatively a lower ionizing path length of electrons and longer ionizing path length of holes in the low bandgap material. In consequence, an appreciably different ionization coefficients of the electrons and holes are obtained at the band edge discontinuities. Excess noise may thus be reduced. As for the second process, it may be noted that with increasing field strength, the dead length becomes comparable to the mean ionization path length and thereby the dead length effect on the avalanche process appears to be quite significant. In the process, the carriers enter the multiplication region with high kinetic energy derived from the strong electric field existing at the sharply peaked band shape at the heterojunction. Such initial energy serves to reduce the dead space followed by the avalanche-inducing carrier.

The dead space effect on the excess noise is considered using a simplified model of Hayat et.al. (Hayat, et.al.2002). Here, the carriers are assumed to be injected with fixed energies in an electric field E, say. Ionisation probability of such injected projectile is set to zero within the limit of the dead length l_0. The probability distribution function (PDF) of the ionization path lengths x of an electron after each collision in the dead space model is described by the following piecewise function as

$$P(x) = \begin{cases} 0 & \text{for } x \le l_0 \\ \alpha^* \exp[-\alpha^*(l_0 - x)] & \text{for } x > l_0 \end{cases} \quad (1)$$

where α^* is the ionization coefficient of electrons in the hard threshold dead space model; the ionization path lengths x are measured from the point of generation of the carriers at the instant of ionization. The multiplication in heterostructure APDs can well be studied by exploiting the eqn.(1) in conjunction with the random path length model proposed by Ong et.al. (Ong et.al., 1998).

Monte Carlo description of motion of electrons :- Transport dynamics of the hot electrons in the strong electric field is simulated by Monte Carlo method considering two dimensional carrier scattering of intervalley optic type. For InP a spherical and non-parabolic band model is used while for InGaAs spherical and parabolic band model is considered. Furthermore, the composition dependent band parameters are introduced into the carrier density expressions (Yokoyama et.al., 1984). Also, it is to be noted that as the scattering of carriers in small devices does not occur instantaneously either in space or in time scale so the only compromise in dealing with such small devices is to use non-stationary carrier transport mechanism. In the MC formalism, the carrier dynamics is described in the phase space taking sample of a flux of 10,000 real particles (in this case, the real electrons) for simulation. It is to mentioned here that the entire MC algorithm (Hockney & Eastwood, *Computer Simulation Using Particles,* NY: McGraw-Hill,1981) consists of two sub-sections e.g. the MC-particle dynamics where the particles are treated as real particles and in the other section the particles are treated as super-particles for particle-mesh force calculation required to set up equation of motion. For estimation of time evolution of the potential and field the potential grid is taken sufficiently dense so as to consider the k-points in the first Brillouin zone extremely close to each other. This consideration is very important in the sense that in short devices the field changes so rapidly that one may miss a significant change of track of the hot electron during its motion and in consequence some information of the carrier transport may be lost. The impact ionization rate λ_{ii} is taken from Keldysh (Keldysh, 1964) model:

$$\lambda_{ii} = C\lambda_{ph}(\varepsilon_{th})\left[\frac{\varepsilon - \varepsilon_{th}}{\varepsilon_{th}}\right]\eta$$

$$\varepsilon_{th} = \gamma\ \varepsilon_g$$

λ_{ph} (ε_{th}) being the phonon scattering rate at the ionization threshold, ε and ε_g being the carrier energy and bandgap energy respectively and C, η, γ are the constants. The Keldysh approximation is exploited by using the threshold and softness coefficients to fit (Spinelli & Lacaita) the measured value of the electron ionization coefficient obtained from the experiment of Bulman et.al. (Bulman et.al.,1985). In the MC formalism, the non-steady state of the ionization process is taken into account by considering the ionization probability to be a function of energy of the primary carriers; this is based on the consideration that the energy is not instantly responsive to the very fast change in the electric field at the heterojunction. For simplicity, the distribution of excess energy (ε- ε_g) of the ionizing carriers after each impact ionization is assumed to be shared equally by the three carriers e.g. one primary electron and two secondary carriers in each of the resulting EHPs. Further, the carrier multiplication is simulated in the MC formalism by random pick up of one primary electron with zero initial speed starting at one end of the multiplication region. The transformation equation to generate random path lengths from the displaced distribution function (1) is given by:

$$l = l_0 - \frac{\ln(r)}{\alpha^*}$$

r being the random number distributed uniformly in {0,1}. The hard threshold α^* is obtained from the probability expression (1) as

$$\alpha^* = \frac{1}{1/\alpha\ -\ l_0}$$

α being the electron ionization coefficient in the continuum theory (McIntyre,1966).
This is obtained in the one particle Monte Carlo method by averaging the distance to impact ionize over a large number of ionizing events and using the relation

$$\alpha = \frac{g_{e-h}}{v_d}$$

where g_{e-h} is the EHP generation rate and v_d is the drift velocity of the electrons. The spatial distribution of points where the ionization events occur are recorded. At the next step, the motion of the generated EHPs from each of these points of ionizations by the primary are studied and noted also where, if any, further ionization has occurred within the avalanche region. If the total number of impact ionizations counted until all the secondary pairs and the original shooting electron leave the avalanche region be N then the multiplication M is given by N + 1 for this first trial. A large number of such trials are made and corresponding M's are noted. Finally, the multiplication noise factor is determined by

$$F = \frac{< M^2 >}{< M >^2}$$

A plot of the ionization coefficients of InP and InGaAs comprising the heterostructure versus the electric field is shown in fig.1. In the simulation, a nominal value of 0.4 μm. is taken as the avalanche width. It is observed that the hard threshold ionization coefficients calculated in the dead length model decrease with decreasing electric field strength as is the case with the MC calculated ionization coefficient in the continuous model of McIntyre. The nature of variation of the ionization coefficient and its independence on the field orientation agree well with the results of Chandramouli et.al. (Chandramouli,1994) where a complex band structure in FBMCis taken into account. It is apparent from the figure that the dead length effect is quite significant in strong electric field. A comparative study on the carrier multiplication individually in component materials of the heterostructure and in the heterostructure itself is made through MC simulation. Interestingly, it is observed from the graphical analysis that the multiplication and hence, in effect, the multiplication noise decreases substantially in heterojunction APDs. This means that the noise in heterojunction

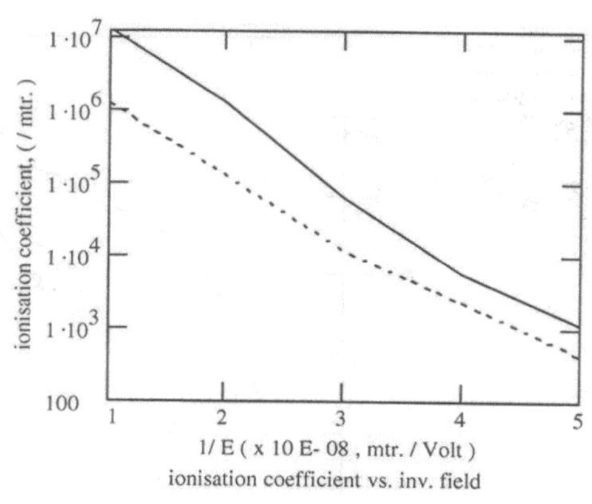

Fig. 2. Ionization coefficient vs. inverse electric field. Solid line is for InP and the dotted line is for the InGaAs

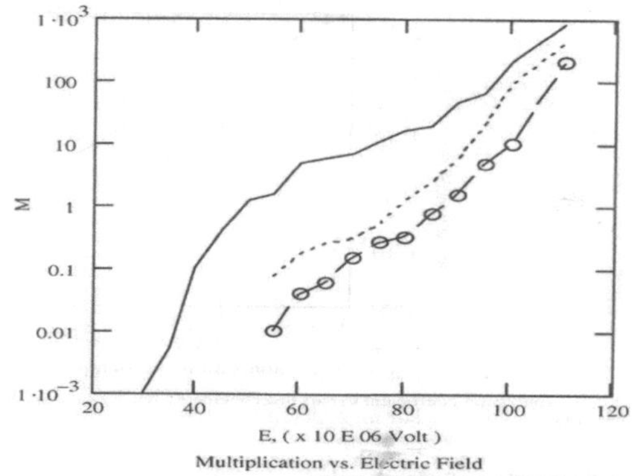

Fig. 3. Multiplication as a function of electric field. The solid line is for InP while the dotted one is for InGaAs and the circled dashed line is for the heterojunction system of InP / InGaAs.

APDs is much less in comparison that in component materials.This simulated result has already been predicted in our theoretical discussion. The dead space effect is quite obvious from the shift of the multiplication curves to the right of the origin. It is also clear that the avalanche field is to be higher in heterojunction to obtain a given magnitude of carrier multiplication. Thus, we arrive at a very important conclusion that the excess noise in APDs can definitely be minimized by using heterojunction at the avalanche region. By inspection of fig.4 It is also found that the noise is more likely to depend on the ionization probability function than on multiplication. Spatial distribution of the ionizing events is suggested to be the reason behind. The same observation is supported in the works of Chandramouli et.al. (Chandramouli et.al. 1994) and of Hayat et.al. (Hayat et.al.2002).

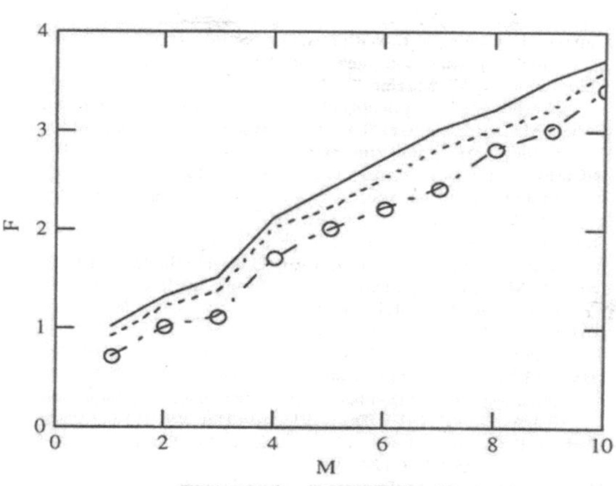

Fig. 4. Excess noise factor varying with the multiplication. Solid line is indicative of the material InP; the dotted line is for the InGaAs while the circled dashed line represents data for the heterojunction.

The probability density function (pdf) of ionization path lengths P (l) of electrons is shown in the fig.4. It shows a rounded off at the peak of the distribution curve while a sharp spike is seen in the distribution obtained in the diplaced exponential model. This sharp peak at the top of the pdf plot indicates that the impact ionization in short heterojunction APDs is more deterministic compared to that in long devices. It also points to the fact that a larger multiplication takes place at the lengths corresponding to the peaks in the P(l) plot. Thus, the dead space effect is validated by calculation of the ionization path length distribution in the displaced exponential model.

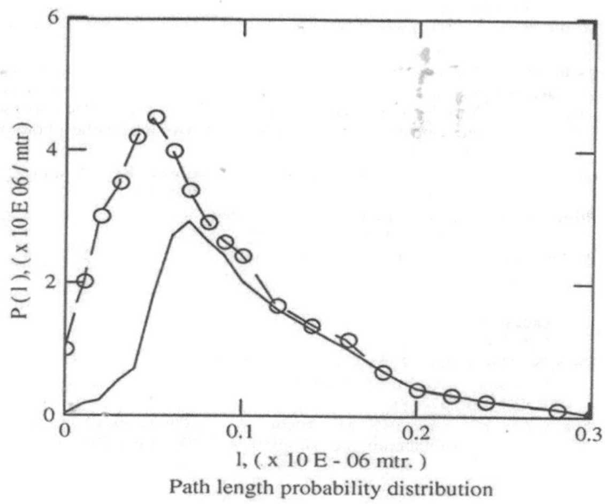

Fig. 5. Plot of P(l) versus l. The upper curve shows the effect in short heterojunction APD and the lower shows that in short one. The ionizing field is set at 65 E 06 V/m.

9. Conclusion

Device technology continues to evolve in response to demand from a myriad of applications that impact our daily lives. Inherent and irresistible noise sources, whatever be its strength or weakness, pose a problem to the high level of operational fidelity of the device. Realisation of absolutely noiseless device is far from reality. What best can be achieved is to fabricate a device with minimum possible noise. Modeling noise sources is important to characterize the noises. A plethora of such models exist in the literature and further new models continue to be introduced. Unfortunately, yet today a single unified theory of noise in devices is not available. The intricacy of 1/f noise still remains a challenge to the area of device research. Puzzling conclusions as to the cause of origin of such noise are being drawn and open up debatable issue. Modification, approximation and etc. are being used time to time to overcome the impasse. Phonon fluctuation or carrier fluctuation or mobility fluctuation - none of them is unique to explain the 1/f noise. In case of the avalanche mode photodiode the conventional continuum theory is not directly applicable to short avalanche photodiode. Lucky drift model and dead space model has improved our understanding of the excess noise in the avalanche diode. The performance of heterojunction APD in the face of noise is substantially improved compared to the homojunction diode. Aggressive downscaling of the electronic and photonic chips embedded with ultra-low dimension devices are much prone to unmanageable noise and poses a threat to the nano-device technology. Future activity in the noise modeling should be dealt with modeling of effects with specific focus related to the device dimensions.

10. Acknowledgement

The author gracefully acknowledges Prof. M.J.Deen, Professor and Senior Canada Research Chair in Information Technology at McMaster University in Hamilton, Canada .The author also acknowledges Prof. A.N. Chakraborti, Ex-Professor and Dean of Technology, Calcutta University, India.

11. References

Brennan K.F.,Wang Y,Teich M.C.,Saleh B.E.A and Khorsandi T, IEEE Trans. Electron Devices, 35, 1456-1467 (1988)

Bufler F.M., Schenk A and Fichtner W, IEEE Trans. Electron Devices, 47(10), 1891(2000)

Bulman G.E, Robbins V.M and Stillman E, IEEE Trans. Electron Devices, 32, 2454- 2466 (1985)

Burgess R.E, Jour.Appl.Phys. 6, 185 (1955)

Burt M. C. , J.Phys.C.,18, L477 (1985)

Capasso F, Tsang W.T and Williams G..F, IEEE Trans. Electron Devices, 30, 381- 390(1983)

Champlin K.S, Jour. Appl. Phys. 30,(7) 1039 – 1050 (1959)

Chandramouli V.C and Maziar C.M, Solid State Electron. 36, 285 (1993)

Chandramouli V.C, Mazier C.M. and Campbell J.C., IEEE Trans. Electron Devices, 41, 648-654 (1994)

du Pre F.K ,Phys.Rev.78,p.615 (1950)

Davies J.H., Hyldgaard P, Hershfield S and Wilkins J.W, Phys.Rev.B46,9620 (1992)

DiCarlo A, Lugli P, Canali C, Malik R, Manfredi M, Neviani A, Zanoni E and G.Zandler, Semicond. Sci.Technol. 13, 858 (1998)

Dunn G.M, Rees G.J, David J.P.R, Plimmer S.A and Herbert D.C, Semicond.Sc.Technol. 12, 111 (1997)

Ghosh A and Ghosh K K , Opt Quant Electron, 40, 439- 446 (2008)

Gonzalez T . et. al . , Phys.Rev.Lett.80,2901 (1998)

Hall R.N., Phys.Rev. 87, 387 (1952)

Hara E.H, Machida S, Ikeda M, Kanbe H and Kimura T, Electron Lett., 17, 150-151 (1981)

Hayat M.M, Saleh B.E.A and Teich M.C. IEEE Trans.Electron Devices, 39, 546 (1992)

Hayat M M, Kwon O.H, Wang S, Campbell J.C, Saleh B.E.A and Teich M. C, IEEE Trans.Electron Devices, 49, 2114- 2123(2002)

Herbert D.C., Semicond. Sci. Technol., 8, 1993- 1998 (1993)

Herve P.J.L and Vandamme L.K.J, Jour. Appl. Phys.77, 5476(1995)]

Hockney&Eastwood, *Computer Simulation Using Particles*, NY: McGraw-Hill (1981)

Hooge F.N, Physica 114B, 391(1982)

Hsu S.T and Whittier R.J., Solid State Electron. 12, 867 (1969)

Hu C, Anselm K.A, Streetman B.G and Campbell J.C, Appl. Phys. Lett., 69, 3734- 3736 (1996)

Jindal R.P and van der Ziel A, Jour. Appl. Phys., 52, 2884 – 2888 (1981)

Johnson J.B, Physical Review, 32, 97 (1928)

Keldysh L.V, Sov.Phys. JETP 21, 1135 – 1144 (1964)

Khintchine A, Math. Annalen 109, 604 (1934)

Kim K and Hess K, Jour. Appl.Phys. 60, 2626(1986)

Kiss L, J.Phys.D, 19, 137 (1986)
Kosina H, Nedjalkov M and Selberherr, IEEE Trans. Electron Devices, 47, 1898 (2000)
Lukaszek W., van der Ziel A and Chenette E.R., Solid State Electron., 19,57 (1976)
Marinov O, Deen M. J., Loukanov V and Velikov V, IEEE Trans Electron Devices, 49 (1), 184 - 186 (2002)
Marsland J.S, Woods R.C and Brownhill C.A , IEEE.Trans.Electron Devices, 139, 1129 (1992)
Marsland J.S., Solid State Electron., 30, 125 (1987)
Mc. Gill T.C, Nicolet M.A, Thornber K.K, Solid State Electron.17, 107 (1974)
McIntyre R.J, IEEE. Trans.Electron Devices, 46 (8), 1623, (1999)
McIntyre R.J., Jour. Appl. Phys., 32, 983- 995 (1961)
McIntyre R.J., IEEE Trans. Electron Devices.13(1),164 (1966)
McIntyre R.J., IEEE Trans. Electron Devices., ED-20, 637- 641 (1973)
McWhorter A.L., *Sc.D. dissertation*, Massachusetts Institute of Technology, Cambridge (1955)
Nougier J.P, *Proc. 6th. Int. Conf. Noise in Physical Systems, Washigton, 1981, Nat. Bureau of Standards*, Publ.NBS-SP-614, p.397 (1981)
Nougier J.P., Gontrand C and Vaissiere J.C., Physica 129B, 580 (1985)
Nyquist H., Phys.Rev., 32, 110 (1928)
Ohashi M, Shiraki K and Tajima K. , Jour. Lightwave Technol. 10, 539 (1992)
Okuto Y and Crowell C.R., Phys.Rev.B.10. 4284 (1974)
Ong D.S, Li K.F,.Rees G.J, David J.P.R and Robson P.N, J.Appl.Phys., 83,.3426 (1998)
Reggiani L, Starikov E, Shiktorov P, Gruzinskis and Varani L, Semicond. Sci. Tech., 12, 141(1997)
Reznikov M, Heiblum M, Shtrikman H and Mahalu D, Phys.Rev.Lett., 75,3340(1995)
Ridley B.K, Jour. Phys. C, 16, 3373(1983)
Rice S.O., Math. Anal. random noise, Bell Syst.Tech.Jour., 23,282 (1945)
Shockley W and Read W.T. Jr. Phys. Rev. 87, 835 (1952)
Shockley W, Copeland J.A., James R.P., *"Quantum Theory of Atoms, Molecules and the Solid State"*,
P.O. Lowdin, ed., Acad.Press, NY., (1966)
Spinelli A and Lacaita A.L, IEEE. Trans. Electron. Devices, 44, 1931 (1997)
Stillman G.E, Cook L.W, Bulman G.E, Tabatabale N, Chin R and Dapkus P.D, IEEE Trans. Electron.Devices, ED-29, 1355- 1371 (1982)
Surdin P.M. Jour. Phys.Radium, 12, 777, (1951)
Susa N, Nakagome H, Mikami O, Ando H and Kanbe H, IEEE Jour.Quantum Electron, 16, 864-870(1980)
Tager A. S, Sov. Phys.- Solid State, 6, 1919- 1925 (1965)
van der Ziel A, Physica,16, 359(1950);
van Kemenade A.J, Harve P and Vandamme L.K.J, Electron Lett. 30. 1338 (1994)
van Vliet C.M, Friedmann A, Zijlstra R.J.J, Gisolf A, van der Ziel A, Jour. Appl. Phys. 46,1814 (1975)
van Vliet K.M, Friedmann A and Rucker L.M, IEEE Trans. Electron.Devices, ED-26, 752 (1979)

Wiener N, Acta Math. 55, 117 (1930)
Yokoyama K, Tomizawa M, and Yoshii A, IEEE Trans. Electron.Devices, 31, 1222- 1229 (1984)

3

Design of Thin-Film Lateral SOI PIN Photodiodes with up to Tens of GHz Bandwidth

Aryan Afzalian and Denis Flandre
ICTEAM Institute Université catholique de Louvain, Louvain-La-Neuve
Belgium

1. Introduction

Short-distance optical communications and emerging optical storage (OS) systems increasingly require fast (i.e with Gigahertz to tens of Gigahertz bandwidth) and integrated Si photodetectors (Csutak et al., 2002; Hobenbild et al., 2003; Zimmermann, 2000). Thin-film SOI integrated devices appear as the best candidate to cope with these high-speed requirements, notably for the 10Gb/s Ethernet standard (Afzalian & Flandre, 2005; 2006.a; Csutak et al., 2002). For such bandwidths design trades-off between speed and responsivity are very severe and require a careful optimization (Afzalian & Flandre, 2006.b). In this context, accurate analytical modeling is very important for insight, rapid technology assessment for the given application, and/or rapid system design. There is however a lack of these accurate models in the literature so that time consuming devices simulations are often the only solution. In (Afzalian & Flandre, 2005), we have proposed such an accurate analytical model for the responsivity of thin-film SOI photodiodes. In here, thorough analytical modeling of AC performances of thin-film lateral SOI PIN photodiodes will be addressed. Speed performances depend on a trade-off between transit time of carriers and a RC constant related to the photodiode and readout circuit combined impedances. We will first focus on the transit time limitation of the thin-film SOI PIN diodes (section 2). Then, we will model the complex diode impedance using an equivalent lumped circuit (section 3). For a lateral SOI PIN photodiode indeed, the usual approximation of considering only the depletion capacitance, C_d, reveals insufficient. Our original model, fully validated by Atlas 2D numerical simulations and measurements, allows for predicting and optimizing SOI PIN detectors speed performances for the target applications in function of technological constraints, in particular their intrinsic length, L_i, which is their main design parameter.

2. Transit time limitation of thin-film SOI PIN diodes

To study the transit time limitation of thin-film SOI PIN diodes, we will elaborate an AC analytical model of a lateral PIN diode under illumination assuming full depletion of the intrinsic region. Applying the drift-diffusion set of equations to the SOI lateral photodiode structure and using a few realistic assumptions, we will calculate DC and AC currents using a perturbation model. From there, we will derive an expression of the transit time -3dB frequency. In a first stage we will neglect multiple reflections in computing the incident optical power. Next, we will generalize the model to take these effects into account. Results given by the analytical model will also be compared to results of numerical simulations and we will derive an expression to calculate the transit time -3dB frequency when the intrinsic length is only partially depleted. Substrate generated charge effects on the transit time frequency will further be investigated and discussed for the first time. These effects are totally neglected in the literature and related to partial AC isolation of the BOX. Specific solutions to this problem adapted to the SOI structure will be proposed.

2.1 General equations

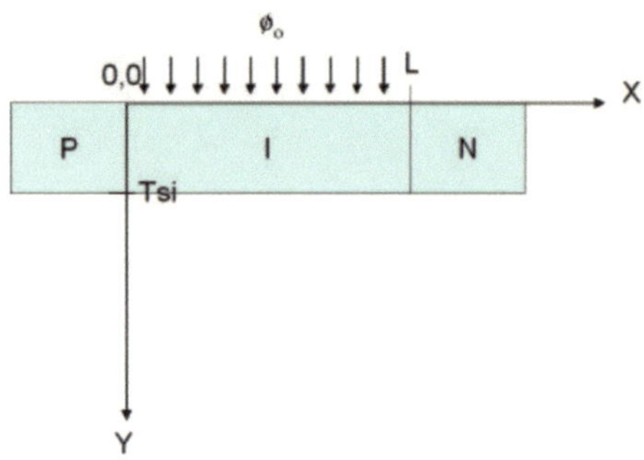

(a) General structure of the lateral PIN photodiode

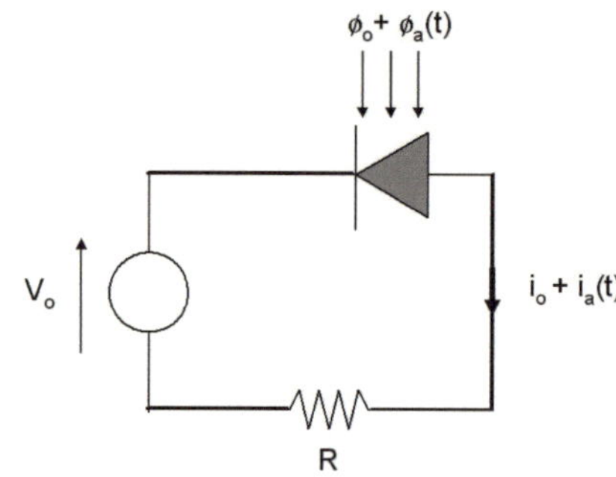

(b) Equivalent scheme of the PIN diode and the external polarization circuit

Fig. 1

a) We first quickly review the general semiconductors equations applied to the particular geometry of lateral PIN diodes (Fig. 1.a). The photodiode is composed of a lowly doped (intrinsic) Si region of length L_i, sandwiched between two highly P+ and N+ doped regions. The Si film of thickness *tsi* is very thin, (typically a few tens to a few hundred of nm). The width of the device is supposed very thick and invariant by translation such that a 2D model is used. The following system is to be solved inside the device (Sze, 1981):

$$\begin{aligned}\frac{\partial E_x(x,y,t)}{\partial x}+\frac{\partial E_y(x,y,t)}{\partial y} &= -\frac{q}{\varepsilon_s}.(p(x,y,t)-n(x,y,t)+\tau(x))\\ \frac{\partial n(x,y,t)}{\partial t} &= \frac{1}{q}\nabla.\overline{J_n}+G-R\\ \frac{\partial p(x,y,t)}{\partial t} &= -\frac{1}{q}\nabla.\overline{J_p}+G-R\end{aligned} \tag{1}$$

In system (1), x is the lateral direction while y is the vertical direction; E is the electric field, q the electron charge, n and p the concentrations of free carriers (electrons and holes resp.) in the device, τ is the doping profile obtained as the difference between the positive N_d and the negative N_a fixed charges $N_d(x)-N_a(x)$ and has been assumed to be a function of x only. G and R are the generation and recombination terms, respectively. Since the concentration of dopant impurities is very low compared to the concentration of Si atoms in the Si crystal, the probability of indirect recombination or generation is very low. Therefore we only consider band-to-band mechanisms and G and R are the same for electrons and holes. J_n and J_p

are the current densities of electrons and holes respectively and are given for a device with homogenous temperature by the drift-diffusion equation (Sze, 1981):

$$\begin{aligned} J_n &= qn\mu_n E + qD_n \nabla n \\ J_p &= qp\mu_p E - qD_n \nabla p \end{aligned} \tag{2}$$

In the right part of equations (2), the first terms are the drift current densities due to the electric field, while the second are related to diffusion. μ_n and μ_p are the mobilities, and D_n and D_p are the well-known diffusion coefficients of the electrons and holes respectively. They are related through the Einstein relationship:

$$D_n = \mu_n \frac{kT}{q} \quad D_p = \mu_p \frac{kT}{q} \tag{3}$$

where k is the Boltzmann constant while T is the absolute temperature of the device. For the mobility, we can consider the electric field dependent model of (Caughey& Thomas, 1967):

$$\mu_n(|E|) = \mu_{no} \cdot \left[\frac{1}{1 + (\frac{\mu_{no} \cdot |E|}{v_{sat}})^2} \right]^2 \tag{4}$$

$$\mu_p(|E|) = \mu_{po} \cdot \frac{1}{1 + (\frac{\mu_{po} \cdot |E|}{v_{sat}})} \tag{5}$$

$$v_{sat} = \frac{2.4 * 10^7}{1 + 0.8 * e^{(\frac{T}{600})}} [cm/s] \tag{6}$$

where $|E|$ is the norm of the lateral electric field and μ_{no} and μ_{po} are the low field electron and holes mobilities respectively. These are strongly influenced by impurity concentration and temperature (Zimmermann, 2000). v_{sat} is the saturation velocity.
b) The illumination is supposed to be uniform over the I-layer($\phi_o(t)$) and zero outside. So the generation term in this layer is only a function of the depth y and the time t and is given by:

$$G(y, t) = \alpha \phi_o(t) e^{-\alpha y} \tag{7}$$

where $\phi_o(t)$ is the number of efficient photons by cm^2 impinging on the device. We can relate it to the illumination power by unit of Area W_{ph} by:

$$\phi_o(t) = W_{ph}(t, \lambda). \frac{\lambda}{hc}. \eta_i \tag{8}$$

where λ is the wavelength of the incident light, h is the Planck constant, c is the speed of light in the vacuum and η_i is the internal quantum efficiency, i.e. the probability that an absorbed photon gives rise to an electron-hole pair.

2.2 Simplifying assumptions in lateral fully depleted thin-film diodes

The system to solve is a two-dimensional problem. However, as we shall see, it can be simplified to a 1D one by using a few realistic assumptions.

1) We will assume the electric field to be constant *vs.* the x-position inside the depletion region as τ is very small compared to its value in the P^+ and N^+ regions and the carrier concentrations are very low: the intrinsic region is supposed to be depleted and the influence of photogenerated carriers on the electric field profile in this absorbing layer can be neglected as they are rapidly swept out by the field.

2) In reverse or low forward bias regime in which the photodiode is biased, diffusion currents are neglected within the intrinsic layer as they are very small with respect to the drift photo-generated currents. Equation (2) simplifies to:

$$J_n = qn\mu_n E \quad J_p = qp\mu_p E \tag{9}$$

3) Charge trapping is neglected at the interface. It is equivalent to say that surface recombination velocities at front ($y = 0$) and back interfaces ($y = t_{si}$) are zero.

4) Only light generation is taken into account. Thermal generation indeed determines dark current, which is very small and of no interest here.

As said earlier, under normal reverse bias condition for the photodiode, the intrinsic region is completely depleted. Only photogenerated carriers have therefore to be accounted for and recombination can be neglected. This is valid as long as the illumination is low, i.e. (Torrese, 2002):

$$\phi_o < min(\mu_p, \mu_n)E.ni \tag{10}$$

However, if the illumination is too strong, recombination will increase and a saturation effect will appear.

Because of hypothesis 3), we can assume current densities only following the x-axis. No current is flowing in the y direction. Because of hypothesis 2), neglecting diffusion currents in the I-layer, this, in turn, implies the electric field to be aligned to the x-axis, too.

We therefore obtain a decoupled problem in the intrinsic layer where the photogeneration takes place: electric field and current density follow the x-axis, while, the generation term is only a function of y. Consequently, the set of equations 1 becomes:

$$\begin{aligned} E(t) &= E_x(t)\hat{a}_x \\ \frac{\partial n(x,y,t)}{\partial t} &= \frac{1}{q}\frac{\partial J_n(x,t)}{\partial x} + G(y) \\ \frac{\partial p(x,y,t)}{\partial t} &= -\frac{1}{q}\frac{\partial J_p(x,t)}{\partial x} + G(y) \end{aligned} \tag{11}$$

2.3 Perturbation model

In order to linearize the transport equations and study the small-signal frequency response of the diode to an optical flux, we will assume that the illumination signal is the sum of a DC time-averaged part and an AC or time-dependent part, small compared to the DC part and use perturbation theory. Expanding equations to the first order, we obtain a system of linear equations. Assuming the AC input signal to be a complex sinusoidal perturbation of infinitesimal amplitude, at only one frequency ω, we know by linear system theory that all other signals will also be complex sinusoidal at frequency ω, i.e. any variables $\psi(x,y,t)$ can be expanded in the form of a dc and ac part:

$$\psi(x,y,t) = \psi_o(x,y) + \psi_a(x,y)e^{j\omega t} \tag{12}$$

where ω is the angular frequency of the modulation.
Note that for the mobilities, we have considered a field dependent model. Since the field is function of time, mobilities will also be function of time. Using a quasi-static model for the mobilities, we find:

$$\begin{aligned} \mu_{p,n}(t) &= \mu_{po,no} + \mu_{pa,na}e^{j\omega t} \\ \mu_{po,no} = \mu_{p,n}(E_o) \qquad \mu_{pa,na} &= \frac{d\mu_{p,n}}{dE}|_{E=(E_o)}.E_a \end{aligned} \tag{13}$$

Combining all together back in equations 11, we can separate these into two distinct sets. One is time-independent, yielding the DC work point, while the other is time-dependent.
The DC set of equations within the depletion region is:

$$\begin{aligned} \overrightarrow{E_o} &= E_o\hat{a}_x \\ (\frac{\partial n_o(x,y)}{\partial x}).\mu_{no}E_o + G_o(y) &= 0 \\ (\frac{\partial p_o(x,y)}{\partial x}).\mu_{po}E_o - G_o(y) &= 0 \end{aligned} \tag{14}$$

while when keeping only the linear terms (small-signal approximation), the AC-system becomes:

$$\begin{aligned} \overrightarrow{E_a} &= E_a\hat{a}_x \\ -(\frac{\partial n_a(x,y)}{\partial x}).\mu_{no}E_o + j\omega n_a(x,y) &= G_a(y) - G_o(y)(\frac{\mu_{na}}{\mu_{no}} + \frac{E_a}{E_o}) \\ (\frac{\partial p_a(x,y)}{\partial x}).\mu_{po}E_o + j\omega p_a(x,y) &= G_a(y) - G_o(y)(\frac{\mu_{pa}}{\mu_{po}} + \frac{E_a}{E_o}) \end{aligned} \tag{15}$$

where generation terms are given by:

$$G_o(y) = \alpha\phi_o e^{-\alpha y} \quad G_a(y) = \alpha\phi_a e^{-\alpha y} \tag{16}$$

2.4 Calculation of the free carriers concentrations

In order to solve this double system of equations we need 2 boundary conditions for each system for the free carriers concentrations. The simplest and quite well accepted conditions in the literature (Torrese, 2002) are to neglect the minority carriers at the edges of the P and N regions (i.e. neglecting diffusion (dark) current. In order to analytically predict the dark current, a third set of equation without generation, but with diffusion and recombination in the intrinsic layer and with the Maxwell-Boltzman boundary conditions can be used (Afzalian & Flandre, 2005)):

$$p_o(L_i) = n_o(0) = 0 \quad p_a(L_i) = n_a(0) = 0 \tag{17}$$

Solving the DC system, we find the following solutions for the free carriers concentrations:

$$n_o = -\frac{\alpha\phi_o e^{-\alpha y}}{\mu_{no} E_o} x \quad p_o = -\frac{\alpha\phi_o e^{-\alpha y}}{\mu_{po} E_o}(L_i - x) \tag{18}$$

The negative sign can surprise at first glance but in fact, the diode being reverse biased, the Electric field is negative (see below). Thus carrier concentrations remain positive everywhere. The negative sign will however imply a negative current, i.e. flowing opposite to the x-axis, as can be expected.
Solving the AC system, we derive:

$$n_a(x,y,\omega) = \frac{G_a(y) - G_o(y)(\frac{\mu_{na}}{\mu_{no}} + \frac{E_a}{E_o})}{j\omega}.(1 - e^{\frac{j\omega}{\mu_{no}E_o}x})$$
$$p_a(x,y,\omega) = \frac{G_a(y) - G_o(y)(\frac{\mu_{pa}}{\mu_{po}} + \frac{E_a}{E_o})}{j\omega}.(1 - e^{\frac{(L_i-x)}{\mu_{po}E_o}}) \tag{19}$$

2.5 Current

From DC and first order AC carriers concentration analytical expressions, expressions for current densities can be derived. From there, currents themselves can be obtained.
The total density of conduction current across the device is the sum of electron density current J_n, hole density current J_p and displacement current density J_d:

$$J = J_n + J_p + J_d \tag{20}$$

The displacement component is proportional to the time variation of the electric field:

$$J_d = \varepsilon_s \frac{\partial}{\partial t} E(x,t) = -\frac{1}{S} C_j \frac{\partial}{\partial t} V_j(x,t) \tag{21}$$

where C_j is the junction capacitance, S its area, and V_j the potential across the junction. The displacement density current is time-dependent and is only significant during transients and at high frequencies (Torrese, 2002). By introducing the perturbation in the continuity equations (9) and then using expressions (18), we get the Oth-order (DC) current. As can be expected, the current density is independent of x:

$$J_o(x,y) = qE_o(n_o(x,y)\mu_{no} + p_o(x,y)\mu_{po}) = -q.L_i\alpha\phi_o e^{-\alpha y} \tag{22}$$

Similarly, but using (19) instead of (18), we get the total 1st-order current density expression:

$$J_a(x,y,\omega) = q[\frac{J_o(x,y).E_a}{qE_o} + E_o(n_a(x,y)\mu_{no} + n_o(x,y)\mu_{na} + p_a(x,y)\mu_{po} + p_o(x,y)\mu_{pa})] + j\omega\varepsilon_s E_a$$
$$= j\omega\varepsilon_s E_a + q\alpha e^{-\alpha y}[-\frac{L_i\phi_o E_a}{E_o} + \frac{E_o}{j\omega}.\langle\mu_{no}.\{\phi_a - \phi_o(\frac{\mu_{na}}{\mu_{no}} + \frac{E_a}{E_o})\}.(1 - e^{\frac{j\omega}{\mu_{no}E_o}x})$$
$$+ \mu_{po}.\{\phi_a - \phi_o(\frac{\mu_{pa}}{\mu_{po}} + \frac{E_a}{E_o})\}.(1 - e^{\frac{j\omega}{\mu_{po}E_o}(L_i-x)})\rangle + \phi_o\{\frac{\mu_{na}}{\mu_{no}}x + \frac{\mu_{pa}}{\mu_{po}}(L_i - x)\}] \tag{23}$$

The DC electric field in the I-part of the diode is due to the external applied voltage, V_{ext} and to the internal P^+N^+ contact potential, ϕ_{PN}. The AC electric field is simply due to the variation of the potential related to the AC photocurrent through the resistor R. According to figure 6.b, we have:

$$V_{ext} = V_o + Ri_o \quad \phi_{PN} = U_t.ln(\frac{N_a.N_d}{n_i^2})$$
$$-\int_O^{L_i} E_o(x)dx = V_{ext} + \phi_{PN} \quad -\int_O^{L_i} E_a(x)dx = Ri_a \tag{24}$$

R is the total equivalent series resistance seen by the diode and can include contributions from diffusion zones, contacts, bias circuit source and the input resistance of the optical circuit front-end stage.
The direction of the positive current has been drawn in order to be related to the sign used in the equations.
E_o is in fact constant in the depleted I region only if $\tau = 0$. In practice this is never the case ($\tau = -N_{a_I} \approx 10^{15}cm^{-3}$, the residual I-doping) and the electric field has a linear profile. In this case, we can use a mean value for E_o given by:

$$E_{o_{cat}} = -\frac{V_{ext} + \phi_{PN}}{L_i} \quad E_o = E_{o_{cat}} + \frac{q.N_{a_I}}{\epsilon_{si}}.\frac{L_i}{2} \tag{25}$$

The DC and AC photo-currents of the device by unit of width (along the Z axis) are obtained by integrating the densities of current along the y-axis and taking them at the limit of the depletion area ($x = 0$ or $x = L_i$):

$$I_o(x) = \int_0^{t_{si}} J_o(x,y)dy \quad i_a(x,\omega) = \int_0^{t_{si}} J_a(x,y)dy \tag{26}$$

Performing the integration we find:

$$I_o(x) = I_o = -q.L_i\phi_o(1 - e^{-\alpha t_{si}}) \tag{27}$$

$$i_a(x,\omega) = j\omega\varepsilon_s E_a.t_{si} + q(1 - e^{-\alpha t_{si}})[-\tfrac{L_i\phi_o E_a}{E_o} + \tfrac{E_o}{j\omega}.\langle\mu_{no}.\{\phi_a - \phi_o(\tfrac{\mu_{na}}{\mu_{no}} + \tfrac{E_a}{E_o})\}.(1 - e^{\frac{j\omega}{\mu_{no}E_o}x})$$
$$+\mu_{po}.\{\phi_a - \phi_o(\tfrac{\mu_{pa}}{\mu_{po}} + \tfrac{E_a}{E_o})\}.(1 - e^{\frac{j\omega}{\mu_{po}E_o}(L_i - x)})\rangle + \phi_o\{\tfrac{\mu_{na}}{\mu_{no}}x + \tfrac{\mu_{pa}}{\mu_{po}}(L_i - x)\}] \tag{28}$$

DC photocurrent, I_o, and static electric field E_o are not related and can be calculated directly using Eq. (27), and Eq. (24). To calculate i_a an iterative method seems best suited, as it depends on E_a, which itself depends on i_a Eq. (24). Starting from E_a equal zero, which also implies the terms describing small-signal mobility equal to zero, we get the small signal photocurrent term directly generated by the variation of flux:

$$i_{ao}(x,\omega) = qE_o\frac{\phi_a}{j\omega}(1 - e^{-\alpha t_{si}})[\mu_{no}(1 - e^{\frac{j\omega}{\mu_{no}E_o}x}) + \mu_{po}(1 - e^{\frac{j\omega}{\mu_{po}E_o}(L_i - x)})] \tag{29}$$

Note that for low frequencies, i.e. frequencies such that the first order expansion of the exponential function is valid, we have a similar expression that for the dc photocurrent:

$$i_{ao_{lf}}(j\omega) = -q\phi_a(1 - e^{-\alpha t_{si}})L_i \quad \omega < \frac{\mu_{po}|E_o|}{L_i} \tag{30}$$

We then calculate the related variation of electric field:

$$E_{ao} = -\frac{R.i_{ao}}{L_i} \tag{31}$$

and iterate this process in Eq. (28) until the required precision is reached. Note that if the equivalent series resistor, R, is very low (ideally zero) equation (29) yields the solution directly without the need for iterative calculations.
Expression (30) also allows us to generalize our model to take into account the multiple reflections in the film by relating AC to DC photocurrent (for wich we have already developped such a model (Afzalian & Flandre, 2005)). Knowing dc photocurrent and modulation ratio k_m between ϕ_0 and ϕ_a, we write:

$$\begin{aligned} i_{ao_{lf}}(j\omega) &= \frac{phi_a}{phi_o}.I_o = k_m.I_o \\ i_{ao}(x,\omega) &= -\frac{i_{ao_{lf}}}{j\omega}.\frac{E_o}{L_i}[\mu_{no}(1 - e^{\frac{j\omega}{\mu_{no}E_o}x}) + \mu_{po}(1 - e^{\frac{j\omega}{\mu_{po}E_o}(L_i - x)})] \end{aligned} \tag{32}$$

We can then rewrite (28) using equation (29) and equation (27):

$$\begin{aligned} i_a(x,\omega) = j\omega\varepsilon_s E_a.L_i.t_{si} + i_{ao}(x,\omega) + i_o.\{\tfrac{E_a}{E_o} + (\tfrac{\mu_{na}}{\mu_{no}} + \tfrac{E_a}{E_o}).\frac{1 - e^{\frac{j\omega}{\mu_{no}E_o}x}}{\frac{j\omega}{\mu_{no}E_o}L_i} \\ +(\tfrac{\mu_{pa}}{\mu_{po}} + \tfrac{E_a}{E_o}).\frac{(1 - e^{\frac{j\omega}{\mu_{po}E_o}(L_i - x)})}{\frac{j\omega}{\mu_{po}E_o}L_i} - \tfrac{1}{L_i}[\tfrac{\mu_{pa}}{\mu_{po}}(L_i - x) + \tfrac{\mu_{na}}{\mu_{no}}x]\} \end{aligned} \tag{33}$$

We have implemented the model on Matlab. We first observed that in Si, with typical value of L_i on the order of μm and illumination power densities of a few mW/cm^2, electric field and mobilities variations only make i_a starting to differ from i_{ao} with huge load resistor values, typically larger than about $1M\Omega$. In this case, however, the frequency response of the detector will be limited by its RC constant, such that in most practical case in Si the calculation of i_{ao} is sufficient to model the transit time behaviour of the lateral PIN diodes.

2.6 Transition frequency

We will now extract a simple analytical expression of the -3dB transition frequency, f_{tr}, from the expression of i_{ao} for $x = L_i$ (cathode electron current). By definition of f_{tr}, we have:

$$\|i_{ao}(j\omega_o)\| = \frac{1}{\sqrt{2}}\|i_{ao_{lf}}\| \quad \omega_o = 2.\pi.f_{tr} \tag{34}$$

For $x = L_i$, expression (32) simplifies to an electron current only:

$$i_{ao}(L_i,\omega) = -\frac{i_{ao_{lf}}}{j\omega}.\frac{\mu_{no}E_o}{L_i}.(1 - e^{\frac{j\omega}{\mu_{no}E_o}L_i}) \tag{35}$$

Injecting equation (35) into eq. (34), we have:

$$\| - \frac{\mu_{no}E_o}{L_i}\frac{1 - cos(\frac{L_i}{\mu_{no}E_o}.\omega_o) - j.sin(\frac{L_i}{\mu_{no}E_o}.\omega_o)}{j\omega_o}\| = \frac{1}{\sqrt{2}} \tag{36}$$

which yields:

$$\omega_o = \frac{\mu_{no}E_o}{L_i}.\sqrt{4.[1 - cos(\frac{L_i}{\mu_{no}E_o}.\omega_o)]} \tag{37}$$

Because the cosine function has a value which range between -1 and 1, we can note:

$$\omega_o = k.\frac{\mu_{no}E_o}{L_i} \quad 0 \leq k \leq 2.\sqrt{2} \tag{38}$$

and solve (38) for:

$$k = \sqrt{4.[1 - cos(k)]} = 2.78 \tag{39}$$

$$f_{tr_n} = f_{tr}(x = L_i) = \frac{k}{2.\pi}.\frac{\mu_{no}E_o}{L_i} \tag{40}$$

We get good agreement when comparing cathode model (eq. (39)) to simulations of both anode and cathode currents as long as the intrinsic length is laterally depleted (i.e. for intrinsic length shorter than about $2\mu m$). Otherwise, carrier diffusion has to be taken into account.

2.7 Carrier diffusion

In our model, we have assumed that the photodiode was laterally depleted, i.e. $L_i < L_{zd}$ and the transit time limit was due to fast drift. L_{zd} is the depletion length and is related to doping and bias voltage (Sze, 1981) The related -3dB frequency, f_{tr}, decreases as L_i^2. However, if L_i becomes greater than L_{zd}, around $2\mu m$ for P_- doping and low voltage operation of actual processes, carriers transit is dominated by a slower diffusion mechanism and the related -3dB frequency, f_{tr}, decreases faster with L_i. On fig. 2, this f_{tr} reduction is observed on the Atlas simulation curve for L_i greater than $2\mu m$, when compared to the fast drift modeled curve that assumes full depletion of the I-region.

In order to estimate the time t_{diff} for the diffusion of electrons through a P region of thickness $L = L_i - L_{zd}$, we can use the equation derived for a time-dependent sinusoidal electron density due to photogeneration in the P layer from the electron diffusion equation (Sarto& Zeghbroeck, 1997; Zimmermann, 2000) and, from there, derive the related -3dB frequency, f_{diff}:

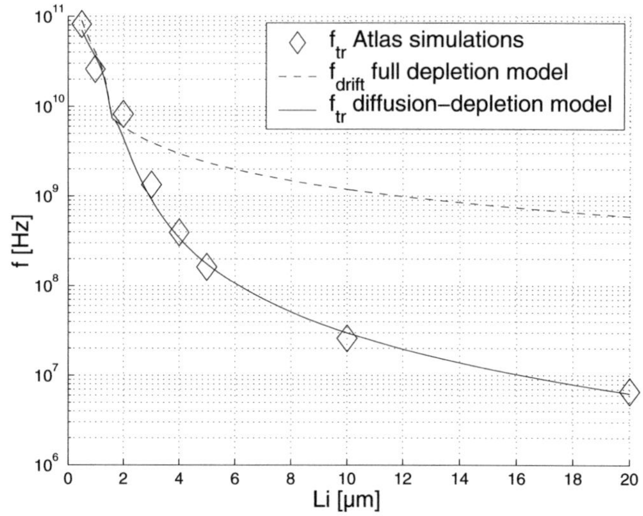

Fig. 2. Comparison of the PIN diode transition frequency given by Atlas simulations, by our model assuming drift only (full depletion hypothesis)(eq. 39), and by our model assuming both drift and diffusion mechanism (eq. 42).

$$t_{diff} = \frac{q.L^2}{2\mu_n.k_B.T} \quad f_{diff} = \frac{k_{diff}}{2\pi.t_{diff}} \tag{41}$$

where k_{diff} is a fitting coefficient.
The -3dB frequency related to the total transit time (drift+diffusion) is then obtained as:

$$f_{tr_{dd}} = (\frac{1}{f_{tr}} + \frac{1}{f_{diff}})^{-1} \tag{42}$$

A value of 2 was obtained for the coefficient k_{diff} by fitting model to numerical simulations (fig. 2).

2.8 Influence of the substrate

Until now in our modeling and numerical simulations we have ignored the effect of the substrate, assuming a perfect or ideal isolation through the buried oxide between the thin-Si film and the Si substrate. This assumption is used in the literature, where it is said that in SOI, unlike in Bulk Si, owing to the BOX, we can avoid the slow vertical diffusion of carriers generated under the depletion region in the substrate.
From an AC point of view, however, the BOX is a capacitor so that at high frequency, carriers photogenerated in the substrate could be mirrored at the front electrodes. In order to investigate this effect we have performed 2D-numerical Atlas simulations of the whole PIN structure, including a $500\mu m$-thick substrate. In current SOI submicron processes, two substrate doping concentrations are most often used. One of them is highly resistive (hr) and has a low substrate P-doping of around $2.10^{12}/cm^3$. The other, the standard resistivity (sr), is P-doped at around $1.10^{15}/cm^3$.
To get an idea of the insulation the BOX can provide we first compare AC photocurrents of a thin-film lateral PIN diode without substrate (SOI ideal case), with a 400nm buried oxide

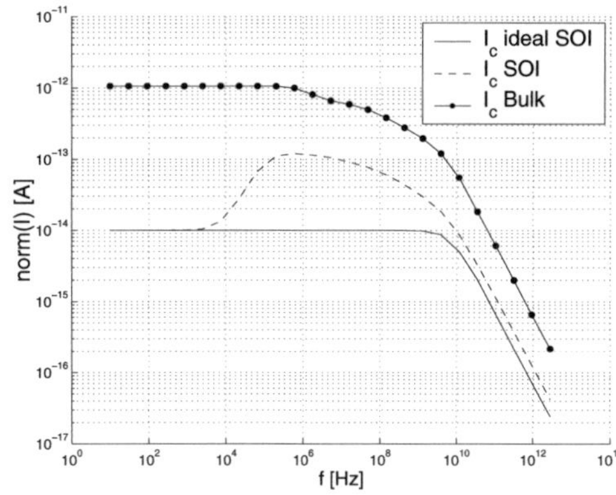

Fig. 3. Comparison of the currents vs. frequency of the PIN diode without substrate (ideal SOI case), with 500μm thick high resistivity substrate (SOI case) and with with 500μm thick high resistivity substrate but without a BOX ("Bulk" case) given by Atlas simulations. $L_i = 2\mu m$, $\lambda = 800nm$, P_{in} dc=1mW/cm^2 ac=0.1mW/cm^2, t_{si}=80nm.

and a 500μm hr Si-substrate (SOI case), and with a 500μm hr Si-substrate but without a buried oxide ("Bulk" case) obtained by numerical simulations (fig. 3). For frequency above a few kHz the BOX does not provide perfect insulation. The worst attenuation factor compared to the bulk case is about a factor 10 in the MHz range. This factor of attenuation, which may be sufficient for practical isolation of thicker SOI materials (t_{si} of few μms) with higher quantum efficiency, seems insufficient for insulating thin film SOI diodes in near IR wavelengths, where their quantum efficiency is only of a few percents.

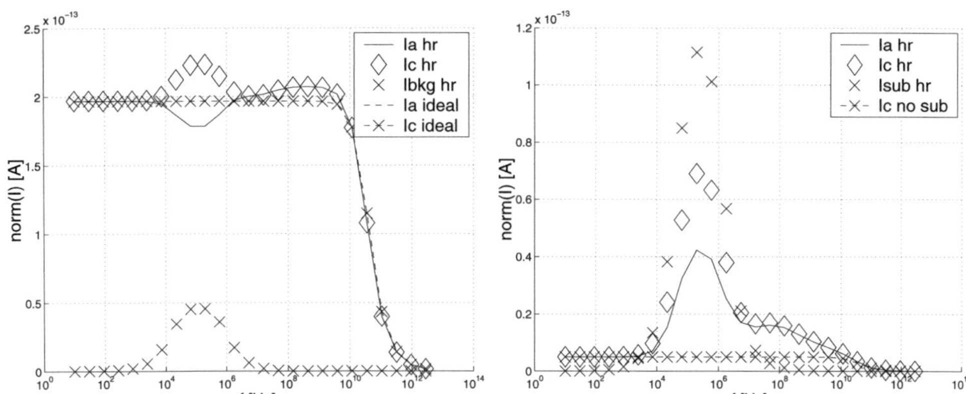

(a) $\lambda = 400nm$ (b) $\lambda = 800nm$

Fig. 4. Comparison of the currents vs. frequency of the PIN diode with hr substrate and without substrate (ideal case) given by Atlas simulations. $L_i = 1\mu m$, P_{in} dc=1mW/cm^2 ac=0.1mW/cm^2

When comparing now AC simulations of PIN diodes with and without substrate, we see that at low frequencies, there is no difference (see fig. 4). The BOX isolates the active thin-film part of the diodes from the charges photogenerated in the substrate by the modulated light source. At higher frequency however, the BOX appears more and more like a short and a capacitive photocurrent originating from the substrate (I_{bkg}) can reach the thin-film. The anode (I_a) and cathode (I_c) currents are influenced by the substrate photogenerated charges. Although this can increase the amplitude of the output photocurrent, this extra photocurrent is a slow diffusive current which will degrade the speed performances of the diodes. At still higher frequency, the number of substrate photogenerated charges that can follow the ac light source signal and diffuse to the thin film on time decreases. This is the cut-off frequency of the substrate generated charges and anode and cathode currents decrease toward the values of the ideal diode case.

The importance of the substrate photogenerated charges depends of course of the wavelength. For wavelength shorter than around 400 nm (see fig. 4.a), this influence can be neglected as most of the light is absorbed in the thin-film region. For higher wavelength (see fig. 4.b), importance of substrate generated charges compared to thin-film generated carriers increases. Simulations show that at a wavelength of 800 nm, the frequency response is strongly influenced.

The peak value and frequency location of the backgate current are influenced by the substrate resistivity. For hr substrate the peak is higher and at a lower frequency which seems worse for high speed application. We explain this as if charges generated in the substrate see two paths to the ground: one impedance through the substrate to the backgate and one impedance through the BOX to the front electrodes. If the resistive impedance through the substrate is lower (sr substrate), the frequency at which the charges can cross the BOX will be higher. The appearance of the backgate current at mid frequency can then be explained by assuming that holes generated in the substrate see a higher impedance through the anode than that electrons undergoes through the cathode, so that the frequency at which holes will flow through the thin-film is higher.

In order to quantify the influence of the substrate photogenerated charges on the temporal response of the SOI photodiodes, we have simulated transient response of PIN diodes with and without substrate for an intrinsic length of $2\mu m$. From our model, these diodes should exibit a transit time frequency of a little less than 10 GHz and then to be available for 10 GBps optical data communication (which is the actual challenge for Si based optical communication).

If the ideal diode shows sufficiently fast temporal behaviour both at 400 and 800nm wavelengths for data train of 0.1 ns (fig. 5.a and 5.b), we can see that the diode with substrate can only be used at short wavelengths, for example 400nm. At this wavelength, as can be expected from the AC simulations, the effect of the substrate is very weak and do not degrade much the speed performance of the diode (fig. 5.a). At 800nm, on the contrary, the slow substrate diffusion current overlap between the adjacent bits and dominates over the photocurrent generated in the thin-film (fig. 5.b) which can make the distinction between zero and one impossible. The so-called long tail response effect is observed.

SOI, owing to its unique structure, can provide specific solutions on top of that available in Bulk to get rid of the slow substrate photogenerated diffusion current at high wavelength.

- The use of PIN SOI diodes on a membrane. This consists in removing the substrate under the PIN diodes by a etching post process which is stopped on the BOX (Laconte et al., 2004). After this removal, as the thin silicon film is now sandwiched between two oxides (front and buried oxide) which both induce compressive stress to the Si film, the Si film can start to buckle. This has been observed on a $500\text{x}500\mu m^2$ lateral PIN diode fabricated in the UCL technology. In (Laconte et al., 2004), to avoid this effect they proposed to use a nitride layer

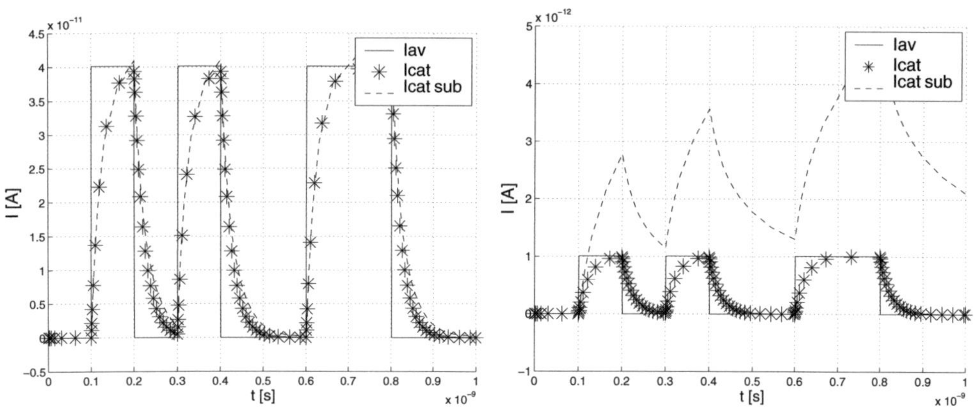

(a) $\lambda = 400nm$ (b) $\lambda = 800nm$

Fig. 5. Comparison between Atlas simulated transient response of the PIN diode with and without substrate to a '0101001100' optical data bit train of 10GBps. $L_i = 2\mu m$, P_{in}'1'=$10mW/cm^2$, P_{in}'0'=$0mW/cm^2$

which add a tensile stress to compensate. We note that this nitride layer can also be used as an anti-reflexion coating.
- Similarly a variant of the SOI technology, the SOS (Silicon on Sapphire) technology in which the Si substrate is replaced by a transparent Sapphire substrate (Apsel& Andreou, 2005) can be used.
- Finally a very promosing solution at high wavelength is to use Germanium on SOI Lateral PIN photodiodes (Koester et al., 2007). Ge is quite compatible with Si integration and is more and more present in MOSFET process for strain silicon devices. The use of ultrathin SOI as substrate for the growing of the Ge layer minimizes the problem of Si diffusion into Ge during thermal annealing steps and allows for an easy co-integration of Ge photodetector with Si circuits. As the absorption length of Ge is only a few hundred of nanometers at 850nm (roughly 50 times less than in Si) owing to its direct bandgap, thin-film (400 nm Ge layer) lateral PIN photodiodes can be fabricated which features similar bandwidth than thin-film SOI photodiodes but with high quantum efficiency. A 10x10 μm with a finger spacing of 0.4μm had a bandwidth of 27GHz at a bias voltage of -0.5V and a quantum efficiency of 30%. The dark current however higher than in a comparable SOI photodetector was still less than 10nA.

3. RC frequency

The diode also exhibits an impedance which combined with the input impedance of the readout circuit leads to a RC -3dB frequency, f_{RC}. In this section we will model the thin-film lateral SOI PIN diode impedance which is mainly capacitive. In what follows, we will first give the complete equivalent lumped circuit we derived in order to model the diode impedance. Then, we will explain the different elements of the circuit and focus on the elements which represent the anode to cathode impedance via the thin film impedance or the ideal diode case, the anode or cathode-to-substrate impedance and the MOS capacitor related to, and finally the coupling impedance between the anode and cathode via the substrate and via the air.

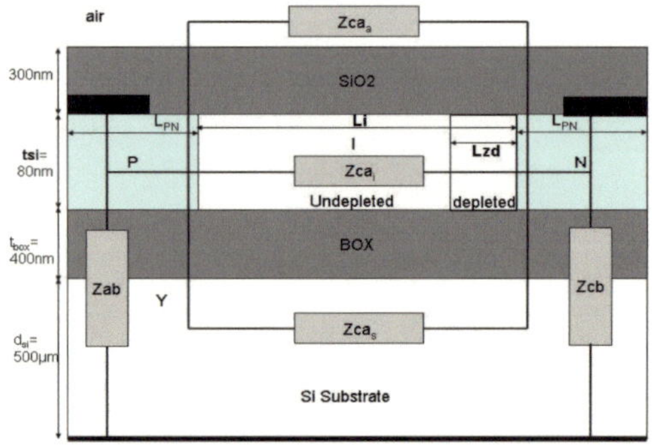

(a) Schematic view of our PIN diode structure and simplified equivalent impedance model

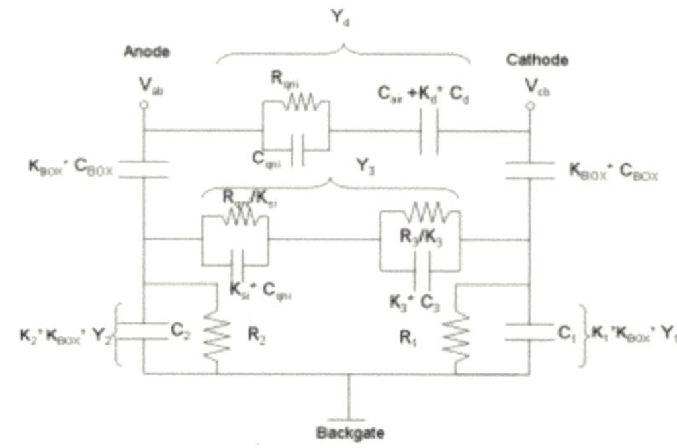

(b) Equivalent model for the calculation of the PIN diode capacitances

Fig. 6

In the general case, when there is a BOX and substrate underneath the thin active Si film, the total cathode or anode impedance Z_{cc} or Z_{aa} of the diode involved in f_{RC} is due to the cathode to anode impedance, Z_{ca}, from which the thin film impedance is just a part, and to the impedance of the N^+ (cathode) or P^+ (anode) region to the substrate, Z_{cb} or Z_{ab} resp. (fig. 6.a). The full impedance behaviour of the diode has to be modeled by the equivalent circuit of fig. 6.b. The different components will be explained in the forthcoming sections. Coefficients K_i are used to take into account the fringing effects which become more and more dominant with down scaling, whereas the admittance cross-sections become smaller and smaller compared to their length.
The value of these fringing factors K_i depends on geometrical dimensions as the length of the diffusion areas, the distance between them, the substrate thickness... Semi-empirical formulation can be derived from microstrip line theory (Garg& Bahl, 1979), (Kirschning&Jansen, 1984).

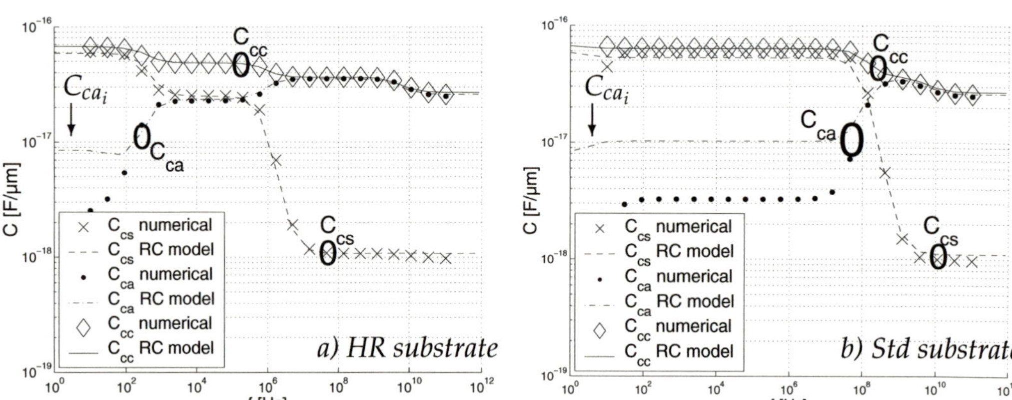

Fig. 7. Comparison of modeled and simulated capacitance by μm width vs. frequency of a) high resistivity (hr) and b) standard resistivity substrates. ST 0.13 μm thin-film SOI diodes. $L_i = 5\mu m$, m=2.

In our case we obtained the K_i factors by fitting model and numerical simulations (see table 1). The 2D numerical simulations were made with the ISE software. We simulated a 2 finger diode (PINIP structure) with full substrate thickness (d_{si}=500μm) and 500μm air layer on top of it to obtain a realistic fringing effect. As can be seen in figure 7.a for highly resistive (hr) substrate (P-doping of $2.10^{12}/cm^3$), the modeled value of the total cathode capacitance C_{cc} between C_{ca} and C_{cb} fairly matches the related simulated curves for frequencies as low as 100Hz. In the case of the standard resistivity (hr) substrate (P-doping of $6.10^{14}/cm^3$) (see figure 7.b) the agreement between modeled and simulated C_{ca} or C_{cb} curves is good only above 100MHz. This can be explained as the low frequency value of C_{ca} tends towards the thin-film capacitance C_{ca_i} (see fig. 7), which depends on the backgate voltage, V_b, and the film conditions. In our model, we have assumed the film as neutral and didn't take into account the influence of V_b. The simulations were performed with a value of V_b of 0V for which the film is in vertical depletion and where C_{ca_i} is reduced. However, the modeling satisfies our high speed purpose and, more over, modeled and simulated total capacitances, C_{cc}, fit very well for all frequencies, for both high and standard resistivity substrate cases.

3.1 The ideal diode impedance

In the ideal case, the diode impedance is only due to the impedance of the thin film region and is dominated upto high frequency by the capacitance of the depletion region C_d. In fact 3 components only are required to model this impedance behaviour versus frequency: C_d and the capacitance, C_{qni}, and resistance, R_{qni} of the quasi neutral part of the I-region if they exists ($L_i > L_{zd}$).

C_d decreases with L_i as long as the I-region is fully depleted, i.e. $L_i < L_{zd}$ and is also proportional to the junction area and to t_{si}, which results in much lower value for thin film SOI than in Bulk. Noting W the width and m the number of fingers of the PIN diode, we have:

$$C_d = m.\frac{\epsilon_{si}.W.t_{si}}{min(L_i, L_{zd})} \tag{43}$$

C_{qni} and R_{qni} determine the cut off frequency, $f_1 = \frac{1}{2\pi R_{qni} C_{qni}}$ where the diode capacitance falls from C_d to $\frac{C_{qni}.C_d}{C_{qni}+C_d} = m.\frac{\epsilon_{si}.W.t_{si}}{L_i}$. From classical semiconductor and circuits theories, noting σ_{qni} the conductivity of the quasi neutral I region, we have:

$$R_{qni} = \frac{L}{m.\sigma_{qni}W.t_{si}} \quad C_{qni} = m.\frac{\epsilon_{si}W.t_{si}}{L} \quad f1 = \frac{\sigma_{qni}}{2\pi.\epsilon_{si}} \simeq 10GHz \tag{44}$$

Fig. 8 shows good agreement between this model and numerical 2D simulations of the cathode to anode capacitive part C_{ca_i} of the impedance Z_{ca_i} of a diode without substrate, defined as:

$$\left(j.\omega C_{ca_i} + \frac{1}{R_{ca_i}}\right)^{-1} \tag{45}$$

R_{ca_i} is the resistive part of the ideal diode impedance in parallel with C_{ca_i} and is totally negligible up to f_1. However this simple model is not sufficient to predict or to simulate the capacitance behaviour of the real PIN diode with a BOX and a substrate underneath.

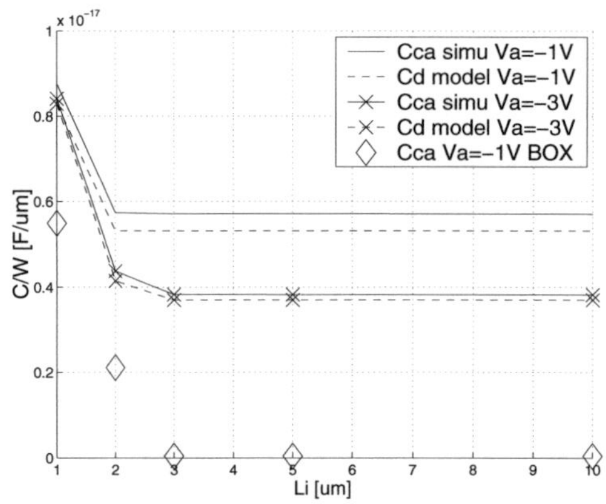

Fig. 8. Comparison of the modeled depletion capacitance Cd with the Atlas simulated cathode to anode capacitance Cca vs. L_i at f=20kHz of thin-film SOI diodes.

3.2 Modeling of the anode or cathode to substrate impedance

We will now focus on the modeling of the terms Y_1 and Y_2 of fig. 6.b related to the impedance of anode or cathode to the substrate. For this purpose we will first study the simpler structure of the N^+ or P^+ region in the Si film and its coupling to the substrate. Model and simulations show that the conclusions we can draw from this simpler structure on the cathode or anode to substrate impedance will stay valid in the general case (i.e. Z_{cb} or Z_{ab}) because the modification of this impedance through the substrate coupling stay negligible when affecting the global anode or cathode impedance (Z_{aa} or Z_{cc} resp.).

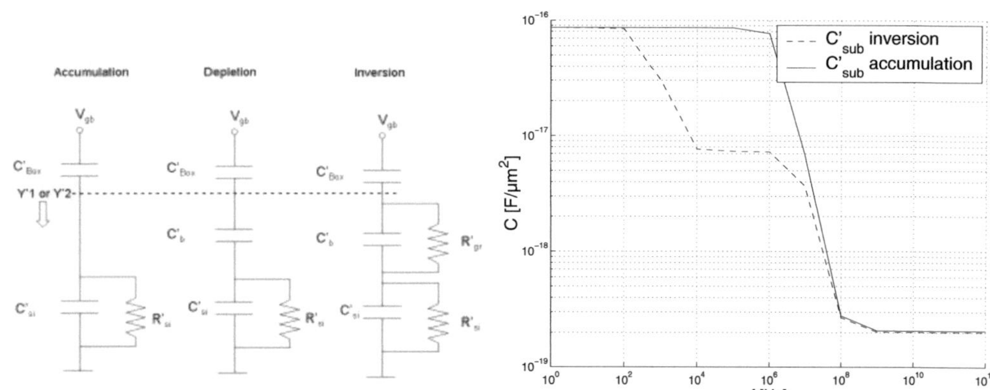

(a) C'_{sub} models (b) C'_{sub} vs. frequency

Fig. 9. a) Equivalent model for the calculation of an N^+ or P^+ diffusion to substrate capacitance (C'_{sub}) in accumulation, depletion and inversion regime and the equivalent Y'1 or Y'2 admittance. b) Modeled C'_{sub} vs. frequency behavior of hr substrates thin-film SOI diodes for strong inversion and accumulation regime.

The anode or cathode to substrate impedance of our simpler case Z_{sub} is in SOI mainly due to a MOS capacitor C_{sub}. Therefore, depending on the electrode (we will call it the gate in the following) to substrate equivalent voltage, $V_{gb_{eq}}$, C'_{sub} can cross three main different regimes: accumulation ($V_{gb_{eq}} < 0$ if p-type substrate), depletion and inversion($V_{gb_{eq}} > 0$).

In fig. 9.a and b, we can see the equivalent circuits for each regime and the evolution of the associated capacitance per unit area C'_{sub} with frequency in inversion and accumulation regimes respectively. $V_{gb_{eq}}$ is related to the actual gate to substrate voltage on contacts, V_{gb}, by (Tsividis, 1999):

$$V_{gb_{eq}} = V_{gb} - \phi_{ms} + \frac{Q'_{BOX}}{C'_{BOX}} \tag{46}$$

Q'_{BOX} and C'_{BOX} are the BOX fixed charge density and capacitance per unit area respectively. ϕ_{ms} is the contact potential or work function difference between gate and substrate and is given for the cathode and anode cases respectively by:

$$\phi_{ms_c} = -Ut.log(\frac{N_s.N_d}{ni^2}) \quad \phi_{ms_a} = Ut.log(\frac{N_s}{N_a}) \tag{47}$$

where N_s is the p-type substrate doping, and N_d and N_a are the cathode and anode doping levels respectively.

In actual processes, under normal (low) voltage operation, the trapped charge density (typ. value of $2 \times 10^{10}.q\ [C/cm^2]$) is usually the dominant term in $V_{gb_{eq}}$ and leads C'_{sub} into the strong inversion regime even for the anode.

In inversion, at very low frequency, any change in the gate-substate voltage V_{gb} (i.e. the cathode- or anode-substrate voltage of the diode) and then in the gate charge, is balanced by a change in the thin inversion charge just underneath the BOX and the capacitance is dominated by the BOX capacitance. Physically, an abundance of electrons exists immediately below the oxide and forms the bottom "plate" of the oxide capacitor, just as an abundance of holes provides that plate in the case of accumulation regime. On the contrary, in the depletion regime, there is no highly conductive inversion or accumulation layer under the BOX, and any change in V_{gb} must be compensated by a change in the depth of the depletion region (X_d) with the surface potential Φ_s and thus V_{gb}. The equivalent capacitance C'_{sub} is then a series combination of the BOX capacitance and the depletion capacitance C'_b and is then lower than C'_{BOX} (Raskin, 1997):

$$X_d = \sqrt{\frac{2\epsilon_{si}.\Phi_s}{q.N_s}} \quad C'_b = \frac{\epsilon_{si}}{X_d} \quad C'_{sub} = \frac{C'_{BOX}.C'_b}{C'_{BOX} + C'_b} \tag{48}$$

For higher frequencies however, the inversion layer charge cannot keep up with the fast changing δV_{gb} and the required charge changes must be provided by covering or uncovering acceptor atoms at the bottom of the depletion region, just as in the case of depletion operation. Again the equivalent capacitance C'_{sub} becomes a series combination of the BOX capacitance and the depletion capacitance C'_b and is then lower than C'_{BOX}.

The relaxation time of minority carriers expresses the inertia of the inversion layer under the oxide layer. Sah and al (Sah et al., 1957) have demonstrated that the finite generation and recombination within the space charge region is the dominant factor in controlling the frequency response of the inversion layer. Hofstein and Warfield (Hofstein& Warfield, 1965) define for the strong inversion regime layer a resistance (R'_{gr}) associated with this generation-recombination U (see fig. 9.a), as follows:

$$R'_{gr} = \frac{\Phi_s}{q.X_d.U} \quad U = \frac{n_i}{\tau_o} \tag{49}$$

where τ_o is the time carrier density fluctuation to decay to its equilibrium concentration by recombination through traps and is typically the order of 10^{-6} sec (Nicollian& Brews, 1982). This equivalent resistance allows one for taking into account the frequency response of the inversion layer in the dark. The relaxation time of the minority carriers is given by $\tau_{rg} = R'_{rg}.C'_b$. For the calculation of X_d in strong inversion, we can consider the classical approximation of Φ_s, the surface potential, pinned to two times the Fermi level (Tsividis, 1999).
For still higher frequencies in the GHz range, the relaxation time of the majority carriers cannot be neglected anymore and can be modeled by a resistance, R'_{si} and a capacitance, C'_{si} which are the substrate silicon resistance and capacitance respectively (Raskin, 1997). Noting d_{si} the Si substrate thickness, we have:

$$R'_{si} = \frac{d_{si}}{\sigma_{si}} \quad C'_{si} = \frac{\epsilon_{si}}{d_{si}} \tag{50}$$

In this range of frequencies, C'_{sub} is then dominated by C'_{si} and is therefore very small which is advantageous for high speed design. The capacitance behaviour of a typical thin film SOI diode in a $0.13\mu m$ PDSOI technology is plotted with standard and high resistive substrate (hr) on fig. 10.a. The diode exhibits total length and width of $50\mu m$ and an intrinsic length of $2\mu m$. The higher the substrate resistivity, the lower the frequency at which this transition happens.

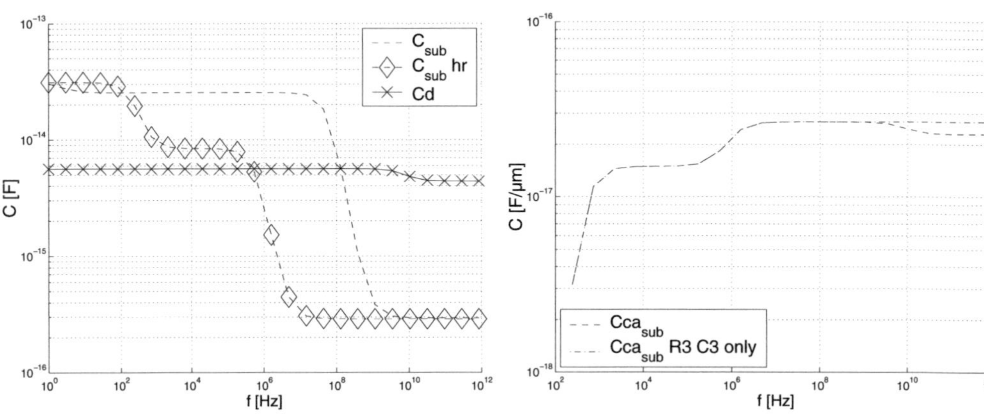

(a) C_{sub} hr and sr vs. C_d (b) Anode to cathode substrate capacitance $C_{ca_{sub}}$

Fig. 10. a) Capacitance vs. frequency behavior of thin-film ST 013 SOI diodes. a) Cathode to substrate capacitance C_{sub}, assumed in the strong inversion regime, for standard and high resistivity (hr) substrates vs. the ideal diode capacitance C_d. ($L_i = 2\mu m$, L_{tot} and W of $50\mu m$ and $L_{PN} = 0.34\mu m$. b) Anode to cathode substrate capacitance by μm width, $C_{ca_{sub}}$, models. hr substrates. $L_i = 5\mu m$,m=2.

If the general behaviour of C_{sub} versus frequency can be now well understood by the model, the plateau values of the model are too low when compared to numerical simulations. This is also pointed out and explained by (Raskin, 1997) when comparing model to measurements. The higher value of the capacitance is due to a fringing field effect. Indeed the length of the diffusions L_{PN} is very small compared to the thickness of the substrate and then the effective area of the capacitor is higher than just $L_{PN}.W$. We then have to use a correction factor, K_1 or K_2 for Y_1 or Y_2. With deep submicron processes, we even have important fringing field effect for C_{BOX}. A coefficient K_{BOX} has then to be introduced. These three coefficients increase with the intrinsic length showing a field confinement effect of the adjacent electrodes. Values for ST 0.13μm process are shown in table 1.

3.3 Coupling effect

Numerical simulations with Atlas or ISE show that the model of the anode to cathode impedance which only take into account the depletion capacitance C_d is too simple. Simulations, indeed, show that the coupling effect through the substrate is dominant at high frequency and therefore cannot be neglected. It shows the same transition frequencies as the capacitances to substrate (fig. 7) and hence is based on similar phenomena than those discussed above. We have to use a new admittance Y_3 as shown in the equivalent model of fig 6.b and from there we can compute $Y_{ca_{sub}}$, the coupling admittance through the substrate. A model was firstly introduced in (Raskin, 1997) to calculate the coupling between coplanar line on SOI substrate only using R3 and C3. The expressions of R3 and C3 are given using the approximation of two infinite lines on a very thick silicon substrate ($t_{si} <<< d_{si}$) (Raskin, 1997), (Walker, 1990) and K3 is a fringing factor.
However, when the diode is not fully depleted ($L_i > L_d$), simulations show a decrease of $C_{ca_{sub}}$ above 10GHz, while this model only shows a constant value (see fig. 10.b). Our explanation is that part of the electric field induced in the substrate is curved upwards and cross again the buried oxide as well as the quasi neutral region. An exact model is quite complex but as the field always see the BOX and a silicon region by adding $Rqni/K_{qni}$ and $K_{qni}.C_{qni}$ we can model the transition with a very good accuracy (see fig. 7). For the expressions of R3 and C3 we derived:

$$R3 = [m.K3.\frac{\pi\epsilon_0\sigma_{si}}{4ln[\frac{\pi.min(L_i,L_{zd})}{K_{BOX}.L_{PN}+t_{si}}+1]}W]^{-1}[\Omega] \quad C3 = m.K3.\frac{\pi\epsilon_0(\epsilon_{rsi})}{4ln[\frac{\pi.min(L_i,L_{zd})}{K_{BOX}.L_{PN}+t_{si}}+1]}W[F] \quad (51)$$

In this formulation, the value of K3 was constant vs. L_i and equal to 5 for L_{PN} of 0.34μm.
For the front coupling through the air, numerical simulations show that the fringing field capacitance through the air, C_{air}, cannot be neglected because the thicknesses of the silicon film and of the electrode, t_{al}, were small compared to L_i. We can assume a formulation to compute this capacitance coupling similar to that used for C3 but with air instead of silicon:

$$C_{air} = m.0.5.\frac{\pi\epsilon_0}{4ln[\frac{\pi.L_i}{L_{PN}+tal}+1]}W[F] \quad (52)$$

We also add the capacitance through the thin film with a fitting coefficient K_d close to unity for L_i small and reducing for increasing values of L_i, for a larger portion of the electric field propagates through the air.
In figure 11.a we see a comparison of the cathode capacitance for standard and high resistive substrates. In the bandwidth of interest for high speed circuits starting from a few hundred of MHz, we see that there is no clear advantages of using a high-resistive substrate.

L_i	K_{BOX}	K_1=K_2	K_3	K_{si}	K_d
1	1.2	3.2	5	0.25	0.9
2	1.8	3.7	5	0.25	0.8
3	1.9	5	5	0.05	0.62
4	1.95	6.5	5	0.05	0.61
5	2	8	5	0.03	0.6
10	2.5	12	5	0.03	0.2

Table 1. Fringing effect coefficients value vs. L_i for ST013 ($L_{PN} = 0.34\mu m$).

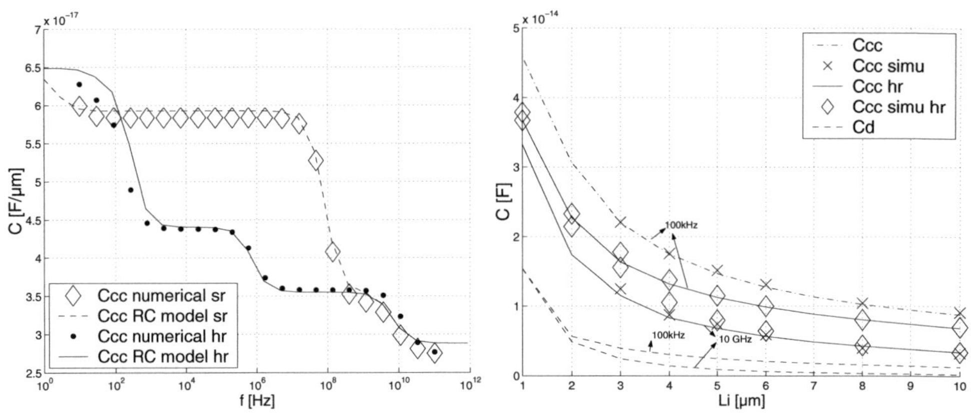

(a) C_{cc} vs. frequency (b) C_{cc} vs. L_i

Fig. 11. Comparison of the cathode capacitance of ST013 thin-film SOI diodes for sr and hr substrates a) by μm width vs. frequency. $L_i = 3\mu m$, m=2. b) vs. L_i of a 50x50 μm^2 ST013 thin-film SOI diodes @100kHz and 10GHz. The ideal case, C_d is also plotted.

In figure 11.b, we show modeled and simulated capacitances of a PIN diode of 50x50 μm^2 vs. L_i at 100kHz and 10 GHz for sr and hr substrates. In all cases, this capacitance mainly decreases with L_i because the number of fingers decreases as well. C_{cc} is also bigger than C_d, the ideal diode case, but keeps same order of magnitude. Again, we observe that, if the value of C_{cc} is lower for hr substrates than for sr ones at 100kHz, there are sensibly equal at 10GHz.

3.4 2nd order effects: reduction of the depletion plateau of C_{sub} with light

For a high resistivity substrate in the usual case of strong inversion, the effect of depletion is more pronounced and makes C_{sub} already low compared to C_d at a still lower frequency of 10kHz (the beginning of the depletion plateau). However this is only true if no light illuminates the depletion region in the substrate. This is the case in the dark (part of N+ and P+ regions covered by metal electrodes) or everywhere at low wavelength (typ. $<$ than 400 nm) where all the light is absorbed in the thin Si-film.
If light is absorbed in the depletion region in the substrate, the positive effect of depletion is firstly reduced because it reduces the surface potential Φ_s and therefore X_d (Grosvalet& Jund, 1967). The plateau value of C_{sub} increases with the power absorbed in this area and then with P_{in}.
Secondly if light is absorbed in depletion region in the substrate, the beginning of the depletion plateau happens at higher frequencies because an extra photogeneration process

speeds up the thermal minority carriers process in the depletion region (Grosvalet& Jund, 1967). Equation 49 has then to be modified in the following way:

$$R'_{gr} = \frac{\Phi_s}{q.X_d.U} \quad U = \frac{n_i}{\tau_o} + g \tag{53}$$

where g is the equivalent or mean generation term in the depletion region.

3.5 Substrate losses

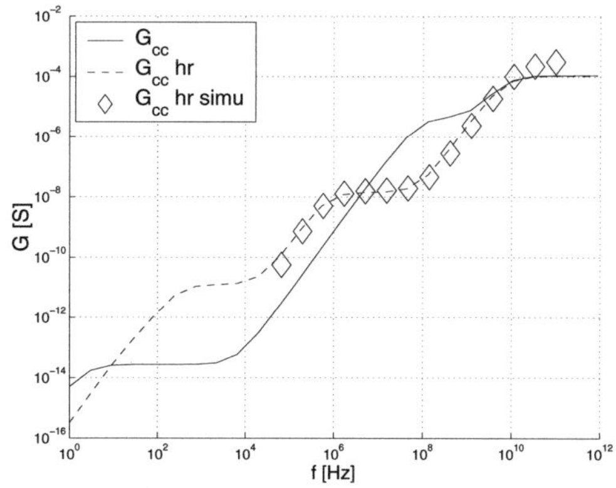

Fig. 12. Comparison of conductance vs. frequency behavior of standard and high resistivity (hr) substrates thin-film SOI diodes ($L_i = 3\mu m$, L_{tot} and W of $50\mu m$ and $L_{PN} = 0.34\mu m$ (ST 013)in strong inversion. For the hr case model is also compared to numerical simulation

The cathode (or anode) impedance has a complex value. If the imaginary part is related to C_{cc}, the real part can be modeled by an equivalent conductance G_{cc} in parallel with C_{cc}. G_{cc} takes into account the signal losses through the substrate. To be negligible, G_{cc}^{-1} has to remain high compared to the next stage equivalent resistor, R, which conditions the current to voltage gain in the bandwidth of interest. For actual SOI processes the bandwidth of interest is in the tens of GHz and R is lower than $1k\Omega$. Fig 12 shows the modeled evolution of G_{cc} for hr and sr SOI substrates. The same transitions than for C_{cc} are appearing. At high frequencies C_{BOX} looks more and more like a short and the losses are increasing. In both cases (hr and sr), however, G_{cc}^{-1} remains at least 10 times larger than R in the 10GHz range. For the hr case, we also compare the modeled G_{cc} curve to that given by the numerical simulations and can note the very good agreement.

3.6 Impedance measurements

In order to further validate our RC model of the PIN photodiodes, on-wafer S parameter measurements were performed. 6 lateral thin-film ungated PIN photodiodes were designed on ST 0.13 μm PD SOI technology with different device parameters (intrinsic length L_i, N^+ and P^+ diffusion lengths L_{pn}, and number of finger m) and with coplanar accesses in order to be able to characterize these devices in a wide range of frequencies. Parameters and a photograph of the realized diodes are shown in Fig. 13.

Most of the diodes were realized using the conservative value of L_{pn}=1.36μm used in the last design rules we receive from ST for lateral photodiodes. One diode was realized using the value of L_{pn}=0.34μm which is the value for the source and drain extension of

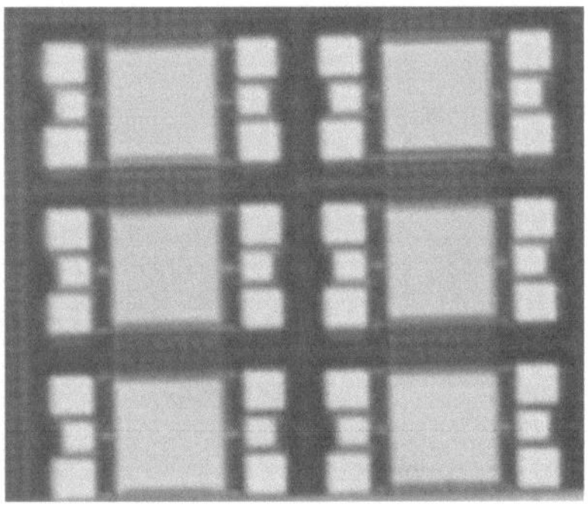

L_i (μm)	L_{pn} (μm)	L_{tot} (μm)	m	W (μm)
1	1.36	249.16	105	250
2	1.36	250	75	250
2	0.34	250.72	107	250
5	1.36	249.4	39	250
10	1.36	251.28	22	250
100	1.36	204.08	2	250

Fig. 13. Photograph and parameters of the PIN photodiodes realized in ST 0.13 μm PD SOI technology.

the MOS transistor in this technology. This last diode wasn't working, certainly because of mask misalignments (a gateless device with N^+ and P^+ contacts is more subject to mask misalignments than a MOS transistor).
AC-capacitances given by numerical simulations (using the parameters $L_{pn} = 1.36\mu m$) and by our model (only readjusting the value of L_{pn}, but leaving unchanged the fringing field coefficients obtained for $L_{pn} = 0.34\mu m$) were in good agreement and no further fitting was necessary. We then compared model and measurements. These measurements were obtained after a calibration to remove the impedance effect of the RF probes and cables used to connect the device to the spectrum analyzer. An open substraction was also performed to remove the impedance of the access pad. This was mainly a capacitive impedance (capacitance of about 90fF). It was relatively negligible compared to the diode impedance, except for that with an intrinsic length of $100\mu m$. A metal path of about $50\mu m$ long, which wasn't removed by de-embedding, remains between the device anode and cathode and their related access pad.
The measurements were done in the 40MHz-40GHz band under illumination or not. For each of these measurements, the DC voltage of the anode was connected to the ground while the DC-voltage of the cathode was successively fixed to 0, 1V and 3V with a bias-T.
In this range of frequency, we observed as expected the transition in the capacitive behavior of C_{cc} between its mid-range value (dominated by the depletion plateau of C_{cb}) to the high range value (dominated by C_{ca} and where C_{cb} is low because the substrate behave like a dielectric) (fig. 14.a). The mid-frequency range plateau value is expected to be influenced by bias, illumination and buried oxide trapped charges since the value of the depletion capacitance which is dominant in this range is strongly dependent on these parameters. This was observed in the measurements as can be seen in fig. 14.b. The high frequency range value is quite unaffected by these parameters as expected since the whole substrate now behaves like a dielectric.
The transition frequency between mid and high range value depends on the substrate doping: The higher the substrate doping, the higher this frequency. From our measurements, we deduce that the substrate doping should be of the order of $1.10^{14}\ cm^{-3}$. This value is, however, higher than the real physical doping of the hr-substrate. This typical effect with SOI hr-substrate (Lederer& Raskin, 2006) is explained by surface conduction in the low resistive

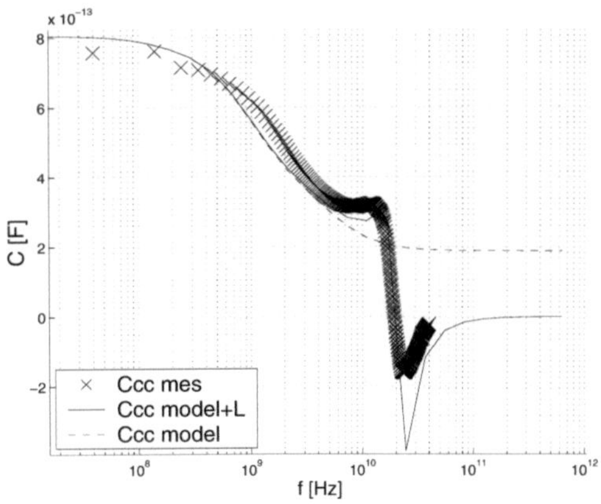

(a) Comparison of measured and modeled Capacitance vs. frequency in the dark

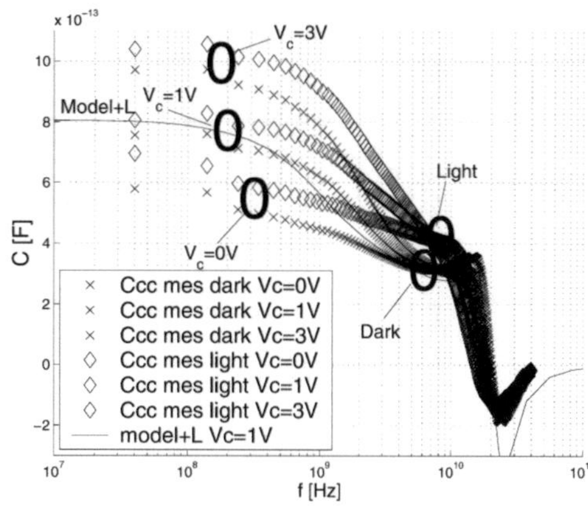

(b) Measured capacitance vs. frequency for different bias and illumination conditions. A modeled curve at Vc=1V in the dark is also shown for comparison

Fig. 14. Thin-film SOI diodes ($L_i = 10\mu m$, L_{tot} and W of about $250\mu m$ and $L_{PN} = 1.36\mu m$. ST 0.13 μm PD SOI technology).

inversion layer that appears just underneath the BOX and the presence of coplanar accesses for the measurements. The impedance of the cathode to the substrate backgate electrode is now in parallel with the impedance of the cathode to the ground plane of the coplanar access via this top substrate inversion layer. It is the latter which dominates the high frequency transition and presents the same kind of RC transition behavior but at a higher frequency because of the lower resistivity of this inversion Si layer.
Finally, a resonance effect appears around the 10GHz range. The imaginary part of Y_{cc} first increases and presents a positive peak, then decreases and presents a negative peak. This is attributed to the self inductance of the metal path between the pads and the electrodes. Indeed by simply adding an inductor of 0.25nH in series with the cathode of the diode, which is a good approximation of having an inductor of 0.125nH in series with the cathode and with the anode if the resonance effect appears in a frequency range where C_{cb} and $C_{ab} < C_{ca}$, our model predicts a very similar behavior (see curves labelled model+L on fig. 14 to fig. 15).
The amplitude of the peak and the frequency at which it happens depend on the capacitor value. It varies, therefore, with L_i and with the inductance value. The higher their LC product, the lower the frequency at which it happens. For usual 50x50μm^2 diodes and shorter metal lines of monolithically integrated diodes and circuits, this effect should not appear. It however have to be kept in mind during the layout phase of the circuits (avoid too long connection lines) and may be checked again and incorporated by simulation after the layout phase. This resonance effect could also be useful, if well controlled, to increase the bandwidth of the system (Gray& Meyer, 1984) as it is done to increase the bandwidth of transimpedance amplifiers (Maxim, 2004).

4. Conclusions

Speed performances of thin-film SOI PIN photodetectors have been investigated in terms of transit time and RC frequency. Our original models, fully validated by 2D numerical simulations and measurements, enable to deeply understand the underlying physical

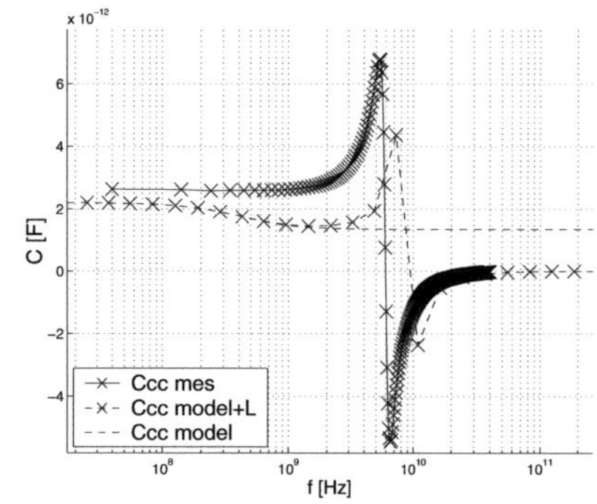

(a) $L_i = 1\mu m$

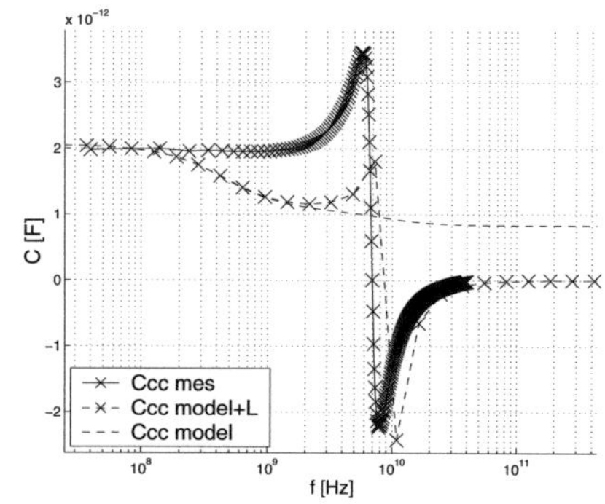

(b) $L_i = 2\mu m$

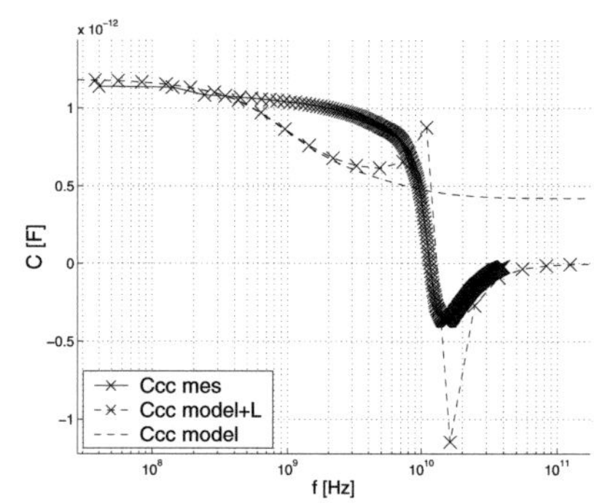

(c) $L_i = 5\mu m$

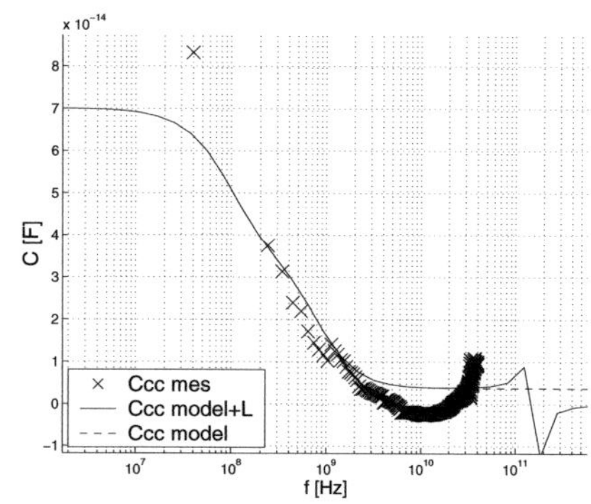

(d) $L_i = 100\mu m$

Fig. 15. Comparison of measured and modeled capacitance vs. frequency of thin-film SOI diodes ($L_i = 1\mu m$, L_{tot} and W of about $250\mu m$ and $L_{PN} = 1.36\mu m$. ST 0.13 μm PD SOI technology).

phenomena and predict and optimize their speed performances for the target applications. Concerning the transit time frequency, our modeling allows one to simply and accurately select the intrinsic length required for a given bandwidth. We showed that as long as the entire I region is laterally depleted, the transit time limit is due to fast drift and the related -3dB frequency, f_{tr}, decreases as L_i^2. If L_i becomes greater than L_d, carriers transit is dominated by a slower diffusion mechanism and the related -3dB frequency decreases faster with L_i. The effectiveness of BOX insulation from the slow substrate photogenerated current and resulting problem of bandwidth degradation due to partial isolation of the BOX in near IR wavelength have been discussed for the first time and solutions have been presented.

Concerning the modeling of the diode impedance, our physical RC model can be implemented in a circuit simulator and allows the co-design and optimization of the photodiode and the readout circuit as a function of design parameters such as the intrinsic length of the diode, L_i. At low frequency, the total cathode capacitor, C_{cc}, is dominated by the cathode to substrate capacitor, C_{cs} which is a MOS capacitor (Raskin, 1997). At higher frequency, C_{cs} reduces below the value of C_d as carriers in the Si substrate cannot follow the ac-signal and the substrate behaves like a dielectric. C_{cc} then also reduces but cannot reach the ideal lowest value of C_d, as at higher frequency the coupling through the substrate (Y_3) between anode and cathode is increased and then C_{ca} increases. Above about 100MHz, C_{cc} of SOI PIN diodes remains,

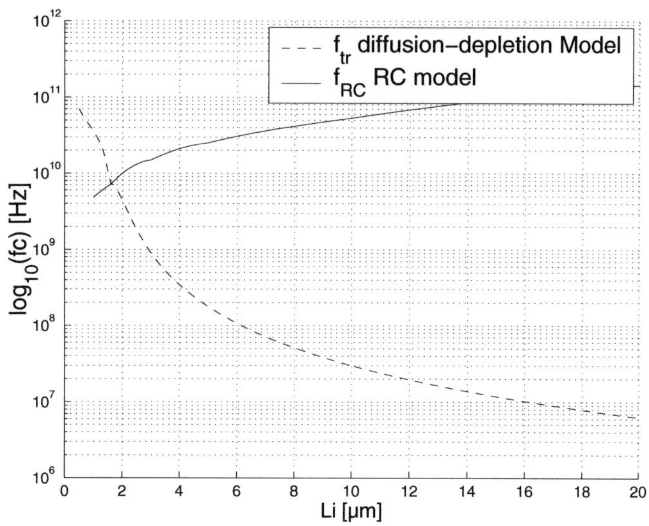

Fig. 16. Evolution of f_{tr} and f_{RC} with Li, bias voltage Vd=-1V, photodiode area A_t=50x50μm^2, typical load resistor of $1k\Omega$.

however, very low compared to the equivalent capacitor of integrated bulk diodes so that, for identical speed performances, we can increase the load resistor and then increase the overall system sensitivity in SOI compared to bulk.
Consequently, the total -3dB frequency combining f_{tr} and f_{RC} shows an optimum vs. L_i (fig. 16) which in thin SOI diodes can reach a few tens of GHz, while the fastest integrated bulk diodes are typically limited to a few GHz only (Zimmermann, 2000). A SOI diode with L_i of $6\mu m$ already fulfills the 250MHz bandwidth requirement for actual Blue DVD specifications under 1V operation, while a $2\mu m$ device is suitable for the 10Gb/s Ethernet standard.

5. References

A. Afzalian, D. Flandre," Physical Modeling and Design of Thin-Film SOI Lateral PIN Photodiodes", *IEEE Transaction on Electron Devices*, Volume 52, No. 6, June 2005, p. 1116-1122.

A. Afzalian, D. Flandre, "Speed performances of thin-film lateral SOI PIN photodiodes up to tens of GHz", 2006 IEEE SOI conference, Oct 2-5, Niagara Falls, New York.

Aryan Afzalian and Denis Flandre, "Monolithically Integrated 10Gb/s Photodiode and Transimpedance Amplifier in Thin-Film SOI CMOS Technology", IEE Electronics Letters, Volume 42, No. 24, 23rd November 2006, pp. 1420-1421.

A.B. Apsel, A.G. Andreou, "A low-power silicon on sapphire CMOS optoelectronic receiver using low- and high-threshold devices", *IEEE Trans. on Circuits and Systems I: Regular Papers*, Vol. 52, No. 2, Feb. 2005, pp. 253-261.

D. M. Caughey and R. E. Thomas, "Carrier mobilities in silicon empirically related to doping and field", *Proc. IEEE 55*, pp. 2192-2193, 1967.

S. Csutak, J. Schaub, W. Wu, R. Shimer, and J. Campbell, "CMOS-Compatible High-Speed Planar Silicon Photodiodes Fabricated on SOI Substrates", *IEEE J. of Quantum Electronics*, vol.38, No. 2, pp. 193-196, Nov. 2002.

R. Garg and I.J. Bahl, "Characteristics of Coupled Microstriplines", *IEEE Transactions on Microwave theory and techniques*, vol.27, No. 7, pp. 700-705, July 1979.

P. Gray, R. Meyer, "Analysis and Design of Analog Integrated Circuits", John Wiley & Sons, 1984.

J. Grosvalet and C. Jund, "Influence of Illumination on MIS Capacitances in the Strong Inversion Region", *IEEE Transactions on Electron Devices*, vol.14, No. 11, pp. 777-780, Nov. 1967.

M. Hobenbild, P. Seegebecht, H. Pless, W. Einbrodt, "High-speed photodiodes with reduced dark current and enhanced responsivity in the blue/uv spectra", *EDMO 2003*, 17-18 Nov. 2003 p. 60-65.

S.R. Hofstein and G. Warfield, "Physical limitations on the frequency response of a semiconductor surface inversion layer", *Solid-State Electronics*, Pergamon Press, vol.8, pp. 321-341, 1965.

M. Kirschning and R.H. Jansen, "Accurate Wide-Range Design Equations for the Frequency-Dependent Characteristic of Parallel Coupled Microstrip Lines", *IEEE Transactions on Microwave theory and techniques*, vol.32, No. 1, pp. 83-90, Jan. 1984.

Steven J. Koester, Clint L. Schow, Laurent Schares, Gabriel Dehlinger, Jeremy D. Schaub, Fuad E. Doany, and Richard A. John, "Ge-on-SOI-Detector/Si-CMOS-Amplifier Receivers for High-Performance Optical-Communication Applications," *J. Lightwave Technol.* 25, 46-57 (2007).

J. Laconte, C. Dupont, D. Flandre, and J.-P. Raskin, "SOI CMOS Compatible Low-Power Microheater Optimization for the Fabrication of Smart Gas Sensors", *IEEE Sensors Journal*, Vol. 4, No. 5, Oct 2004, pp 670-680.

D. Lederer and J.-P. Raskin, "Bias effects on surface crosstalk in HR SOI substrates", *Proceedings of the 6th Topical Meeting on Silicon Microwave Integrated Circuits for RF Systems*, San Diego, January 2006, pp. 8-11.

A. Maxim, "A 10Gb/s SiGe transimpedance amplifier using a pseudo-differential input stage and a modified Cherry-Hooper amplifier", *Symposium on VLSI Circuits 2004*, 17-19 June 2004, pp. 404 - 407.

E.H. Nicollian and J.R. Brews, *MOS Physics and Technology*, Bell Laboratories, Murray Hill, New Jersey, John Wiley and Sons, New-York, 1982.

J.-P. Raskin, "Modeling, Characterization and Optimization of MOSFET's and Passive Elements for the Synthesis of SOI MMIC's", Ph.D thesis, *Universite Catholique de Louvain, Laboratoire d'Hyperfrequences*, Louvain-La-Neuve, Belgium, Dec. 1997.

C.T. Sah, R. Noyce and W. Shockley, "Carrier Generation and Recombination in PN Junctions and PN Junction Characteristics", *Proceedings of the IRE*, pp. 1228-1243, September 1957.

A. Sarto, B. Van Zeghbroeck, "Photocurrents in a Metal-Semiconductor-Metal Photodetector" *IEEE J. of Quantum Electronics*, Vol. 33, No. 12, Dec. 1997, pp. 2188-2194.

S.M. Sze, "The Physics of Semiconductor Devices", *New York: Wiley*, 1981.

G. Torrese, "Ultra-Wide Bandwith Photodetectors for Optical Receivers", Ph.D thesis, *Universite Catholique de Louvain, Laboratoire d'Hyperfrequences*, Louvain-La-Neuve, Belgium, Feb. 2002.

Y. Tsividis, *Operation and modeling of the MOS transistor*, Mc Graw-Hill, 1999.

C.S. Walker, *Capacitance, inductance and crosstalk analysis*, Artech House, Boston, 1990.

H. Zimmermann, "Integrated Silicon Opto-electronics", Springer, Berlin 2000.

4

Modeling and Optimization of Three-Dimensional Interdigitated Lateral p-i-n Photodiodes Based on $In_{0.53}Ga_{0.47}As$ Absorbers for Optical Communications

P Susthitha Menon, Abang Annuar Ehsan and Sahbudin Shaari
Universiti Kebangsaan Malaysia, Bangi, Selangor
Malaysia

1. Introduction

Access networks such as Fiber-to-the-Home based on passive optical networks (FTTH-PON) are experiencing a paradigm shift where these 'last-mile' networks are experiencing the need to provide converged services to the end-user at home. Triple-play services such as data and voice operating at the optical wavelength, λ=1310 nm as well as video (λ=1550 nm) at a minimal speed of 2.5 Gbps are demanded to achieve an all-optical-network revolution (Kim, 2003; Lee & Choi 2007). It is estimated that in 2011, there will be 10.3 million FTTH households in the USA alone (Lee & Choi 2007). Thus, there arises a need to produce optical components which can be fabricated easily and in a cost-effective manner to cater for this ever-increasing demand.

The development of interdigitated lateral p-i-n photodiodes (ILPP) based on $In_{0.53}Ga_{0.47}As$ (InGaAs) absorption layer can be achieved using cheap and easy CMOS fabrication techniques such as diffusion and ion implantation. This can cater for the ever increasing demand of fiber-to-the home passive optical access networks (FTTH-PON) operating at a minimal speed of 2.5 Gb/s. The InGaAs ILPP which converts optical signals to electrical signals in the optical receiver has advantages compared to other photodiode structures because it has a high-resistance intrinsic region thus reducing Johnson noise, has low dark currents, permits a large detection area along with a low device capacitance, and can be monolithically integrated with planar waveguides or other devices. This chapter summarizes our key results on the modeling, characterization and optimization of ILPP based on $In_{0.53}Ga_{0.47}As$ absorbers for optical communications. A three dimensional model of ILPP InGaAs operating at the optical wavelength, λ of 1.55 μm was developed using an industrial-based numerical software with a proposed fabrication methodology using spin-on chemicals. New parameters for three different carrier transport models were developed and the proposed design was characterised for its dark and photo I-V, responsivity, -3dB frequency and signal-to-noise ratio (SNR) values. Statistical optimization of the InGaAs ILPP model was executed using fractional factorial design methodology. Seven model design factors were investigated and a new general linear model equation that relates the responsivity to significant factor terms was also developed (Menon, 2008; Menon et al.,2008a; Menon et al., 2008b; Menon et al. 2009)

2. Application, fabrication and simulation of lateral P-I-N photodiodes

2.1 The application of ILPP

Else than optical communication systems, InGaAs-based photodiodes are also used in optical measurement systems such as for high precision length measurement, light patterns, spectrum analyzer, speed measurement in luminuous flow as well as in imaging applications. Other application include high speed sampling, optoelectronic integrated circuits (OEIC), high speed device interconnects, optoelectronic mixers and also for microwave single sideband modulation.

Photodiode design structures can be categorised based on the illumination direction, detection mechanism and the structure itself. The illumination direction of a photodiode can be classified into two; the vertical illuminated photodiode (VPD) and the edge illuminated photodiode which is known as the waveguide photodiode (WGPD). VPDs are the preferred photodiodes for OEICs due to its planar structure but the layers in VPDs are grown epitaxially using complex fabrication methods. Meanwhile WGPDs overcome the limitation of the bandwidth-efficiency product in VPDs because the electrical transit of carriers are perpendicular to the optical propagation direction. Photodiodes can also be classified based on the detection mechanism such as the avalanche photodiode (APD) or the p-n junction photodiode where the former has sensitivity limits of 5-10 db higher than the latter due to the multiplication region at the absorption layer hence producing high gains.

The structure of photodiodes can be divided to p-i-n types or metal-semiconductor-metal (MSM) types (inclusive of Schottky photodiodes). Vertical p-i-n photodiodes consist of one p+ doped layer at the topmost region followed by the absorption (intrinsic) layer in the middle and finally the n+ dope region at the bottom. MSM photodiodes have metal fingers deposited on the semiconductor and photons are detected via collection of electron-hole pairs that experience drift due to the presence of the electric field between the metal fingers (Zhao, 2006).

The internal gain mechanism in APDs makes it suitable to be used in long distance fiber optic transmission systems. However, the impact ionization produces additional noise in APDs and reduces the signal-to-noise ratio (SNR). Else than that, APDs are costly, hence they are normally not utilized in medium and short-haul optical communication systems (Huang, 2003). MSM photodiodes have lower capacitances compared to VPDs for the same amount of device active area. It is also a planar structure and can be integrated wasily with MESFET-based pre-amplifiers. However, the fabrication of MSM photodiodes is not compatible with CMOS processes where ohmic junctions are preferred compared to metal junctions (Menon, 2005). MSM photodiodes also have a larger dark current compared to p-i-n photodiodes (Koester et al., 2006).

In this chapter, we will discuss about the formation a p-i-n photodiode structure which is a combination of WGPD and MSM photodiodes and is known as the interdigitated lateral p-i-n photodiodes (ILPP). It can be given surface or edge illumination such as in the WGPD but the arrangement of the p, i and n region are in a planar form. Interdigitated electrodes such as those used in MSM photodiodes reduces the device capacitance and increases the area of optical absorption. The planar structure eases monolithic integration compared to a vertical structure (Koester et al., 2006). Moreover, the ILPP can be fabricated using standard CMOS processes such as diffusion or ion implantation. Fig. 1 shows the differences between different photodiode structures.

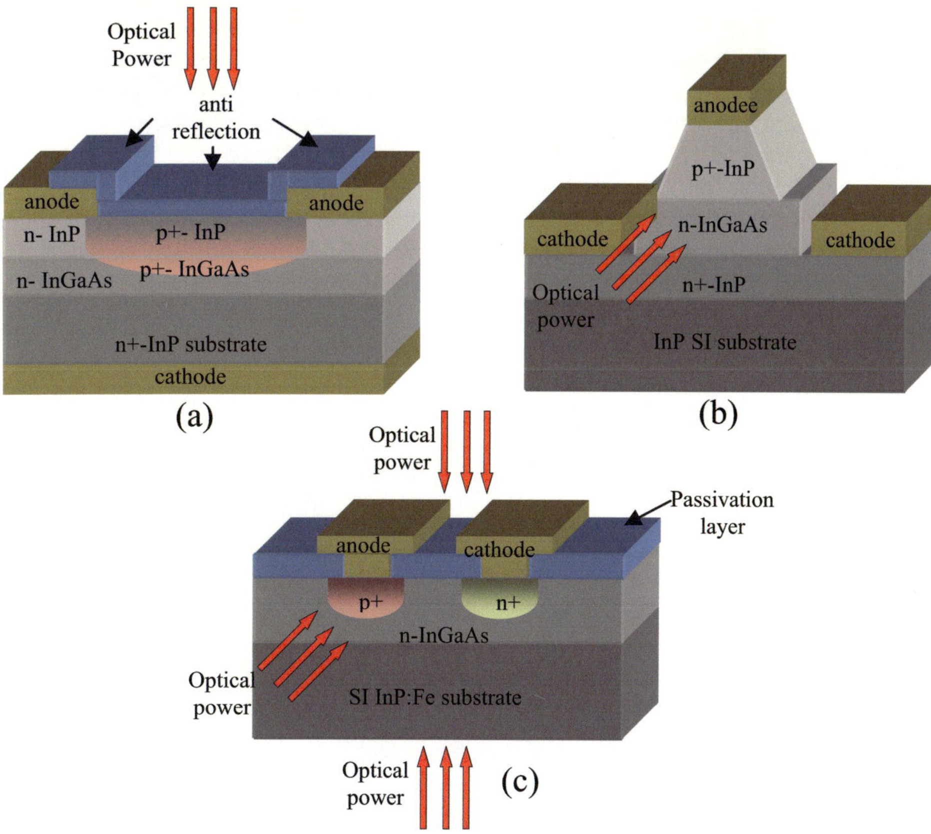

Fig. 1. Various p-i-n photodiode design structures based on InP substrate for (a) VPD, (b) WGPD and (c) ILPP

2.2 III-V Material: In(0.53)Ga(0.47)As

At optical wavelengths of 1.55 µm, III-V semiconductor materials are normally used because the energy gap can be modified according to the intended wavelength by changing the relative composition of the material which is lattice-matched to the substrate. Three basic alloy systems which are useful for telecommunication applications are AlGaSb, InGaAs and HgCdTe. Although all three ternary materials can be used for the development of an ILPP, InGaAs remains as the preferred choice due to the available technology for laser, LED and diodes fabrication developed from this material as well (Tsang, 1985). Indium gallium arsenide (InGaAs) is a III-V material that consist of indium, gallium and arsenide components. It is used in high power and high frequency optoelectronics applications due to the high electron and hole saturation velocity (~6 x 10^6cm/s and ~4.6 x 10^6 cm/s) as well as the high absorption coefficient (0.65 μm^{-1} at λ=1.55 µm). The electron mobility in InGaAs is 1.6 times higher than GaAs and is 9 times higher compared to Si (Sze, 2002). The energy gap of InGaAs which is 0.75 eV at optical wavelength of 1550 nm makes it a suitable material to be used as a detector in fiber optic communication systems both at 1300 and 1550 nm wavelengths.

The indium content in InGaAs determines the two dimensional charge carrier density. The optical and mechanical properties of InGaAs can be modified by changing the ratios of indium and gallium to form $In_xGa_{1-x}As$. InGaAs-based devices is normally developed on indium phosphide (InP) substrates which has a energy gap of 1.35 eV. To match the lattice constant of InP dan to avoid mechanical strain, the commonly chosen composition is $In_{0.53}Ga_{0.47}As$ where the cut-off wavelength is at 1.68 µm. By increasing the ratio of In compared to As, the cut-off wavelength can be increased upto 2.6 µm. The lattice constant and cut-off wavelength for alloys that produce InGaAs is shown in Fig. 2.

A high electron mobility transistor (HEMT) utilizing InGaAs channels is one of the fastest transistors which can achieve speeds upto 600 GHz. InGaAs is a popular material in infra-red photodiodes and is replacing Ge as a photodiode material mainly because of the low dark current whereas in APDs, the multiplication noise in the active multiplication region based on InGaAs is much lower compared to Ge. Therefore, the technology and applications based on InGaAs material is wide and its usage as the absorbing layer in an ILPP is most appropriate for the current trends (Menon, 2008).

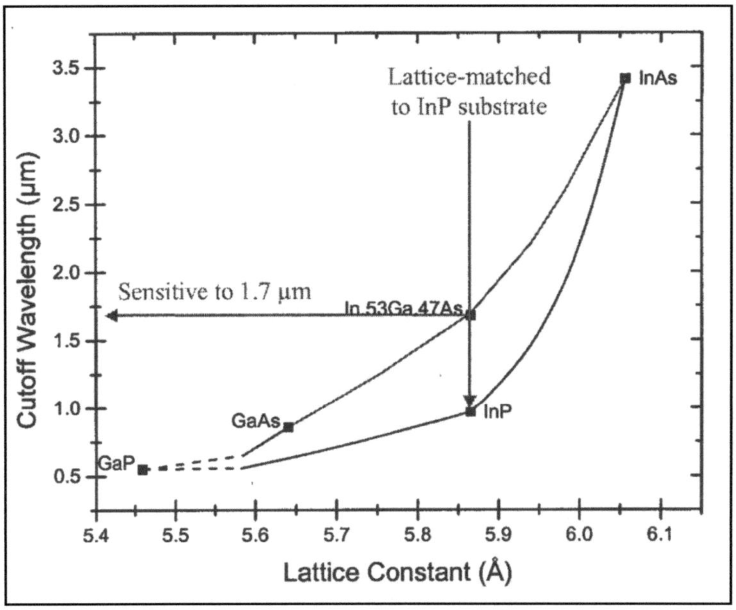

Fig. 2. Lattice constant and cut-off wavelength for alloys that produce InGaAs (Source: Goodrich, 2006)

2.3 Review of fabricated InGaAs/InP-based ILPP

The ILPP structure has been developed on many different substrates utilizing different material as the absorbing layer depending on the respective wavelengths. These include silicon (Schow et al., 1999), silicon-on-SOI (silicon-on-insulator) (Li et al., 2000), Ge-on-SOI (Koester et al., 2006) and GaAs (Giziewicz et al. 2004).

ILPP on InGaAs/InP substrates were developed by Lee et al (1989) which was integrated with an InP-based JFET amplifier. The p+ well was formed using ion implantation using Mg+ at 25 keV and dosage of 1×10^{14} cm^{-2}. The thickness of the InGaAs absorption layer was 2 μm. The width and length of the fingers were 2 μm and 47 μm respectively. Distance between the p+ and n+ fingers was 3 μm and the total active area space was 50 μm x 50 μm. Optical sensitivity of -29 dBm was achieved at a bit rate of 560 Mbit/s. Leakage current and capacitance was measured to be 1.5 μA and 130 fF. Responsivity value achieved was 0.56-0.6 A/W at optical wavelength of 1.3 μm.

Diadiuk & Groves (1985) produced a lateral p-i-n photodiode on a semi-insulating (SI) InGaAs substrate developed using the Liquid Phase Epitaxy (LPE) technique. Metal contacts consisting of AuZn and NiGeAu were deposited on the InGaAs layer with thicknesses of 2-4 μm. Dopant from the metal electrodes will diffuse into the semiconductor during the alloying process and subsequently p+ and n+ junctions are formed. Interdigitated electrode structure without an anti-reflecting layer has finger lengths of 300 μm and width of 100 μm where distance between fingers is 3 - 20 μm. The capacitance was ~ 18 pF (electrode distance 3 μm) whereas the quantum efficiency is ~40% at λ=1.24 μm. Breakdown voltage was at 50 V and the leakage current was < 1 nA at bias voltage of 0.9 V. The response time in the InGaAs substrate was 50 ps (FWHM) where f_{-3dB}=0.4/50 ps = 8 GHz for devices with finger distances of 3 μm.

A lateral p-i-n photodiode utilizing InGaAs as the absorption layer on InP substrate was developed by Yasuoka et al. (1991). Monolithic integration was achieved using a coherent receiver with waveguide coupling with a pair of interdigitated lateral p-i-n photodiodes. The -3dB frequency was at 2 GHz and quantum efficiency was 85 % at λ=1540 nm at a bias voltage of 5 V. The p+ well was formed using Zn diffusion via a SiN_x layer to form the p+ InP junction whereas the n-InP cap layer acted as the n+ region. Dark current value was < 10 nA until bias voltages of 12 V and the capacitance was 0.3 pF at V = 5V.

A planar p-i-n photodetector based on InGaAs substrate was fabricated using self-aligned contact technique (Tiwari et al., 1992). The p-type and n-type contacts were formed using W (Zn) and $MoGe_2$ metalurgy. Bandwidths exceeding 7.5 GHz and responsivity of 0.53 A/W was achieved by this device with bias voltage of 5 V and optical wavelength of 1.3 μm. Jeong et al. (2005) developed an InGaAs-based ILPP comprising of p+ and i regions to be integrated with a two terminal heterojunction phototransistor (2T-HPT). The photodiode which was fabricated using epitaxial methods with InGaAs absorption layer thickness of 800 nm achieved bandwidths of 100 MHz due to surface leakage current phenomena at the exposed InGaAs surface layer. Responsivity was at 0.21 A/W for a device size of 20 μm x 20 μm. Lateral p-i-n photodiodes based on InGaAs/InP usually achieve quantum efficiencies of 50-90%, responsivity of ~ 1 A/W and bandwidth upto ~60 GHz (Saleh et al., 1991).

2.4 Spin-on chemical fabrication method of InGaAs/InP-based ILPP

Normally, p-type dopants are incorporated into InGaAs or InP layers using a closed-tube system where the substrate and the dopant powder source are placed in an ampoule (Ho et al., 2000; Feng & Lu, 2004). Air in the capsule is released until low pressure is achieved and the capsule is sealed. Then the capsule is placed in a furnace and heat evaporates the dopant source and produces dopant vapour that will diffuse into the substrate. The diffusion is uniform due to the non-existence of air in the capsule. However, the costly capsule as well as the vacuum pump causes this process not to be preferred.

Solid phase diffusion (SPD) is a doping technique without rays and examples of SPD include dopant diffusion from doped epitaxial layers, doped oxide layers and spin-on dopants (SOD). The SOD technique is a common SPD technique where a conformal layer is spun-on the surface of the substrate. This dopant source consists of oxide powder mixed with a solvent. Substrates which are covered with dopants are heated to evaporate the solvent and leave a doped oxide layer which is compatible with the substrate surface. Next, the substrate is placed in a deposition tube where annealing treatment will release the dopant from the oxide and aim it into the substrate. SODs have the potential to produce high uniformity and productivity. Common problems that would arise are such as dopant spreading in the oxide layer, thickness variation especially during the early stages and cost related to additional processes such as etching and annealing (Zant, 2000).

The SOD technique have been used to produce the p+ and n+ wells in silicon substrates (Gangopadhyay et al. 2003; Oh et al. 2004), n+ wells in germanium layers (Posthuma et al. 2007) as well as the p+ and n+ wells in GaAs substrates (Filmtronics, 2006a; Filmtronics, 2006b). In addition, the SOD diffusion technique was used previously in the development of a silicon-based lateral p-i-n photodiode in our laboratory (Ehsan & Shaari, 2001; Menon, 2005).

In InGaAs/InP-based substrates, the p+ wells were produced using the SOD technique by several researchers (Lange et al. 2000; Kamanin et al. 1996 & Lauterbach 1995). The quality of the p+ junctions is at par with those produced using the closed ampoule technique (Lange et al. 2000). However, based on the literature survey, the SOD technique has never been used to produce the n+ junction in InGaAs/InP substrates. Normally, n-InGaAs is produced using silicon as a dopant (Si_2H_6 in H_2) which is diffused into the InGaAs layer during the epitaxial layer growth process (Murray et al. 2003). Another option is to dope the InP epitaxial layer with sulphur or Si (Dildey et al, 1989). Therefore, in this project, the SOD technique is used for the first to time to produce both the p- and n-type wells in the absorbing InGaAs layer in an ILPP device (Menon et al. 2010).

2.5 Review of simulated InGaAs/InP-based ILPP

The simulation of optoelectronic devices such as lateral p-i-n photodiodes require accurate modeling based on semiconductor physics principles. Generally, semiconductor device equations can be solved using two methods; a fully numerical approach or an analytical approach. For models that are used in circuit simulation, the latter is more suitable because the computation time can be reduced whereas the former approach is more appropriate in simulation of optoelectronics devices simulation. The numerical approach can be divided to Monte-Carlo (MC) based simulation or solving the approximation of the drift-diffusion (DD) equation where a solution for the Boltzmann transport equation can be obtained (Konno et al., 2004). Nowadays, there are many device simulators which are developed using the DD equation solution.

In the modeling the ILPP device, various methods have been used before this. Konno et al. (2004) has used the analytical model in the Fourier space as well as numerical method to characterize a silicon-based p-i-n photodiode. The analytic analysis for the frequency response for an InGaAs-based p-i-n photodiode was undertaken by Sabella & Merli (1993). A numerical analysis on the non-linear response of InGaAs-based photodiode at high illumination conditions was also performed previously (Dentan & Cremoux, 1990). A numerical model for an InGaAs-based p-i-n photodiode which utilizes the pulse response equation was developed by Cvetkovic et al. (2000). Software such as SPICE and MATLAB has also been utilized previously to perform numerical simulation on p-i-n photodiodes (Parker, 1988; Loo, 2007).

Semiconductor device software such as MEDICI was used by Li et al. (2000) to simulate an ILPP device based on silicon. Numerical software such as ATLAS from Silvaco Inc. can be used to develop two and three dimensional models of semiconductor devices. Prior to this, this software was used to develop semiconductor devices such as an avalanche photodiode

(Lee et al., 2004), a uni-travelling photodiode (UTC-PD) (Srivastava & Roenker, 2003), a vertical photodiode (Jacob et al. 2005) and a silicon-based lateral p-i-n photodiode (Menon, 2005).

In this work, Silvaco Atlas was used, for the first time, to develop a three dimensional model of an interdigitated lateral p-i-n photodiode based on InGaAs. The 3D analysis takes into consideration the Poisson equation, electron and hole continuity equation, the concentration-dependent minority carrier lifetime model, the concentration-dependent carrier mobility model, Shockley-Read-Hall (SRH) recombination model, Auger recombination model, optical recombination model and the Fermi-Dirac statistics model. Curve fitting methodology using MATLAB was used to obtain parameters of empirical equations used to derive the minority carrier lifetime and the carrier mobility. The developed model was characterized for its responsivity, -3dB frequency, I-V, C-V and the signal-to-noise ratio (SNR) (Menon, 2008; Menon et al.,2008a; Menon et al., 2008b; Menon et al. 2009, Menon et al. 2010).

2.6 Statistical modeling

The design of experiment (DOE) was created by the British scientist, Sir R. A. Fisher in the 1920-es. The fractional factorial design (FFD) is one the methodologies in DOE design based on statistical consideration that brings about meaningful information about the effects of design parameters on the device characteristics. Other famous statistical methods include the Monte-Carlo and the worst-case statistical method. The advantages of FFD are it can save experimental time because the number of simulation experiment which needs to be executed can be reduced with the assumption that all variables have no interactions.

The FFD technique has been used before in both device simulated modeling as well as in device fabrication. Sipahi & Sanders (2002) used the FFD technique to investigate parameters that affect the simulated model of a low noise amplifier whereas Yuan et al. (2005) applied the FFD technique with a resolution IV in the simulation of their silicon-based chip packaging model. Jacob et al. (2006) used the full factorial design to optimise the characterisation of design parameters in the simulation of a photodiode for imaging applications.

Therefore, the simulated ILPP model was optimised prior to the fabrication process. Seven design factors at two levels each i.e., the thickness of the InGaAs absorption layer, the distance between the finger electrodes, the junction depth, the width of the interdigitated finger electrodes, the bias voltage and the input optical power were investigated. A resolution IV fractional factorial design (Montgomery, 2005) was executed to identify the effects of these factors on the key characteristics of an ILPP which are the responsivity, the -3dB frequency and SNR where a general linear model for all three characteristics was developed respectively using FFD.

3. Theoretical modeling

3.1 Material and model parameters

Semiconductor devices can be modelled in two ways; the first is through the determination of the electrical properties at the device's terminal based on data fitting of empirical results or via the second method; analysis of the carrier transportation processes which occurs in the device. Both these methods were utilized in this research based on the physical model of the device which can determine the terminal characteristics and the carrier transport behaviour within the device. The physical model of the device was developed based on the description of the substrate material properties and the carrier transport physics. A semiconductor device software, ATLAS from Silvaco International was used simulate the ILPP. It is a numerical simulator which uses differential equations that defines the device physics at different locations within the device. The performance analysis is achieved via self-consistent solution of basic equations that define semiconductor equations such as the Poisson's equation, carrier continuity equations and the current density equations in two or three dimensions. The solution to these equations is based on the device structure, its geometry and the boundary limits which is determined by the electrical contacts and the bias voltage (Silvaco, 2004). The three dimensional analysis takes into account the following physical models; concentration-dependent minority carrier lifetime model, concentration and temperature-dependent mobility model, parallel field mobility, Shockley-Read-Hall (SRH) recombination model, Auger recombination model, optical generation/radiative recombination model and Fermi-Dirac statistics. The material and model parameters were taken from periodical literature (Silvaco, 2004; Srivastava & Roenker, 2003; Adachi, 1992; Datta et al., 1998). Table 1 provides a summary of the material parameters used in this modelling (Menon et al. 2010). The complete list of model and material parameters can be obtained from (Menon, 2008)

Parameter	**Symbol**	**$In_{0.53}Ga_{0.47}As$**
Energy gap	E_g (eV)	0.734
Electron effective mass	m_e^*/m_o	0.033
Hole effective mass	m_h^*/m_o	0.46
Light hole	m_{lh}^*/m_o	0.01
Heavy hole	m_{hh}^*/m_o	0.46
Real index		3.6
Imaginary index		0.08

Table 1. Summary of material parameters used in the modelling (Menon, 2008; Menon et al. 2010)

3.2 Concentration-dependent minority carrier lifetime model

In ATLAS (Silvaco, 2004), the minority electron and hole lifetimes are given by

$$\tau_n = \frac{TAUN0}{1 + \frac{N}{NSRHN}} \tag{1}$$

$$\tau_p = \frac{TAUP0}{1 + \frac{N}{NSRHP}} \tag{2}$$

where N is the total impurity concentration and *NSRHN* and *NSRHP* are the critical doping concentration above which impurity scattering dominates. Experimental data of electron lifetime (Tashima et al., 1981) in p+-InGaAs was fitted into a simple exponential form as follows (Conklin et al., 1995)

$$\tau_n(n\sec) = 10^{\beta-\gamma \log N_A} \tag{3}$$

where N_A is the p+-InGaAs doping and parameter values of β=12.6, γ=0.73 for $N_A > 8x10^{17}$ /cm^3 were obtained. For smaller base doping than $8x10^{17}$ /cm^3, a constant lifetime of 0.3 ns was assumed. Similarly for the n+ InGaAs, the hole lifetime is a function of doping and can be fit to the empirical expression (Datta et al., 1998)

$$\tau_p(n\sec) = 10^{\beta-\gamma \log N_D} \tag{4}$$

where N_D is the n+ InGaAs doping and the fit parameters are β=22.4, γ=1.2 for $N_D > 8x10^{17}$ /cm^3. For smaller base doping than $8x10^{17}$ /cm^3, a constant lifetime of 10 ns was assumed. The empirical models for the concentration dependent minority carrier lifetime (Eq. (3) and Eq. (4)) were fit into the ATLAS Eq. (1) and Eq. (2) as shown in Fig. 3 (a) and Fig. 3 (b). Values for the fit are given in Table 2 (Menon et al. 2010).
The negative differential mobility model of Barnes et al. (1996) was used to account for the carrier drift velocity that peaks at some electric field before reducing as the electric field increases. The model is given by (Silvaco, 2004):

$$\mu(E) = \frac{\mu_0 + \dfrac{v_{sat}}{E}\left(\dfrac{E}{E_{crit}}\right)^{\gamma}}{1 + \left(\dfrac{E}{E_{crit}}\right)^{\gamma}} \tag{5}$$

where v_{sat} is the carrier saturation velocity, E_{crit} is the critical electric field, E_0 and γ are constants and μ_0 is the low-field carrier mobility. The model parameters used for InGaAs are shown in Table 2 as well.

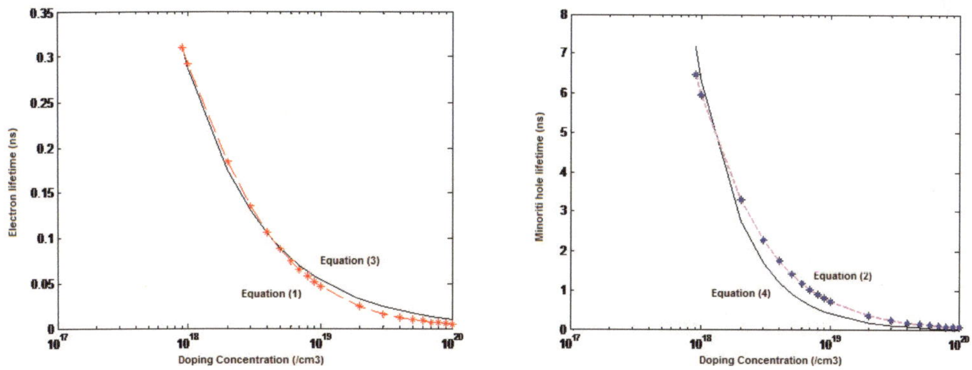

(a) (b)

Fig. 3. Fit of (a) concentration dependent minority electron lifetime in p+ InGaAs and (b) concentration dependent minority hole lifetime in n+ InGaAs (Source: Menon et al. 2010)

Parameters	Symbol	Units	Electrons	Holes
Concentration-dependent minority carrier lifetime model				
Lifetime	τ_0	Ns	0.6985	31.02
Total impurity conc.	N_{SRH}	$1/cm^3$	7.134×10^{17}	2.37×10^{17}
Parallel electric field-dependent mobility model				
Saturation velocity	v_{sat}	cm/sec	2.5×10^7	5×10^6
Fit parameter	Γ		4	1
Critical electric field	E_{crit}	V/cm	3000	4000

Table 2. Summary of concentration-dependent minority carrier lifetime and parallel electric field-dependent mobility model parameters (Source: Menon et al. 2010)

3.3 Low-field carrier mobility models

The electron and hole mobilities in $In_{0.53}Ga_{0.47}As$ (InGaAs) as a function of doping concentration and temperature are important parameters for device design and analysis. The default low field mobility parameters for InGaAs carriers in commercial device simulation packages are given by linear interpolations from the binary compounds of GaAs and InP (Silvaco, 2004) which do not describe the dependency of carrier mobility on doping concentration or temperature. Therefore, in this chapter, we attempt to provide the fitted parameters of a concentration- and temperature-dependent carrier mobility model to accurately characterize an InGaAs ILPP (Menon et al., 2008a).

The dependence of electron and hole low-field mobilities on doping and temperature (Caughey & Thomas, 1967) for electrons and holes is given by Eq. (6) and Eq. (7) respectively.

$$\mu_n = mu1n.caug.\left(\frac{T_L}{300K}\right)^{alphan.caug} + \frac{mu2n.caug.\left(\frac{T_L}{300K}\right)^{betan.caug} - mu1n.caug.\left(\frac{T_L}{300K}\right)^{alphan.caug}}{1+\left(\frac{T_L}{300K}\right)^{gamman.caug}.\left(\frac{N}{ncritn.caug}\right)^{deltan.caug}} \tag{6}$$

$$\mu_p = mu1p.caug.\left(\frac{T_L}{300K}\right)^{alphap.caug} + \frac{mu2p.caug.\left(\frac{T_L}{300K}\right)^{betap.caug} - mu1p.caug.\left(\frac{T_L}{300K}\right)^{alphap.caug}}{1+\left(\frac{T_L}{300K}\right)^{gammap.caug}.\left(\frac{N}{ncritp.caug}\right)^{deltap.caug}} \tag{7}$$

where N is the total impurity doping (cm^{-3}) and T_L is the lattice temperature (K). *mu1n.caug*, *mu1p.caug*, *mu2n.caug*, and *mu2p.caug* are the minimum and maximum mobilities of electrons and holes respectively (cm^2/V-s). The fitting parameters are *alphan.caug*, *alphap.caug*, *betan.caug*, *betap.caug*, *gamman.caug*, *gammap.caug*, *deltan.caug* and *deltap.caug* whereas *ncritn.caug* and *ncritp.caug* are the critical doping densities (cm^{-3}) for electrons and holes, respectively above which ionized impurity scattering becomes dominant (Menon et al., 2008a).

The mobility data has been fit to Eqs. (6) and (7) by Datta et al. (1998) for InGaAs at T_L=300K where *mu1n.caug*=3372, *mu1p.caug*=75, *mu2n.caug*=11599, *mu2p.caug*=331, *ncritn.caug*=8.9×10^{16}, *ncritp.caug*=1×10^{18}, *deltan.caug*=0.76 and *deltap.caug*=1.37. In this work we have obtained the remaining fitting parameters of Eqs. (6) and (7) using curve-fitting methodology where we obtained *alphan.caug*=0.437, *alphap.caug*=0.9222, *betan.caug*=1.818, *betap.caug*=1.058, *gamman.caug*=2.526 and *gammap.caug*=7.659.

A comparison between these fitted results versus the calculated carrier mobility from (Sotoodeh et al., 2000; Arora et al., 1992; Chin et al., 1995) as well as some experimental Hall data (Lee & Forrest, 1991; Ohtsuka et al., 1988; Pearsall, 1981) is shown in Fig. 4(a) till Fig. 4 (c) for T=77K, 100K and 200K. Fig. 4(d) shows the electron mobility as a function of temperature in InGaAs where calculated electron mobility from this work is compared to the experimental Hall data from Takeda et al. (1981). A very good agreement is obtained for temperatures >150K (Menon et al., 2008a).

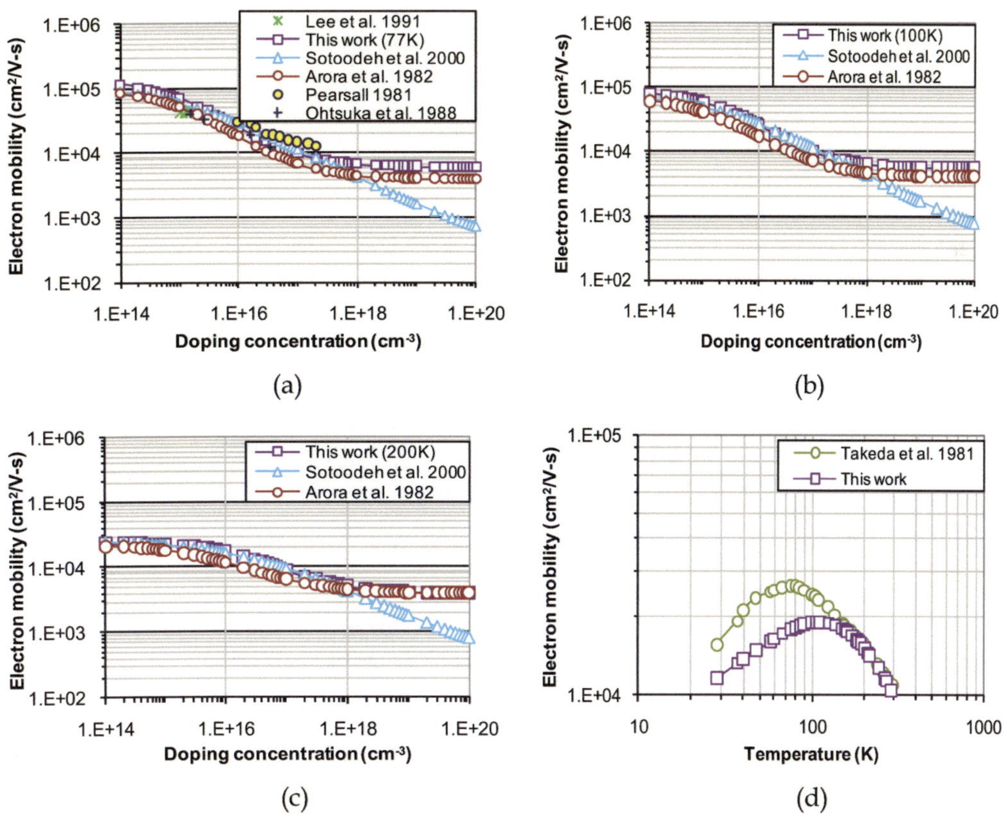

Fig. 4. Electron mobility in $In_{0.53}Ga_{0.47}As$ as a function of (a) doping at T_L=77K, (b) T_L=100K, (c) T_L=200K and (d) temperature at N=1.5e^{16}cm^{-3} (Source: Menon et al., 2008a)

The hole mobilities for InP-based material are similar to those seen for GaAs and AlGaAs (Datta et al., 1998). Fig. 5 (a) till Fig. 5 (c) show the fitted results versus calculated hole mobility for T=77K, 100K and 200K. Fig. 5(d) shows the hole mobility as a function of temperature in InGaAs. Good agreement is obtained for temperatures ≥200K.

Carrier mobility decreases sharply when doping density is increased for low doping densities (less than 1e18 cm^{-3}). For high doping densities, the mobility tends to decrease more slowly and shows a saturated trend. Similarly, for low operating temperatures (<100K), the carrier mobility tends to increase with increment in temperature. However, above 100K, the mobility shows a downward bowing trend as temperature is increased. Therefore, it has been proven that the fitted parameters are reliable and match available experimental or theoretical data. These carrier mobility equations were used in the development of an ILPP based on InGaAs absorption layer.

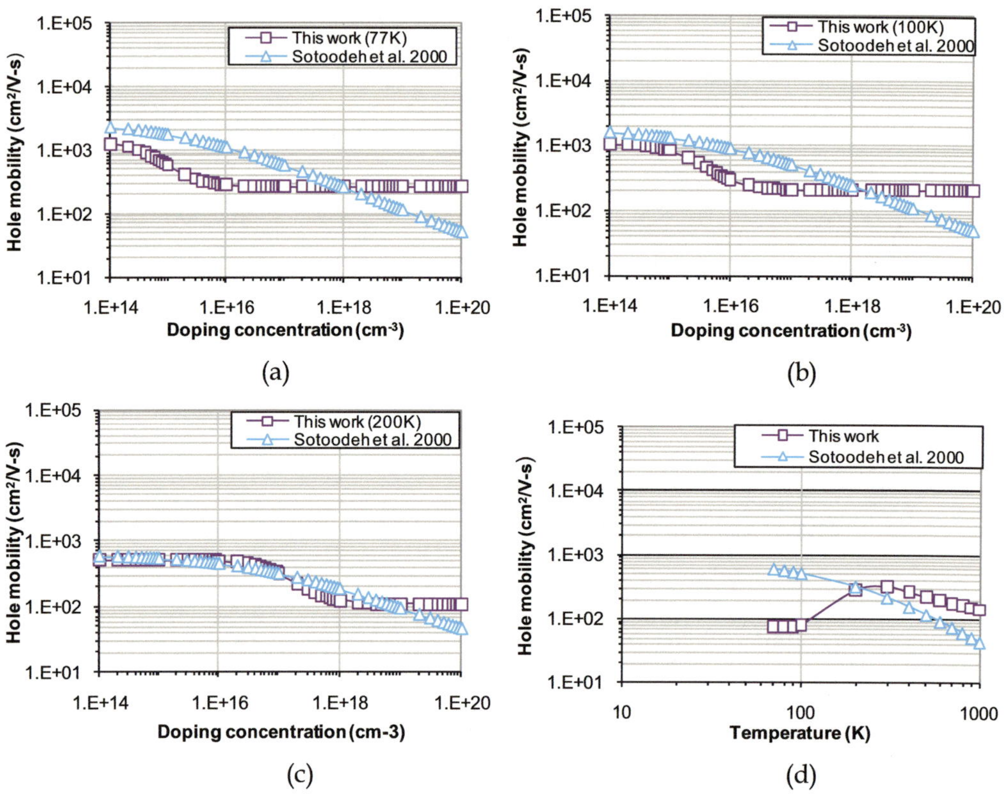

Fig. 5. Hole mobility in $In_{0.53}Ga_{0.47}As$ as a function of (a) doping at T_L=77K, (b) T_L=100K, (c) T_L=200K and (d) temperature at N=1e^{17}cm^{-3} (Source: Menon et al., 2008a)

4. Numerical modeling

4.1 Device material selection

InGaAs is specified as the absorbing material in ATLAS by setting the mol fraction of quaternary material $In_{1-x}Ga_xAs_yP_{1-y}$ where x=0.43 and y=1 to form $In_{0.53}Ga_{0.47}As$. It is used as the absorbing layer with a thickness of 3 µm and at this depth, 83% of the optical power will be absorbed by the device based on the InGaAs absorption coefficient, α=6070 /cm at λ=1.55 µm. The absorbing layer is also given an n-type background doping of 1e11 cm^{-3} with a uniform doping profile.

The p+ wells in the ILPP device will be formed using zinc SOD hence the junction parameters were obtained from available experimental data. Kamanin et al. (1996) formed a p+ junction in InGaAs using thin film zinc-based polymer diffusion at a temperature of 500°C for 30 minutes to obtain junction depth of 0.8 µm and dopant surface concentration of ~8x10^{18} cm^{-3}. Similarly, in the ILPP model, the junction depth of the p+ wells was selected to

be 0.8 μm with a surface doping level of $4x10^{18}$ cm^{-3}. The n+ wells in InGaAs is proposed to be formed using selenium-doped SOD. Penna et al. (1985) performed ion implantation of Se into InGaAs to form a junction of ~0.4 μm deep. Alternatively, selenium-doped SOD have been used on GaAs to obtain junctions with a depth of 1.3 μm and surface concentration of $6x10^{18}$ cm^{-3} (Filmtronics, 2006). In this ILPP model, the junction depth of the selenium-doped n+ wells were set to be 0.8 μm with a surface concentration of $1x10^{19}$ cm^{-3} to produce a uniform electric field between the alternating junctions.
Spin-on glass (SOG) will be used as the passivation layer for InGaAs. It will serve to protect the junction surfaces as well as for planarizing the device. In this model, a 0.1-μm thick SiO_2 was used to reflect the presence of SOG on top of the SOD-doped InGaAs absorbing layer. Finally, the alternating interdigitated fingers were modelled as gold-based.

4.2 Design of device structure

The electrode finger width/ spacing and length are 1 μm and 50 μm respectively. The device's active area is 41 x 5 x 50 μm^3 with a total of 10 pairs of interdigitated electrodes. The junction depth for both the p+/n+ wells are 0.8 μm respectively and lateral diffusion per well is 0.3 μm. The compensation ratio θ (N_A/N_D) is set at 0.1 where donor concentration, N_D=1e^{19} cm^{-3}. Fig. 6 shows the potential of the InGaAs ILPP three-dimensional model upon illumination of an optical beam with spectral width of 41 μm, optical spot power of 10 W/cm2 and wavelength, λ=1.55 μm.

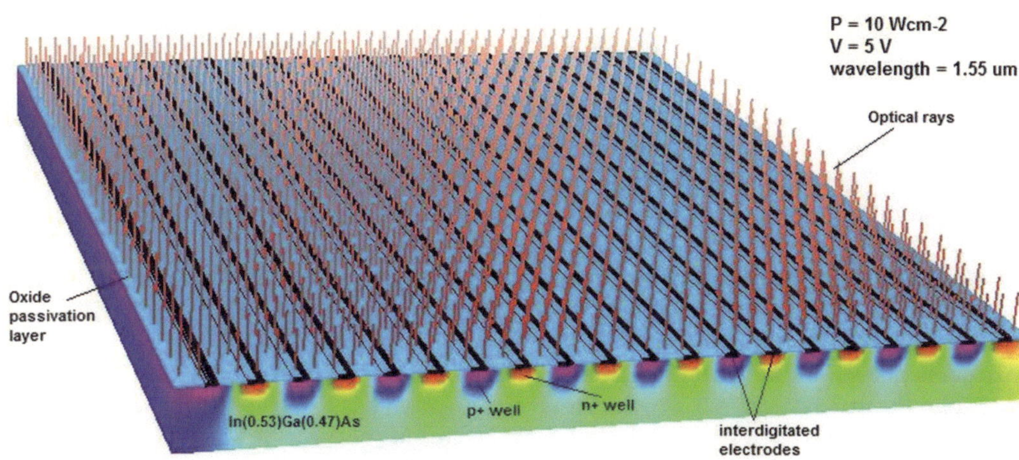

Fig. 6. Potential within the InGaAs ILPP 3D model upon illumination of an optical beam

4.3 Characterization equations

The ILPP dark current, I_D is given by:

$$I_D = I_{SAT}(e^{\frac{qV_A}{k_B T}} - 1) \tag{8}$$

where I_{SAT} is the reverse saturation current, q is the electron charge, V_A is the applied bias voltage, k_B is the Boltzmann constant and T is the absolute temperature in Kelvin. Illuminating the photodiode with optical radiation, shifts the I-V curve by the amount of photocurrent (I_P). Thus, the total current I_T is given by $I_T = I_D + I_P$.

The ILPP responsivity, R is calculated using:

$$R = \frac{I_T}{I_S}\left(\frac{\lambda}{1.24}\right) \tag{9}$$

where I_S is the source photocurrent and λ is the optical wavelength.
The simulator calculates the real (I_R) and imaginary (I_I) component current values for every equivalent AC frequency value. Hence, the -3dB frequency (f_{-3dB}) is calculated using the following equation:

$$f_{-3dB} = 20 * \log\left(\frac{I_R}{I_{R_0}}\right) \tag{10}$$

where I_{R0} is the real component current at low AC frequencies which is normally a constant value. Finally, the ILPP signal-to-noise ratio (SNR) is calculated using the following equation

$$SNR = \frac{\left\langle i_p^2 \right\rangle}{2q(I_p + I_D)B + 4k_BTB / R_L} \tag{11}$$

where I_P is the average photocurrent, B is the bandwidth and R_L is the load resistance set to be as 50 Ω (Menon et al. 2010).

4.4 Device characterization results

A cross section of the 3D device is shown in Fig. 7 (a) portraying the net doping within the device. Dark current value at 5 V was measured to be 21 nA and is much higher than that achieved by conventional InGaAs VPDs (in pA values) (Huang et al., 2007) due to the absence of a capping layer such as InP to reduce the surface leakage current. However, the modelled device's dark current is comparable to conventional ILPP that have been fabricated before as portrayed in Fig. 7 (b). The ideality factor, n was measured to be ~1 and the series as well as dynamic resistances were measured to be 43 Ω and ~238 MΩ respectively. Breakdown voltage was >40 V.
The capacitance values recorded at a bias voltage of 5V was 2.87 nF and this value is much higher than the capacitance values achieved by conventional ILPP devices due to the smaller intrinsic region width (1 μm in this design versus 3 μm in (Yasuoka et al., 1991)) and longer electrode fingers in the current design (50 μm in this design versus 20 μm and 47 μm in (Tiwari et al., 1992) and (Lee et al. 1989)). The C-V results are shown in Fig. 8(a).
Dark and photo-IV curves for the optical beam at λ=1.55 μm and P=dark (0), 1, 5, 10, 50, 100 and 200 Wcm^{-2} is shown in Fig. 8 (b). At operating voltage of 5V, the photocurrent increased from 0.011 mA (P=1 Wcm^{-2}) to 2.28 mA (P=200 Wcm^{-2}).
Fig. 9(a) is the responsivity curve of the modelled device at P=10 Wcm-2, V=5V and the wavelength is swept up from 0.75 μm until 1.75 μm. In optical communication networks, data signals are usually transmitted at λ=1.31 μm whereas video signals are transmitted at λ=1.55 μm. At both these wavelengths, the responsivity was measured to be 0.55 A/W and 0.56 A/W respectively which is equivalent to an external quantum efficiency of 44 %. These values are comparable to the experimentally developed InGaAs ILPP devices but are much smaller than VPDs due to the electrode shadowing effect in ILPP designs. Fig. 9(b) shows

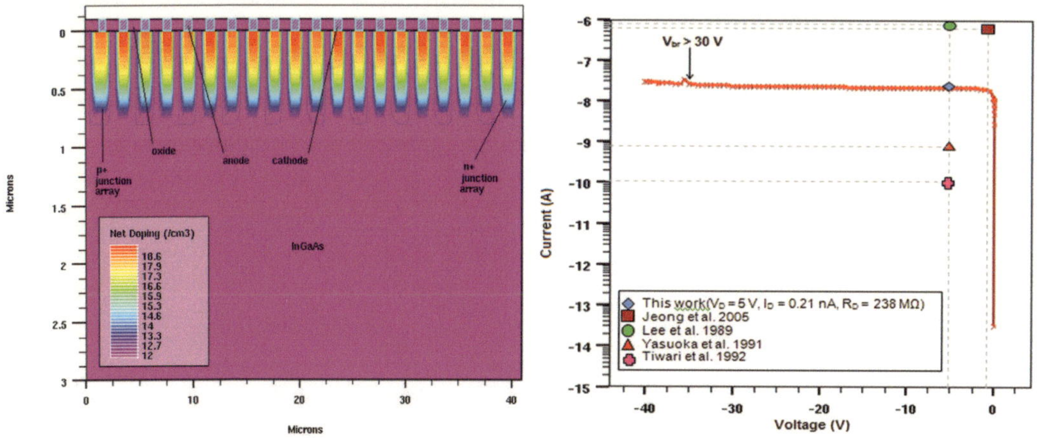

(a) (b)

Fig. 7. The ILPP's (a) cross section of the 3D device portraying the net doping within the device and (b) dark current trend (Source: Menon et al. 2010)

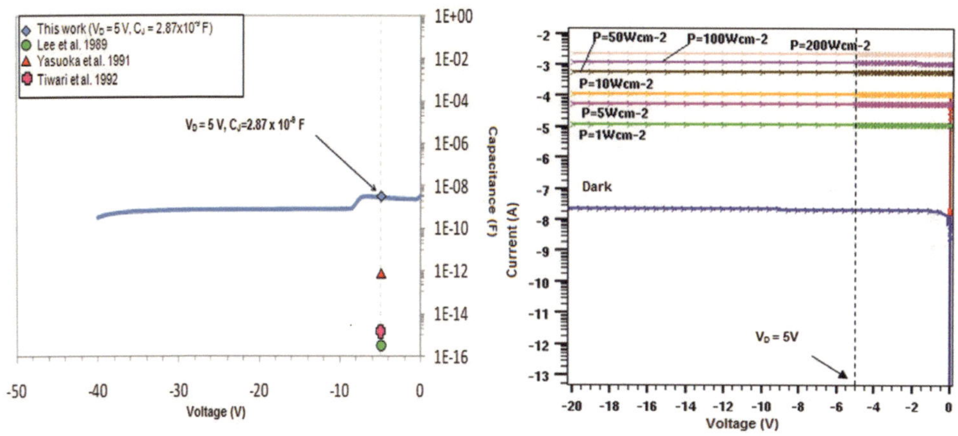

(a) (b)

Fig. 8. The ILPP's (a) C-V trend and (b) dark and photo-IV curves for the optical beam at λ=1.55 μm and P=dark (0), 1, 5, 10, 50, 100 and 200 Wcm^{-2} (Source: Menon et al. 2010)

the -3dB frequency of 8.93 GHz achieved by the model and it is 16% higher than conventional ILPP prototypes (Yasuoka et al., 1991; Tiwari et al., 1992; Lee et al. 1989; Jeong et al., 2005) mainly due to the smaller intrinsic region width utilized in this design.

The dark current noise is 0.06 fA/√Hz, quantum noise is 0.33 nA/√Hz and Johnson noise is 2.96 pA/√Hz with load resistance of 50 Ω where the Johnson noise is the highest noise contributor. The device SNR was calculated to be ~36 dB and dynamic range ranges from -16 dBm until 17.9 dBm (Menon et al. 2010).

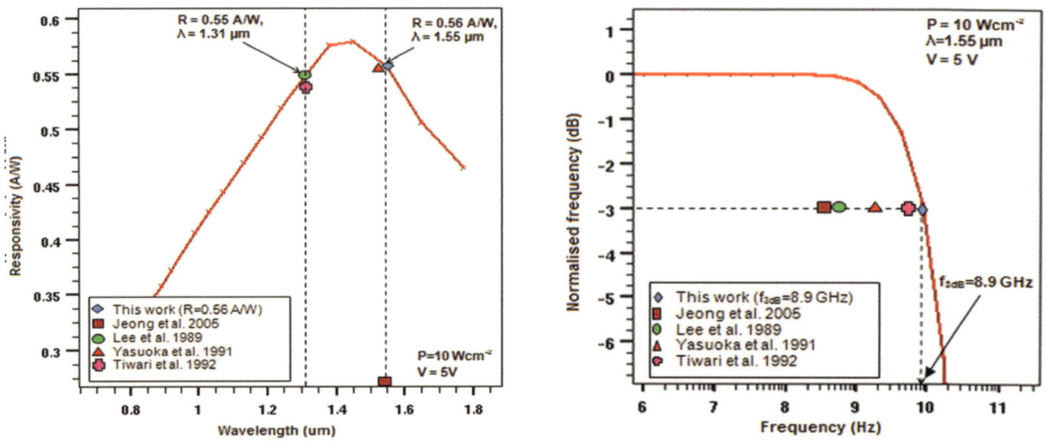

(a) (b)

Fig. 9. The ILPP's (a) responsivity curve of the modeled device at P=10 Wcm-2, V=5V and (b) the -3dB frequency (Source: Menon et al. 2010)

5. Statistical modeling

5.1 Fractional Factorial Design

Fractional factorial design (FFD) was used to identify the factors that affect the device responsivity significantly. Next, the significant factors were used to develop a general linear model to predict the responsivity of different ILPP models. In this research, seven factors i.e. InGaAs absorbing layer thickness (T), finger width (FW), finger spacing (FS), junction depth (JD), finger length (FL), bias voltage (V) and optical beam power (P) were investigated, each of which were tested at two levels. A one-quarter fractional factorial design (resolution IV) comprising of 32 runs (Montgomery, 2001) was carried out to obtain information on the effects of the investigated factors. Fig. 10 displays the ILPP model where the chosen factors are highlighted. Table 3 lists the factors and their respective values which were used in the DOE. A well-known statistical software, Minitab was used to obtain the statistical results (Menon et al. 2008b).

The normal probability plots and the pareto chart for the device responsivity are shown in Fig. 11 (a) and Fig. 11 (b). The significant factors which include interactive factors are highlighted in red in the normal probability plots. Significant or active effects are larger and further away from the fitted line than inactive effects which tend to be smaller and centered around zero, the mean of all the effects. The pareto charts display the absolute value of the effects.

In the normal probability plot for the device responsivity, the most significant factor that affects this response is the InGaAs thickness (A), followed by the finger width (B), finger spacing (C) and the interaction factor between InGaAs thickness and finger width (A*B). These significant factors prove that when the absorbing layer thickness is increased, the absorbed optical power, $P(x)$ at a depth of x increases according to the equation $P(x)=P_0(1-e^{1-a(x)})$where P_0 is the incident optical power and a is the absorbing coefficient. Decrement in the electrode finger width (FW) and increment in the electrode finger spacing (FS) increases the total illumination area from the top of the device hence increasing the total generated photocurrent within the device and subsequently increases the device responsivity.

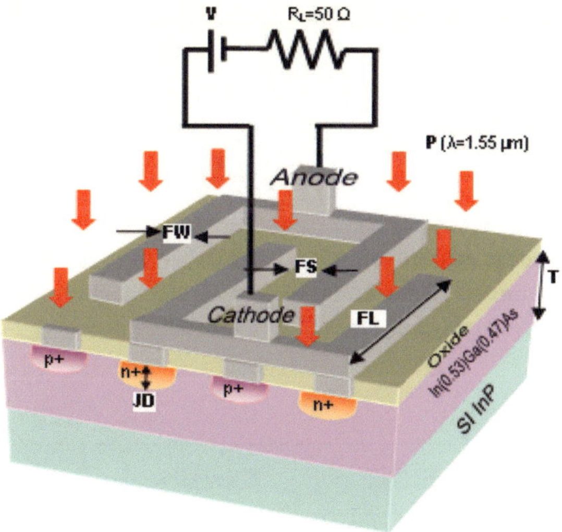

Fig. 10. Schematic diagram of the ILPP model. The chosen factors are highlighted in the diagram.

Variable (Code)	Factor name	-1 Level (Low)	+1 Level (High)
A(T)	InGaAs thickness (μm)	1	3
B(FW)	Finger width (μm)	1	3
C(FS)	Finger spacing (μm)	1	3
D(JD)	Junction depth (μm)	0.4	0.8
E(V)	Voltage (V)	2	5
F(P)	Beam power (Wcm^{-2})	1	10
G(FL)	Finger length (μm)	20	50

Table 3. Fractional factorial design factors and values.

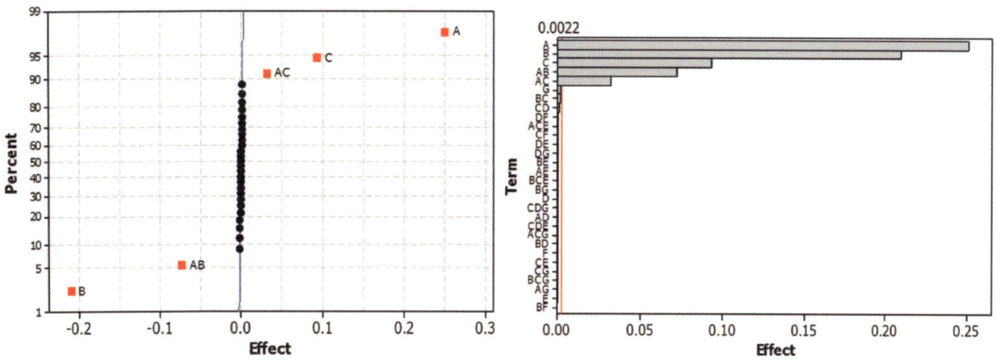

(a) (b)

Fig. 11. (a): Normal probability plot for the responsivity. The factors highlighted in red are significant and (b) Pareto chart for the responsivity displaying absolute values of the factor effects in descending order.

Next, the significant factors for the device responsivity was used to develop a reduced model at a confidence level of 95%. This was done by screening out the insignificant effects from the full model and evaluating the fit of the new reduced model using analysis of variance (ANOVA). The main effects as well as significant two-way interaction effects which are significant gives a *p*-value. If $p<0.05$, then the effect or term is significant whereas if $p>0.05$, then the terms are insignificant and hence can be excluded from the reduced model. From Fig. 16 and Fig. 17, the new reduced model will now comprise of the main effects (A, B and C) as well as two-way and three-way interactive factors which include these main effects. Table 4 lists the analysis of variance for the device responsivity using the factorial fit from the reduced model (Menon et al., 2009).

Term	Effect	Coefficient	*p*-value
Constant		0.3564	0.000
T	0.2506	0.1253	0.000
FW	-0.2095	-0.1047	0.000
FS	0.0930	0.0465	0.000
T*FW	-0.0722	-0.0361	0.000
T*FS	0.0324	0.0162	0.000
FW*FS	-0.0017	-0.0008	0.034
T*FW*FS	-0.0009	-0.0005	0.221

Table 4. Analysis of variance for responsivity (S=0.002, R^2=99.9%, R^2(adj)=99.9%).

All the terms have a *p*-value of <0.05 except the last term (T*FW*FS) where the *p*-value is 0.221 deeming it insignificant. The *S*, R^2 and adjusted R^2 are measures of how well the model fits the data where *S* represents how far the standard distance data values fall from the regression line, R^2 describes the amount of variation in the observed response values and adjusted R^2 is a modified R^2 that has been adjusted for the number of terms in the model. For a given fit, the lower the value of *S* and the higher the values of R^2 and adjusted R^2, the better the equation predicts the response. In this model, values of *S*, R^2 and adjusted R^2 are 0.002 and 99.9% respectively proving that a robust model for predicting the InGaAs ILPP responsivity has been established. Next, the coefficients of each significant term is used to construct a regression or analytic equation representing the relationship between the device responsivity and the design factors. The regression equation which defines the responsivity of the InGaAs ILPP is as follows (Menon et al., 2009):

$$y_{(resp)} = 0.3564 + 0.1253(T)_c - 0.1047(FW)_c \\ +0.0465(FS)_c - 0.0361(T)_c(FW)_c \\ +0.0162(T)_c(FS)_c - 0.0008(FW)_c(FW)_c \tag{12}$$

where X_c is the factor value in coded units and it is related to the actual factor value X_a by

$$X_a = \frac{X_c - \left[\frac{(X_H + X_L)}{2}\right]}{\frac{(X_H - X_L)}{2}} \tag{13}$$

where X_L and X_H are the factor values at the low level and high level as given in Table 1. Eq. (13) can be rearranged to obtain the value of X_c:

$$X_c = \frac{(X_H + X_L)}{2} + \left\{\frac{(X_H - X_L)}{2}\right\} X_a \tag{14}$$

The coded values for all the factors which defines the device responsivity is calculated and is given as follows:

$$(T)_c = 2 + (T)_a \tag{15}$$

$$(FW)_c = 2 + (FW)_a \tag{16}$$

$$(FS)_c = 2 + (FS)_a \tag{17}$$

Eqs. (15) to (17) are replaced into Eq. (12) to obtain the general linear model which defines the responsivity of an InGaAs ILPP in uncoded units.

$$y_{(resp)} = 0.143106 + 0.163188(T)_a - 0.0327433(FW)_a \\ +0.0138567(FS)_a - 0.0351578(T)_a(FW)_a \\ +0.0171634(T)_a(FS)_a - 0.000096781(FW)_a(FS)_a \tag{18}$$

where T_a, FW_a, $FS_a \neq 0$.

5.2 Model verification

Eq. (18) was used to recalculate the responsivity of the numerical models used in the 32 runs of the fractional factorial DOE and the comparative results between the simulated and calculated values as well as the error ratios are displayed in Fig. 12. Good correlation is observed between the two values and the error ratios are less than 3% for all the 32 models. Table 5 lists the factor values of some InGaAs ILPP designs from previous experimental work. The responsivity of these devices were recalculated using Eq. (18) and error ratios between 16% to 27% were obtained between the actual and calculated responsivity values. The results are displayed in Fig. 13. The high error ratios could be attributed to the drift-diffusion model used in the simulation for ILPP devices whereas the actual devices were fabricated using different techniques where carrier transport model may vary. The simulated model also does not take into consideration fabrication defects and reflects an ideal ILPP device. Eq. (18) is a new analytic equation which can be used to predict the responsivity of InGaAs ILPP as a function of the device design factors prior to fabrication.

No	T (μm)	FW (μm)	FS(μm)	Reference
1	1.7	1	3	Yasuoka et al., 1991
2	1.4	20	2	Tiwari et al., 1992
3	2	2	3	Lee et al., 1989

Table 5. Factor values from periodical literature

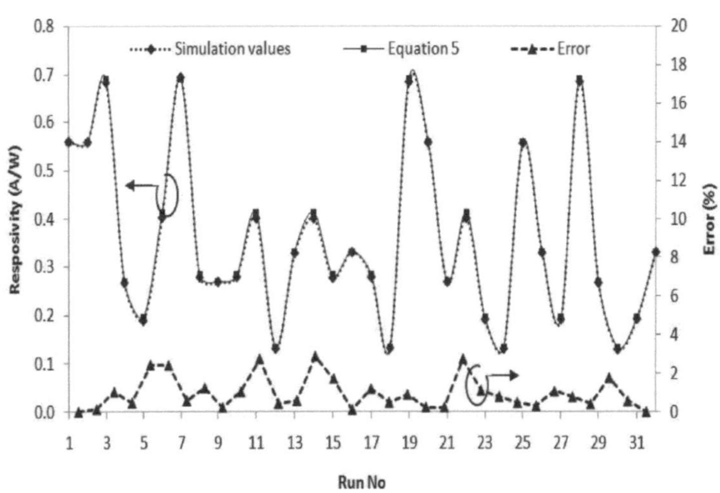

Fig. 12. Comparitive results between the simulated and calculated responsivity values from Eq. (18) as well as the error ratios.

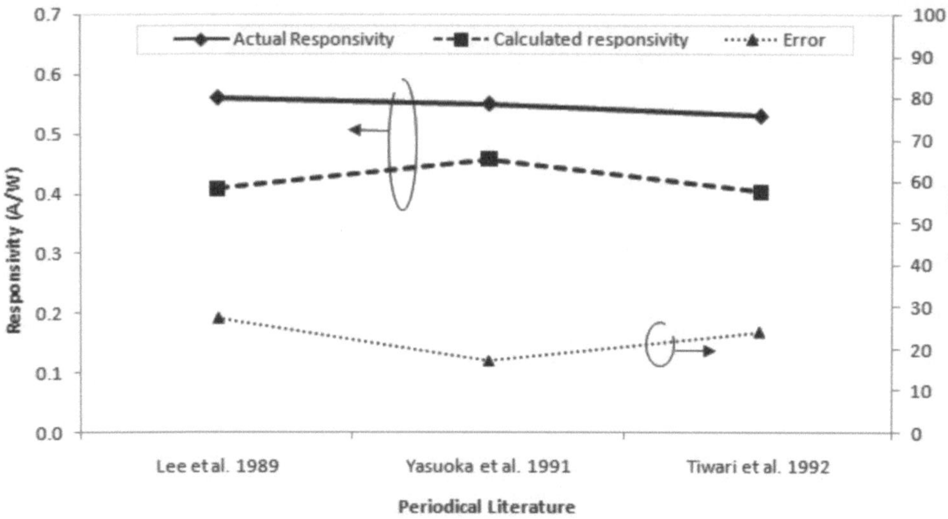

Fig. 3. Comparison between the actual responsivity versus calculated responsivity values using Eq. (18) for past experimentally developed devices.

5.3 Statistical optimization

A statistically optimized model for the InGaAs ILPP device was obtained by specifying the target range values that would like to be attained for each device characteristic. This is shown in Table 6. The optimized design factors that must be chosen in order to achieve the optimal target characteristic values as stipulated in Table 6 are given in Table 7. These optimized design factors can be used in the fabrication of InGaAs-based ILPP devices in the future.

Characteristics	Units	Low Target Value	High Target Value	Optimal Target Value
Responsivity	A/W	0.5	1	0.68
-3dB frequency	GHz	5	10	7.43
SNR	dB	10	50	12.11

Table 6. Target and optimal characteristic values obtained statistically

Variable (Code)	Factor name	-1 Level (Low)	+1 Level (High)	Optimal Target Value
A(T)	InGaAs thickness (μm)	1	3	3
B(FW)	Finger width (μm)	1	3	1
C(FS)	Finger spacing (μm)	1	3	3
D(JD)	Junction depth (μm)	0.4	0.8	0.8
E(V)	Voltage (V)	2	5	5
F(P)	Beam power (Wcm^{-2})	1	10	1.14
G(FL)	Finger length (μm)	20	50	20

Table 7. Target design factors for a statistically optimized InGaAs ILPP device

6. Conclusion

A novel interdigitated lateral p-i-n photodiode (ILPP) model utilizing $In_{0.53}Ga_{0.47}As$ as the absorbing layer was developed numerically and optimized statistically using fractional factorial methodology. Seven model factors were investigated and an analytical expression to predict the device responsivity was defined. Comparison between the simulated and calculated responsivity values yielded error ratios of less than 3%. Finally, a statistically optimized InGaAs ILPP model with -3dB frequency of 7.5 GHz, responsivity of 0.61 A/W and SNR of 20 dB was developed at an operating voltage of 5 V, wavelength of 1.55 μm and optical input power of 10 Wcm^{-2}. The modeled device provides a cheap and easy solution to cater for the increasing demand of FTTH-PON users

7. Acknowledgement

The authors would like to thank the Malaysian Ministry of Science, Technology and Innovation (MOSTI), the Malaysian Ministry of Higher Education (MOHE) and Universiti Kebangsaan Malaysia (UKM) for sponsoring this project under grants IRPA 03-02-02-0069-EA231 and UKM-OUP-NBT-27-119/2010

8. References

Adachi, S. (1992). *Physical properties of III-V Semiconductor Compounds: InP, InAs, GaAs, GaP, InGaAs and InGaAsP*, New York: Wiley Interscience.

Arora, N. D., Hauser, J.R. & Roulstan, R. J. (1982). Electron and hole mobilities in silicon as a function of concentration and temperature. *IEEE Transactions on Electron Devices* ED-29: 292.

Barnes, J. J., Lomax, R. J. & Haddad, G. I. (1996). Finite element simulation GaAs MESFET with lateral doping profiles and submicron gates. *IEEE Transactions on Electron Devices* ED-23: 1042-1048.

Caughey, D. M. & Thomas, R. E. (1967). Carrier mobilities in silicon empirically related to doping and field. *Proceedings of the IEEE*, 2192-2193.

Chin, W. L., Osotchan, T. & Tansley, T. L. (1995). Electron mobility in In(0.53)Ga(0.47)As as a function of concentration and temperature. *Microelectronics Journal* 26: 653-657.

Conklin, T., Naugle, S., Shi, S., Roenker, K. P., Frimel, S. M. Kumar, T & Cahay, M. M. (1996). Inclusion of tunneling and ballistic transport effects in an analytical approach to modeling of NPN InP based heterojunction bipolar transistors. *Superlattice Microstructures* 18: 1-12.

Cvetkovic, M., Matavulj, P., Radunovic, J. & Marincic, A. (2000). An InGaAs P-I-N photodiode model: Description and implementations in the analysis of the 1.55 μm lightwave system. *Journal of Optical Communications* 31(3): 35-39.

Datta, S., Shi, S., Roenker, K. P., Cahay, M. M. & Stanchina, W.E. (1998). Simulation and design of InAlAs/InGaAs pnp heterojunction bipolar transistors. *IEEE Transactions on Electron Devices*, 45(8): 1634-1643.

Dentan, M & Cremoux, B. D. (1990). Numerical simulation of the nonlinear response of a p-i-n photodiode under high illumination. *Journal of Lightwave Technology* 8: 1137-1144.

Diadiuk, V. & Groves, S. H. (1985). Lateral photodetectors on semi-insulating InGaAs and InP. *Applied Physics Letters* 46(2): 157-158.

Dildey, F., Treichler, R., Amann, M.-C., Schier, M. & Ebbinghaus, G. (1989). Replacement of magnesium in InGaAs/InP heterostructures during zinc diffusion. *Applied Physics Letters* 55(9): 876

Ehsan, A. A., Shaari, S. (2001). *Fabrikasi dan pencirian diodfoto planar P-I-N silikon untuk OEIC.* Technical Report No MTF 004. Universiti Kebangsaan Malaysia.

Feng, S. & Lu, C. (2004). Influence of InP cap layer on photo-responsivity of InP/InGaAs PIN detector. *Proceedings of the 7th International Conference on Solid-State and Integrated Circuits Technology*, 2332-2334.

Filmtronics, Incorporated. (2006a). *Experimental Selenium Film Se-965*. Datasheet. USA.

Filmtronics, Incorporated. (2006b). *Experimental Zinc Zn-980*. Datasheet. USA.

Gangopadhyay, U., Park, S., Kim, K., Park, J., Kim, D. & Yi, J. (2003). Large area monocrystalline silicon solar cell using SOD. *Proceedings of 3rd World Conference on Photovoltaic Energy Conversion*, 1431-1434.

Giziewicz, W., Prasad, S. & Fonstad, C. G. Jr. (2004). Lateral p-i-n photodetectors fabricated in a standard commercial GaAs VLSI process. *Proceedings of IEEE on Sensors*, 284-287

Goodrich Corporation. (2006). What is InGaAs?. Application Note 4110-0039: 1-3. http://www.sensorsinc.com/GaAs.html

Ho, C.-L., Wu, M.-C., Ho, W.-J. & Liaw, J.-W. (2000). Comparison between planar InP/InGaAs/InP pin photodiodes with symmetrical and asymmetrical doping profiles. *IEEE Proceedings on Optoelectronics* 147(2): 109-113.

Huang, Z. (2003). *Multi gigahertz InGaAs/InP inverted MSM photodetectors for photoreceiver and waveguide applications*. Ph.D Thesis. Georgia Institute of Technology, Atlanta.

Jacob, B., Witzigmann, B., Klemenc, M. & Petit, C. (2005). A TCAD methodology for high-speed photodetectors. *Solid -State Electronics* 49: 1002-1008.

Jeong, T.W.; Iiyama, K.; Takamiya, S. (2005). Two terminal InP/InGaAs heterojunction phototransistor with lateral photodiode as sensing section. *International Conference on Indium Phosphide and Related Materials*, 250-253.

Kamanin, A. V., Mokina, I. A., Shmidt, N. M., Busygina, L. A., & Yurre, T. A. (1996). Polymer diffusants in III-V semiconductor compounds technology. *Proceedings of the Eighth International Conference on Indium Phosphide and Related Materials IPRM '96*, 334-337

Kim, K. S. (2003). On the evolution of PON-based FTTH systems. *Information Sciences* 149: 21-30.

Koester, S. J., Schaub, J. D., Dehlinger, G & Chu, J. O. (2006). Germanium-on-SOI infrared detectors for integrated photonic applications. *IEEE Journal of Selected Topics in Quantum Electronics* 12(6): 1489-1502.

Konno, K., Matsushima, O., Navarro, D. & Miura-Mattausch, M. (2004). High frequency response of p-i-n photodiodes analyzed by an analytical model in Fourier space. *Applied Physics Letters* 96(7): 3839-3844.

Lange, M. J., Dixon, & Olfsen, G. H. (2000). p-n junction formation in 3- and 4-inch indium gallium arsenide epitaxial wafers using a doped glass diffusion source. *Conference on Lasers and Electro-Optics (CLEO 2000)*, 351-352.

Lauterbach, C. (1995). Zinc diffusion in InP from spin-on films of various zinc concentrations. *Semiconductor Science and Technology* 10: 500-503.

Lee, B., Yoon, H., Hyun, K. S., Kwon, Y. H. & Yun, I. (2004). Investigation of manufacturing variations of planar InP/InGaAs avalanche photodiodes for optical receivers. *Microelectronics Journal* 35(8): 635-640.

Lee, C. D & Forrest, S. R. (1991). $In_{0.53}Ga_{0.47}As$/InP heterojunction with low interface defect densities. *Journal of Applied Physics* 69: 342.

Lee, C.-H. & Choi, K.-M. (2007). Fiber to the home. *IEEE Proceedings*. 937-938

Lee, W. S., Kitching, S. A. & Bland, S. W. (1989). Monolithic integration of fully ion-implanted lateral GaInAs pin detector/InP JFET amplifier for 1.3-1.55 μm optical receivers. *Electronics Letters* 25(8): 522-523.

Li, R., Schaub, J. D., Csutak, S. M., & Campbell, J. C. (2000). A high-speed monolithic silicon photoreceiver fabricated on SOI; *IEEE Photon. Technol. Lett.* 12:1046-1048.

Loo, E. C. H. 2007. *Pembangunan diodfoto planar silicon dan InGaAs dengan elektrod H_N dan H_P*. Ph.D. Thesis. Universiti Kebangsaan Malaysia.

Menon, P. S. (2005). *Pembangunan diodfoto planar p-i-n silikon (Development of silicon-based p-i-n photodiode)*. MSc Thesis. Universiti Kebangsaan Malaysia.

Menon, P. S., Kandiah, K., Ehsan, A. A. & Shaari, S. (2009). The development of a new responsivity prediction model for In(0.53)Ga(0.47)As interdigitated lateral PIN photodiode. *Journal of Optical Communications*, 30, 2009: 2-6.

Menon, P. S., Kandiah, K. & Shaari, S. (2008a). Concentration and temperature-dependent low-field mobility model for In0.53Ga0.47As interdigitated lateral PIN PD. *IEICE Electronics Express* 5(9): 303-309.

Menon, P. S., Kandiah, K. & Shaari, S. (2008b). SNR prediction model of an In(0.53)Ga(0.47)As interdigitated lateral p-i-n photodiode. *Proceedings of the 2008 IEEE International Conference on Semiconductor Electronics*, ICSE2008 (Johor Bahru), 292-296.

Menon, P. S., Kandiah, K., Ehsan, A. A. & Shaari, S. (2010). Concentration-dependent minority carrier lifetime in an In(0.53)Ga(0.47)As interdigitated lateral PIN photodiode based on spin-on chemical fabrication methodology. *International Journal of Numerical Modelling: Electronic Networks, Devices and Fields.* DOI: 10.1002/jnm.792

Menon, P. S. (2008). *Pembangunan diodfoto planar p-i-n In(0.53)Ga(0.47)As dengan elektrod berselang-seli (Development of an In(0.53)Ga(0.47)As interdigitated lateral p-i-n photodiode).* Ph.D Thesis. Universiti Kebangsaan Malaysia.

Montgomery, D.C. (2005). *Design and analysis of experiment.* 6th Edition. New York: John Wiley.

Murray, S. L., Newman, F. D., Wilt, D. M., Wanlass, M. W., Ahrenkiel, P., Messham, R. & Siergiej, R. R. (2003). MOCVD growth of lattice-matched and mismatched InGaAs materials for thermophotovoltaic energy conversion. *Semiconductor Science and Technology* 18: 202-208.

Oh, J., Im, K., Ahn, C-G., Yang, J-H., Cho, W-J., Lee, S. & Park, K. (2004). Ultra shallow and abrupt n+-p junction formations on silicon-on-insulator by solid phase diffusion of arsenic from spin-on-dopant for sub 50 nm Si metal-oxide-semiconductor devices. *Materials Science and Engineering B* 110(2): 185-189.

Ohtsuka, K., Ohishi, T., Abe, Y., Sugimoto, H., Matsui, T. & Ogata, H. (1988). High purity $In_{0.53}Ga_{0.47}As$ layer grown by liquid phase epitaxy. *Journal of Crystal Growth* 89: 391.

Parker, D.G. (1988). The theory, fabrication and assessment of ultra high-speed photodiodes. *GEC Journal of Research* 6(2): 106-117.

Pearsall, T. P. (1981). Alloy scattering effects and calculated mobility in n-type $Ga_{0.47}In_{0.53}As$. *Electronics Letters* 17: 169.

Penna, T., Tell, B., Liao, A. S. H., Bridges, T. J. & Burkhardt, G. (1985). Ion implantation of Si and Se donors in $In_{0.53}Ga_{0.47}As$. *Journal of Applied Physics* 57(2): 351-354.

Posthuma, N. E., van der Heide, J., Flamand, G. & Poortmans, J. (2007). Emitter formation and contact realization by diffusion for germanium photovoltaic devices. *IEEE Transactions on Electron Devices* 54(5): 1210-1215.

Sabella, R. & Merli, S. (1993). Analysis of InGaAs P-I-N photodiode frequency response. *IEEE Journal of Quantum Electronics* 29:906-916.

Saleh, B. E. A. & Teich, M.C. (1991). *Fundamentals of photonics.* USA: John Wiley & Sons Incorporated.

Schow, C.L., Li, R., Schaub, J.D. & Campbell, J.C. (1999). Design and implementation of high-speed planar Si photodiodes fabricated on SOI substrates. *IEEE Journal of Selected Topics in Quantum Electronics* 35(10): 1478-1482.

Sipahi, L.B. & Sanders, T. J. (2002). Modeling, simulation and comparative analysis of RF bipolar and MOS low noise amplifiers for determining their performance dependence on silicon. *Proceedings of the fifth international conference on modeling and simulation of Microsystems MSM2002*, 155-158.

Srivastava, S. & Roenker, K.P. (2003). Numerical modeling study of the InP/InGaAs uni-travelling carrier photodiode. *Solid -State Electronics* 48: 461-470.

Sze, S.M. 2002. *Semiconductor Devices, Physics and Technology.* 2nd Edition. New York: John Wiley & Sons.

Tiwari, S., Burroughes, J., Milshtein, M. S., Tischler, M. A. & Wright, S. L. (1992). Lateral Ga(0.47)In(0.53)As and GaAs p-i-n photodetectors by self-aligned diffusion. *IEEE Photonics Technology Letters* 4(4): 396-398.

Tsang, W. T. (1985). Semiconductors and Semimetals. In *Lightwave Communications Technology.* New Jersey: Academic Press.

Silvaco International. (2004). *ATLAS User's Manual.* 10th Edition. USA: SILVACO International Incorporated.

Sotoodeh, M., Khalid, A. H. & Rezazadeh, A. A. (2000). Empirical low-field mobility model for III-V compounds applicable in device simulation codes. *Journal of Applied Physics* 87: 2890-2900.

Takeda, Y., Littlejohn, M. A. & Hauser, J. R. (1981). Electron Hall mobility calculation and alloy scattering in $In_{0.53}Ga_{0.47}As$. *Electronics Letters* 17: 377.

Tashima, M. M., Cook, L. W. & Stillman, G. E. (1981). Room temperature electron diffusion lengths in liquid phase epitaxial InGaAsP and InGaAs. *Applied Physics Letters* 39: 960-961.

Yasouka, N., Sanada, T., Hamaguchi, H., Makiuchi, M., Mikawa, T., Kuramata, A., Wada, O., Deri, R.J. (1991). High-speed monolithic coherent optical receiver integrated on InP substrate. *Electronics Letters* 27(22): 2020-2022.

Yuan, C-A., Han, C. N., Yew, M-C.,Chou, C-Y. & Chiang, K-N. (2005). Design, analysis and development of novel three-dimensional stacking WLCSP. *IEEE Transactions on Advanced Packaging* 28:387-396.

Zant, P. V. (2000). *Microchip fabrication.* USA: McGraw-Hill.

Zhao, X. (2006). *Carrier transport in high-speed photodetectors based on two-dimensional-gas.* Ph.D Thesis. Drexel University.

5

Simulation of Small-pitch High-density Photovoltaic Infrared Focal Plane Arrays

Mikhail Nikitin[1], Albina Drugova[2],
Viacheslav Kholodnov[2] and Galina Chekanova[1]
[1]*Federal State Unitary Enterprise ALPHA*
[2]*Institute of Radio Engineering and Electronics Russian Academy of Sciences*
Russia

1. Introduction

Scanning and starring photovoltaic infrared focal plane arrays (PV IRFPAs) based on ternary alloys $Hg_{1-x}Cd_xTe$ (Whicker, 1992; Triboulet & Chatard, 2000; Baker & Maxey, 2001; Norton, 2002; Kinch, 2007) and binary compound InSb and its alloys (Glozman et al., 2006) are considered as the most sensitive, flexible and perspective for detection of infrared radiation in spectral ranges 1.5-2.7 μm Short-Wave IR (SWIR), 3-5.5 μm Mid-Wave IR (MWIR), 8-14 μm Long-Wave IR (LWIR) and longer than 14 μm Very Long-Wave IR (VLWIR). Those FPAs are updated and improved continuously and move gradually from linear arrays such as 288×4 (TDI); 480×(4-8) (TDI); 768×8 (TDI) pixels to mid-format (sub-TV and TV) including but not limited 64×64; 320×256; 384×288; 640×512 pixels and finally to megapixel format (High Definition TV) like 1280×768; 1280×1024 pixels and more. Nowadays all manufacturers offer LWIR PV FPA with peak wavelength $\lambda_p \approx 8.5 \pm 0.5$ μm. It means that scanning thermal imagers (TI) based on old LWIR photoconductive (PC) linear arrays ($\lambda_p \approx 11$ μm) covers 8-14 μm atmospheric "window" of transparency totally whereas TI based on LWIR PV FPA with $\lambda_p \approx 8.5 \pm 0.5$ μm covers left (shorter) part of that "window" only. As the result TIs based on LWIR PC linear arrays ($\lambda_p \approx 11$ μm) allow adequate visualizing of cold landscape (scene) with temperatures as low as minus 60 ^{0}C. Thermal Imagers based on LWIR PV FPA with $\lambda_p \approx 8.5 \pm 0.5$ μm can visualize adequately cold landscape at scene temperatures higher than minus 30 ^{0}C (even higher than minus 20 ^{0}C). Full replacement of scanning type TI by starring type TI will take place when extended LWIR PV FPA with λ_p shifted to 10-11 μm at T_{op}=80-100 K will become affordable. Megapixel high performance IRFPA having extended spectral covering with λ_p=10-11 μm at T_{op}=80-100 K could be preferable to create future TI systems.

Increasing of array format along with improvement in performance is general development trend in IRFPA technology. It is accompanied inevitably by decreasing of pixel size and pixel pitch to minimal size reasonable from point of view of infrared physics to provide the best resolution and producing comfortable imaging with electro-optic (EO) system. Pitch in small-pitch PV IRFPA can be equal to from 10 μm to 20 μm. PV arrays based on InSb and its alloys or $Hg_{1-x}Cd_xTe$ alloys are fabricated often on single layer (substrate) that is common for all pixels of array.

Implementation of large format high performance PV IRFPAs covering above mentioned spectral ranges both single-color and multi-color requires comprehensive simulation of photodiodes (PD) performance depending on base material layers properties, interfaces parameters, array topology, array design and operating conditions. Analysis of MWIR and LWIR PD performance at operating temperatures from 77 K to 100 K and higher is needed also due to strong tendency to use so called HOT (higher operating temperature) mode for lowering weight and power consumption in perspective TIs with cryogenically cooled megapixel IRFPAs.

Perhaps novel $Hg_{1-x}Cd_xTe$ FPAs will be based on photodiodes with p-n junction opposite to usually used n^+-p junction. PD with optimal p-n junction could have lower dark current value than same size n^+-p junction. It is desirable for adequate multiplexing of PD arrays to Silicon Read-out Integrated Circuits (ROICs).

2. Key aspects of IRFPA performance requiring simulation

1. Simulation of IR photodiodes detectivity and responsivity depending on cut-off wavelength, type of junction: n^+-p junction or p-n junction and operating temperatures from 77 K to 100 K and higher.
2. How does recombination rate at nearest interface to PD absorber impact on PD dark current?
3. Development of theoretical approach producing analytical expressions for collection of photogenerated charge carriers in small-pitch infrared PV arrays enabling optimization of array topology for reaching the best resolution, good filling factor and minimal cross-talking.

Due to small thickness of layers in epitaxial heterostructure interfaces are located close to active regions of p-n junction and hence generation-recombination processes at interfaces can impact on value of current flowing through junction. In high-density arrays with thin common layer, collection length of photogenerated charge carriers will exceed pixel pitch as a rule. It means that each pixel can collect excess charge carriers generated far from PD's p-n junction border. Therefore optimization of resolution, filling factor and cross-talking level of small-pitch high-density PV FPA requires complete estimation of photocurrent generation in neighbor PD pixels depending on pixel and array design, material properties and operating conditions. In two technologically viable 2D IRFPA architectures: front-side illuminated High-Density Vertically Integrated Photodiode (HDVIP) or ("Loop-hole") and backside illuminated flip-chip bonded via In-bumps to Si-ROIC are used special guard rings or grids to solve a. m. problems. Therefore development of theoretical simulation describing analytically collection of photogenerated charge carriers in small-pitch infrared PV arrays seems useful.

3. Simulation of LWIR $Hg_{1-x}Cd_xTe$ PD with small sensitive area

3.1 Photodiode models and simulation approach

Simulation was done for front-side illuminated LWIR $Hg_{1-x}Cd_xTe$ photodiode based on n^+-p or p-n junction. Performance of LWIR photodiodes ($Hg_{0.785}Cd_{0.215}Te$ and $Hg_{0.766}Cd_{0.234}Te$) was estimated at operating temperatures 77 K and 100 K. Evaluation was performed at reverse bias 0.05 V because every real $Hg_{1-x}Cd_xTe$ PD array multiplexed to Silicon Read-out Integrated Circuit (ROIC) is operated under reverse bias.

Upper limit of PD performance was calculated under assumption that diffusion current is prevailing component of dark current in PD pixel at low reverse bias. Photocurrent excited by background radiation was taken into account as well because its value is competitive to dark (diffusion) current. Tunnel current is controlled mainly by total absorber doping and in calculations its value was considered many times lower than diffusion current value at reverse bias 0.05 V. Currents due to generation in space charge region of p-n junction and surface (interface) shunting were ignored. Interface shunting elimination can become the hardest task to solve. Surface (interface) recombination acts as generator of minority charge carriers into absorber region of either n^+-p or p-n junction and at high rates it can enlarge seriously dark current value, especially when p or n absorber region is thin (shorter than diffusion length of minority charge carriers). For simplicity surface recombination rate was taken low (negligible) - 10^2 cm/sec and high (infinitive) - 10^7 cm/sec.

3.2 PD performance: simulation formalism

Let's take photodiode with n-p junction as a model and consider contribution of quasi-neutral n-side and p-side of photodiode to dark current and background current.
Depletion current per unit volume from the n-side for a planar one-side photodiode is given by expression:

$$J_p(-W_n) = J_p^F(-W_n) + J_p^D(-W_n) \tag{1}$$

Density of background current from n-side is described by formula:

$$J_p^F(-W_n) = \eta \times q \times F \times \exp(-\gamma \times W_1) \times \left[\frac{\gamma^2 \times L_p^2}{1-\gamma^2 \times L_p^2}\right] \times \{1\} \tag{2}$$

$$\{1\} = 1 + \frac{1}{\gamma \times L_p} \times \frac{\frac{D_p}{L_p} \times sh\frac{W_1}{L_p} + S_p \times ch\frac{W_1}{L_p} - (\gamma \times D_p + S_p) \times \exp(\gamma \times W_1)}{\frac{D_p}{L_p} \times ch\frac{W_1}{L_p} + S_p \times sh\frac{W_1}{L_p}} \tag{3}$$

Density of dark current from n-side is described by formulae:

$$J_p^D(-W_n) = -q \times \frac{D_p}{L_p} \times \Delta p_{ne}(-W_n) \times \left[\frac{\frac{D_p}{L_p} \times sh\frac{W_1}{L_p} + S_p \times ch\frac{W_1}{L_p}}{\frac{D_p}{L_p} \times ch\frac{W_1}{L_p} + S_p \times sh\frac{W_1}{L_p}}\right] \tag{4}$$

$$\Delta p_n(-W_n) = p_{ne}\left(\exp\left(\frac{q \times V}{kT}\right) - 1\right) \tag{5}$$

Contribution to responsivity from n-side of photodiode:

$$S_{J\lambda}^{N} = \eta \times 0.8 \times 10^{4} \times \lambda_{co} \times \exp(-\gamma \times W_1) \times \left[\frac{\gamma^2 \times L_p^2}{1 - \gamma^2 \times L_p^2} \right] \times \{1\} \tag{6}$$

$$\lambda_{co}(\mu m) = \frac{h \times c}{E_g(eV)} \tag{7}$$

Depletion current per unit volume from the p-side for a planar one-side photodiode is given by expression:

$$J_n(W_p) = J_n^F(W_p) + J_n^D(W_p) \tag{8}$$

Density of background current from p-side is described by formula:

$$J_n^F(W_p) = \eta \times q \times F \times \exp(-\gamma \times W_1) \times \left[\frac{\gamma^2 \times L_n^2}{(\gamma \times L_n)^2 - 1} \right] \times \{2\} \tag{9}$$

$$\{2\} = 1 - \frac{1}{\gamma \times L_n} \times \frac{S_n \times ch\frac{W_3}{L_n} + \frac{D_n}{L_n} \times sh\frac{W_3}{L_n} + \exp(-\gamma \times W_3) \times [-S_n + \gamma \times D_n]}{S_n \times sh\frac{W_3}{L_n} + \frac{D_n}{L_n} \times ch\frac{W_3}{L_n}} \tag{10}$$

Density of dark current from p-side is described by formulae:

$$J_n^D(W_p) = -q \times \frac{D_n}{L_n} \times \Delta n_{pe}(W_p) \times \frac{\frac{D_n}{L_n} \times sh\left(\frac{W_3}{L_n}\right) + S_n \times ch\left(\frac{W_3}{L_n}\right)}{\frac{D_n}{L_n} \times ch\left(\frac{W_3}{L_n}\right) + S_n \times sh\left(\frac{W_3}{L_n}\right)} \tag{11}$$

$$\Delta n_p(W_p) = n_{pe}\left[\exp\left(\frac{q \times V}{kT}\right) - 1 \right] \tag{12}$$

Contribution to responsivity from p-side of photodiode:

$$S_{J\lambda}^{P} = \eta \times 0.8 \times 10^{4} \times \lambda_{co} \times \exp(-\gamma \times W_1) \times \frac{\gamma^2 \times L_n^2}{1 - \gamma^2 \times L_n^2} \times \{2\} \tag{13}$$

Here:
$-W_n$ - coordinate of depletion region border on n-side; W_p - coordinate of depletion region border on p-side; W_1 - thickness of quasi-neutral n-side; W_3 - thickness of quasi-neutral p-side; q - electron charge; $\eta = 1 - r$ - quantum efficiency; γ and r - absorption and reflection coefficients; F - background radiation flux density; D_n, D_p - diffusion coefficient for electrons and holes properly; L_n, L_p - diffusion length for electrons and holes properly; S_n, S_p - surface recombination rate for electrons and holes properly; λ_{co} - cut-off wavelength. Majority and minority charge carrier concentrations are defined (Blakemore, 1962)

In n-side:

$$n = n_e + n_{bgr}\,;\, p_n = p_{ne} + n_{bgr}\,;\; n_e = \frac{N_d}{2} + \frac{\left(N_d^2 + 4n_i^2\right)^{1/2}}{2}\,;\; n_{bgr} = p_{bgr} = g_{bgr} \times \tau_{eff} \quad (14)$$

In p-side:

$$p = p_e + n_{bgr}\,;\, n_p = n_{pe} + n_{bgr}\,;\; p_e = \frac{N_a}{2} + \frac{\left(N_a^2 + 4n_i^2\right)^{1/2}}{2}\,;\; n_{bgr} = p_{bgr} = g_{bgr} \times \tau_{eff} \quad (15)$$

Where:
n_e and p_e - equilibrium electron and hole concentrations; N_d / N_a donor/acceptor dopant concentration; n_i – intrinsic carrier concentration; $n_{bgr} = p_{bgr}$ – average concentration of excess charge carriers generated by infrared background flux; $g_{bgr} = \eta \times \gamma \times F$ – excess charge carriers generation rate by background flux; τ_{eff} - resulting excess charge carriers' lifetime.
Energy gap value $E_g(x,T)$ in eV is determined by formula (Laurenti et al., 1990), where x is composition of $Hg_{1-x}Cd_xTe$:

$$E_g = -0.303 \times (1-x) + 1.606 \times x - 0.132 \times x \times (1-x) + \{3\} \quad (16)$$

$$\{3\} = \frac{\left[6.39 \times (1-x) - 3.25 \times x - 5.92 \times x \times (1-x)\right] \times 10^{-4} \times T^2}{11 \times (1-x) + 78.7 \times x + T} \quad (17)$$

Intrinsic charge carriers concentration in $Hg_{1-x}Cd_xTe$ is given by expression (Schmit, 1970):

$$n_i = 4.293 \times \cdot 10^{14} \times (1.093 - 0.296x + 0.442 \times 10^{-3} \times T) \times T^{3/2} \times E_g^{3/4} \times \exp\left(-\frac{E_g}{2kT}\right) \quad (18)$$

In pure non-compensated $Hg_{1-x}Cd_xTe$ material there are two band-to-band processes which control total recombination rate: radiative recombination and Auger recombination due to transitions A1 and/or A7 (Kinch et al, 1973; Gelmont, 1980; Gelmont 1981; Kinch, 2007):

$$\frac{1}{\tau_R} = \frac{n_e + p_e + n_{bgr}}{2n_i \times \tau_{Ri}}\,;\; \frac{1}{\tau_{A1}} = \frac{(n_e + n_{bgr}) \times (n_e + p_e + n_{bgr})}{2 \times n_i^2 \times \tau_{A1}^i}\,;\; \frac{1}{\tau_{A7}} = \frac{(p_e + n_{bgr}) \times (n_e + p_e + n_{bgr})}{2 \times n_i^2 \times \tau_{A7}^i} \quad (19)$$

$$\tau_{Ri} = 7 \times 10^8 \times (1+\mu)^{3/2} \times \frac{1}{E_g n_i} \times \left(\frac{T}{77}\right)^{3/2}\,;\; \tau_{A1}^i = \frac{1}{7.2 \times 10^{13}} \times \frac{1}{E_g} \times \left(\frac{E_g}{kT}\right)^{3/2} \times \exp\left[(1+2\mu) \times \frac{E_g}{kT}\right];$$

$$\tau_{A7}^i = 3.69 \times 10^{-16} \times \mu^{-5/2} \times \frac{E_g}{kT} \times \exp\left[(1+\mu) \times \frac{E_g}{kT}\right];\; \mu = (m_e / m_{hh})\,. \quad (20)$$

Resulting excess charge carriers' lifetime equals to:

$$\frac{1}{\tau}=\frac{1}{\tau_R}+\frac{1}{\tau_{A1}}+\frac{1}{\tau_{A7}} \qquad (21)$$

Iteration procedure was used to calculate n_{bgr} (5): $\tau_{eff}\Big|_{n_{bgr}=n_{bgr}^{(i)}} = \tau_{eff}\Big|_{n_{bgr}=n_{bgr}^{(i-1)}}$, $i=1, 2, \ldots$ k, $n_{bgr}^{(0)}=0$. Convergence took place at number of iteration $k \le 10$.

The following noise sources were taken into account:

- Johnson-Nyquist thermal noise of PD's dynamic resistance;
- Background current shot noise;
- Dark current shot noise.

Noise currents densities are taken at preselected reverse bias V_b (typically 0.01-0.1 V).

$$\delta I^2=\frac{4kT}{R_{dV}}\Delta f+2\times q\times(J_\Sigma^{Ff}\times A_{Ff}+J_\Sigma^{D}\times A_d)\times\Delta f \qquad (22)$$

Total density of noise current:

$$I_{sh}=\sqrt{\delta I^2} \qquad (23)$$

Here:

A_d - geometrical area of photodiode's p-n junction; A_{Ff} - collection area of photogenerated current in photodiode ("light capture" area); Δf - operative bandwidth; R_{dV} - resistance of photodiode at preselected reverse bias V, J_Σ^{Ff} - total background current, J_Σ^{D} - total dark current.

$$\frac{1}{R_d}=A_d\times\frac{q^2}{kT}\times\left[-p_n\times\frac{D_p}{L_p}\times\frac{\frac{D_p}{L_p}\times sh\frac{W_1}{L_p}+S_p\times ch\frac{W_1}{L_p}}{\frac{D_p}{L_p}\times ch\frac{W_1}{L_p}+S_p\times sh\frac{W_1}{L_p}}-n_p\times\frac{D_n}{L_n}\times\frac{\frac{D_n}{L_n}\times sh\frac{W_3}{L_n}+S_n\times ch\frac{W_3}{L_n}}{\frac{D_n}{L_n}\times ch\frac{W_3}{L_n}+S_n\times sh\frac{W_3}{L_n}}\right] \qquad (24)$$

$$R_{dV}=R_d\times\exp\left(-\frac{q\times V}{kT}\right) \qquad (25)$$

First term in curly brackets determinates contribution of n-side to resistance of photodiode at reverse bias and second term the same of p-side.

Impact of surface recombination rate on charge carriers concentration and currents densities was accounted correctly.

Total density of background current:

$$J_\Sigma^{Ff}=J_n^{Ff}(-W_n)+J_p^{Ff}(W_p) \qquad (26)$$

Total density of dark current:

$$J_\Sigma^{D}=J_n^{D}(-W_n)+J_p^{D}(W_p) \qquad (27)$$

Let's assume for simplicity that:

$$A_d = A_{Ff} = A \tag{28}$$

Density of total current through photodiode will be sum of two terms:

$$J_{FfD} = J_{\Sigma}^{Ff} + J_{\Sigma}^{D} \tag{29}$$

Detectivity is calculated following to standard expression:

$$D^* = \frac{S_J \times A \times \Delta f}{(\delta I^2)^{1/2}} = \frac{S_J}{\left(\frac{4kT}{R_{dV} \times A} + 2q \times J_{FfD}\right)^{1/2}} \tag{30}$$

$$Ff = k_f \times 2\pi \times c \int_{2\times10^{-4}}^{\lambda_{co}} \frac{1}{\lambda^4} \times \frac{d\lambda}{\exp\left(\frac{h \times c}{\lambda \times kT_{Ff}}\right) - 1} \tag{31}$$

Here: $k_f = \sin^2(\Theta/2)$ where Θ - full solid angle within that background and signal radiation comes in sensitive area of photodiode.

3.3 LWIR PD performance: calculation results

We have done calculations for model photodiodes based on asymmetric n^+-p or p-n junction always used in practice. Data used in calculation are presented in Table 1.

	PD with n^+-p junction		PD with p-n junction	
Operating temperature, T (K)	77	100	77	100
$Hg_{1-x}Cd_xTe$ absorber composition, x (mol. fr.)	0.234 / 0.215	0.234 / 0.215	0.234 / 0.215	0.234 / 0.215
Energy gap, E_g (eV)	0.138 / 0.104	0.144 / 0.112	0.138 / 0.104	0.144 / 0.112
Cut-off wavelength, λ_{co} (μm)	9.0 / 11.9	8.6 / 11.1	9.0 / 11.9	8.6 / 11.1
Peak wavelength, λ_p (μm)	≈ 8.1 / ≈ 10.5	≈ 7.7 / ≈ 10	≈ 8.1 / ≈ 10.5	≈ 7.7 / ≈ 10
Absorption coefficient (Blue, 1964), γ (cm^{-1})	3×10^3	3×10^3	3×10^3	3×10^3
Quantum efficiency, η	0.7	0.7	0.7	0.7
Junction area, A (μm × μm)	20 × 20	20 × 20	20 × 20	20 × 20
Junction regions doping, n and p (cm^{-3})	$n^+=10^{17}$ $p=10^{16}$	$n^+=10^{17}$ $p=10^{16}$	$p=5\times10^{16}$ $n=10^{15}$	$p=5\times10^{16}$ $n=10^{15}$
Junction regions thickness, t (μm)	$t(n^+) = 0.5$ t(p-absorber) = 4-40	$t(n^+) = 0.5$ t(p-absorber) = 4-40	t(p) = 0.5 t(n-absorber) = 4-40	t(p) = 0.5 t(n-absorber) = 4-40
Electron mobility, μ_n ($cm^2/(V\times sec)$)	1.9×10^5	1.29×10^5	1.9×10^5	1.29×10^5
Hole mobility, μ_p ($cm^2/(V\times sec)$)	600	390	600	390
Reverse bias value, V_b (V)	-0.05	-0.05	-0.05	-0.05
Surface recombination rate, s (cm/sec)	10^2 10^7	10^2 10^7	10^2 10^7	10^2 10^7

Table 1. Data used for estimation of small-size $Hg_{0.766}Cd_{0.234}Te$ and $Hg_{0.785}Cd_{0.215}Te$ photodiodes performance

Calculation results are presented on Fig. 1-6. Typically discussed photovoltaic case (V_b =0) has been studied as well.
Obtained results presented on Fig. 1-6 say that extended LWIR PD with p-n junction will be potentially of 4-5 times lower dark current value than PD with n+-p junction at T_{op}=77 K and 2 times lower at T_{op}=100 K. As the result it is hoped that decrease in D* value with elevating of operating temperature up to 100 K will be moderate in the case of PD with p-n junction opposite to significant decreasing observed on LWIR PD with n+-p junction as it presented on Fig. 1-6. Calculated detectivity at reverse bias 0.05 V is higher than in the case of zero bias (photovoltaic mode). Formalism of R_0A product is not suitable for the case of LWIR PD arrays multiplexed to Silicon ROIC.

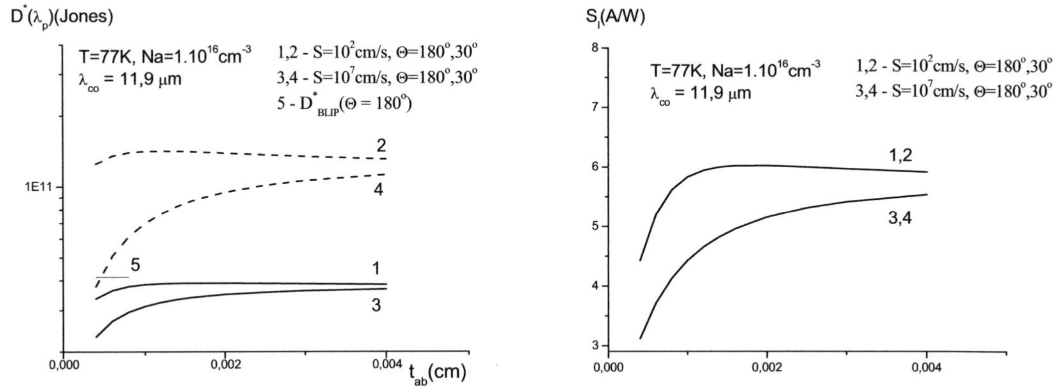

Fig. 1. Calculated peak detectivity $D^*(\lambda_p)$ and peak responsivity $S_I(\lambda_p)$ of $Hg_{0.785}Cd_{0.215}Te$ photodiodes with n^+-p junction versus thickness of p-absorber t_{ab} at FOV=180^0 - (1 and 3) and FOV=30^0 - (2 and 4). Surface recombination rate s=10^2 cm/sec (1 and 2) and s=10^7 cm/sec (3 and 4). Operating temperature 77 K. Background temperature equals to 293 K. Doping of p-absorber p_{77}=10^{16} cm^{-3}, n^+-p junction area 20 μm × 20 μm

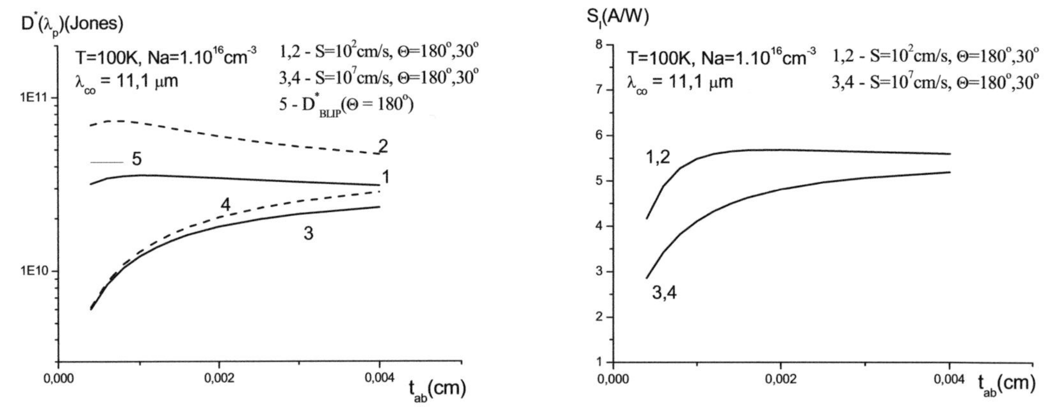

Fig. 2. Calculated peak detectivity $D^*(\lambda_p)$ and peak responsivity $S_I(\lambda_p)$ of $Hg_{0.785}Cd_{0.215}Te$ photodiodes with n^+-p junction versus thickness of p-absorber t_{ab} at FOV=180^0 - (1 and 3) and FOV=30^0 - (2 and 4). Surface recombination rate s=10^2 cm/sec (1 and 2) and s=10^7 cm/sec (3 and 4). Operating temperature 100 K. Background temperature equals to 293 K. Doping of p-absorber p_{77}=10^{16} cm^{-3}, n^+-p junction area 20 μm × 20 μm

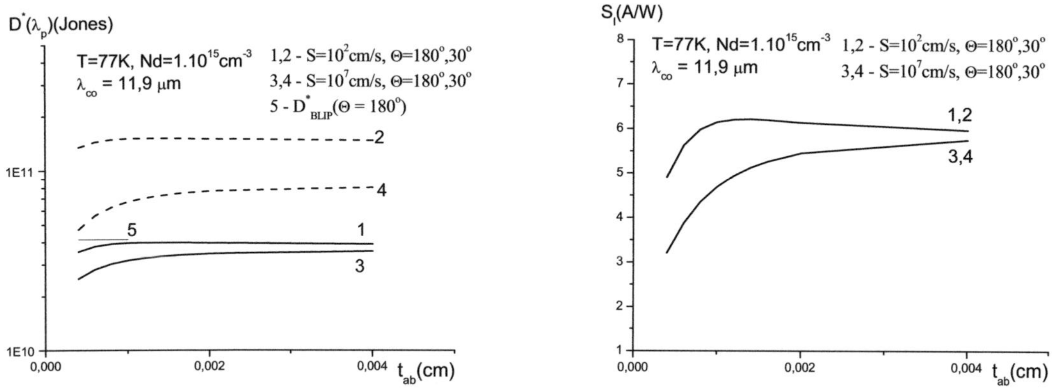

Fig. 3. Calculated peak detectivity $D^*(\lambda_p)$ and peak responsivity $S_I(\lambda_p)$ of $Hg_{0.785}Cd_{0.215}Te$ photodiodes with p-n junction versus thickness of n-absorber t_{ab} at FOV=180^0 - (1 and 3) and FOV=30^0 - (2 and 4). Surface recombination rate s=10^2 cm/sec (1 and 2) and s=10^7 cm/sec (3 and 4). Operating temperature 77 K. Background temperature equals to 293 K. Doping of n-absorber n_{77}=10^{15} cm^{-3}, p-n junction area 20 μm × 20 μm

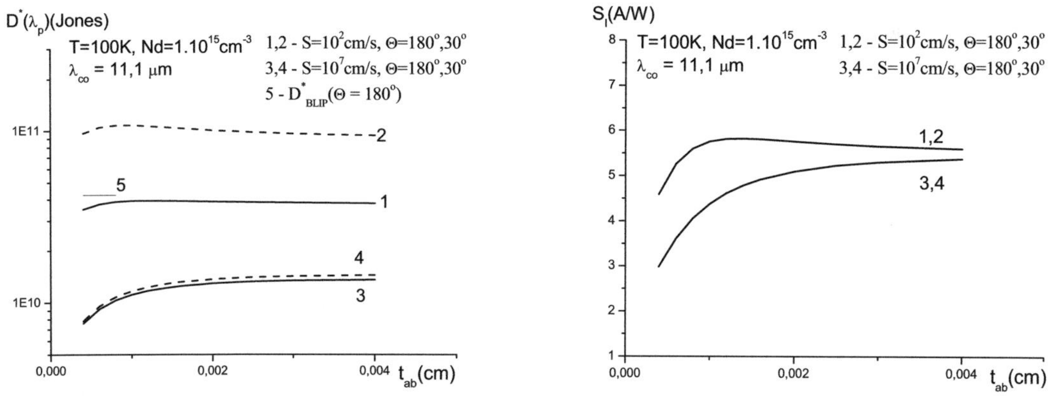

Fig. 4. Calculated peak detectivity $D^*(\lambda_p)$ and peak responsivity $S_I(\lambda_p)$ of $Hg_{0.785}Cd_{0.215}Te$ photodiodes with p-n junction versus thickness of n-absorber t_{ab} at FOV=180^0 - (1 and 3) and FOV=30^0 - (2 and 4). Surface recombination rate s=10^2 cm/sec (1 and 2) and s=10^7 cm/sec (3 and 4). Operating temperature 100 K. Background temperature equals to 293 K. Doping of n-absorber n_{77}=10^{15} cm^{-3}, p-n junction area 20 μm × 20 μm

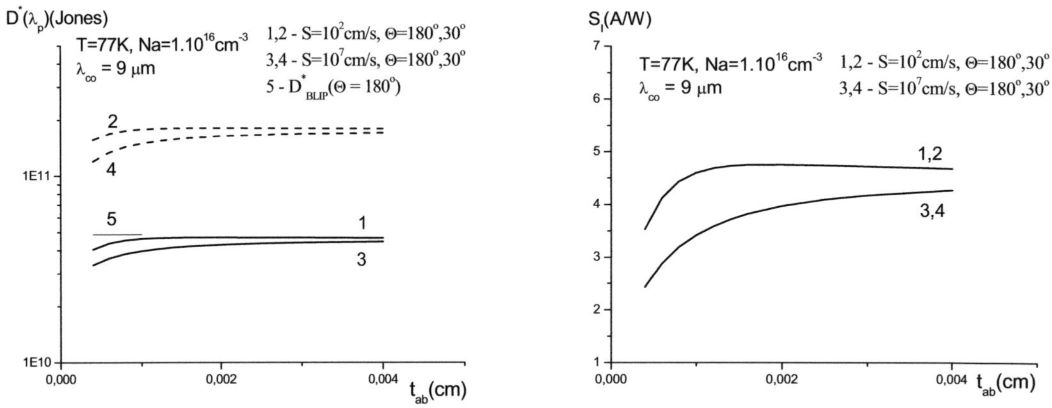

Fig. 5. Calculated peak detectivity $D^*(\lambda_p)$ and peak responsivity $S_I(\lambda_p)$ of $Hg_{0.766}Cd_{0.234}Te$ photodiodes with n^+-p junction versus thickness of p-absorber t_{ab} at FOV=180^0 - (1 and 3) and FOV=30^0 - (2 and 4). Surface recombination rate s=10^2 cm/sec (1 and 2) and s=10^7 cm/sec (3 and 4). Operating temperature 77 K. Background temperature equals to 293 K. Doping of p-absorber $p_{77}=10^{16}$ cm^{-3} , n^+-p junction area 20 μm × 20 μm

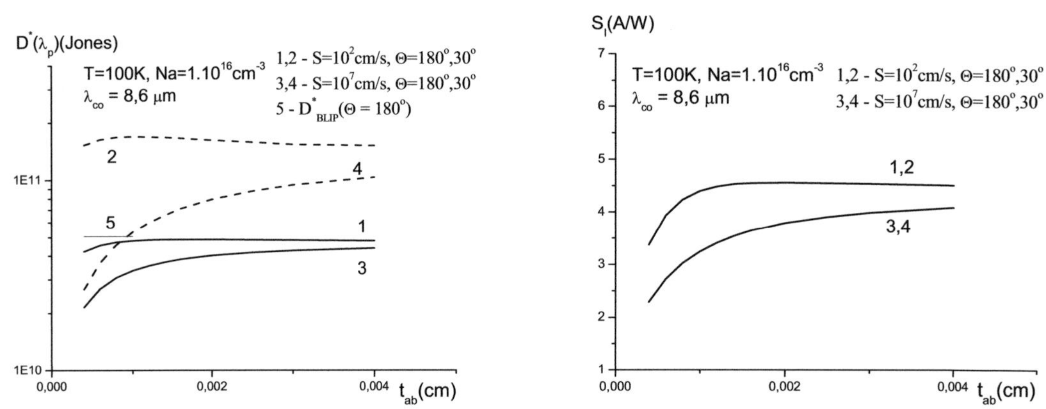

Fig. 6. Calculated peak detectivity $D^*(\lambda_p)$ and peak responsivity $S_I(\lambda_p)$ of $Hg_{0.766}Cd_{0.234}Te$ photodiodes with n^+-p junction versus thickness of p-absorber t_{ab} at FOV=180^0 - (1 and 3) and FOV=30^0 - (2 and 4). Surface recombination rate s=10^2 cm/sec (1 and 2) and s=10^7 cm/sec (3 and 4). Operating temperature 100 K. Background temperature equals to 293 K. Doping of p-absorber $p_{77}=10^{16}$ cm^{-3} , n^+-p junction area 20 μm × 20 μm

4. Surface recombination impact on currents in LWIR $Hg_{1-x}Cd_xTe$ photodiode

4.1 Approach and formalism

Cross-section of model photodiode (pixel) is shown on Fig. 7.

Dependences of dark and background currents in reverse-biased LWIR $Hg_{1-x}Cd_xTe$ photodiode on surface recombination rate S at back surface of p base ($t = LP$) were studied.

Basing on parameters of considered photodiode let's assume that:

1. Hole current inflowing into space charge region is negligible.
2. Generation-recombination current in space charge region is negligible.

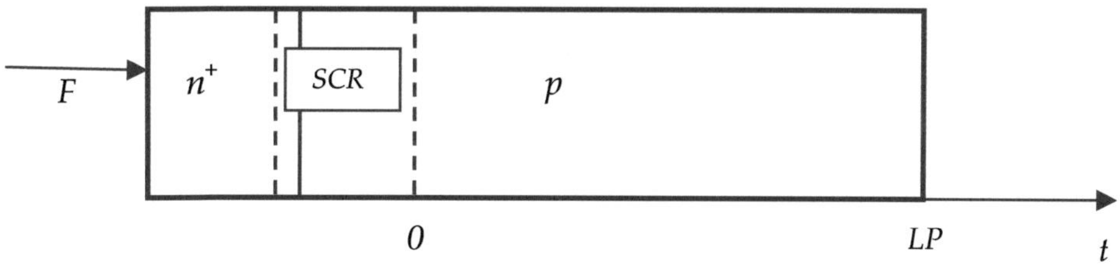

Fig. 7. Cross-section of model photodiode pixel. Here: n^+ is n^+ - region of $n^+ - p$ junction; p is base region common for all pixels of PV array. SCR is space-charge (depletion) region of $n^+ - p$ junction. Front surface of photodiode is irradiated by background photon flux F that is absorbed and generates photocurrent in photodiode. Zero point on t -axis means the boundary between space-charge region and quasi-neutral part of p base region. Point $t = LP$ is coordinate of p base region back surface

Concentration profiles of non-equilibrium dark and background generated charge carriers in p base versus t coordinate were analyzed theoretically in reversed-biased $Hg_{1-x}Cd_xTe$ photodiode at different Field-Of-View (FOV) and surface recombination rate S values. It is shown that growth of concentration of non-equilibrium dark charge carriers near SCR depends significantly on S that differs essentially from behavior of non-equilibrium background generated charge carriers. It gives in the result high growth of dark current with increasing of surface recombination rate. At the same time background current is varied low. Calculations based on obtained analytic expressions were done at temperature $T = 77$ K.

Continuity equation of electron current in p base of photodiode is defined by expression:

$$\frac{\partial i_n(t)}{\partial t} + q \times g(t) - q \times R_n = 0 \tag{32}$$

Where, $i_n(t)$ - electron current density, $g(t)$ – specific (per cubic centimeter) photogeneration rate of electron-hole pairs which is defined by formula:

$$g(t) = \gamma \times \eta \times F \times \exp(-\gamma \times t) \tag{33}$$

Where:

$R_n = R_p = \Delta n / \tau$ - specific band-to-band recombination rate of non-equilibrium electrons and holes; $\Delta n = n_d + n_{bgr}$ and τ - non-equilibrium electrons and holes concentrations and lifetime; n_d and n_{bgr} - concentration of non-equilibrium dark and background radiation generated charge carriers.

Dark and background generated currents flowing through photodiode were calculated at short-circuit mode of operation under low reverse-biased $V_b \le 0.05$ V.

Boundary conditions of the task are stated as follows:

$$i_n(LP) = -q \times S \times \Delta n(LP)\ ;\ \Delta n(0) = n_p \times \left\{ \exp\left[\frac{qV}{kT}\right] - 1 \right\} \tag{34}$$

Where,
S - surface recombination rate of non-equilibrium minority charge carriers (electrons) at back surface of photodiode p base (at coordinate $t = LP$; $\Delta n(0)$ - non-equilibrium charge carriers concentration at the boundary between space charge region and quasi-neutral part of p base region; n_p - concentration of equilibrium minority charge carriers (electrons) in p base and V - bias across space charge region of photodiode that is independent on illumination.
Total current I flowing through photodiode in considered conditions is formed by electrons inflowing into space charge region from quasi-neutral part of p base region:

$$I = I_n(0) = A_{pd} \times i_n(0) \tag{35}$$

Where:
A_{pd} - area of photodiode where current is formed. Please note that for the case of photodiode sensitive area and area of photodiode where current is formed are matched.
Let's assume that there is no built-in electric field in quasi-neutral parts of $n^+ - p$ junction. Solving equation (32) in diffusion approximation we find that:

$$i_n(0) = i_{bgr}(0) + i_d(0) \tag{36}$$

$$i_{bgr}(0) = q \times \eta \times Q_n = q \times \eta \times F \times \left(\frac{\gamma^2 \times L_n^2}{\gamma^2 \times L_n^2 - 1} \right) \times \left(1 - \frac{M1P}{\gamma \times L_n} \right) \tag{37}$$

$$M1P = \frac{\left(\frac{D_n}{L_n}\right) \times sh\left(\frac{LP}{L_n}\right) + S \times ch\left(\frac{LP}{L}\right) + (\gamma \times D_n - S) \times \exp(-\gamma \times LP)}{\left(\frac{D_n}{L_n}\right) \times ch\left(\frac{LP}{L_n}\right) + S \times sh\left(\frac{LP}{L_n}\right)} \tag{38}$$

$$i_d(0) = -q \times \frac{D_n}{L_n} \times \Delta n(0) \times \left[\frac{\left(\frac{D}{L_n}\right) \times sh\left(\frac{LP}{L_n}\right) + S \times ch\left(\frac{LP}{L_n}\right)}{\left(\frac{D_n}{L_n}\right) \times ch\left(\frac{LP}{L_n}\right) + S \times sh\left(\frac{LP}{L_n}\right)} \right] \tag{39}$$

Where:
$i_{bgr}(0)$ and $i_d(0)$ - background and dark components of electron current density $i_n(0)$;
D_n and L_n - electrons' diffusion coefficient and ambipolar diffusion length of charge carriers in p base defined via ambipolar diffusion coefficient of electrons in p base;
Q_n - collection coefficient of non-equilibrium photogenerated charge carriers in p base and LP - thickness of quasi-neutral part of p base region.
Concentration profiles of non-equilibrium dark $n_d(t)$ and photogenerated $n_{bgr}(t)$ charge carriers are defined by expressions (40) and (41) properly:

$$n_d(t) = \Delta n(0) \times \left[\frac{D_n \times ch\left(\frac{LP-t}{L_n}\right) + S \times L_n sh\left(\frac{LP-t}{L_n}\right)}{D_n \times ch\left(\frac{LP}{L_n}\right) + S \times L_n sh\left(\frac{LP}{L_n}\right)} \right] \quad (40)$$

$$n_{bgr}(t) = \left(\frac{\gamma \times \eta \times F \times \tau}{1 - \gamma^2 \times L_n^2} \right) \times \left[\exp(-\gamma \times t) + M2 \right] \quad (41)$$

$$M2 = \frac{(\gamma \times D_n - S) \times sh\left(\frac{t}{L_n}\right) - S \times sh\left(\frac{LP-t}{L_n}\right) - \frac{D_n}{L_n} \times ch\left(\frac{LP-t}{L_n}\right)}{S \times sh\left(\frac{LP}{L_n}\right) + \frac{D_n}{L_n} \times ch\left(\frac{LP}{L_n}\right)} \quad (42)$$

4.2 LWIR PD currents: calculation results

Data used in calculation are given in Table 2.

Data used for estimation of dark and background generated currents in small-size $Hg_{0.776}Cd_{0.224}Te$ photodiode:

	PD with n^+-p junction
Operating temperature, T	77 K
$Hg_{1-x}Cd_xTe$ absorber composition, x	0.224
Energy gap, E_g	0.12 eV
Cut-off wavelength, λ_{co}	10.3 μm
Peak wavelength, λ_p	≈ 9.2 μm
Absorption coefficient, γ	3×10^3 cm^{-1}
Quantum efficiency, η	0.7
Photodiode collection area, A_{pd}	20 μm × 20 μm = 4×10^{-6} cm^2
Thickness of quasi-neutral part of p-base, LP	10 μm=10^{-3} cm
Junction regions doping, n and p	$n^+=10^{17}$ cm^{-3}; $N_A= p=5\times10^{15}$ cm^{-3}
Bias across space charge region, V_b	-0.05 V
Minority charge carriers lifetime in p-base, τ	7.95×10^{-8} sec
Electron mobility, μ_n	1.67×10^5 cm^2/(V×sec)
Hole mobility, μ_p	600 cm^2/(V×sec)
Electron diffusion coefficient, D_n	1.15×10^3 cm^2/sec
Hole diffusion coefficient, D_p	4.14 cm^2/sec
Ambipolar diffusion length, L	48 μm

Table 2. Data used for estimation of dark and background generated currents in small-size $Hg_{0.776}Cd_{0.224}Te$ photodiode

Developed approach (32) - (42) was applied to calculate non-equilibrium dark and background generated concentration of minority charge carriers in *p* base and dark and background generated currents flowing through small-size $Hg_{0.776}Cd_{0.224}Te$ photodiode at low reverse bias.

Calculated dependences of non-equilibrium dark $n_d(t)$ and background generated $n_{bgr}(t)$ concentration of minority charge carriers in p base on surface recombination rate S and cold shield Field-Of-View (FOV) are shown on Fig. 8. As it is seen from Fig. 8 calculated non-equilibrium dark concentration of minority charge carriers at back boundary of p base $t = LP$ increases up to two orders in comparison with concentration at SCR boundary $t = 0$ with growing S. At the same time background generated concentration of minority charge carriers varies not so significantly in a few times only.

Respectively dark I_d and background generated I_{bgr} currents are varied with growing S analogously to variation of non-equilibrium dark and background generated concentrations of minority charge carriers (Fig. 9). To do comparison of $I_d(S)$ and $I_{bgr}(S)$ dependencies more convenient we present on Fig. 9 graphs in arbitrary units as well. Every curve is specified to its minimum. Minimum value of I_d responds S = 10^2 cm/sec and for I_{bgr} it responds S = 10^7 cm/sec. It is obvious that dark component varies in a few orders and background component near to constant.

Physical reason of that result becomes clear if we address to Fig. 8. Independently to surface recombination rate value at back surface of p base gradient of concentration of background generated minority charge carriers is practically the same near SCR (near $t = 0$). But gradient of concentration of non-equilibrium dark minority charge carriers increases rapidly with increasing S. Proper currents are proportional to gradients of proper concentrations at $t = 0$. Therefore background current is varied slightly and dark current increases significantly when surface recombination rate grows.

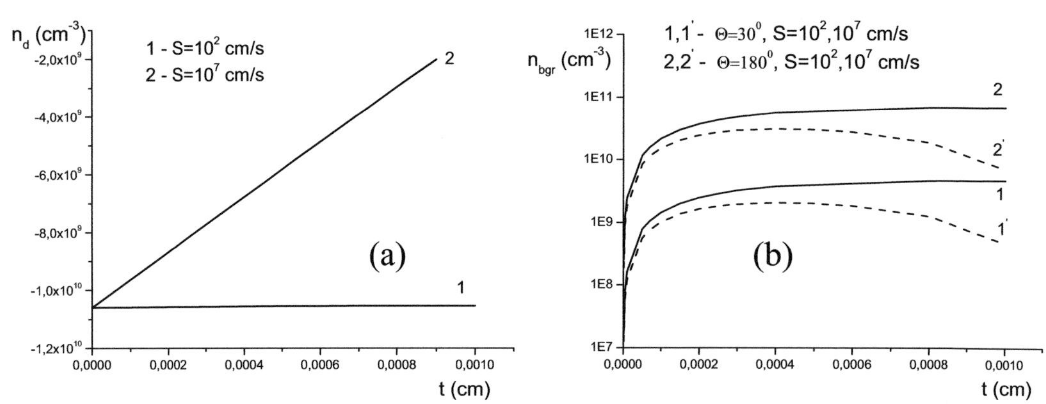

Fig. 8. Calculated concentration profiles of non-equilibrium dark $n_d(t)$ - (a) and background generated $n_{bgr}(t)$ - (b) minority charge carriers in quasi-neutral p base versus thickness t of p base at different surface recombination rate S and cold shield formed Field-Of-View (Θ) in $Hg_{1-x}Cd_xTe$ (x=0.224) photodiode described by data given in Table 2

The reason of different reaction of non-equilibrium dark and background generated charge carriers' concentration profiles on surface recombination rate's variation is as follows. In accepted conditions major share of infrared radiation is absorbed in part of p base joining to space charge region (nearby point $t = 0$). Thickness of that absorbing part is a few times smaller than total thickness LP of p base. Again thickness of p base is almost order of value less than ambipolar diffusion length $L_n \approx 10^{-2}$ cm. In addition background concentration in zero point ($t=0$) is always equal to zero i.e. $n_{bgr}(0)=0$. As the result concentration profile of photogenerated charge carriers nearby to point $t=0$ is formed preferably by their photogeneration with subsequent extraction into SCR. On the other hand due to disparity $LP << L_n$ extraction of dark minority carriers into SCR takes place from whole thickness of p base where they have existed initially (at $V_b=0$). Furthermore value of concentration $n_d(0)=\Delta n(0)<0$ is fixed according to expression (34) by applied bias and algebraic value $n_d(LP)\leq 0$ grows with increasing of S. In other words ratio $n_d(LP)/n_d(0)$ is raised. This entire means that gradient of concentration of non-equilibrium dark minority charge carriers along axis t grows with increasing of S (Fig. 8a).

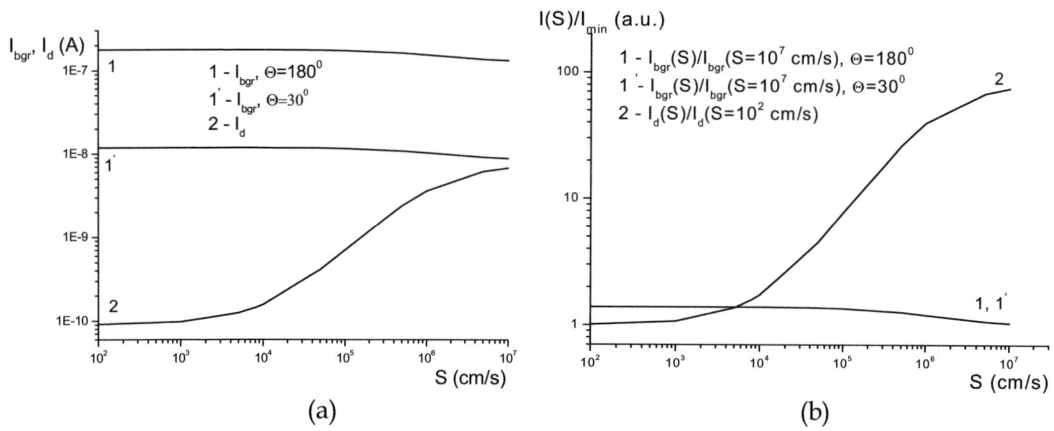

Fig. 9. Dark I_d - (a) and background generated I_{bgr} - (b) currents versus S in $Hg_{1-x}Cd_xTe$ (x=0.224) photodiode described by data given in Table 2. On graph (a) currents are given in absolute units and on graph (b) – in arbitrary units when curves (a) are specified to minimum photocurrent values

5. Photocurrent generation and collection in small-pitch high-density IRFPA

Theoretical approach was developed for the case of front-side illuminated IRFPA based on regular structure of n^+-p junctions enlaced by n_{gr}^+ - guard ring around, Fig. 10.

5.1 PV IRFPA design model

Cross-section of model PD array fragment (pixel) is shown on Fig. 10.

5.2 Photocurrent generated by sideways δ-shaped light beam

For estimation purpose let's consider one-dimensional (along line A) $n_{gr}^+-p-n_m^+-p-n_{gr}^+$ fragment (Fig. 10) of model PD array illuminated by δ-shaped light beam perpendicularly to surface of array, where n_m^+ is n^+ - region of n^+-p junction, n_{gr}^+ is n^+ - guard ring around n^+-p junction and p is layer (substrate) common for all pixels of PD array. Pixel is area including n^+-p junction and limited by guard ring (Fig. 11). Model array fragment is symmetrical regarding n_m^+ - region (Fig. 11). For simplicity word photocurrent will mean further photocurrent generated by pixel illuminated by proper light. Photocurrent generated in pixel is calculated at short-circuit between lead V and Ground (Fig. 11).

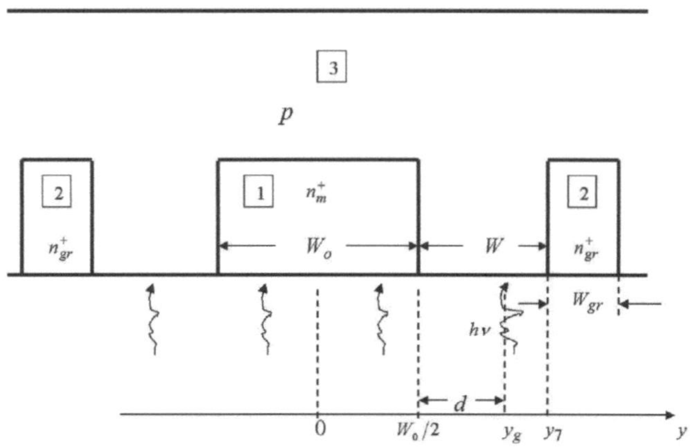

Fig. 10. Cross-section of model PD array fragment (pixel). 1 - n^+_m is n^+ - region of $n^+ - p$ junction with width W_0 ; 2 - n^+_{gr} is n^+ - guard ring with width W_{gr} ; 3 - p is thin layer (substrate) common for all pixels of PD array. Spacing between periphery of $n^+ - p$ junction and guard ring is marked as W . Front surface of array is irradiated by photon flux $h\nu$ (δ - shaped light beam or uniform flux or spotlight) that is absorbed and generates photocurrent

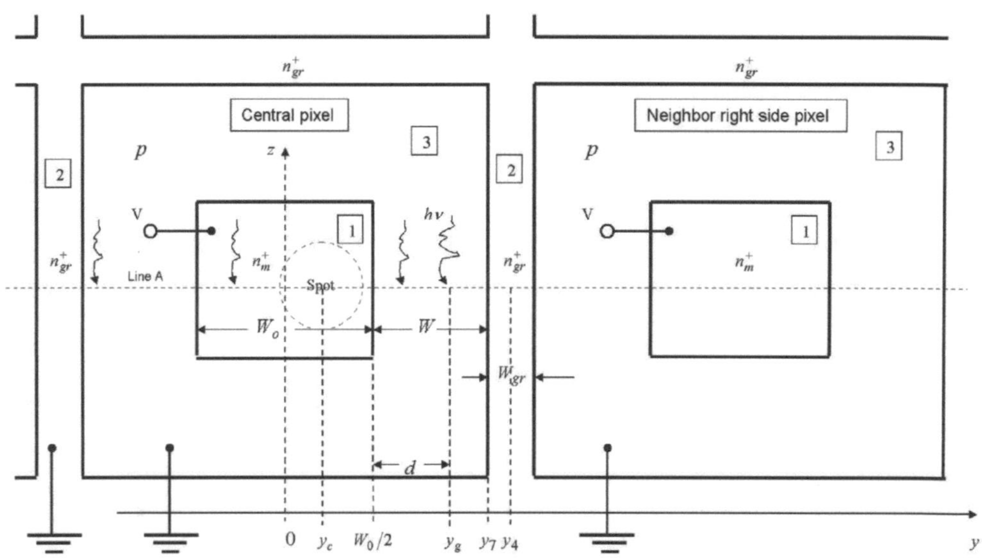

Fig. 11. Front view of model PD array fragment. 1 - n^+_m is n^+ - region of $n^+ - p$ junction with width W_0 ; 2 - n^+_{gr} is n^+ - guard ring with width W_{gr} ; 3 - p is thin layer (substrate) common for all pixels of PD array. Spacing between periphery of $n^+ - p$ junction and guard ring is marked as W . Front surface of array is irradiated by photon flux $h\nu$ (δ -shaped light beam or uniform flux or spotlight) that is absorbed and generates photocurrent in pixel. One-dimensional consideration is developed along line A (illumination moves along that line). Common p thin layer and n^+_{gr} - guard ring grid are grounded. Photocurrent generated in pixel is calculated between Ground and V diode lead connected to n^+_m - region of $n^+ - p$ junction

Let's assume:
Recombination rates of excess electrons and holes are equal to each other.

$$R_n = R_p = \frac{\Delta n}{\tau} \tag{43}$$

Where: R_n and R_p - recombination rates, Δn - concentration and τ - lifetime of excess electrons and holes.
Drift of excess charge carriers in electric field in p - region is negligible.
Band-to-band photogeneration of charge carriers at point $y = y_g$, i.e. specific rate of photogeneration is described by formula:

$$g(y) = G_\delta \times \delta(y - y_g) \tag{44}$$

Where: $\delta(y - y_g)$ - delta-function and G_δ - total photogeneration rate of charge carriers.
In analyzed conditions distribution of $\Delta n(y)$ in p - region is defined by diffusion equation:

$$D \times \frac{\partial^2 \Delta n}{\partial y^2} - \frac{\Delta n}{\tau} = -G_\delta \times \delta(y - y_g) \tag{45}$$

Where: D - coefficient of ambipolar diffusion.
Do solve equation (45) in intervals $W_o / 2 < y \le y_g$ and $y_g \le y \le y_7 \equiv W_0 / 2 + W$ assuming boundary conditions:

$$\Delta n(W_o / 2) = n_p \times \left[\exp\left(\frac{qV}{kT} \right) - 1 \right] \text{ and } \Delta n(y_7) = 0 \tag{46}$$

And stitching conditions are:

$$\Delta n(y_g - 0) = \Delta n(y_g + 0) \text{ and } D \times \left(\left. \frac{\partial \Delta n}{\partial y} \right|_{y=y_g+0} - \left. \frac{\partial \Delta n}{\partial y} \right|_{y=y_g-0} \right) = -G_\delta \tag{47}$$

Where: n_p - concentration of equilibrium minority charge carriers (electrons) in p - region.
Condition (46) means continuity of excess charge carriers' concentration, and condition (47) is derived relation resulted from integration of equation (45) in neighborhood of point $y = y_g$. Photocurrent value I_{ph}^{δ} at $y = W_0 / 2$ is defined by formula:

$$I_{ph}^{\delta} = q \times G_\delta \times K \tag{48}$$

Where: K - coefficient of one-sided sideways photoelectric conversion defined as:

$$K = \frac{sh[(W - d) / L]}{sh(W / L)}. \tag{49}$$

Where: $L = \sqrt{D \times \tau}$ - ambipolar diffusion length of charge carriers.

Graph of K versus normalized distance d/W between δ-shaped light beam and periphery of n_m^+ - region of $n^+ - p$ junction is presented on Fig. 12.
If sideways δ-shaped light beam illumination is symmetrical in relation to n^+ - region of $n^+ - p$ junction (i.e. junction is illuminated from left and right sides, Fig. 10) then total photocurrent value will be two times higher than got from expression (48).

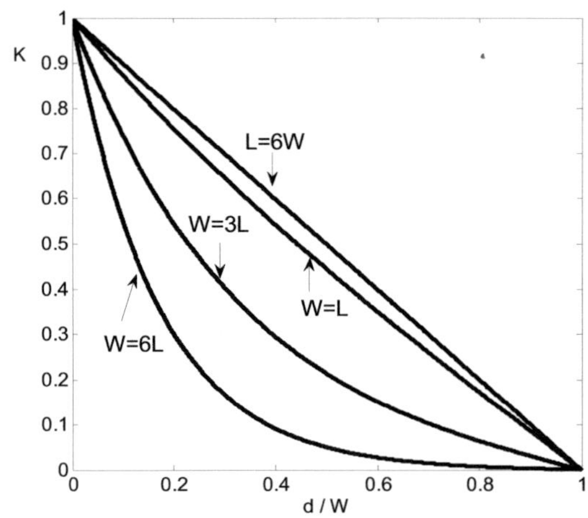

Fig. 12. Dependence of one-sided sideways photoelectric conversion coefficient K on normalized distance d/W between δ-shaped light beam and periphery of n_m^+ - region

5.3 Photocurrent generated by uniform sideways and front illumination

To calculate photocurrent value I_{ph}^{lat} under symmetrical regarding n_m^+ - region sideways illumination we need integrate expression (48) with respect to y between $W_o/2$ and W and than multiply result by coefficient 2.
In the case of uniform illumination ($G_\delta(x) = const$) we get:

$$I_{ph}^{lat} = q \times G_{2W} \times K_{tot}^{lat} . \tag{50}$$

Where: G_{2W} - total sideways photogeneration rate (taking into account both left and right sides) is defined as:

$$G_{2W} = G_\delta \times 2W \tag{51}$$

And sideways photoelectric conversion coefficient K_{tot}^{lat} if defined by:

$$K_{tot}^{lat} = \frac{L}{W} \times th\left(\frac{W}{2L}\right) . \tag{52}$$

Assuming that photoelectric conversion coefficient is equal to 1 under front-side illumination we can write photocurrent value I_{ph}^{fr} in this case as follows:

$$I_{ph}^{fr} = q \times G_\delta \times W_0 . \tag{53}$$

As it follows from expressions (50) - (53) ratio of photocurrents generated by $n^+ - p$ junction under uniform sideways and front-side illumination is defined by:

$$R \equiv \frac{I_{ph}^{lat}}{I_{ph}^{fr}} = 2 \times \frac{L}{W_o} \times th\left(\frac{W}{2L}\right) = 2 \times a_0 \times th\left(\frac{1}{2a}\right) = a_0 \times Y \qquad (54)$$

$$a_o = L / W_o \ , \ a = L / W \text{ and } Y = 2 \times th\left(\frac{1}{2a}\right). \qquad (55)$$

Graph of calculated universal dependence $Y = 2 \times th\left(\frac{1}{2a}\right)$ versus L / W is given on Fig. 13.

Herein:

$$R = a_o \times Y(L / W). \qquad (56)$$

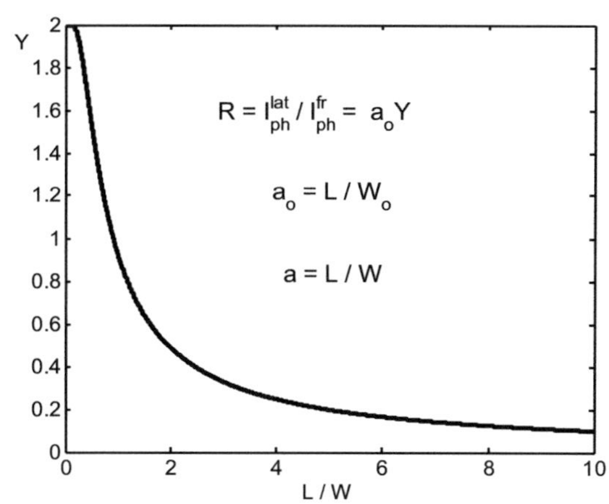

Fig. 13. Graph of universal dependence $Y = 2 \times th\left(\frac{1}{2a}\right)$ versus L / W following to (55)

5.4 Photocurrent generated by moving small-diameter uniform spotlight

Basic relation (48) allows estimating of photocurrent I_{ph} variation when small diameter (D_{spot}) uniform spotlight is moving along surface of PD array.

To calculate photocurrent value we need integrate expression (48) with respect to y within uniformly illuminated region except guard ring region (W_{gr}). Further we will limit consideration by condition (57):

$$D_{spot} \leq W_o . \qquad (57)$$

Within uniform spotlight area dependence of photocurrent I_{ph} on spot center position y_c will be described by formulae given further.

Case (a): Gap between n_m^+ - region border and n_{gr}^+ - guard ring is higher than spot diameter:

$$W \geq D_{spot} \,. \tag{58}$$

Generation of photocurrent when spot illuminates right half of central pixel.
Let's mark $I_{ph}^{(c)}$ photocurrent generated in central pixel when spot moves within interval $-W - W_0/2 \leq y \leq W_0/2 + W$.
1a. Spot center moves within the interval:

$$0 \leq y_c \leq y_1 \equiv W_0/2 - r \,. \tag{59}$$

In this case spot is located within n_m^+ - region of $n^+ - p$ junction totally. Photocurrent $I_{ph}^{(c)}$ is frontal only that is:

$$I_{ph}^{(c)} = I_{ph}^{fr} = q \times G_\delta \times D_{spot} \,. \tag{60}$$

2a. Spot center moves within the interval:

$$y_1 \leq y_c \leq y_2 \equiv W_0/2 + D_{spot}/2 \,. \tag{61}$$

Spot light is appearing on the side of n_m^+ - region and at $y_c > y_2$ get it away.
In the interval (61) we get:

$$\frac{I_{ph}^{(c)}(y_c)}{q \cdot G_\delta} = F_1(y_2 - y_c, y_3 - y_c) \equiv y_2 - y_c + \frac{L}{sh(W/L)} \times \left[ch\left(\frac{W}{L}\right) - ch\left(\frac{y_3 - y_c}{L}\right) \right] \tag{62}$$

$$y_3 \equiv (W_0/2) + W - D_{spot}/2 \tag{63}$$

3a. Spot center moves within the interval:

$$y_2 \leq y_c \leq y_3 \,. \tag{64}$$

Spotlight is located totally between n_m^+ - and n_{gr} - regions, therefore $I_{ph}^{fr} = 0$ and

$$\frac{I_{ph}^{(c)}(y_c)}{q \cdot G_\delta} = F_2(y_7 - y_c) \equiv 2L \times \frac{sh(D_{spot}/2L)}{sh(W/L)} \times sh\left(\frac{y_7 - y_c}{L}\right) . \tag{65}$$

Case (a_1): Let's impose some condition - width of guard ring is narrower than spotlight diameter:

$$W_{gr} < D_{spot} \,. \tag{66}$$

4a_1. Spot center moves within the interval:

$$y_3 \leq y_c \leq y_5 \equiv (W_0/2) + W + D_{spot}/2 \,. \tag{67}$$

Spotlight gets away gradually from considered central pixel. Photocurrents generated in central pixel and neighbor right side pixel will be equal to each other when y_c will coincide to mid y_4 of right side guard ring (68):

$$y_4 \equiv (W_0/2) + W + (W_{gr}/2). \tag{68}$$

In the interval (67):

$$\frac{I_{ph}^{(c)}(y_c)}{q \cdot G_\delta} = F_3(y_5 - y_c) \equiv 2L \times \frac{sh^2[(y_5 - y_c)/2L]}{sh(W/L)}. \tag{69}$$

5a. Spot center moves beyond coordinate y_5

$$y_c \geq y_5. \tag{70}$$

In this case spotlight leaves central pixel entirely and no photocurrent will be generated

$$I_{ph}^{(c)}(y_c) = 0. \tag{71}$$

Generation of photocurrent when spot illuminates left half of neighbor right side pixel. Photocurrent generation in right side pixel $I_{ph}^{>}$ will take place when edge of spotlight appears in that pixel, i.e. at condition (72):

$$y_c \geq y_6 \equiv (W_0/2) + W + W_{gr} - D_{spot}/2. \tag{72}$$

It means that till spot's edge hasn't reach periphery of right side pixel and no photocurrent is generated

6a. $$y_c \leq y_6;\; I_{ph}^{>}(y_c) = 0. \tag{73}$$

Photocurrent $I_{ph}^{>}(y_c)$ and $I_{ph}^{(c)}(y_c)$ values are symmetrical about mid line of guard ring region y_4, i.e.:

$$I_{ph}^{>}(y_c) = I_{ph}^{(c)}(2y_4 - y_c). \tag{74}$$

Therefore we do have the following cases:

7a₁. $$y_6 \leq y_c \leq y_{11} \equiv (W_0/2) + W + W_{gr} + D_{spot}/2;\; \frac{I_{ph}^{>}}{q \times G_\delta} = F_3(y_c - y_6). \tag{75}$$

8a₁. $$y_{11} \leq y_c \leq y_{10} \equiv (W_0/2) + 2W + W_{gr} - D_{spot}/2;\; \frac{I_{ph}^{>}(y_c)}{q \cdot G_\delta} = F_2(y_c - y_9). \tag{76}$$

Where:

$$y_9 = (W_0/2) + W + W_{gr}. \tag{77}$$

9a₁. $$y_{10} \leq y_c \leq y_{12} \equiv (W_0/2) + 2W + W_{gr} + r;\; \frac{I_{ph}^{>}(y_c)}{q \times G_\delta} = F_1(y_c - y_{10}, y_c - y_{11}). \tag{78}$$

10a₁. $$y_{12} \le y_c \le y_8 \,;\ \frac{I^{>}_{ph}(y_c)}{q \times G_\delta} = D_{spot} \,. \quad (79)$$

Where distance between centers of n_m^+ - regions of central and right side pixels:

$$y_8 = W_0 + 2W + W_{gr} \,. \quad (80)$$

Generation of photocurrent when spot illuminates left half of central pixel.
Let's mark photocurrent at negative and positive coordinate y_c as $I_{ph-}(y_c)$ and $I_{ph}(y_c)$ properly. Values $I_{ph-}(y_c)$ and $I_{ph}(y_c)$ are the same in respect to zero point $y_c = 0$, i.e.

$$I_{ph-}(y_c) = I_{ph}(-y_c) \,. \quad (81)$$

Therefore we do have the following cases:

11a. $$-y_1 \le y_c \le 0 \,;\ I_{ph-}(y_c) = q \times G_\delta \times D_{spot} \,. \quad (82)$$

12a. $$-y_2 \le y_c \le -y_1 \,;\ I_{ph-}(y_c) = q \times G_\delta \times F_1(y_2 + y_c, y_3 + y_c) \,. \quad (83)$$

13a. $$-y_3 \le y_c \le -y_2 \,;\ I_{ph-}(y_c) = q \times G_\delta \times F_2(y_7 + y_c) \,. \quad (84)$$

14a. $$-y_5 \le y_c \le -y_3 \,;\ I_{ph-}(y_c) = q \times G_\delta \times F_3(y_5 + y_c) \,. \quad (85)$$

15a. $$y_c \le -y_5 \,;\ I_{ph-}(y_c) = 0 \,. \quad (86)$$

Generation of photocurrent when spot illuminates right half of neighbor left side pixel.

16a. $$-y_6 \le y_c \le 0 \,;\ I_{ph-}(y_c) = 0 \,. \quad (87)$$

17a₁. $$-y_{11} \le y_c \le -y_6 \,;\ I_{ph-}(y_c) = q \times G_\delta \times F_3(-y_c - y_6) \,. \quad (88)$$

18a₁. $$-y_{10} \le y_c \le -y_{11} \,;\ I_{ph-}(y_c) = q \times G_\delta \times F_2(-y_c - y_9) \,. \quad (89)$$

19a₁. $$-y_{12} \le y_c \le -y_{10} \,;\ I_{ph-}(y_c) = q \times G_\delta \times F_1(-y_c - y_{10}, -y_c - y_{11}) \,. \quad (90)$$

20a₁. $$-y_8 \le y_c \le -y_{12} \,;\ I_{ph-}(y_c) = q \times G_\delta \times D_{spot} \,. \quad (91)$$

Case (b): Gap between n_m^+ - region border and n^+ - guard ring is less than spot diameter:

$$W \le D_{spot} / 2 \,. \quad (92)$$

Generation of photocurrent when spot illuminates right half of central pixel.

21b. $$0 \le y_c \le y_1 \,;\ I^{(c)}_{ph} = I^{fr}_{ph} = q \times G_\delta \times D_{spot} \,. \quad (93)$$

22b. $$y_1 \le y_c \le y_3\,;\ \frac{I_{ph}^{(c)}(y_c)}{q \cdot G_\delta} = F_1(y_2 - y_c, y_3 - y_c)\,. \tag{94}$$

In interval (96) part of spot is located in n_m^+ - region but spot edge does not reach guard ring. Case (b_1): Let's impose some condition:

b_1. $$W_{gr}/2 \le (D_{spot}/2 - W)\,. \tag{95}$$

23b_1. $$y_3 \le y_c \le y_2\,;\ \frac{I_{ph}^{(c)}(y_c)}{q \cdot G_\delta} = F_4(y_2 - y_c) \equiv y_2 - y_c + L \times th\left(\frac{W}{2L}\right). \tag{96}$$

24b_1. $$y_2 \le y_c \le y_5\,;\ \frac{I_{ph}^{(c)}(y_c)}{q \cdot G_\delta} = F_3(y_5 - y_c)\,. \tag{97}$$

25. $$y_5 \le y_c \le y_8\,;\ I_{ph}^{(c)} = 0\,. \tag{98}$$

Generation of photocurrent when spot illuminates left half of neighbor right side pixel.

26. $$0 \le y_c \le y_6\,;\ I_{ph}^{>}(y_c) = 0\,. \tag{99}$$

27b_1. $$y_6 \le y_c \le y_{10}\,;\ \frac{I_{ph}^{>}(y_c)}{q \times G_\delta} = F_3(y_c - y_6)\,. \tag{100}$$

28b_1. $$y_{10} \le y_c \le y_{11}\,;\ \frac{I_{ph}^{>}(y_c)}{q \times G_\delta} = F_4(y_c - y_{10})\,. \tag{101}$$

29b_1. $$y_{11} \le y_c \le y_{12}\,;\ \frac{I_{ph}^{>}(y_c)}{q \times G_\delta} = F_1(y_c - y_{10}, y_c - y_{11})\,. \tag{102}$$

30b_1. $$y_{12} \le y_c \le y_8\,;\ \frac{I_{ph}^{>}(y_c)}{q \cdot G_\delta} = D_{spot}\,. \tag{103}$$

Generation of photocurrent when spot illuminates left half of central pixel.

31. $$-y_1 \le y_c \le 0\,;\ I_{ph-}(y_c) = q \times G_\delta \times D_{spot}\,. \tag{104}$$

32b. $$-y_3 \le y_c \le -y_1\,;\ I_{ph-}(y_c) = q \times G_\delta \times F_1(y_2 + y_c, y_3 + y_c)\,. \tag{105}$$

33b_1. $$-y_2 \le y_c \le -y_3\,;\ I_{ph-}(y_c) = q \times G_\delta \times F_4(y_2 + y_c)\,. \tag{106}$$

34b_1. $$-y_5 \le y_c \le -y_2\,;\ I_{ph-}(y_c) = q \times G_\delta \times F_3(y_5 + y_c)\,. \tag{107}$$

35. $$-y_8 \le y_c \le -y_5\,;\; I_{ph-}(y_c) = 0\,. \quad (108)$$

Generation of photocurrent when spot illuminates right half of neighbor left side pixel.

36. $$-y_6 \le y_c \le 0\,;\; I_{ph-}(y_c) = 0\,. \quad (109)$$

37b1. $$-y_{10} \le y_c \le -y_6\,;\; I_{ph-}(y_c) = q \times G_\delta \times F_3\left(-y_c - y_6\right). \quad (100)$$

38b1. $$-y_{11} \le y_c \le -y_{10}\,;\; I_{ph-}(y_c) = q \times G_\delta \times F_4\left(-y_c - y_{10}\right). \quad (111)$$

39b1. $$-y_{12} \le y_c \le -y_{11}\,;\; I_{ph-}(y_c) = q \times G_\delta \times F_1\left(-y_c - y_{10}, -y_c - y_{11}\right). \quad (112)$$

40b1. $$-y_8 \le y_c \le -y_{12}\,;\; I_{ph-}(y_c) = q \times G_\delta \times D_{spot}\,. \quad (113)$$

5.5 LWIR PD array: calculation of photocurrent collection profiles

Data used in calculation of photocurrent generated in small-pitch high-density $Hg_{0.776}Cd_{0.224}Te$ PD array are given in Table 2. Junction regions thickness t was taken t(n^+) = 0.5 µm and t(p-absorber) = 6 µm. Surface recombination rate 10^2 cm/sec.

Developed approach (57) - (113) was applied to calculate photocurrent generated in small-pitch $Hg_{0.776}Cd_{0.224}Te$ PD array. Calculated dependences of photocurrent I_{ph} generated by spotlight in $Hg_{1-x}Cd_xTe$ (x=0.224) PD array are shown on Fig. 14 and ratio of photocurrents generated at uniform frontal and sideways illumination can be estimated easily from Fig. 14. It is seen clearly that developed approach allows analytical estimation of photocurrent generation in different close-packed PD arrays. Following to dependence presented on Fig. 13 contribution of photocurrent generated by sideways uniform illumination to total photocurrent of pixel can be too much high at not reasonable ratios between L, W and W_0. Dependences of photocurrent value I_{ph} are calculated as function of spot center position coordinate y_c for central and neighbor pixels of array. Condition $y_c = 0$ means that in start position Zero of coordinate system and spot center are matched. Length (distance) is given in units D_{spot} (spot diameter). Photocurrent is calculated in units $q \times G_\delta \times D_{spot}$. It is accepted in calculation that width of n_m^+ - region of $n^+ - p$ junction W_o = 20 µm; width of n_{gr}^+ - guard ring W_{gr} = 5 µm; spot diameter D_{spot} = 15 µm; operating temperature $T_{op} = 77K$; ambipolar diffusion length in p layer $L = 48$ µm. Spacing between periphery of $n^+ - p$ junction and guard ring W = 20 µm (a) and W = 5 µm (b). Photocurrent in central, neighbor right-side and neighbor left-side pixels are presented on graphs by solid curves, dashed curves and dash-and-dot curves properly

6. Conclusion

We have attempted to develop some general approach for simulation MWIR and LWIR PD IRFPA including estimation of major electro-optical parameters. Estimations have shown that extended LWIR $Hg_{1-x}Cd_xTe$ PD with p-n junction will be potentially of 4-5 times lower dark current value than PD with n^+-p junction at T=77 K and 2 times lower at T=100 K. Additionally extended LWIR $Hg_{1-x}Cd_xTe$ PD with p-n junction will be seriously lower sensitive to operating temperature increasing than PD with traditional n+-p junction. We have shown that surface recombination rate value at back surface of thin *p* absorber can have serious effect on dark current in small-size LWIR $Hg_{1-x}Cd_xTe$ PD. We have developed analytical expressions describing collection of photogenerated charge carriers in small-pitch IRFPA for practical cases: uniform and small-size spotlight illumination.

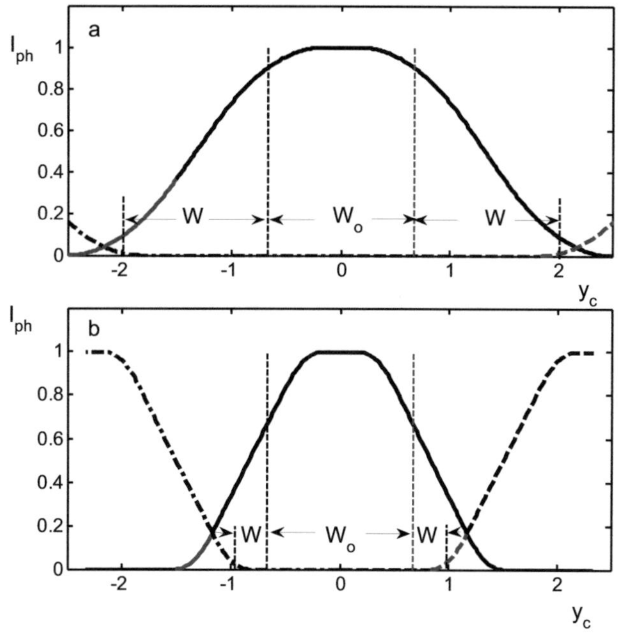

Fig. 14. Graphs of photocurrent generated in $Hg_{1-x}Cd_xTe$ (x=0.224) PD array following to expressions (57)-(113)

7. References

Whicker, S. (1992). "New technologies for FPA dewars", *Proceedings of SPIE, 1683*, pp. 102-112, ISBN 9780819408488, August 1992, SPIE Press, Bellingham, Washington

Triboulet, P. & Chatard, J.-P. (2000). "From research to production: ten years of success", *Proceedings SPIE 4130*, pp. 422-440, ISBN 9780819437754, December 2000, SPIE Press, Bellingham, Washington

Baker, I. & Maxey, C. (2001). "Summary of HgCdTe 2D array technology in the UK", *Journal of Electronic Materials*, Vol. 30, No. 6, (June 2001) 682-689, ISSN 0361-5235

Norton, P. (2002). "HgCdTe infrared detectors", *Opto-Electronics Review*, Vol. 10, No. 3, (September 2002) 159-174, ISSN 1230-3402

Kinch, M. (2007). *Fundamentals of Infrared Detector Materials*, SPIE Press, ISBN 978-0-8194-6731-7, Bellingham, Washington

Glozman, A.; Harush, E.; Jacobsohn, E.; Klin, O.; Klipstein, Ph.; Markovitz, T.; Nahum, V.; Saguy, E.; Oiknine-Schlesinger, J.; Shtrichman, I.; Yassen, M.; Yofis, B. & Weiss, E. (2006). "High performance InAlSb MWIR detectors operating at 100 K and beyond", *Proceedings SPIE 6206*, pp. 6206M, ISBN 9780819462602, May 2006, SPIE Press, Bellingham, Washington

Kinch, M.; Brau, M. & Simmons, A. (1973). "Recombination mechanisms in 8-14 μ HgCdTe", *J. Appl. Phys.*, Vol. 44, No. 4, (April 1973) 1649-1663, ISSN 0021-8979

Gelmont, B. (1980). "Auger recombination in narrow band-gap semiconductors", *Sov. Semicond. Phys. & Tech.*, Vol. 14, No. 11, (November 1980) 1913-1917, ISSN 1063-7826

Gelmont, B. (1981). "Auger recombination in narrow band-gap p-type semiconductors", *Sov. Semicond. Phys. & Tech.*, Vol. 15, No. 9, (September 1981) 1316-1319, ISSN 1063-7826

Blue, M. (1964). "Optical Absorption in HgTe and HgCdTe", *Physical Review A*, Vol. 134, No. 1, (January 1964) 226-234, ISSN 1050-2947

Laurenti, J.; Camassel, J.; Buchemadou, A.; Toulouse, B.; Legros, R. & Lusson, A., (1990). "Temperature dependence of the fundamental absorption edge of mercury cadmium telluride", *J. Appl. Phys.*, Vol. 67, No. 10, (May 1990) 6454-6460, ISSN 0021-8979

Schmit J., (1970). "Intrinsic carrier concentration of $Hg_{1-x}Cd_xTe$ as a function of x and T using k-p calculations, *J. Appl. Phys.*, Vol. 41, No. 7, (June 1970) 2876-2879, ISSN 0021-8979

Blakemore, J. (1962). *Semiconductor Statistics*, Pergamon Press, ISBN 0-486-49502-7, New York, New York

Part 2

Silicon Devices

6

Methodology for Design, Measurements and Characterization of Optical Devices on Integrated Circuits

G. Castillo-Cabrera[1,3], J. García-Lamont[2] and M. A. Reyes-Barranca[3]
[1]*Superior School of Computing (ESCOM), National Polytechnic Institute (IPN),*
[2]*Institute of Basic Science and Engineering, CITIS, Hidalgo State University,*
[3]*Electrical Engineering Department, SEES, CINVESTAV-IPN,*
Mexico

1. Introduction

The main application of optical devices is image processing which is a research field still in study for a wide variety of applications, such as video digital cameras for entertainment use, pattern recognition based in artificial neural networks, real time object tracking, clinical uses for repair by stimulation parts of visual system and artificial vision for application in silicon retinas, among others. So, it is important to evaluate the performance of available integrated photo-sensor devices used in these applications, considering issues as noise, resolution, processing time, colour, etc. Actually, there are several technologies available for integration of photo devices, commonly CCD, BiCMOS and GaAs. Although all of them are usually applied in image acquisition systems, there are still some performance aspects that should be optimised, as voltage levels, leakage currents, high fabrication costs, etc., so research is still being done to overcome these limitations. Standard CMOS integrated circuit technology is also an attractive alternative, since devices like phototransistors and photodiodes can be implemented as well. The foremost advantage of CMOS devices is its availability in standard technology. It should be mentioned that this technology has also some limitations but since fabrication of CMOS integrated circuits has low costs, exploration of the potential of new technologies for image processing is still an interesting field. Besides, algorithms can be implemented along for tasks such as border detection (space vision), movement detection (space-time vision), image enhancement (image processing vision) and pattern classification or recognition (neuro-fuzzy vision).
Considering the state of the art (Aw & Wooley 1996; Storm & Henderson, 2006; Theuwissen, 2008), as well as clinic approaches (Zaghloul, & Boahen, 2004), in this work, a chip was designed and fabricated, with two possible photo-sensor structures: *p+/N-well/p-substrate*, for phototransistors and *N-well/p-substrate*, for photodiodes, through the standard 1.5μm AMI's- , *N-Well* technology. In the future, it is the intention to design a second chip that must include electronics for image processing with pulse frequency modulation (PFM), once the characterization gives enough information about the performance of the stages studied. A complete description is given.

2. Devices Involved, type of structures

After the CCDs, the new generations of optical devices are based in standard CMOS technology. Experimental study based is here presented, about two typical structures in the field of art, namely, phototransistor and photodiodes which were designed and fabricated through the standard 1.5µm AMI's- technology. Technically those are known as "structures P+/N-Well/P-substrate" and "structures N-Well/P-substrate" respectively, which are presented by Fig. 1.

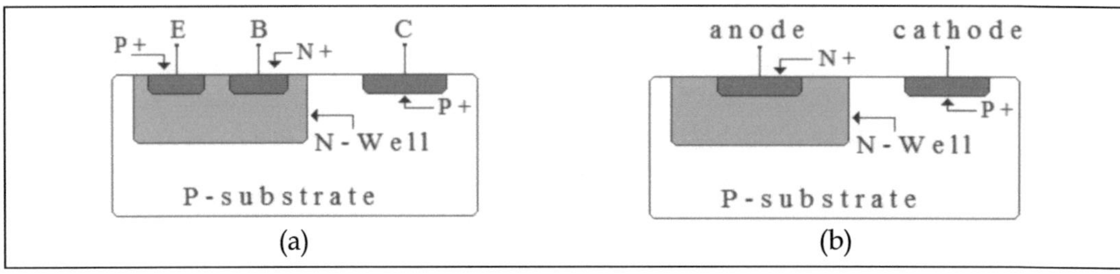

Fig. 1. Optical devices, (a): P+/N-Well/P-substrate, (b): N-Well/P-substrate

In the former, Fig. 1(a), P+/N-Well/P-substrate, the P+/N-Well is being an active junction as well as N-Well/P-substrate junction. An active junction is one in which two semiconductors with different conductivity, "p" and "n" type, are joined and electrically interacting. N-Well is a diffused region n-type on substrate. P+ on N-Well is an implanted material and also serves as low resistive ohmic contact. N+ on N-Well is an n-type implanted region and solely is used as low resistive ohmic contact. P+ on P-substrate is an implanted p+-type material and is used as low resistive ohmic contact. Terminals E, B and C are Emitter, Base and collector respectively in the phototransistor.

In the last one, Fig. 1(b), N-Well/P-substrate, there is only one active junction. N+ and P+ are implanted regions which does low resistive ohmic contact with N-Well and P-substrate respectively.

Both, structures P+/N-Well/P-substrate and N-Well/P-substrate, symbols are presented by the Fig. 2.

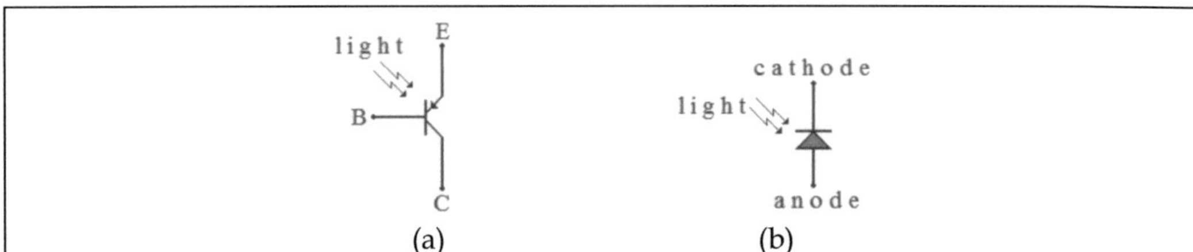

Fig. 2. Symbols for optical structures (a): P+/N-Well/P-substrate, (b): N-Well/P-substrate

3. Circuital architecture of pixel for characterization

3.1 Components of architecture

Fig. 3 presents the pixel architecture which has resulted efficient for optical devices characterization. It consists in optical device, four transistors, a source of current and buffer for readout. The common source amplifier consists, in M1 transistor and current source Isc, and it is used to handle the photocurrent. M2 is row select transistor and is not part of amplifier strictly speaking, however its position play an important role on this architecture, as will be shown in section 4. Photocurrent, from optical device, is integrated at the parasitic capacitance of the p-channel transistor M1, between node 2 and substrate, which is tied to ground. Assuming that photocurrent is constant, the relation of integrated voltage can be obtained by using the relation $q = C \cdot V$ in the parasitic capacitance.

$$V_{int} = \frac{1}{C}\int_0^{\Delta t} i dt \qquad (1)$$

where

$$i = \frac{dq}{dt} \qquad (2)$$

MSHUT along with signal VSHU controls the exposition time Δt. MREST along with signal VRES have as function to reset the nodes 1 and 2 at level Vreset. Optical devices can be P+/N-Well/P-substrate or N-Well/P-substrate structures. BUFFER OUTPUT provides power to avoid disturbance during readout.

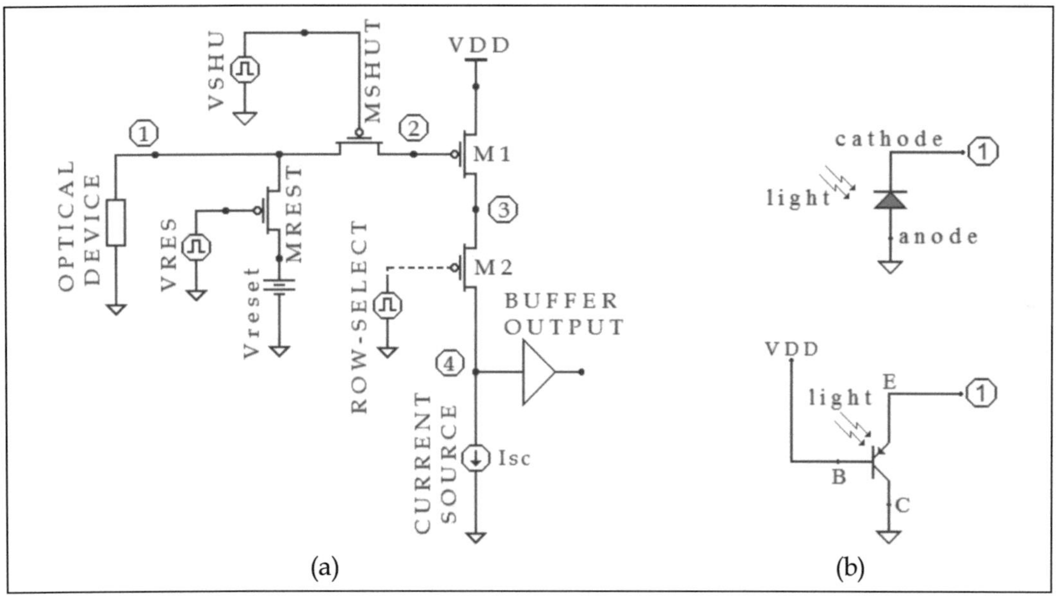

Fig. 3. (a) Architecture of pixel, (b) Optical device

Transistors MSHUT and MREST operate as switches in order to integrate the photocurrent generated in the photo-sensor at a given time and operation frequency. Fig. 4 shows the waveforms of their respective gate voltages. Integration time Δt, takes place while MSHU is on and MRES is off. During this time, the photocurrent is converted to voltage at the gate of transistor M1 (node-2 in Fig. 1). The relation between the integrated voltage and photocurrent is expressed in Eq. (1), assuming that the photocurrent is uniform in time. The DC voltage source, Vreset, is the reference voltage from which integration of the photocurrent is carried out.

3.2 Signals of control

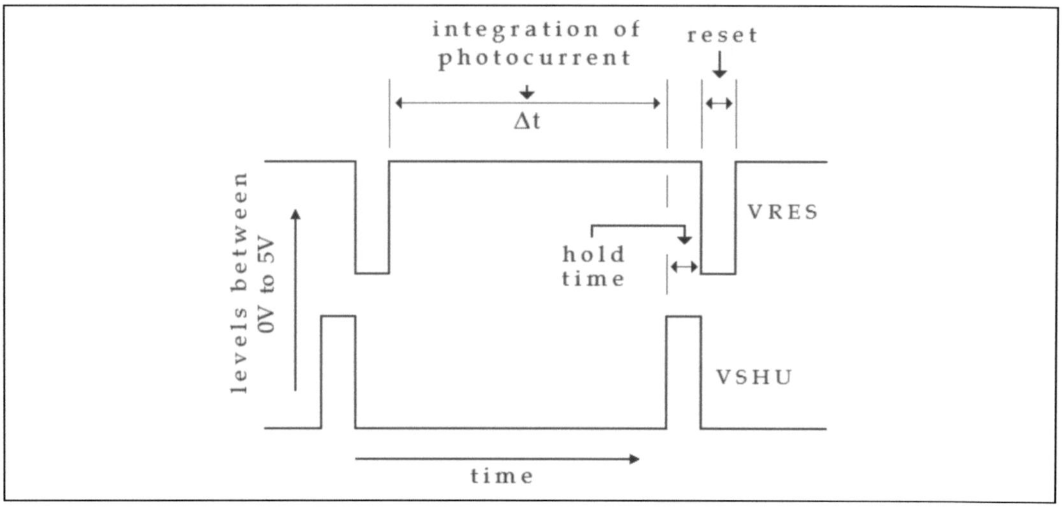

Fig. 4. Waveform of signals shutter and reset

$$I_{ph} = C_T \left(\Delta V_{\text{int}} / \Delta T \right) \tag{3}$$

where:

I_{ph} : photocurrent from the photo-device.

C_T : capacitance at node-2

ΔV_{int} : integrated voltage in C_T $(\Delta V_{\text{int}} = \Delta V_{measured} / Av)$

ΔT : integration time

Av : amplifier's gain

In addition, Vreset sets the quiescent point of the amplifier. Fig. 4 shows the input voltage pulses applied to the gate of MRES and MSHUT, respectively.

Since the objective of this work is to propose a methodology for the characterization of photo-devices and pixel architectures, the design was made considering those parameters affecting the performance of pixels and the way they will be measured. The design was carried out with the more simple architecture. Unlike what was reported with current amplifiers in the current-mode readout configuration (Philipp et al 2007), the amplifier is configured for voltage-mode operation in this work. From results obtained with this design, useful information can be processed and analysed considering a specific application, regarding factors such as the spectral response and silicon integration area of *p+/N-well/p-substrate* phototransistors and *N-well/p-substrate* photodiodes, as well as integration time, integrated voltage, transistors' aspect ratio and reference voltage used, for instance. Considerations about technology parameters, is important also in the definition of the dependence of the response and the architecture operation upon the photo-device structure.

Here, we present the electronic design, layout, simulations and some measurements on the fabricated chip. This prototype was made using the 1.5µm AMI's technology. The chip contains five pairs of phototransistors and five pairs of photodiodes, with one instance of each pair covered with metal for dark current characterization purposes. Results reported here are only from those devices having an area of (9µm)x(9µm). Bigger photo-devices were also measured, but as they saturated the amplifier, no useful results were obtained. A drawing is given in the Fig. 5 which is showing the fabricated array.

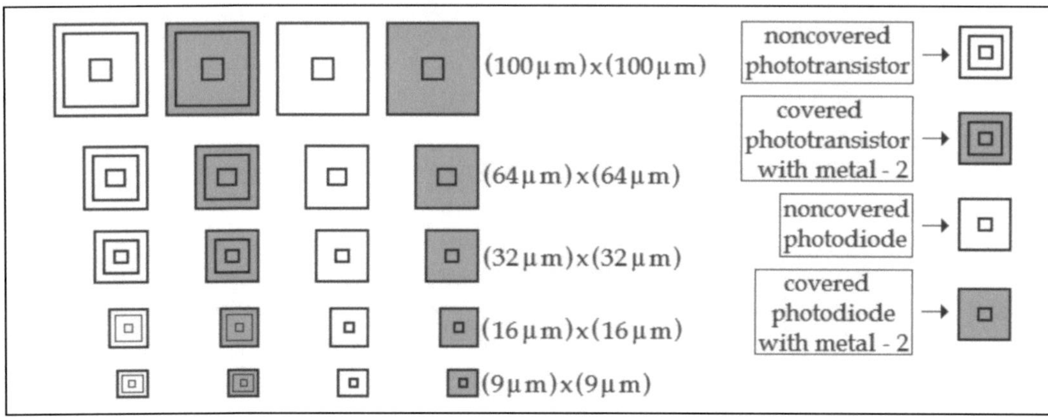

Fig. 5. Drawing for the array of optical devices

4. Circuit analysis

Now, doing reference to the Fig. 3, some design criteria for the circuit are defined. (1) Transistors MREST and MSHUT are used as switches, so their sizes can be drawn with minimum dimension features allowed by the technology. (2) Channel modulation effects must be avoided with M1, therefore its channel length should be at least five times the minimum dimension allowed. (3) For proper operation of the amplifier, it is recommended to choose a stable current source for biasing, thus, a cascode configuration was selected and designed for sourcing 20µA. (4) By adjusting the current source and Vreset the voltage gain of the amplifier it is varied, giving a useful degree of freedom for the characterization of the architecture and different devices, including the possibility another kind of not optical integrated sensors. (5) Transistor M2 it is inserted into the amplifier since it is a standard way for selecting row within an array. However, the role of M2 on the amplifier here proposed must be analyzed. (6) Finally, a standard buffer circuit is used for provide of power to the output.

Fig. 6 (a) shows the schematic of the cascode current source used for biasing the amplifier. Transistors involved in the amplifier are M1, M2, M4 and M6. Fig. 6 (b) shows the equivalent circuit for the amplifier and the cascode current source, used to find a mathematical relationship between the voltage gain and the size of M2.

From Figure 4(b), r_s is the channel resistance of M2; here, subscripts 01, 04 and 06 are identifiers for transistors M1, M4 and M6, respectively. A circuit analysis gives the following expression for the output voltage, V_{out} :

$$r_{01}i_1 + r_s i_2 + r_{04}i_3 + r_{06}i_2 = 0 \tag{4}$$

$$i_2 - i_1 = g_{m1}v_{sg1} \tag{5}$$

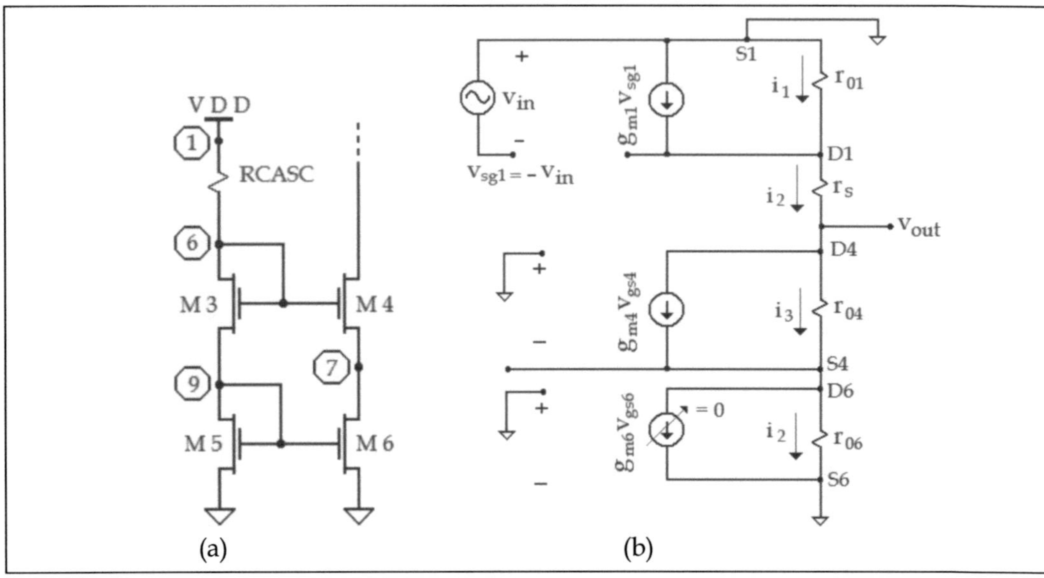

Fig. 6. Cascode source current (a) schematic, (b) equivalent circuit

$$i_2 - i_3 = g_{m4} v_{gs4} \tag{6}$$

$$-v_{out} = r_{01} i_1 + r_s i_2 \tag{7}$$

$$v_{out} = r_{04} i_3 + r_{06} i_2 \tag{8}$$

Equations from (4) up to (8) are mesh equations from which the next expression is obtained:

$$R_0 i_2 = v_{out} \tag{9}$$

Where

$$R_0 = r_{04} + g_{m4} r_{06} r_{04} + r_{06} \tag{10}$$

$$-i_1 + i_2 = g_{m1} v_{sg1} \tag{11}$$

$$r_{01} i_1 + r_s i_2 = -v_{out} \tag{12}$$

And by using (11) and (12), i_2 can be obtained:

$$i_2 = \frac{r_{01} g_{m1} v_{sg1} - v_{out}}{r_s + r_{01}} \tag{13}$$

From (9) and (10), the voltage gain of the amplifier is deduced:

$$R_0 \left(\frac{r_{01} g_{m1} v_{sg1} - v_{out}}{r_s + r_{01}} \right) = v_{out} \tag{14}$$

Defining the next ratio:

$$K = \frac{R_0}{r_s + r_{01}} \tag{15}$$

$$r_{01} g_{m1} v_{sg1} K - v_{out} K = v_{out} \tag{16}$$

And since

$$v_{sg1} = -v_{in}$$
$$Av = \frac{v_{out}}{v_{in}} \tag{17}$$

Finally the voltage gain is obtained:

$$Av = -\frac{K}{K+1} r_{01} g_{m1} \tag{18}$$

When $K >> 1$, the gain can be approximated to:

$$Av \cong -r_{01} g_{m1} \tag{19}$$

This is only possible if r_s is sufficiently small, from (15). So, assuming that $r_{01} = r_{04} = r_{06}$ from (10) and (11), the size of M2 must be such that r_s will result very small, compared with r_{01}, r_{04} and r_{06}. With an iterative procedure, the aspect ratio W / L of M2 was made large enough such that r_s does not have a strong influence over the amplifier's operation (Baker *et al* 2005). So, for the technology used the calculated aspect ratio for M2 was, $W = 64.8 \mu m$ and $L = 2.4 \mu m$. Then, considering these design outlines, the operation of the circuit based on M2 can be traduced in a convenient performance evaluation. As a result of the above design considerations, this basic analysis of the equivalent circuit reveals that the role of the row-select transistor is important for the proper operation of the amplifier. The voltage gain in (19) can be estimated using the following expressions (Baker *et al* 2005):

$$r_{01} = \frac{1}{\lambda \cdot I_D} \tag{20}$$

$$g_{m1} = \sqrt{2(KP)\tfrac{W}{L} I_D} \tag{21}$$

Using values of KP and λ, from the 1.5μm AMI technology, the maximum voltage gain was estimated as: $Av = 35dB$.

4.1 Simulation

Once the sizes of transistors used in the pixel were calculated, simulations with PSPICE were made to confirm the behavior of the circuit. Fig. 7 shows the gain range that can be achieved with the amplifier, going from 10dB to 32dB. Beside this, Input voltage, which is provided with Vreset, goes from 2.2V up to 3.5V, taken as parameter RCASC in the source current, Fig. 6(a).

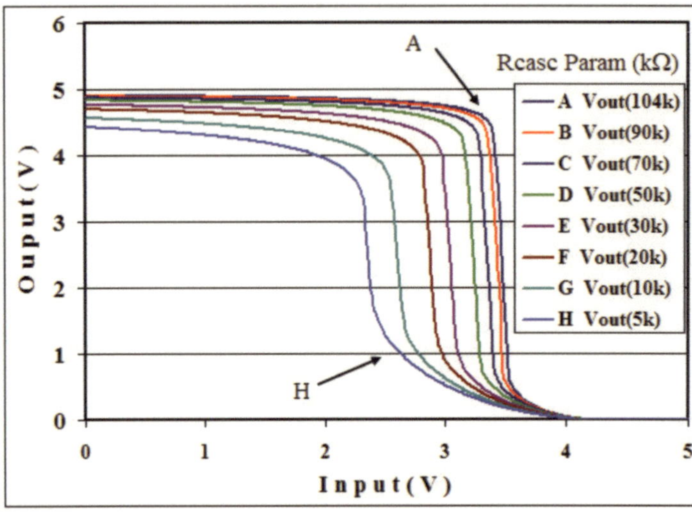

Fig. 7. Transfer function simulated

In order to evaluate temporal response, input voltage was adjusted to 3.5V, which belong to the gain of 32dB. Fig. 8 shows a simulated temporal response. Time of integration is of 0.9ms and time of reset is 0.1ms, according with the waveforms of VSHU and VRES seen in the Fig. 4. Level of photocurrent for simulation was taken of 10pA, from Reginald-Krishna's model (Perry 1996).

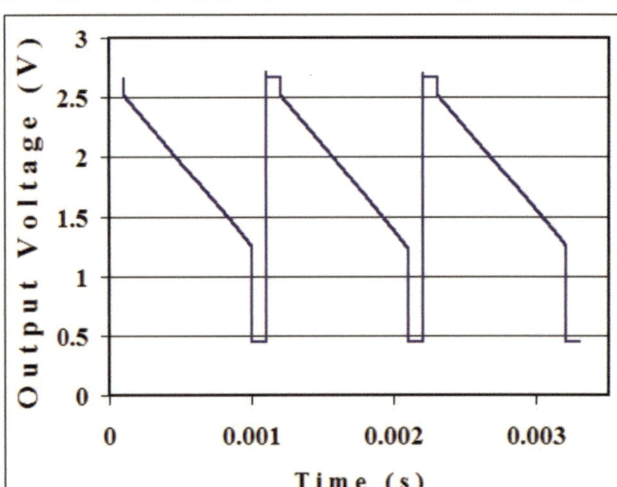

Fig. 8. Temporal response structure P+/N-Well/P-substrate

4.2 Layout

Cross section and layout are given in Fig. 9 and 10, for P+/N-Well/P-substrate and N-Well/P-substrate structures, respectively.

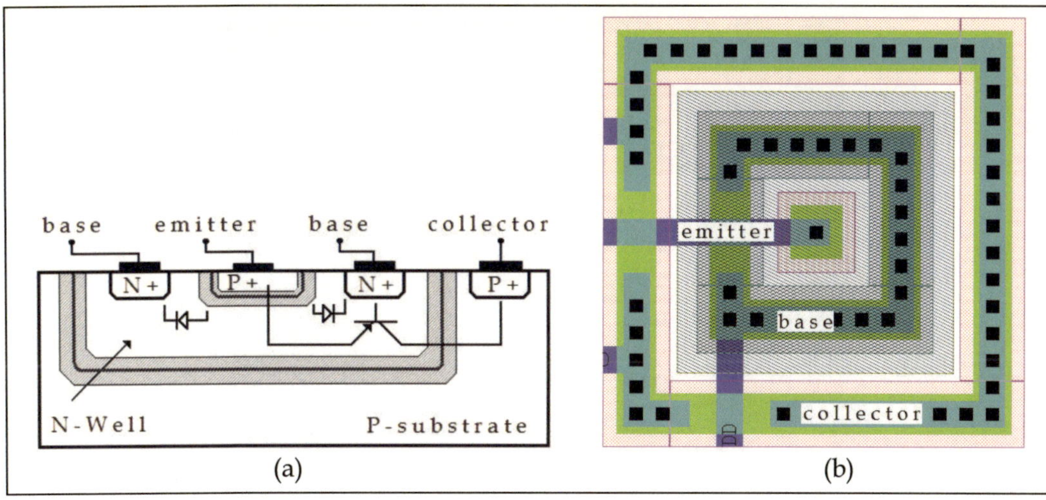

Fig. 9. P+/N-Well/P-substrate structure (a) Cross section (b) Layout

We can see two junctions in the Fig. 9. As it was been established in the Fig. 3, terminal of "emitter" is periodically reset at CD level of Vreset, which is also the amplifier´s point of operation. Terminal of "base" is tied to VDD which is power supply. "Collector" is tied to ground. Both junctions are biased in reverse way. Base-emitter junction is to (5V-Vreset) and base-collector junction to (5V-0V).

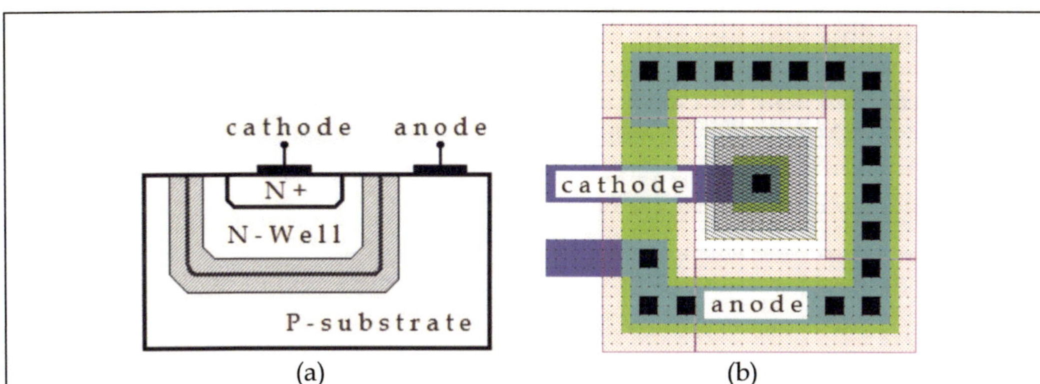

Fig. 10. N-Well/P-substrate structure (a) Cross section (b) Layout

Photodiode, structure N-Well/P-substrate given in Fig. 10, only has two terminals. Terminal of cathode is reset to Vreset periodically and anode is tied to ground. So, it is biased in reverse way.

The phototransistor's "base" has the same dimensions of the photodiode's anode ring, (9μm)x(9μm). So, both has the same active surface. The photocurrent generated is collected by the phototransistor's emitter and the same happens with the cathode of the photodiode.

5. Results

An array, which has been drawn in the Fig. 5, was fabricated and it is shown in the Fig. 11.

Fig. 11. Fabricated array of devices, microphotography

At the bottom of each column we can see a block, which comprise both, cascode source current Isc and BUFFER OUTPUT. First left column has different size of not covered phototransistors, P+/N-Well/P-substrate structures. Second column of left to right consists of P+/N-Well/P-substrate structures, similar sizes that first left column but covered with metal 2 (process AMI of 1.5μm). Third and fourth columns of left to right are not covered and covered N-Well/P-substrate structures respectively, photodiodes. Transfer function measured, of amplifier is given in the Fig. 12.

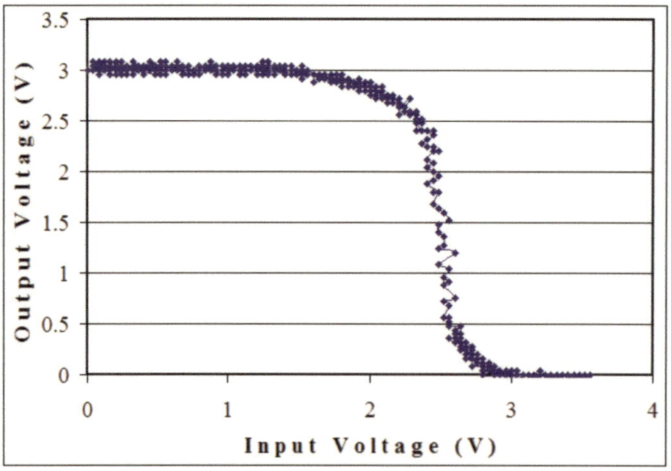

Fig. 12. Experimental transfer function of amplifier

Experimental transfer function of amplifier is quite fitted to the criteria design. Fig. 12 shows one of these functions. During the procedure of calibration, in a first set of measurements it was saw a strong response in the case photodiodes. So, in order to carry out measurements, the amplifier gain was set at 10dB in case of photodiodes measurements, while phototransistors at 32dB, in order to have a good reading without saturated response. It is clear that response of the photodiode, shown in Fig. 13 tend to be much larger than phototransistors, Fig. 14. This is an indication that the integrated current within the photodiode is higher compared to that of the phototransistor, even with the same incident illumination power.

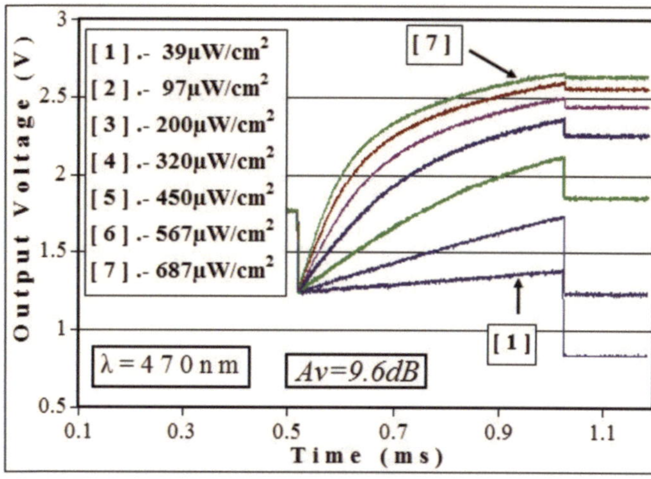

Fig. 13. Measurements of the temporal response in the N-Well/P-substrate structures (photodiodes)

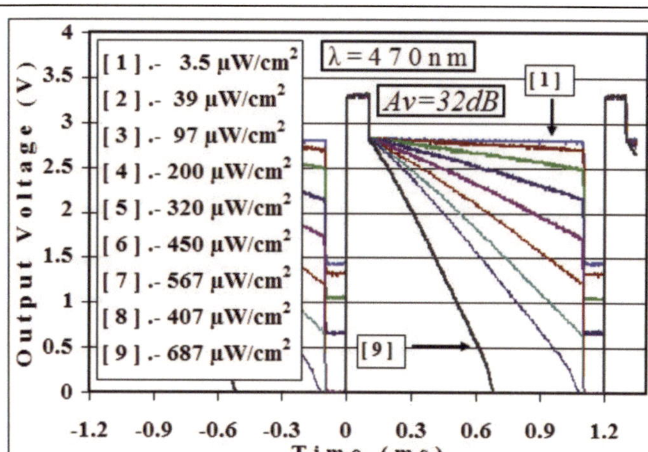

Fig. 14. Measurements of the temporal response in the P+/N-Well/P-substrate structures (phototransistors)

This is confirmed with the plots shown in Fig. 15 and 16, where the photocurrent was estimated as a function of wavelength. The difference between each other is about one order of magnitude. These responses were carefully obtained adjusting the gain of the amplifier and with experimental data of sensitive surface, capacitance in the node 2 and integration time. The photocurrent was estimated using equation (3) and similar data from Figure 13 and 14. Here, the technology spread can be seen also as five chips were measured giving some dispersion from each photo-device measured. As result, a display logarithmic is shown in Fig. 17. A monochromator HILGER & WATTS and an ISA lamp were used. Characteristic spectral for that lamp, used in this work, is shown in the Fig. 18.

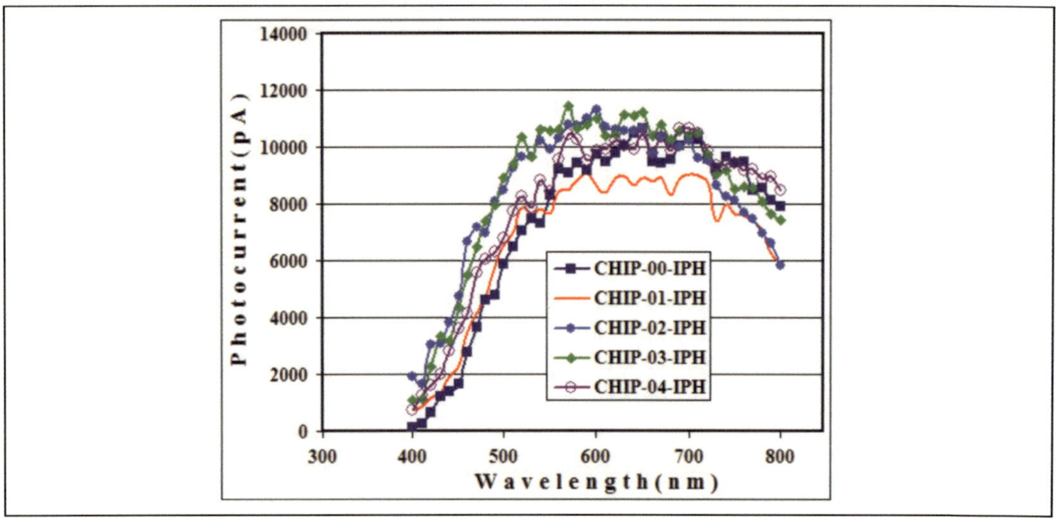

Fig. 15. Spectral response of N-Well/P-substrate structures (photodiodes)

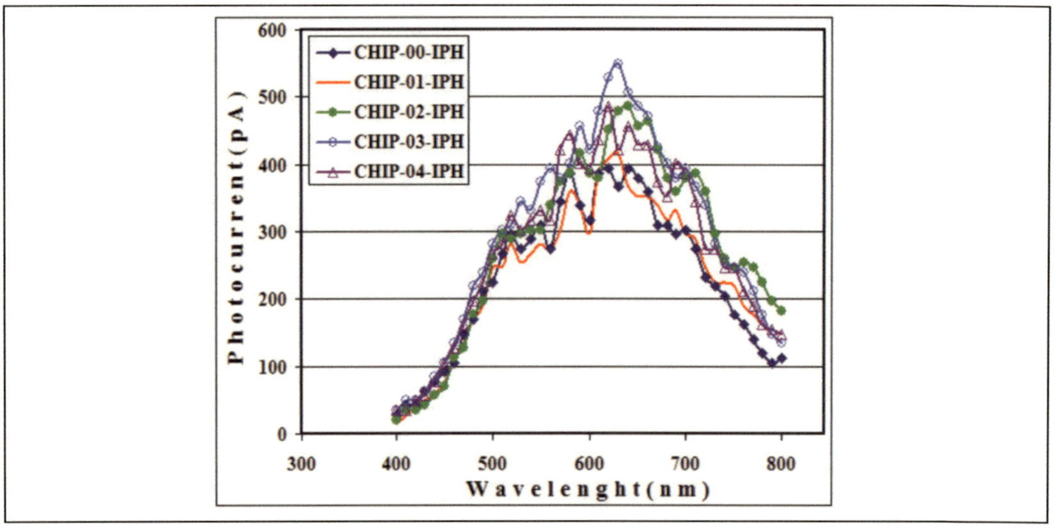

Fig. 16. Spectral response of P+/N-Well/P-substrate structures (phototransistors)

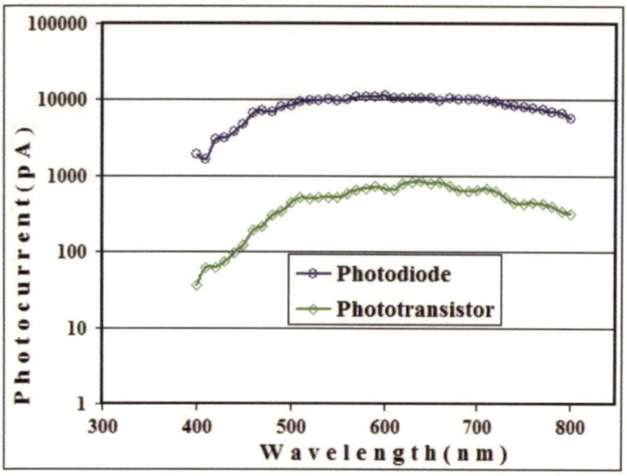

Fig. 17. Spectral response in logarithmic scale of P+/N-well/P-substrate (phototransistors) and N-Well/P-substrate (photodiodes), structures.

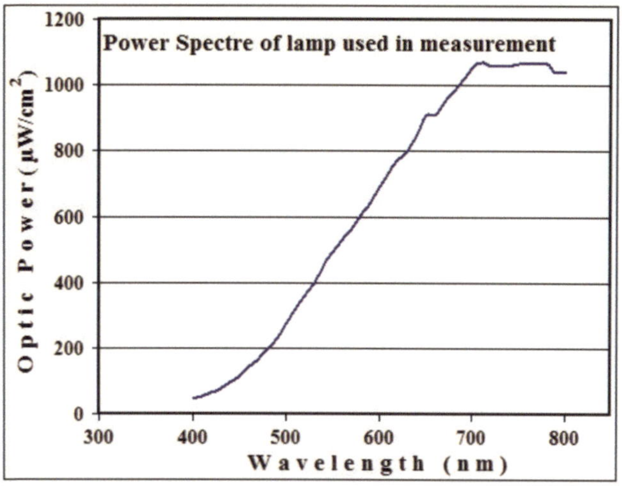

Fig. 18. Power spectre of lamp ISA used for measurements

A kind of small oscillations can be observed from Fig. 15 and 16. It has been suggested these oscillations are due to the Fabry-Perot interference (Lee *et al* 2007, Liang *et al* 2001). For both (9µm)x(9µm) phototransistors and photodiodes, the spectral response was measured. Fig. 17 presents a comparison in magnitude between the spectral responses of phototransistors and photodiodes. Again, it can be seen that the photocurrent from the photodiode is higher than the photocurrent from the phototransistor. Fig. 19 shows a comparison among the spectral response of the phototransistor used in this work (with 500 µW/cm^2 optical power) and the reported Reginald-Krishna's model (Perry 1996) using two other alike photodiode structures.

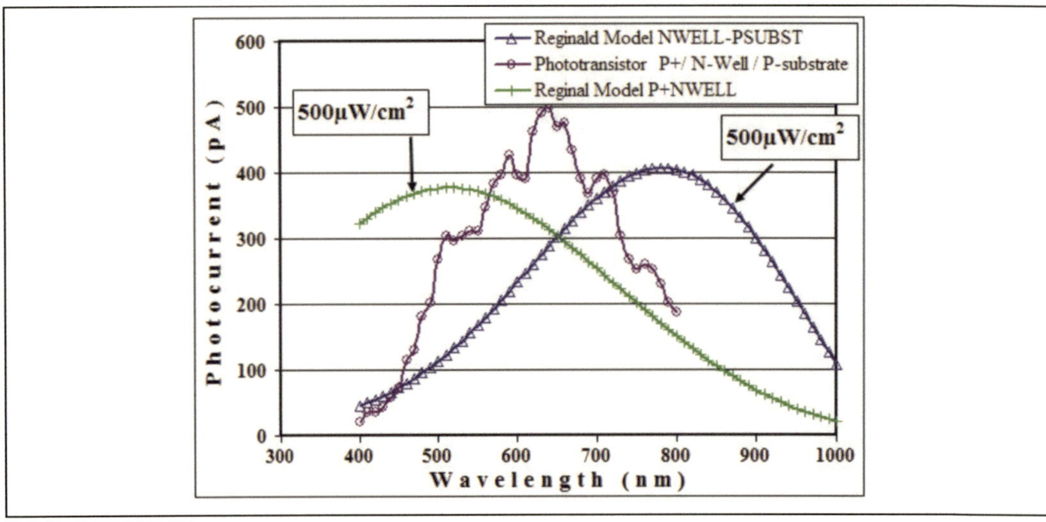

Fig. 19. Spectral response P+/N-Well/P-substrate vs. Reginald-Krishna's model of P+NELL and NWELL-PSUBST (Perry 1996).

It should be noted that the experimental photo-response of the P+/N-Well /P-substrate structure (phototransistor) has its maximum just where the junctions NWELL-PSUBST and P+NWELL models overlap. The model's plots are ideal optical response of the indicated junctions, so they do not include parasitic currents generated by additional events as crosstalk, for instance.

Crosstalk is a problem that can be present with neighbouring illuminated pixels that collect reflected or refracted light through lateral paths. Ideally, this should be avoided to minimize the degradation of pixels. Crosstalk can increase if integration density of pixel arrays is increased as technology shrinkage the size of devices. Then, the experimental curve is shifted toward the right from the ideal P+NWELL curve. It should be remembered that the emitter-base junction is the active one in the phototransistor. Fig. 13 (a) and (b) show photosensitivity and quantum efficiency respectively, for the measured phototransistors. The experimental photocurrent density was evaluated with Eq. (22).

$$J_{ph} = \frac{C_T \cdot \Delta V_{in}}{A_{ph} \cdot \Delta T} \tag{22}$$

Where:

J_{ph} : photo-sensor current density.

C_T : capacitance at node-2.

ΔV_{in} : integrated voltage at node-2.

A_{ph} : sensitive area of the phototransistor.

ΔT : integration time according to Figure 2.

The photosensitivity was evaluated using the following expression:

$$S_\lambda = \frac{J_{ph}}{P_{opt}} \tag{23}$$

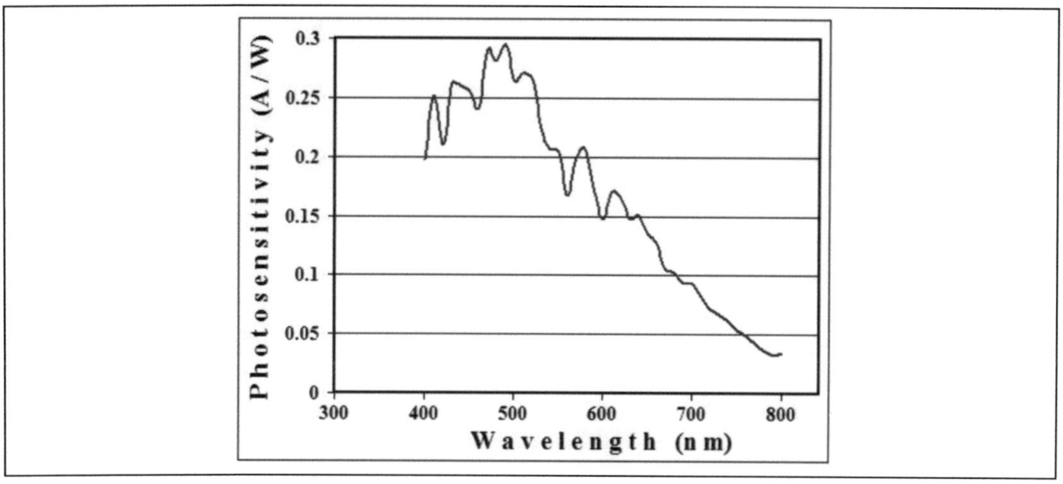

Fig. 20. Photosensitivity of P+/N-Well/P-substrate structure

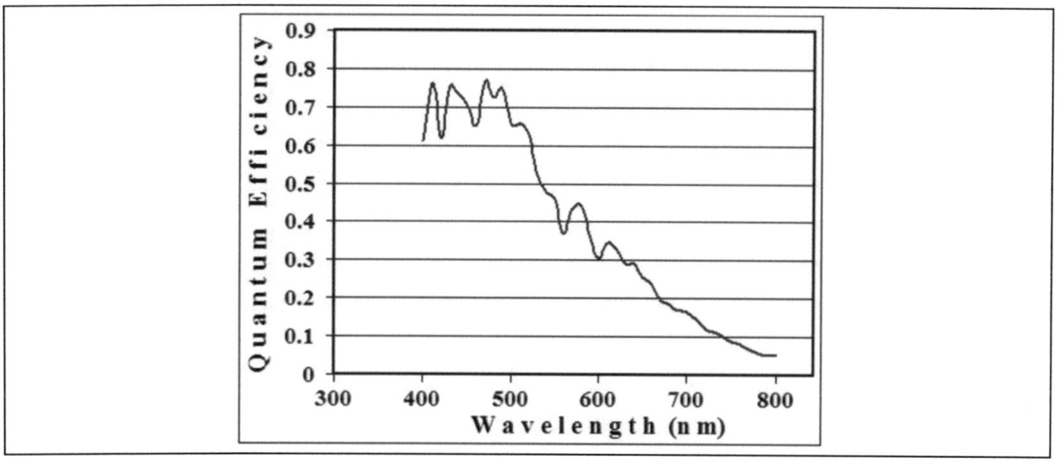

Fig. 21. Quantum efficiency of P+/N-Well/P-substrate

Where

P_{opt}: optical power density from Figure 11(b).

The quantum efficiency QE was evaluated with:

$$QE = S_{\lambda}\frac{h \cdot c}{\lambda \cdot q} = 1240\frac{S_{\lambda}}{\lambda} \tag{24}$$

and

$h \cdot c / \lambda$: energy of photons, with λ in µm.

q: charge of electrons.

6. Discussion

Temporal response shown in Fig. 13 and 14 were made using a fixed wavelength with optical power as a parameter. After several test over the photo-devices, 470 nm was selected for measurements since this wavelength gave the maximum sensitivity, as can be seen from Fig. 20. With the structures used as photo-devices, a strong difference of the spectral photo-response between phototransistors and photodiodes can be identified. It can be seen from Fig. 17 that for same illumination conditions, the response of photodiodes is higher than the response of phototransistors, by almost one order of magnitude. Moreover, measurements were made also over structures covered with a layer of metal, and the results are shown in Fig. 22, 23 and 24 for phototransistors and photodiodes, respectively. In the case of phototransistors, it is seen that the response of covered devices is almost 85% weaker than that of a not covered device.

In the case of photodiodes, this difference is lower about 30%, what is due to a big substrate leakage current by the properties of the N-well/p-subs structure, which may have other contributions adding to carriers directly generated by photons from inside of the photodiodes' area. However, all of the measured current is stimulated by photons as we can see from Fig. 24(b). Furthermore, it can be suggested that the N-well/p-subs structure used

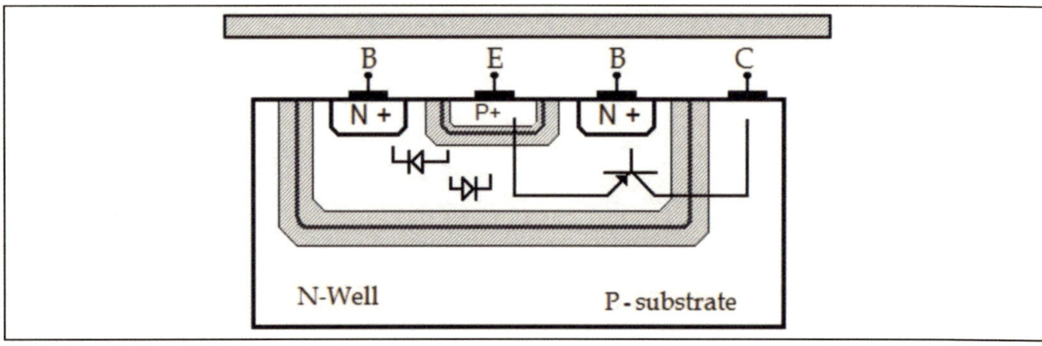

Fig. 22. Cross section of covered structure P+/N-Well/P-substrate structure, by metal 2 (phototransistor)

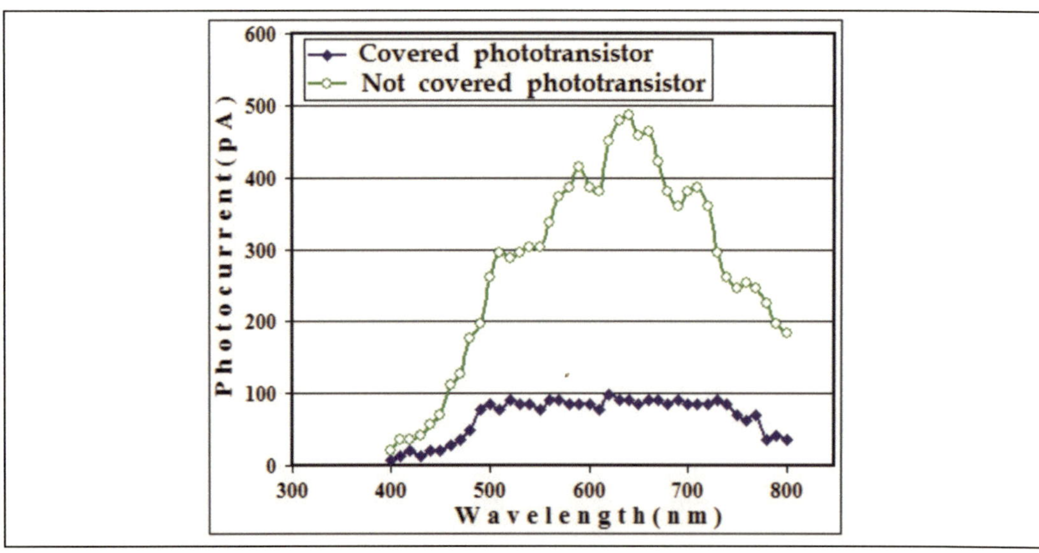

Fig. 23. Response of covered and not covered phototransistors

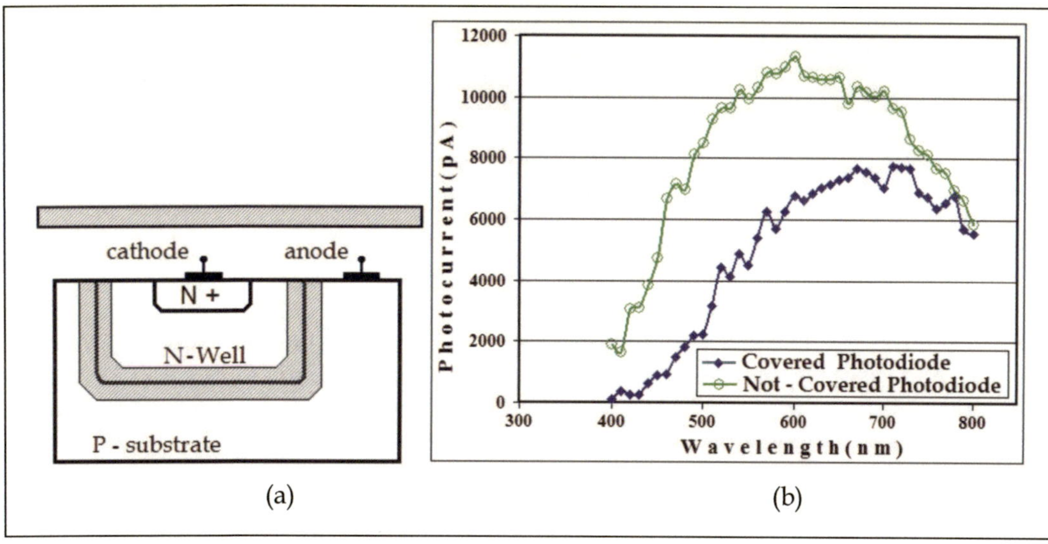

Fig. 24. Photodiode, structure N-Well/P-substrate: (a) cross section, (b) response on covered and not covered device

as a photodiode is not too efficient due to crosstalk, degrading pixel's characteristics as the dynamic range and increasing the fixed pattern noise.

It is reported that crosstalk is a mechanism that is pronounced at longer wavelengths (Lee 2008) since light with these characteristics can go deep into silicon having a high probability to be reflected by the different layers present in the structure of the pixel. This can be confirmed with Fig. 24(b), where it is seen that the photocurrent of a covered photodiode increases as wavelength is increased. Furthermore, comparing Figs. 23 and 24(b), it can be concluded that the base-collector junction of the phototransistor operates as a barrier for carriers generated by light that penetrates beyond the surface region into the substrate. Since the photodiode has not this additional junction, it is collecting extra carriers, thus inconveniently reducing the difference between the response of covered and not covered devices. This should be considered also when designing pixel architectures, providing the pixel with surrounding materials with low dielectric constants and index of refraction, as long as the technology allows it. Otherwise, junction barriers as base-collector in phototransistors can play a similar role. Due to the importance of this mechanism over the performance of pixels, following an explanation is given regarding crosstalk.

6.1 Crosstalk mechanisms

We have focused in the characterization and analysis of likely mechanisms that could contributes to the crosstalk on structures "P+/N-Well/P-substrate" (phototransistors) and "N-Well/P-substrate" (photodiodes). On these kinds of photo-devices, crosstalk has been defined and classified in two main mechanisms: (a) optical crosstalk and (b) electrical crosstalk (Brouk 2002, Kang 2002, Tabet 2002). Here, we introduce an additional classification of crosstalk, as is shown in the Fig. 25 and 26. The first classification is the lateral crosstalk mechanism which includes lateral optical crosstalk, and lateral electrical crosstalk. The second one is the vertical crosstalk mechanisms.

Crosstalk mechanisms	
Lateral crosstalk mechanisms	Lateral optical crosstalk mechanisms
	Lateral electrical crosstalk mechanisms
Vertical crosstalk mechanisms	There aren't vertical optical crosstalk mechanisms
	Vertical electrical crosstalk mechanisms

Table 1. Crosstalk mechanisms classification

6.2 Lateral crosstalk mechanisms

Lateral optical crosstalk is due to light traveling laterally among the layers up to the junction (near the surface of the device) acting as a waveguide, as is shown in the Fig. 25. Lateral electrical crosstalk is the phenomenon whereby photons generate carriers in the "**near to the surface region**". That phenomenon has its origin in short wavelength light, mainly under of 650nm. Both, optical and lateral electric crosstalk, are present either in, phototransistors or photodiodes. However, in phototransistors this contribution is as a photocurrent collected by the base contact, which is tied to VDD, hence masking the effect. This is a first reason whereby the photo-current is larger in photodiodes than in phototransistors. Fig. 25(a) shows the way in which both, optical crosstalk and lateral electric crosstalk mechanisms, affect the response of phototransistors and Fig. 25(b) shows the corresponding for photodiodes. The effect is very strong since the current comes from all directions.

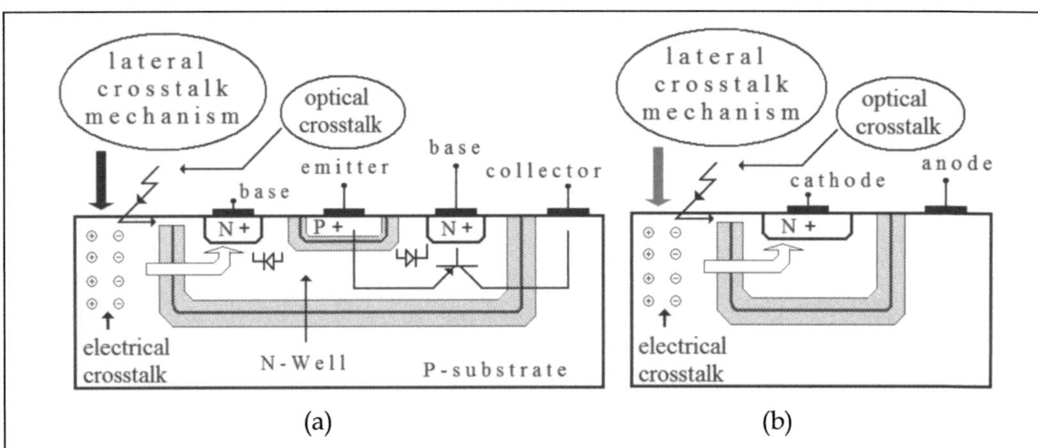

Fig. 25. Lateral crosstalk mechanisms in phototransistor

6.3 Vertical crosstalk mechanisms

Vertical crosstalk mechanism is shown in Fig. 17. It originates only due to electrical crosstalk, so it is called vertical electrical crosstalk. In the case of phototransistors, carriers generated along the substrate, as well as behind and outside the "N-Well", are collected by the base contact since it is connected to a higher voltage than the emitter (see Fig. 17(a)). So, only majority carriers are collected by the base.

In this case, diffusion of minority carriers is present as leakage current. Hence, little or no contribution to the spectral response of phototransistors is due to vertical electrical crosstalk.

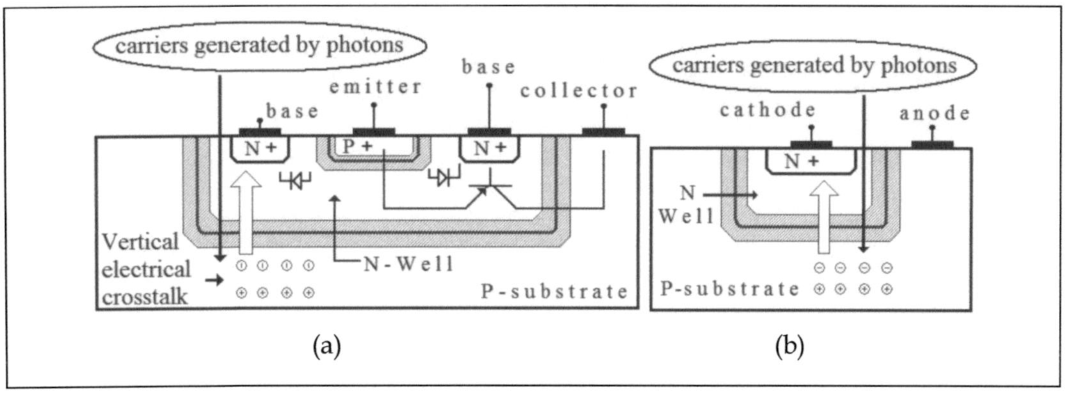

Fig. 26. Vertical crosstalk mechanisms in photodiode, (a) Phototransistors (b) photodiodes

Vertical electrical crosstalk effect in photodiodes is illustrated in Fig. 26(b). Carriers generated by photons behind and outside the N-Well contribute to the spectral response of the photodiode with the leakage current coming from the substrate. Carriers generated deep in the substrate are due to longer wavelengths. This component of leakage current has a very strong effect over photodiodes but this is not the case for phototransistors, so the difference appreciated in Fig. 17 can be attributed to this.

7. Conclusions

An architecture was proposed, from which characterization of photo-devices can be made, giving useful information for the performance evaluation of junction structures available in CMOS standard technologies. An adjustable gain amplifier, with a gain range of 10dB - 32dB, was configured allowing different biasing and operating points for photo-response measurement of different devices. Good agreement between simulated and experimental transfer function of the amplifier was obtained. The row-select transistor, M2, plays an important role in the operation of the amplifier. It was found that the aspect ratio of this transistor should be high in order to have a small channel resistance and to ensure an adjustable gain property to the amplifier. On the other hand, phototransistors (p+/N-well-/p-subs) and photodiodes (N-well/p-subs) were characterized for a 1.5μm technology, but the same methodology can be used with other silicon foundries. Structures have a maximum quantum efficiency of about 0.7 and a maximum sensitivity of almost 0.3A/W. Besides, photodiodes made with an N-well/P-subs junction, have shown a strong substrate leakage current contribution due to crosstalk that can affect parameters such as dynamic range and fixed pattern noise. So, depending on the features added to the architecture and the technology available, photodiodes may not be a good choice for image sensor arrays.

8. References

Albert J.P. Theuwissen. (2008). CMOS image sensors: State-of-the-art. *Solid-State Electronics,* Vol. 52, (2008) page numbers (1401-1406), ISSN: 0038-1101

Chye Huat Aw and Bruce A. Wooley (1996). A 128x128-pixel Standard CMOS Image Sensor with Electronic Shutter. *IEEE Journal of Solid-State Circuit,* Vol. 31, No.12, December 1996 page numbers 1922-1930, ISSN: 0018-9200

Graeme Storm, Robert Henderson, J. E. D. Hurwitz, David Renshaw, Keith Findlater, and Matthew Purcell. Extended Dynamic Range From a Combined Linear-Logarithmic CMOS Image Sensor. *IEEE Journal of Solid-State Circuits,* Vol. 41, No. 9, September. 2006, page numbers 2095-2106, ISSN: 0018-9200.

Kareem A. Zaghloul, and Kwabena Boahen. (2004). Optic Nerve Signals in a Neuromorphic Chip I: Outer and Inner Retina Models. *IEEE Transactions on biomedical engineering,* Vol. 51, No. 4, (April 2004) page numbers (657-666), ISSN: 0018-9294.

Ralf M. Philipp, David Orr, Viktor Gruev, Jan Van der Spiegel, and Ralph Etienne-Cummings. (2007). Linear Current-Mode Active Pixel Sensor. *IEEE Journal of Solid-State Circuits,* Vol. 42, No. 11, (November 2007) page numbers (2482-2491), ISSN: 0018-9200.

R. J. Baker, Harry W. Li and David E. Boyce (2005). *CMOS, Circuit Design, Layout, and Simulation.* IEEE PRESS, of the Institute of Electrical and Electronics Engineers, Inc. ISBN: 0-7803-3416-7. 345 East 47th Street, New York, NY 10017-2394.

R. J. Perry and Krishna Arora (1996),. Using PSPICE to simulate the photoresponse of ideal CMOS Integrated Circuits Photodiodes. *Proceedings of the IEEE Southeastcon Bringing Together Education, Science and Technology,* (1996) page numbers (374-380), ISSN 0-7803-3088-9

Tae Gyoung Lee, Won Nam Kang, Young Ju Park, Eun Kyu Kim (2007). Fabry-Perot Interference Characteristics of the Photoluminescence in Nanoclustered SiNx:H Thick Films. *Journal of the Korean Physical Society,* Vol. 50, No. 3, (March 2007) page numbers (581-585).

Wenjie Liang, Marc Bockrath, Dolores Bozovic, Jason H. Hafner, M., Tinkham and Hongkun Park (2001). "Fabry-Perot interference in a nanotube electron waveguide". *Journal of Nature,* Vol. 411, No. 7, (2001) page numbers (665-669), ISSN

Lee, Ji Soo and Mouli, Chandra (Boise, ID, US), Image Sensors with optical trench, United States Patent 7315014, January 2008.
url: "http://www.freepatentsonline.com/7315014.html"

I. Brouk, Y. Nemirovsky, S. Lachowicz, E. A. Gluszak, S. Hinckley, and K. Eshraghian (2002). Characterization of crosstalk between CMOS photodiodes. *Solid-State Electronics,* Vol. 46, (month and year of the edition) page numbers (first-last), ISSN

In Man Kang (2002) "The simulation of the crosstalk between Photodiodes Fabricated Using the 0.18μm CMOS Process", Semiconductor Materials and Devices Laboratory, School of Electrical Engineering and Computer Science, Seoul National University, Republic of Korea.SMDL Annual Report 2002. URL of the website is:
http://smdl.snu.ac.kr/Research/annual/annual2002/2002pdf/ann_kim_2002.PDF.
http://smdl.snu.ac.kr/Research/annual/annual2002/

M. Tabet (2002). Double Sampling Techniques for CMOS Image Sensors. Doctor of Philosophy thesis, University of Waterloo Electrical and Computer Engineering. Waterloo, Ontario, Canada, 2002.

7

Performance Improvement of CMOS APS Pixels using Photodiode Peripheral Utilization Method

Suat U. Ay
University of Idaho
USA

1. Introduction

Charge-coupled device (CCD) technology had been leading the field of solid-state imaging for over two decades, in terms of production yield and performance until a relatively new image sensor technology called active pixel sensor (APS) (Fossum, 1993), using existing CMOS facilities and processes, emerged as a potential replacement in the early 1990s. While CMOS APS technology was originally considered inferior, continuous improvements in cost, power consumption (Cho et al., 2000), dynamic range (Gonzo et al., 2002), blooming threshold, readout scheme and speed (Krymsky et al.,1999), low supply voltage operation (Cho et al., 2000), large array size (Meynants, 2005), radiation hardness (Eid et al., 2001), and smartness have achieved performance equal to or better than CCD technology (Agranov et al., 2005; Krymsky et al., 2003).

Electro-optical performance of a photodiode (PD) type APS pixel is directly related to physical properties of photodiode diffusion layer. Doping concentration, junction depth, junction grading, biasing conditions, and physical shape of the photodiode diffusion layer determine the pixel full-well capacity, which is one of the main performance benchmarks of the PD-APS pixel. Pixel full-well capacity is related to sensitivity, charge capacity, charge saturation, dynamic range, noise performance, and the spectral response of the pixel (Theuwissen, 1995). Pixel dynamic range versus full well capacity for different pixel noise levels could be plotted as shown on Fig. 1. Thus, increasing full well capacity is desirable.

In this chapter, so called photodiode peripheral utilization method (PPUM) is introduced addressing performance improvement of photodiode type CMOS APS pixels, (Ay, 2008). PPUM addresses the improvement of the metrics full well capacity and spectral response especially in blue spectrum (short wavelength). First, identification of junction and circuit parasitics and their use in improving the full-well capacity of a three-transistor (3T) PD-APS pixel through photodiode peripheral capacitance utilization is discussed. Next, spectral response improvement of PD-APS pixels by utilizing the lateral collection efficiency of the photodiode junction through PPUM is discussed. The PPUM method and its proposed benefits were proven on silicon by designing a multiple-test-pixel imager in a 0.5μm, 5V, 2P3M CMOS process. Measurement results and discussions are presented at the end of the chapter.

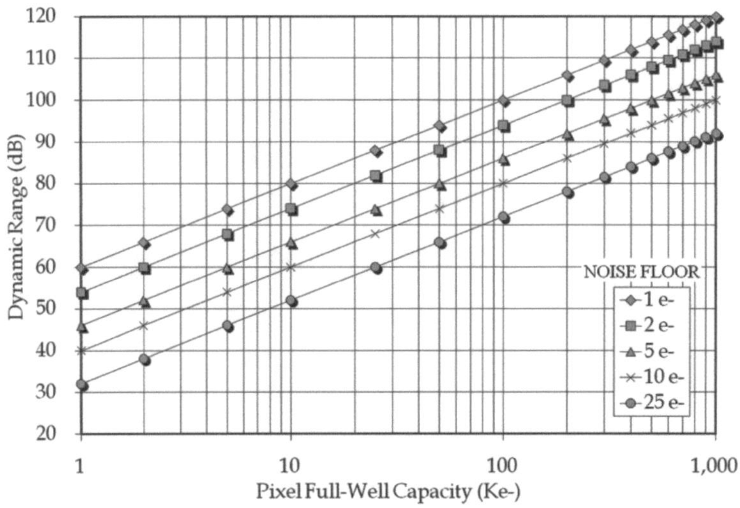

Fig. 1. Pixel dynamic range versus full well capacity and noise floor.

2. Photodiode Peripheral Utilization Method (PPUM)

The theory behind the photodiode peripheral utilization method (PPUM) is that, if the pixel pitch is restricted to a certain size, then pixel full-well capacity could be increased by opening holes in the photodiode's diffusion. These diffusion holes could be used to increase photodiode parasitic capacitance, by increasing the perimeter capacitance of the photodiode for certain process technologies shown on Fig 2. Diffusion holes also can increase spectral response of a photodiode by utilizing lateral collection of charges converted close to the semiconductor surface at the edges of photodiode, (Fossum, 1999; Lee and Hornsey, 2001).

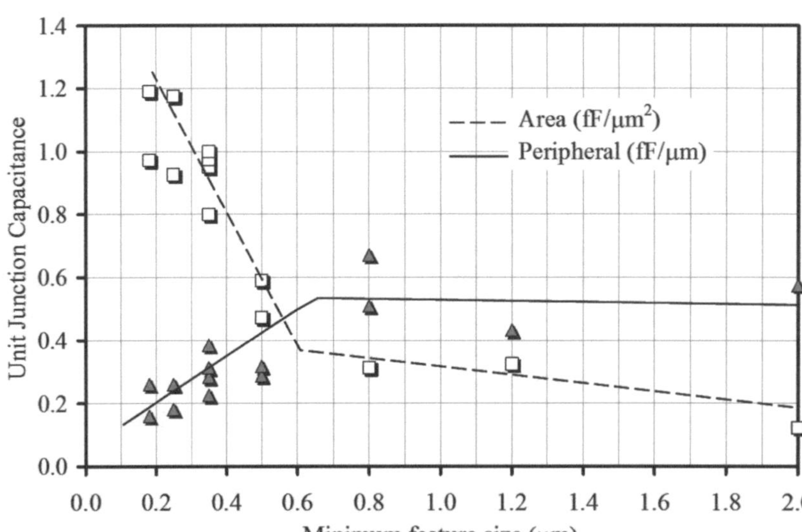

Fig. 2. Unit junction capacitance of CMOS processes, (Ay, 2004).

A reverse-biased PN-junction diode is used in photodiode (PD) type CMOS APS pixels as a photon conversion and charge (electron) storage element. The total capacitance of the photodiode diffusion layer determines key pixel performance parameters. For example, wide-dynamic-range pixels require large pixel full-well capacity and low readout noise. Photodiode full-well capacity is comprised of two components: bottom plate (area) and side wall (peripheral) junction parasitic capacitance. Designer controls the size of the photodiode diffusion bottom plate, while peripheral junction depth and doping concentration are process and technology dependent. The photodiode's unit area junction capacitance (C_A) and unit peripheral junction capacitance (C_P) are given in the following equations, (Theuwissen 1995), including technology and design parameters, for the first-order capacitance that contributes to total well capacity.

$$C_{PD} = C_A \cdot A + C_P \cdot P \tag{1}$$

$$C_{PD} = \frac{C_{J0A} \cdot A}{\left[1 - \frac{V_{PD}}{\Phi_B}\right]^{MJ}} + \frac{C_{J0SW} \cdot P}{\left[1 - \frac{V_{PD}}{\Phi_{BSW}}\right]^{MJSW}} \tag{2}$$

where

C_A, C_P unit area junction capacitance and unit peripheral junction capacitances, respectively;

C_{J0A}, C_{J0SW} unit zero-bias area and peripheral junction capacitances, respectively;

A, P area and peripheral of the photodiode regions, respectively;

Φ_B, Φ_{BSW} built-in potential of area and side-wall junctions, respectively;

M_J, M_{JSW} junction grading coefficients of area and side-wall junctions, respectively;

V_{PD} photodiode junction voltage.

Other parasitic capacitances due to the reset and readout transistors in pixel contributing to total photodiode junction capacitance are shown in Fig. 3. for a three-transistor (3T) PD-APS pixel. These parasitic capacitances contribute to total pixel capacitance differently in different modes of pixel operation, (Ay, 2004). Right after photodiode reset and during scene integration periods, overlap capacitances C_{O1} and C_{O2} and gate-to-body capacitance of the

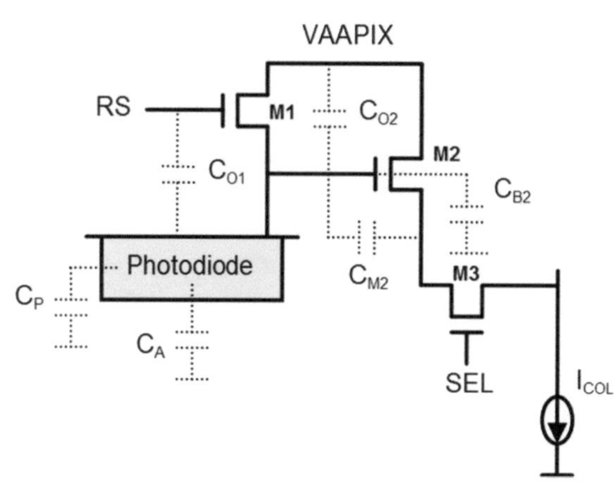

Fig. 3. Parasitic capacitances of photodiode type CMOS APS pixel.

readout transistor M2 (C_{B2}) add to the total photodiode capacitance. During a readout period, miller capacitance C_{M2} and overlap capacitances C_{O1} and C_{O2} contribute to the total photodiode capacitance. Contribution of pixel circuit parasitic capacitances is described by the following equations during imaging (3) and readout (4):

$$C_{par,imaging} = \left[W_{M1} \cdot L_{OL,M1} + W_{M2} \cdot \left[L_{M2} - L_{OL,M2}\right]\right] \cdot C_{OX} \tag{3}$$

$$C_{par,read} = \left[\frac{2}{3} \cdot W_{M2} \cdot \left[L_{M2} - 2 \cdot L_{OL,M2}\right] \cdot \left[1 - G\right]\right] \cdot C_{OX} \tag{4}$$

$$+\left[W_{M1} \cdot L_{OL,M1} + W_{M2} \cdot L_{OL,M2} \cdot \left[2 - G\right]\right] \cdot C_{OX}$$

where

W_1, W_2 channel width of the reset and source-follower transistors, respectively;

L_{OL1}, L_{OL2} channel overlap length of the reset and source-follower transistors, respectively;

C_{OX} unit oxide capacitance,

G pixel source follower gain factor.

C_A and C_P of a few CMOS process technologies, with minimum feature sizes 2.0µm–0.18µm, is shown in Fig. 2., (Ay, 2004). Unit-area capacitance is larger for deep sub-micron devices with a minimum feature size <0.5µm, due to the increased channel-stop doping-level (for better device isolation, higher diffusion doping concentrations, and shallower junction depths) (Packan, 2000). Thus, peripheral junction capacitance could be better utilized in processes that have equal or more unit peripheral junction capacitances than in processes with <0.5µm feature sizes, by opening holes in the photodiode region. As will be shown in the next sections, this will not only improves the total full-well capacity of the pixel, but also improves the spectral response for detecting short wavelength photons.

3. Photodiode lateral collection improvement

The photosensitive element in APS pixels, the photodiode (PD), works in charge integration-mode where pixels are accessed at the end of a time interval called the integration period. When it is accessed, photodiode is read and then cleared for next scene integration. Fig. 4. shows the cross-section of a PN-junction photodiode formed in a CMOS process; the photodiode is reverse-biased and formed by using the shallow N+ doped, drain-source diffusion of an NMOS device. A bias voltage applied to the N+ region forms a depletion region around the metallurgical PN-junction, which is free of any charge because of the electrical field. Any electron-hole pairs generated in this region see the electrical field as shown in the AA′ cross-section view of the photodiode in Fig.4. Electrons move in the opposite direction of the electric field (toward the N+ region), while holes move toward the P-region. As a result, electrons are collected in a charge pocket in the N+ region, while holes are recombined in the substrate. This type of photodiodes has been widely used in CMOS and early CCD-type image sensors as a photo conversion and collection element.

There are two issues associated with using the N+ drain/source diffusion of an NMOS transistor as photosensitive element. First is the dark current induced by stress centres around the diffusion, (Theuwissen 1995). These stress centres are formed during the field

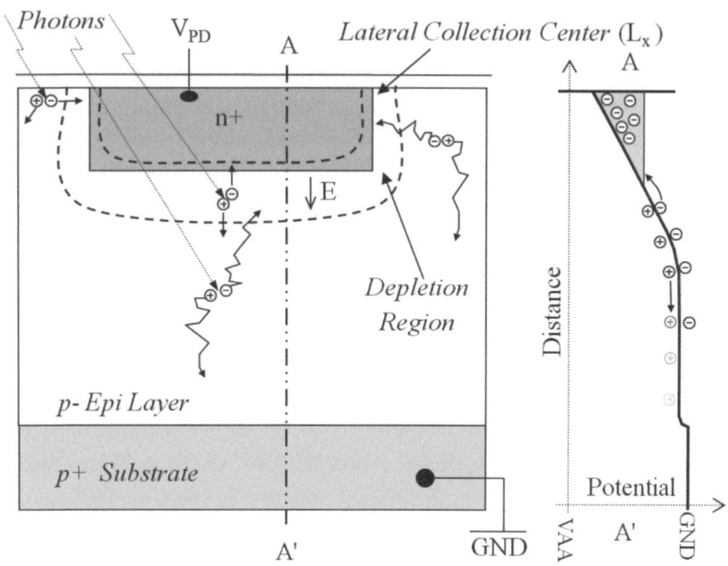

Fig. 4. a)Cross-section and b)potential-well diagram of a PN-photodiode.

oxide (FOX) formation in standard CMOS processes. The second issue is the surface-related dark current generated from the work function difference between the N+ diffusion surface and overlaying isolation oxide layer. This second one causes surface recombination centers and defects. Both of them absorb photo-generated electron-hole pairs close to the surface, resulting in quantum loss at shorter wavelengths. As a result, silicon photodiodes show less sensitivity in the blue spectrum (<400nm. Most blue photons are collected through lateral diffusion of the carriers generated on or in the vicinity of a photodiode peripheral—known as peripheral photoresponse or lateral photocurrent (Lee et al., 2003). Thus, increasing lateral collection centers or peripheral length of a photodiode potentially improves collection efficiency for short-wavelength photons (Fossum, 1999; Lee et al., 2001) as it is depicted in Fig. 5. This method was adopted for UV photodiode devices in P-well CMOS processes (Ghazi et al.,2000).

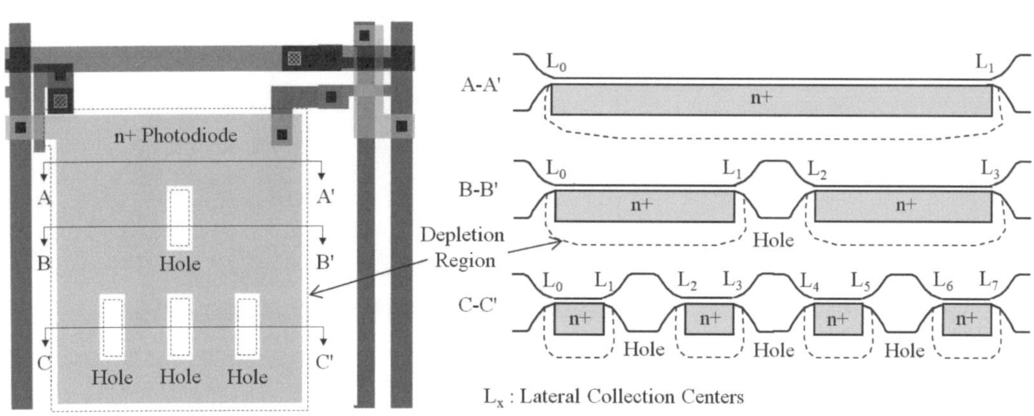

Fig. 5. Improving lateral collection by increasing photodiode peripheral for blue photons.

4. CMOS pixel design using PPUM

There are many ways to test CMOS imaging pixels using test vehicles. Some uses product grade imager platforms to test not only the performance of the imaging pixels, but also their performance in final product environment. Some uses very small array of dumb pixels to measure basic characteristics of the pixel under investigation. A commonly used architecture is called fully flexible open architecture (FFOA) that composes of sample and hold circuits, correlated double sampling (CDS) and differential delta sampling (DDS) circuits, and source follower amplifiers (Nixon et al, 1996; Mendis et al., 1997). Simple FFOA architecture gives very reliable and predictable signal path characteristics. It also allows multiple pixel types with different sizes to be integrated on the same chip.
A test imager was designed containing reference and pixels utilizing PPUM as proof of concept. The reference or baseline three-transistor (3T) photodiode type (PD) APS reference pixel (REF) is shown in Fig. 6. It was designed to normalize measurement results of the test pixels with diffusion holes. A fairly large pixel size of 18μm × 18μm was chosen. It has circular-looking photodiode diffusion region for reducing overall dark current. Row select and reset signals were drawn on top of each other using horizontal metal-2 and metal-3 lines, and metal-1 was used on the vertical direction for routing pixel output and supply signals.
The reference photodiode diffusion area and peripheral were 141.7μm2 and 44.6μm, respectively. Unit area and peripheral capacitance of the photodiode's N+ diffusion layer in used process were 0.25fF/μm2 and 0.22fF/μm, respectively. Total pixel capacitance was calculated by including the Miller contribution of the source-follower transistor (M2) and other parasitic capacitances from equations (3) and (4). Miller contribution to the total photodiode capacitance at 0.75 source-follower gain was calculated to be 1.1fF; peripheral junction capacitance made up of 20 percent of the total photodiode capacitance, and the total calculated photodiode capacitance was about 47.5fF.

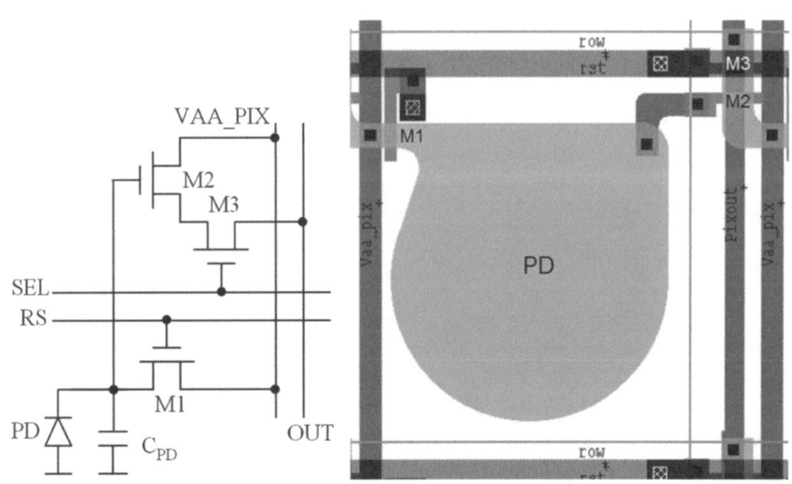

Fig. 6. 3T CMOS APS reference pixel (REF) a) schematic, b) layout.

Four test pixels with a number of circular diffusion openings were designed to model the peripheral utilization effect on pixel performance, with layouts shown in Fig. 7. Pixels have

1.6μm-diameter circular holes on the photodiode diffusion with the same base as the reference pixel (REF). The total number of circular diffusion holes was 17, 14, 11, and 7 for pixel layouts called c17, c14, c11, and c7, respectively. Holes were randomly placed on the reference design. Again, the circular shape was chosen for holes to reduce stress-related dark current.

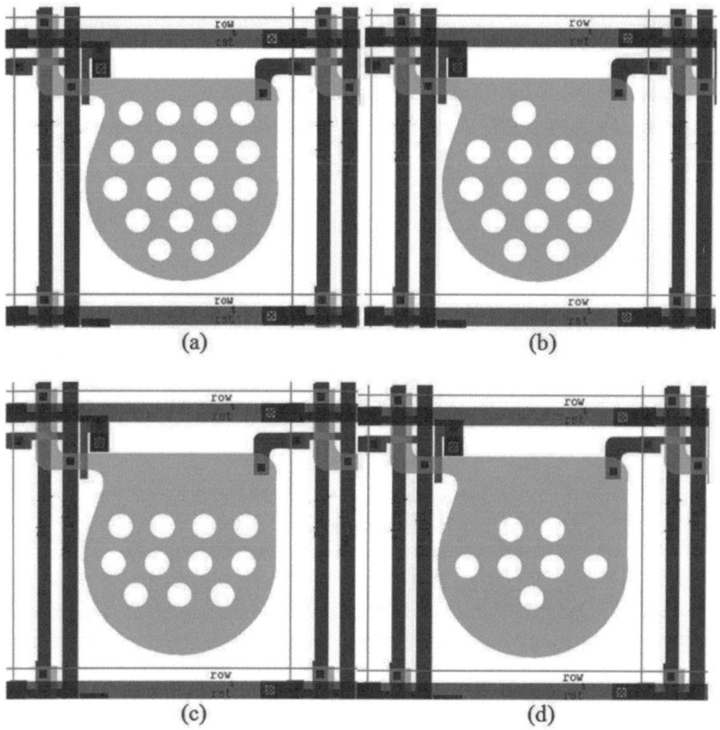

Fig. 7. Test pixels with circular openings; a) c17, b) c14, c) c11, d) c7

5. CMOS APS imager design

All test pixels were placed in the same imager to compare performance under common imaging and environmental conditions. Single-channel serial-readout architecture was adopted to pass all pixel signals through the same signal path for accurate comparison of the effects (Ay et al., 2002). Imagers were composed of a 424x424 pixel array, row decoder and drivers, timing generators, digital and analog buffers, a column analog signal processor (ASP), a column decoder and multiplexer, and a single, global readout channel. The pixel array was divided in to 16 different subsections with 106 x 106 pixel arrays, with different pixel designs in each subsection. A shift-register type decoder was used in the column, too. Decoder control signals were generated in the timing generator block separately for frame operation. A pseudo-differential charge amplifier and sample-and-hold circuits were used in the global readout block. Chip outputs were in differential analog signals (SIG) and reset (RST). Signal analog-to-digital conversion used an analog-frame-grabber card.

A detailed schematic of the prototype imager's analog signal chain is shown in Fig. 8. Each column contains a PMOS source follower, two sample-and-hold capacitors and a number of switches. A PMOS source follower was used for level shifting and signal amplification. Column signals were read during column time through single channel, pseudo-differential charge amplifiers and buffered for off-chip analog-to-digital conversion.

Fig. 9. shows a microphotograph of the prototype imager. The prototype was designed in 0.5μm, 5V, 2P3M CMOS process, and different test pixel quadrants could be recognized on the pixel array with the naked eye. Table 1 provides specifications for the prototype imager. Global charge amplifier gain was adjusted so that the gain-loss in pixel and column source followers balanced to achieve unity gain from pixel-to-chip output. Operating at 5 Mp/s readout speed, the prototype achieved a 30-frame per second (FPS) frame rate. A 5V supply was used and the total power consumption of the chip was <200mW. Noise floor of the readout channel was 850μV.

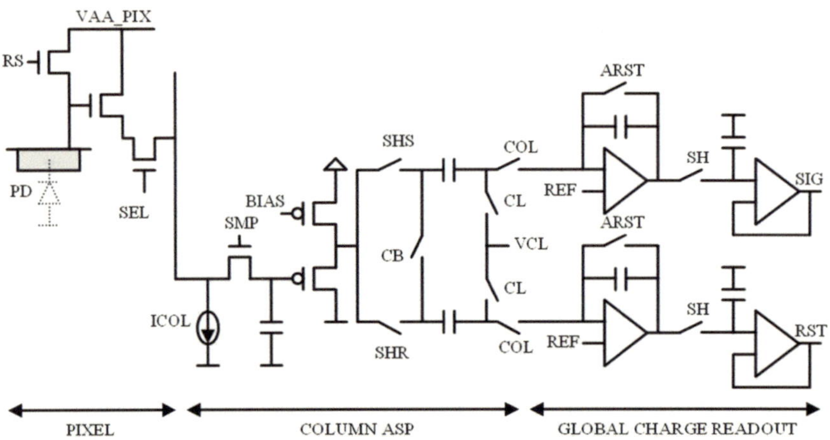

Fig. 8. Analog signal chain from pixel to chip output.

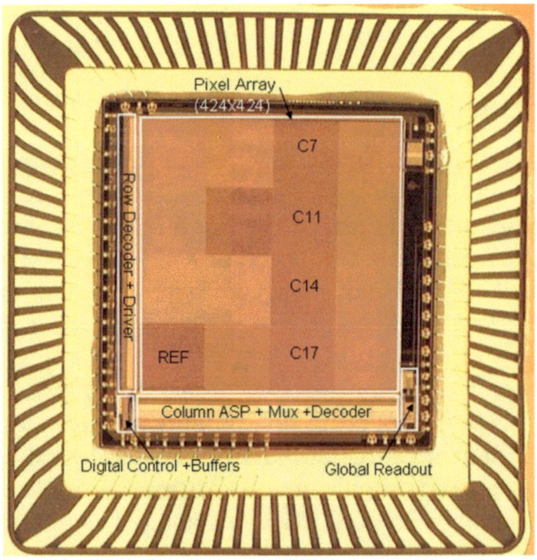

Fig. 9. Micrograph of Prototype CMOS APS imager chip.

Array Size	424 x 424
Pixel Size	18 μm x 18 μm
Pixel Type	3T Photodiode APS
Technology	0.5 μm CMOS (2P3M)
Output Format	Differential Analog
Frame Rate	30 FPS
Gain (photodiode to chip output)	1.0 V/V
Noise Floor	850 μVolt
Fixed Pattern Noise (FPN)	< 0.25 (% Vsat)
Power Supply (VAA,VAA_PIX)	5 Volt
Power Consumption	< 200 mWatts
Package	PGA 84L
Chip Size	9.75 mm x 9.75 mm

Table 1. Design specifications of the prototype CMOS imager

6. Measurement results

Electrical and optical characteristics of reference and circular-opening test pixels measured under the same environmental and imaging conditions, (Ay, 2004). Having them integrated on same focal plane array make these measurements more manageable and easy.

6.1 Reference pixel measurements

Dark current was measured at room temperature. It was 10.63 mVolt per second. This equals to 3155 e-/sec with the measured conversion gain of 3.37 μVolt per electrons. Measured photon transfer curve of the reference pixel is shown in Fig. 10. Total measured pixel capacitance was 47.5 fF as oppose to the calculated value of 46.5fF. Measured pixel full-well capacity was 508Ke- with 1.714V effective photodiode voltage.

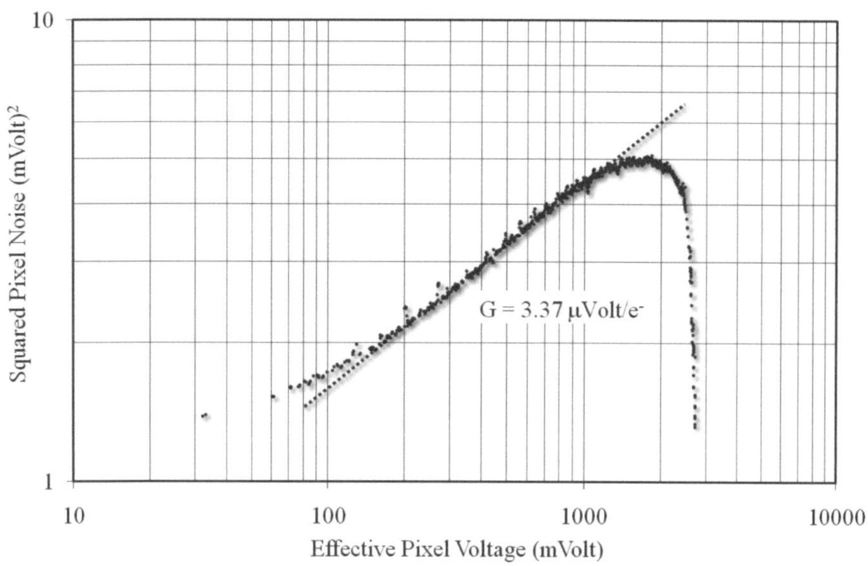

Fig. 10. Measured photon transfer curve of the reference pixel (REF1).

Measured light sensitivity was 2.44 Volt/Lux*sec while peak quantum efficiency was 48.55 percent at 500nm as shown in Fig.11. At 400nm, quantum efficiency of the reference pixel was 23.4 percent. Dynamic range of the reference pixel was around 66.4 dB because of the higher noise floor measured. Rest of the measurement and calculations are listed in Table 2 for the reference pixel (REF1). All measurements of the test pixels with diffusion holes are normalized with the reference pixel characteristics.

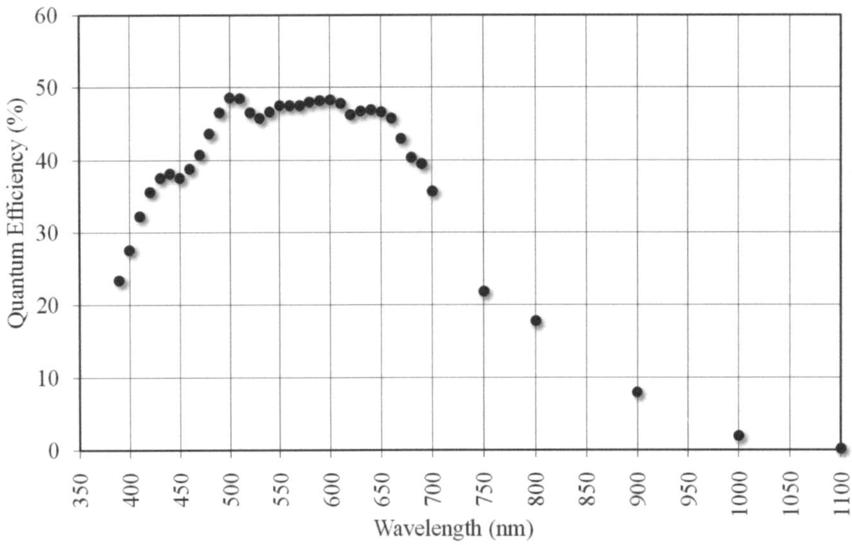

Fig. 11. Measured quantum efficiency of the reference pixel (REF1).

Parameter	Measured	Calculated	Unit
Sensitivity	2.44		Volt/lux.sec
Light Saturation	1.07		lux.sec
Saturation Voltage	2.74		Volt
Quantum Efficieny	23.41		at 390nm
	47.44		at 550nm
	48.55		peak
Conversion Gain	3.370	3.446	μV/e-
Full Well	1.714	1.714	Volt
	508,736	497,530	e-
Pixel Capacitance	47.54	46.49	fF
Dark Current	10.63		mVolt/sec
	3155		e-/sec
Dynamic Range	66.39	79.67	dB

Table 2. Calculated and measured parameters of the reference pixel (REF1).

6.2 Conversion gain and pixel full-well capacity measurements

Conversion gain and the full-well saturation voltage of the reference and test pixels were measured to determine pixel well capacity. Measurement results are shown in Fig. 12. Pixel well capacity increases with proper utilization of the photodiode peripheral junction, by using the holes on the photodiode diffusion region. Conversion gain of the pixel reduces with increased pixel capacitance, and the pixel full-well capacity increases.

It was observed that a linear correlation between total peripheral capacitance and pixel full-well capacity exist, because C_A is almost equal to C_P in the process used. The area loss was compensated for by the peripheral increase, by a factor of 2.5. Because the radius of the opening was set to 0.8µm, and the opening peripheral was (p = 2πr) 5.027µm while the area was (a = πr2) 2.01µm2. A factor of four could easily be achieved by choosing an opening radius of approximately 0.5µm. However, reducing diameter results in depletion region overlap, and lowers peripheral capacitance and utilization.

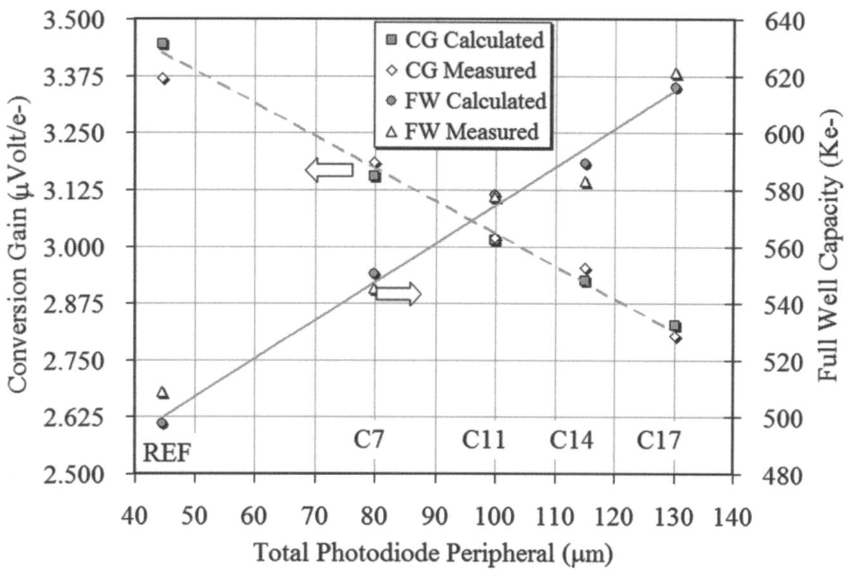

Fig. 12. Conversion gain and full-well capacity of the pixels.

6.3 Quantum Efficiencies (QE)

Quantum efficiencies (QE) of the reference (REF) and test pixels were measured by using a very stable light source, a monochromator, and a calibrated photodiode. Measurement was performed between 390nm and 700nm, with 10nm steps. QE measurement results for reference (REF) and test pixels (c17, c14, c11) are shown in Fig. 13. In the figure, QE difference between the reference pixel and a test pixel with 17 openings (c17), normalized by reference QE, was also plotted. Spectral response improvement was observed with an increased number of openings on the photodiode. The most improvement was achieved at the shorter wavelengths and large number of openings, which is more visible in Fig. 14.

Blue photons generated as electron-hole pairs close to the surface of the silicon were collected better laterally at close surroundings of the photodiode area. By adding circular openings these lateral collection areas were increased, which leads to a better QE response at shorter wavelengths. However, deep-penetrating photon collection probability did not increase as much as that of surface photons, giving less improvement in longer wavelengths.

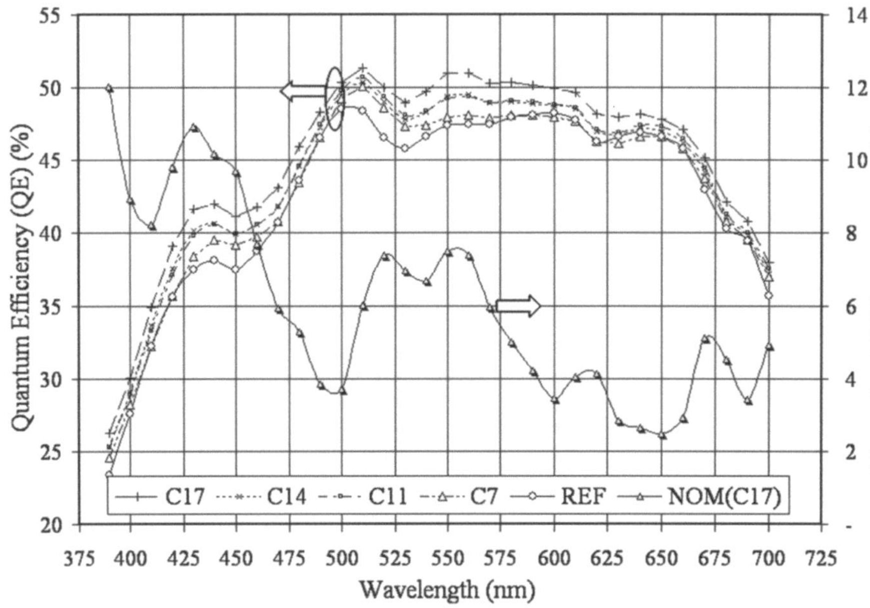

Fig. 13. Quantum efficiency of the test and reference pixels.

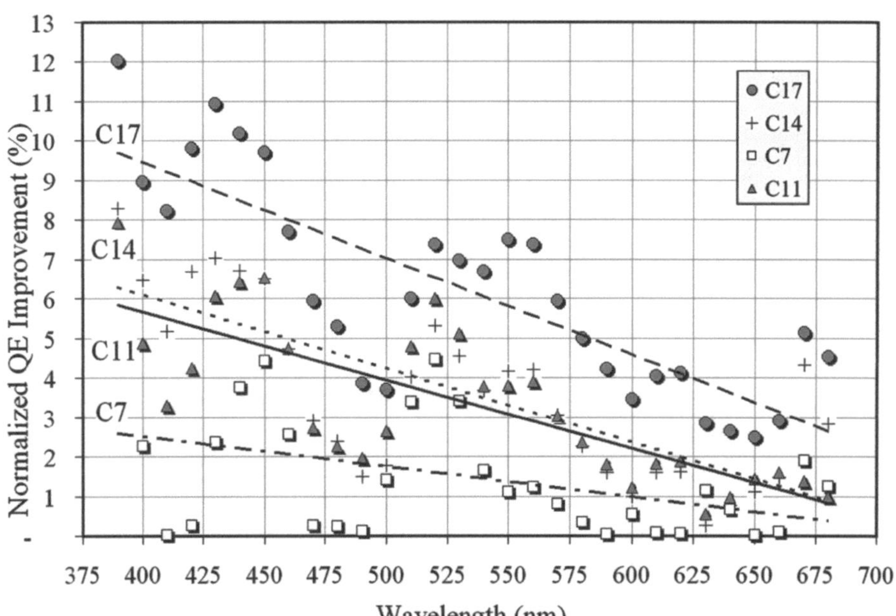

Fig. 14. Quantum efficiency improvement trends of test pixels.

6.4. Dark current

Measured dark current for the reference pixel was 10.63 mV/s at room temperature, or 3155 e-/s with the measured conversion gain of 3.37μV per electron. More dark current was observed from the test pixels with longer peripherals than the reference pixel, as shown in Fig. 15. Dark current, in terms of electrons per second, increases by approximately one-third of the reference dark electrons when the photodiode peripheral doubles (assuming the surface dark current effect was neglected).

In reality, measured dark current has two components, surface dark current and stress-center- related dark current. Surface dark current is related to the area of the photodiode, while stress-center-based dark current is related to the peripheral region. Opening a hole on a photodiode region reduces the surface contribution and increases the peripheral contribution on the total dark current. It is possible to determine the contribution of these two components of the dark current by designing fixed-area and varying-peripheral test pixels.

Dark current also increases noise floor effectively working against the gain achieved by PPMU method. In current design this contribution was not observed because the readout noise was larger than the dark current shot noise in low light condition. Measured dynamic range was around 66dB due to the higher readout channel noise. Dark current electrons add up on pixel capacity, yet, their contribution is less that 0.5% of the full well in worst case.

6.5 Sensitivity

Sensitivity of the test and reference pixels were measured with a very sharp green (550nm ±20nm) bandpass filter at 175ms integration time. Measurement results are shown in Figure 13. Sensitivity's correlation with pixel capacity was extracted by fixing the light wavelength, pixel fill factor, and integration time. It was observed that the higher the pixel capacity, the lower the sensitivity was, for an inverse correlation. A 20 percent increase in pixel capacity causes a 17 percent decrease in pixel sensitivity between reference and test pixels with 17 openings, as shown in Fig. 16.

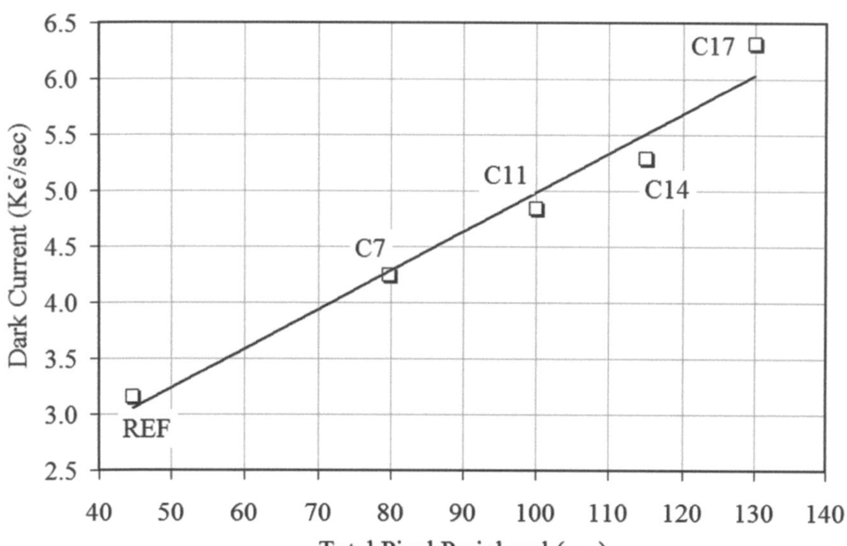

Fig. 15. Measured dark current rates of reference and test pixels.

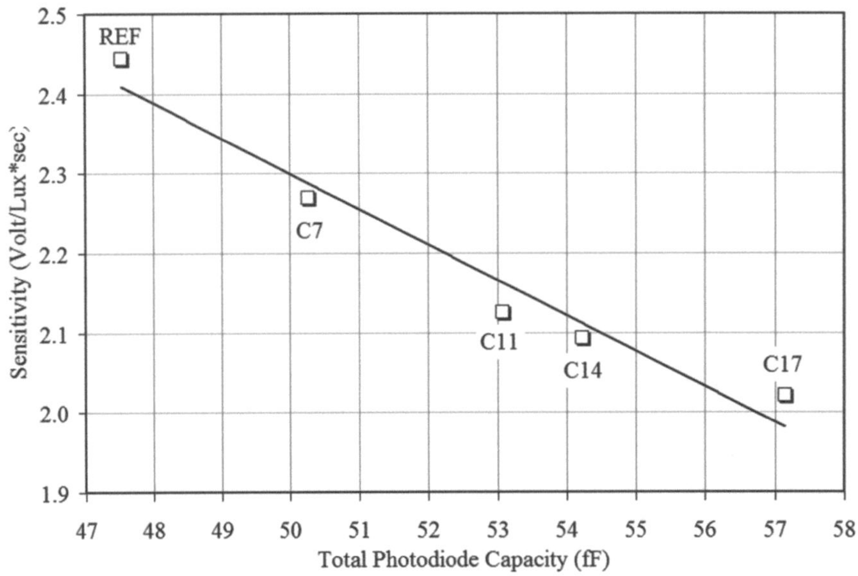

Fig. 16. Measured sensitivity of the reference and test pixels.

7. Conclusion

Photodiode-type CMOS APS pixels' quantum efficiency was improved by opening number of circular holes on the photodiode diffusion area of a prototype imager. A method called photodiode peripheral utilization method (PPUM) was developed to accommodate pixel performance improvement in a fixed size pixel. Utilizing PPUM, four test pixels with 7, 11, 14, and 17 circular openings, and a reference pixel (REF), were designed, fabricated, and tested in a prototype APS imager made with a 0.5μm, 5V, 2P3M CMOS process. Measured pixel characteristics are summarized in Table 3.

Pixel Design	C17	C14	C11	C7	REF	Unit
PD Area	107.5	113.5	119.6	127.6	141.7	μm^2
PD Peripheral	130.1	115.0	99.9	79.8	44.6	μm
Dark Current	6.31	5.29	4.84	4.25	3.15	Ke-/sec.
Conversion Gain	2.80	2.95	3.02	3.19	3.37	μVolt/e-
Quantum Efficieny	26.22	25.34	25.26	24.58	23.41	% @ 390nm
	51.00	49.41	49.24	47.97	47.44	% @ 550nm
	51.31	50.35	50.72	50.05	48.55	% @ peak
QE improvement	12.0	8.3	7.9	5.0	0.0	$\%QE_{REF}$@390nm
	7.5	4.2	3.8	1.1	0.0	$\%QE_{REF}$@550nm
Sensitivity	2.02	2.09	2.13	2.27	2.44	Volt/Lux.sec
Pixel Full-Well Depth	621.4	583.1	577.6	545.2	508.7	Ke-
Pixel Full-Well Improvement	22.1	14.6	13.5	7.2	0.0	$\%FW_{REF}$
Pixel Capacitance	57.1	54.2	53.1	50.3	47.5	fF

Table. 3. Key measured pixel parameters and improvements

8. References

Agranov, G., Gilton, T., Mauritzson, R., Boettiger, U., Altice, P., Shah, J., Ladd, J., Fan, X., Brady, F., McKee, J., Hong, C., Li, X., Patrick, I. (2005). Optical-electrical characteristics of small, sub-4µm and sub-3µm pixels for modern CMOS Image Sensors, *Proceedings of IEEE Workshop on CCDs and Advanced Image Sensors*, R-31, Tokyo, Japan, June 2005.

Ay, S. U., Lesser, M., Fossum, E. R., (2002). CMOS Active Pixel Sensor (APS) Image sensor for Scientific Applications, *in Survey and Other Telescope Technologies and Discoveries, Proceedings of SPIE*, vol. 4836, pp. 271-278, 2002, ISBN: 9780819446152.

Ay, S. U. (2004). *Design Issues and Performance of Large Format Scientific CMOS Image Sensors*, University of Southern california, Ph.D. Thesis, OCLC:65169363.

Ay, S. U. (2008). Photodiode peripheral utilization effect on CMOS APS pixel performance, *IEEE Transactions on Circuit and Systems I*, vol.55, no.6, pp.1405-1411, ISSN 1549-8328.

Cho, K.-B., Krymsky, A. I., Fossum, E. R. (2000). A 1.2 V Micropower CMOS Active Pixel Image Sensor for Portable Applications, *Digest of Technical Papers IEEE International Solid-State Circuits Conference (ISSCC)*, pp.114-115, ISSN 78035853-8, San Francisco, CA, 7-9 February 2000.

Eid, E.-S., Spagnuolo, R., Chan, T. Y., Fossum, E. R., Tsai, R. H., Deily, J., Byers Jr., W. B.; Peden, J. C., (2001). Design and characterization of ionizing radiation-tolerant CMOS APS image sensors up to 30 Mrd (Si) total dose, *IEEE Transactions on Nuclear Science*, vol. 48, no. 6, Dec. 2001, pp. 1796-1806, ISSN 0018-9499

Fossum, E. R., (1999). *Quantum Efficiency Improvements in Active Pixel Sensors*, US Patent No. 6,005,619, 1999.

Fossum, E. R. (1993). Active Pixel Sensors: Are CCD's Dinosaurs, *in Charge-Coupled Devices and Solid State Optical Sensors III, Proceedings of SPIE*, vol. 1990, Feb. 1993, pp. 2-14, ISBN: 9780819411334.

Ghazi, A., Zimmermann, H., Seegebrecht, P. (2000). CMOS Photodiode With Enhanced Responsivity for the UV/Blue Spectral Range, *IEEE Transaction on Electron Devices*, vol. 49, no. 7, July 2000, pp.1124-1128, ISSN: 0018-9383.

Gonzo, L., Stopa, D., Simoni, A., and Dalla Betta, G. (2002). Novel CMOS Image sensor with a 132-db dynamic range, *IEEE Journal on Solid State Circuits*, vol. 37, no. 12, December 2002, pp. 1846–1852, ISSN 0018-9200.

ITRS (1999). *The International Technology Roadmap for Semiconductors*, Semiconductor Industry Association, San Jose, CA, 1999.

Krymsky, A. I., VanBlerkom, D., Andersson, A., Bock, N., Mansoorian, B., Fossum, E. R. (1999). A high speed, 500 frames/s, 1024×1024 CMOS active pixel sensor, *Digest of Technical Papers 1999 Symposium on VLSI Circuits*, pp. 137-138, ISBN: 4930813956, Kyoto , Japan, 17-19 June 1999.

Krymski, A. I., Khaliullin, N., Rhodes. H. (2003). A 2e- Noise 1.3 Megapixel CMOS Sensor, *Proceedings of IEEE Workshop on CCDs and Advanced Image Sensors*, pp13-18, Elmau, Bavaria, Germany, May 15-17, 2003.

Lee, J. S., Hornsey, R.I, (2001). CMOS Photodiodes with Substrate Openings for Higher Conversion Gain in Active Pixel Sensors, *Proceedings of IEEE Workshop on CCDs and Advanced Image Sensors*, pp. 173-175, Crystal Bay, Nevada, June 2001.

Lee, J. S., Hornsey, R. I., Renshaw, D. (2003). Analysis of CMOS Photodiodes—Part II: Lateral Photoresponse, *IEEE Transaction on Electron Devices*, vol. 50, no. 5, May 2003, pp. 1239-1245, ISSN:00189383.

Mendis, S., Kemeny, S. E., Gee, R. C., Pain, B., Kim, Q., Fossum, E. R., (1997). CMOS active pixel image sensors for highly integrated imaging systems, *IEEE Journal of Solid-State Circuits*, vol. 32, no. 2, February 1997, pp. 187-197, ISSN: 00189200.

Meynants, G., Dupont, B., Witvrouwen, N., Wolfs, B., Schippers, G., Maher, K., Dierickx, B., Lee, B., Arnzen, D., Lee S. (2005). A 9 Megapixel APS-size CMOS image sensor for digital still photography, *IEEE Workshop on CCDs and Advanced Image Sensors*, R-40, Tokyo, Japan, June 2005

Nixon, R. H., Kemeny, S. E., Pain, B., Staller, C. O., Fossum, E. R., (1996). 256x256 CMOS active pixel sensor camera-on-a-chip, *IEEE Journal of Solid-State Circuits*, vol. 31, no. 12, December 1996, pp. 2046-2050, ISSN:00189200.

Packan, P.A., (2000). Scaling Transistors into the Deep-Submicron Regime, *MRS Bulletin*, Volume 25, No. 6, p. 18, June 2000.

Theuwissen, A.J.P. (1995). *Solid-State Imaging with Charge-Coupled Devices*, Kluwer Academic, ISBN:0792334566, Boston.

8

Color-Selective CMOS Photodiodes Based on Junction Structures and Process Recipes

Oscal T.-C. Chen and Wei-Jean Liu
National Chung Cheng University
Taiwan

1. Introduction

In the manufacture process of image sensors, it is commonly used to fabricate photodiodes with sensing different colors such as red, green and blue by means of color filters. Since each photodiode only senses a specific color, the back-end image processing mechanism is employed to restore the original color image. However, existence of color filters not only degrades photo-responses of photodiodes but it also makes the fabrication process complicated, thus increasing the fabrication cost.

The image sensor can be partitioned into two parts of the front-end photodiode array and the back-end signal processing circuit where its architecture can be depicted in Fig. 1. Each photodiode is connected to an amplifier that transfers the captured image signal into an electrical signal. Additionally, an overall photodiode array is established by using a pattern with red, green and blue photodiodes as shown in Fig. 1 according to the human perceptual principle [1], [2]. As for the signal processing part, it comprises a decoder, timing-control unit, compensating and synthesizing circuit and so on. The conventional color photodiode is displayed in Fig. 2, which is built by a layer of the light-filtration material as well as the general standard process, to yield the desired color. Such a fabrication method makes a photodiode having a large response toward a specific wavelength and lessening responses of the unneeded wavelengths. Let's take the commonly-used red, green and blue filters as an example. These three color photodiodes exhibit the photo-responses shown in Fig. 3. It is apparent that peak wavelengths of spectral responses appear in regions of red (670nm), green (550nm) and blue (470nm). However, such a fabrication method has the following drawbacks.

1. The fabrication process requires the extra steps in addition to the original standard process. That is to say, the extra several masks are required when color filters are added on top of the commonly-used standard fabrication process, which increase the manufacture cost.
2. Most of incident light is absorbed and reflected during passing through the color filter that decreases the photo-response of a photodiode. Accordingly, one extra process of micro lens is included for the commonly-used color filter to focus incident light with the purpose of increasing the induced current.
3. In order to meet the requirement of sensing different colors, the process must provide different color filters that increase the degree of difficulty on fabrication process. To integrate with micro lens, the fabrication process should be modified in extra when

there is a larger degree of alteration for the sensing area of a photodiode that yields a better response under a specific curvature radius of the lens.

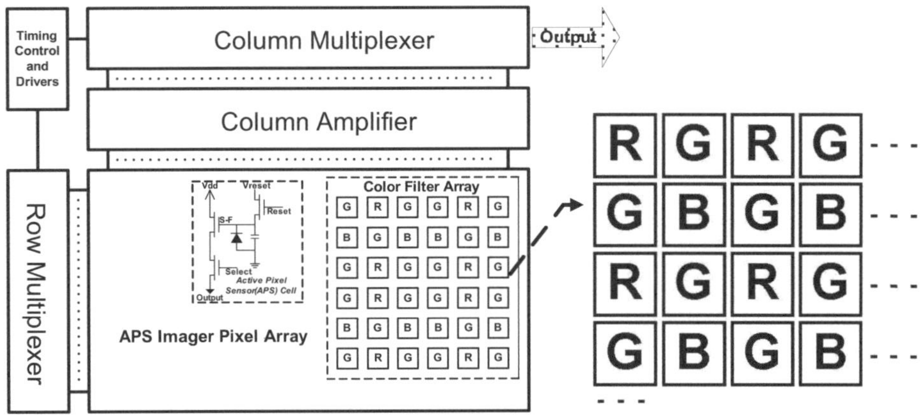

Fig. 1. Architecture of a CMOS image sensor with a color pixel array arranged in the form of the Bayer's pattern [1].

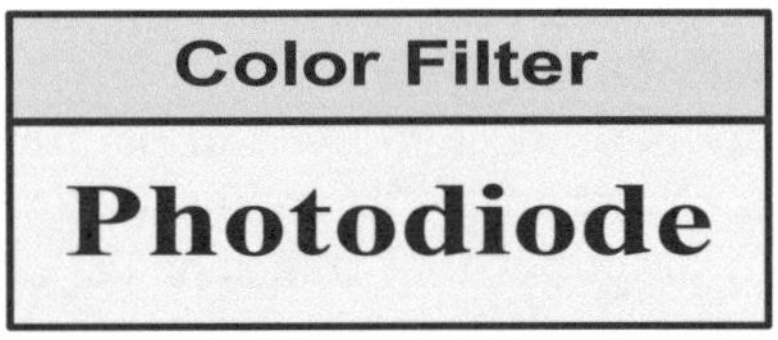

Fig. 2. Simplified structure of the conventional photodiode with a color filter.

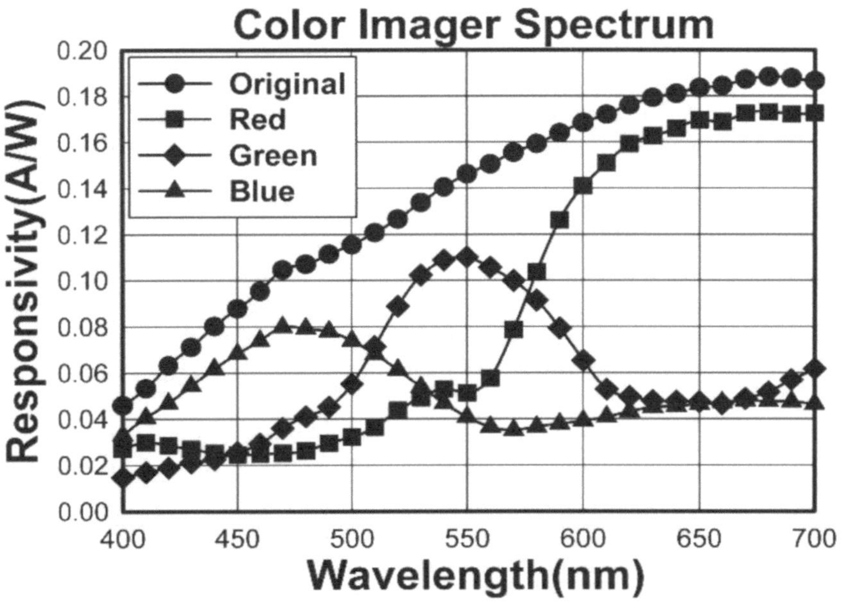

Fig. 3. Commonly-used red, green and blue filters and their corresponding spectral responses.

Despite of the abovementioned three drawbacks, the color filter should be designed by taking into consideration photo-sensing characteristics of the material itself because each photo-sensing material has non-uniform responses toward different wavelengths of incident light. On the other hand, the back-end color compensation circuit is developed with considering characteristics of color filters, mainly the transmission ratio of incidence light, which makes the overall design highly complicated. From the abovementioned issues, it is understood that there still are many drawbacks in the conventional fabrication process which needs to be improved.

To overcome these disadvantages, Chang *et al.* proposed a color-sensing phototransistor with a hetero-junction structure based on the hydrogenated amorphous silicon (a-Si:H) [3]. By controlling the voltage drop from 13V to 1V between a collector and an emitter in a phototransistor, the peak wavelength varies from 610nm to 420nm. Tasi *et al.* proposed a structure employing a change in the external bias polarity of the collector and emitter of a phototransistor. The peak wavelength varies from 600nm to 450nm under a biased voltage from -2V to 2V [4]. Since a-Si:H provides advantages of high photosensitivity in a visible light region and low-cost fabrication, it is extensively employed in the bias-controlled photo-sensing devices [3]-[10]. However, a-Si:H suffers from the Staebler-Wronski effect which results in light-induced instability, and it is not suited to the application of circuit implementation [11]. To overcome this drawback, some researchers proposed the technology of the Thin Film on ASIC (TFA) to combine a-Si:H and crystalline silicon, which are utilized in photo-sensing and electric devices, respectively [12], [13]. This technology successfully provides both advantages of high sensitivity and high integration from a-Si:H and CMOS process, respectively. Moreover, such a technology also successfully replaces color filters by controlling biased voltages.

Recently, some researchers investigated color CMOS image sensors using the standard CMOS process without color filters [14]-[18]. For example, Findlater *et al.* employed a double junction photodiode to approach a color CMOS image sensor [15]. Particularly we also developed a CMOS photodiode by shorting a p-n junction to approach multiple spectral responses in a single CMOS photodiode [16]. Moreover, color CMOS image sensors implemented by the standard CMOS process without color filters were successfully realized [14], [17], [18]. It indicates that a color CMOS image sensor can be implemented with lower cost and higher integration than the conventional ones.

In this chapter, the spectral responses of various CMOS photodiodes are explored and compared. Additionally, the proposed model is employed to investigate the reasons of spectral responses varied with different CMOS photodiodes. An epitaxial layer that can lower the crosstalk among pixels and degrade the spectral response in the infrared region is usually utilized in CMOS image sensors [19]. Moreover, the thickness and doping concentration of the epitaxial layer can be adjusted to achieve the desired spectral response for specific image sensors without color filters [20], [21]. Hence, the epitaxial affect on the spectral responses is also examined. Since the spectral response of a-Si:H photodiodes can be easily varied by controlling biased voltages, the biased voltage effect on the spectral responses of CMOS photodiodes is also investigated. Finally, the design methodology is presented to address a specific spectral response on CMOS photodiodes without color filters.

2. Structures and applications of various CMOS photodiodes

The standard CMOS process is employed to fabricate photodiodes in this work. In the current CMOS process, there exist several types of CMOS compatible photodiodes [22].

Figure 4 depicts cross sections of the process. Generally, there exists four types of CMOS photodiodes: n+/p-/p-sub, n-/p-sub, p+/n-/p-sub and p+/n- shown in Figs. 4(a) and 4(b) under the general and epitaxial wafers, respectively. To meet the demands of various applications, photodiodes in CMOS image sensors have been designed using different process recipes and junction structures. Particularly, a CMOS photodiode with an n-/p-sub or p^+/n- junction is often applied to CMOS image sensors [15], [23]-[26]. Since the spectral response of silicon covers visible light and infrared regions, n-/p-sub and p^+/n-/p-sub photodiodes can be used in commercial CMOS image sensors. Owing to the junction between n- and p-substrate deeper than that between p^+ and n-, the n-/p-sub junction is more responsive in the infrared region (>800nm) than the p^+/n- junction. Accordingly, the n-/p-sub photodiode can serve as an infrared detector as well as a visible light detector [27], [28].

A p^+/n-/p-sub photodiode with a particular recipe that enables the whole p^+ layer to be fully depleted, can function like a pinned photodiode to improve the sensitivity at short wavelengths and reduce the thermal noise at its surface [29]. Two p-n junctions, p^+/n- and n-/p-sub, of the p^+/n-/p-sub photodiode can form a pnp phototransistor, which yields a high gain through a transistor action [30]-[32]. An epitaxial layer that can lower the crosstalk among pixels is usually utilized in CMOS image sensors. Particularly, the epitaxial may degrade quantum efficiency of the CMOS photodiode with p-epi/p+sub in the infrared region [33]. Additionally, the thickness and doping concentration of the epitaxial layer can be adjusted to achieve the desired quantum efficiency for specific image sensors without color filters [34].

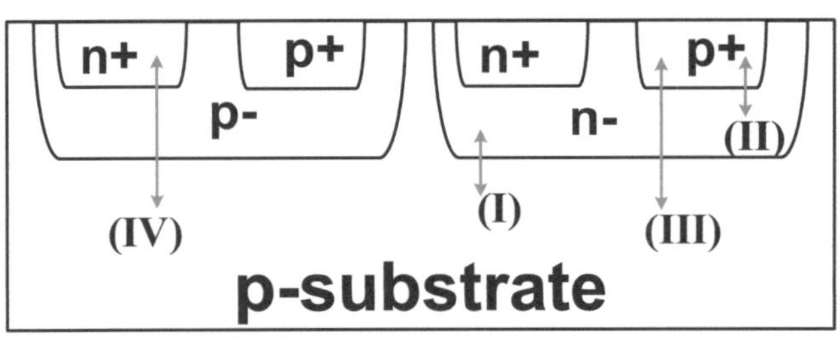

(a)

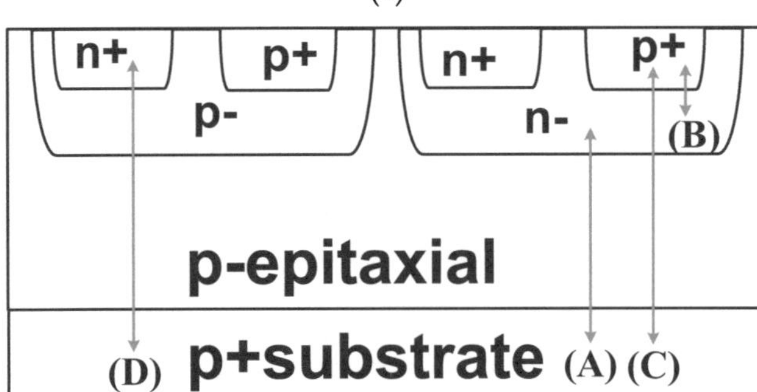

(b)

Fig. 4. Simplified structures of the standard CMOS processes with four types of photodiodes in (a) General wafer and (b) Epitaxial wafer.

3. Analyses and measured results of common-used CMOS photodiodes

There are two main factors, the structure and recipe of a photodiode and the reverse biased voltage, affecting spectral responses of CMOS photodiodes. In this section, these two factors are investigated to understand variations of spectral responses.

A. Effects from photodiode structure and recipe

CMOS photodiodes are studied in terms of the depth of the p-n junction, doping concentration and substrate type. The measured spectral responses of four CMOS photodiodes in Fig. 4(a) under a zero biased voltage are depicted in Fig. 5 to interpret the variations caused by depth of the p-n junction and doping concentration. In this figure, spectral responses of n-/p-sub and p^+/n-/p-sub are similar except for those in the short wavelength region. The reason for this phenomenon is that the excess minority carriers in the device surface, which are excited by incident light with short wavelengths, recombine rapidly owing to heavy doping in the p^+ layer. The surface recombination velocity is significantly influenced by process factors such as surface roughness, surface contamination and oxidation temperature, some of which are not easily controlled [35], [36]. Omitting reflection coefficients, Fig. 6 depicts the simulated spectral responses of the n-/p-sub under different surface recombination velocities. Since excess minority carriers in the device surface are excited by incident light with short wavelengths, and the recombination probability of these excited carriers increases with surface recombination velocity, the degraded amount of spectral response in short wavelengths is larger than that for long wavelengths.

Curve (IV) in Fig. 5 depicts that the spectral response of n^+/p- photodiode reduces significantly in the long wavelength region. The reason of this phenomenon is caused by absorption coefficient. The relationship between absorption coefficient and light wavelength for silicon material can be approximately formulated as [37]

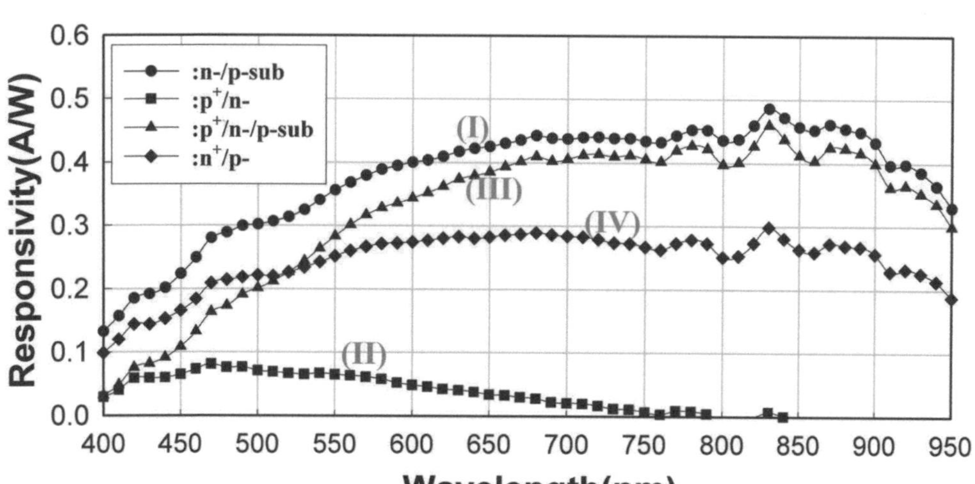

Fig. 5. Measured spectral responses of four CMOS photodiodes in Fig. 4(a) under zero biased voltage.

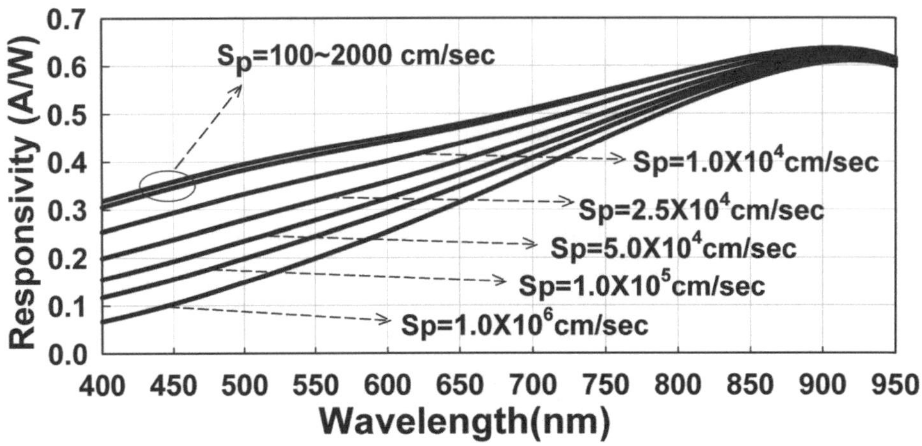

Fig. 6. Simulated spectral responses of n-/p-sub under different surface recombination velocities where Sp indicates a surface recombination velocity.

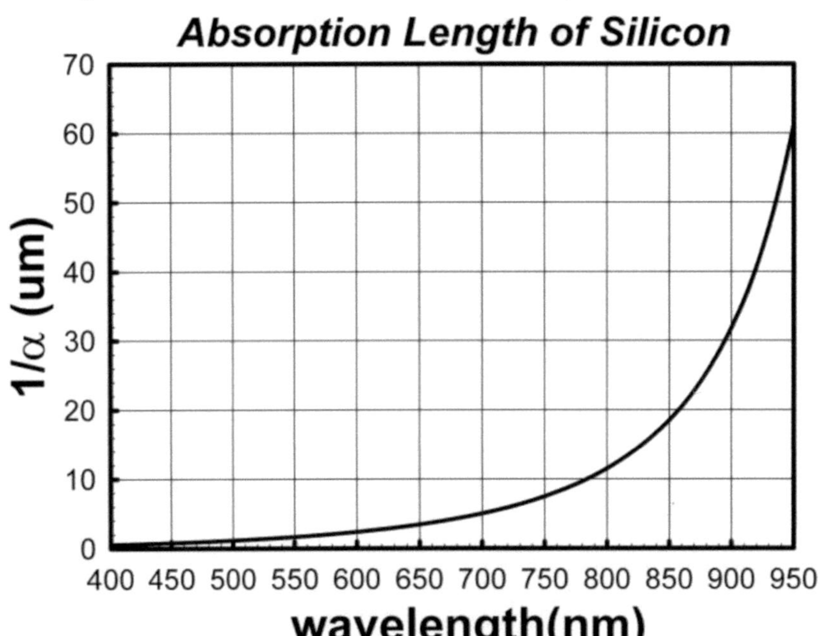

Fig. 7. Simulated result of the relationship between the incident light wavelength and absorption length.

$$\alpha = \left(\frac{84.732}{\lambda} - 76.417 \right)^2 . \tag{1}$$

The reciprocal of absorption coefficient is absorption length that indicates the length where the photon flux decays to 1/e of the previous one. Figure 7 shows the simulated result of the relationship between the incident light wavelength and absorption length. The excess minority carriers need diffuse to the boundary of the space-charge region and drift over the space-charge region to become an effective photocurrent. The p-n junction depth of n^+/p-photodiode is about several hundreds of nanometers. Most of the induced excess minority carriers excited by incident light with long wavelengths can not diffuse to the p-n junction to generate photocurrents. Hence, the spectral response of n^+/p- photodiode is small in the long wavelength region. Additionally, curve (II) in Fig. 5 is apparently different from the other three ones. In this structure, all of the induced excess minority carriers excited in the p-substrate can not generate photocurrents. The photocurrent is produced only by the excess minority carriers excited by short wavelengths in the shallow region. Hence, comparing the spectral responses of other three photodiodes, the spectral response of this one is significantly small. Additionally, the peak wavelength of the spectral response apparently shifts to the short wavelength region (460nm). In fact, since this $p^+/n-/p$-sub photodiode comprises two p-n junctions of p^+/n- and n-/p-sub. The total spectral response is formed by adding the spectral response of these two p-n junctions. The spectral response from one of two p-n junctions can be effectively acquired by shorting the other p-n junction, as shown in Fig. 8. As a p-n junction is shortened, the electron-hole pairs recombine to make the junction current become zero such that there exist two kinds of shorting connection manners in this photodiode in Figs. 8(a) and (b) to yield two photo-responses in curves (II) and (III) of Fig. 5, respectively.

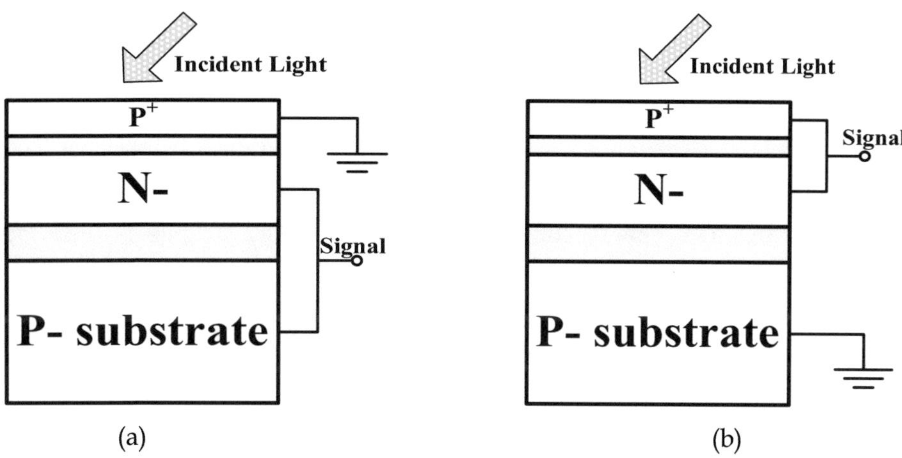

Fig. 8. Connection manners of the $p^+/n-/p$-sub photodiode for deriving different spectral responses: (a) Curve (II) and (b) Curve (III) in Fig. 5 [16].

The measurement results of the CMOS photodiodes shown in Fig. 4(b) is illustrated in Fig. 9. Comparing Figs. 5 and 9, the spectral responses of photodiodes with p-epitaxial/p+substrate are much smaller than those with p-substrate in a long wavelength region. This difference is because the diffusion length of the minority carriers in p-substrate is as high as several hundred micrometers. Accordingly, most excess minority carriers in the p-substrate region generate the induced photocurrent. For the p-epitaxial/p+substrate structure, although the diffusion length of the minority carriers in the p-epitaxial region can also be as high as several hundred micrometers, most minority carriers in this region are transferred from p-epitaxial to p+substrate via drifting or diffusion. However, as the diffusion length of minority carriers in p+substrate is only several micrometers, most minority carriers in p+substrate are rapidly recombined so that the minority carriers cannot form an induced photocurrent. Hence, Fig. 9 also depicts that spectral responses begin to decay dramatically in the infrared region. Figure 10 displays the simulations associated with the excess minority carrier densities in p-substrate and p-epitaxial/p+substrate. From this figure, the excess minority carrier density in the p-substrate region is significantly higher than that in the p-epitaxial/p+substrate region. Particularly, the thickness and doping concentration of epitaxial can be adjusted to obtain the desired spectral response for applications of digital color image sensors without color filters [20], [21].

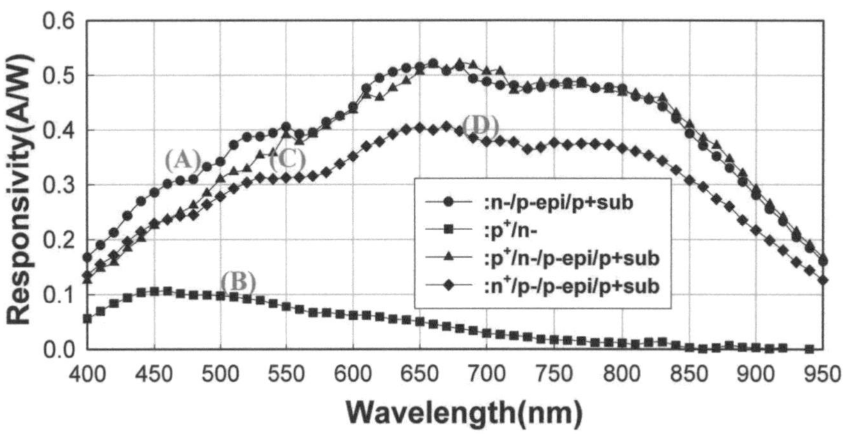

Fig. 9. Measured spectral responses of four CMOS photodiodes in Fig. 4(b) under zero biased voltage.

B. Effects from reverse biased voltages

In practice, to facilitate charge collection, a reverse biased voltage is often applied to photodiodes. The aforementioned measurement operations are conducted at a zero reverse biased voltage. In the following, the effect of the reverse biased voltage on the spectral response is investigated. The n-/p-sub, p^+/n-/p-sub and p^+/n- photodiodes in Fig. 4(a) and n-/p-epi/p-sub photodiode in Fig. 4(b) are employed to observe the influence of reverse biased voltages on spectral responses of CMOS photodiodes with different structures. Figure 11 displays the measured spectral responses of these four photodiodes at reverse biased voltages of 0V, -1V, -3V and -5V. The spectral responses does not apparently increase for n-/p-sub, p^+/n-/p-sub and n-/p-epi/p-sub photodiodes under the absolute reverse biased voltage from 0V to -5V. In the 1-D analysis, the reverse biased voltage only influences the vertical width of the space-charge region. Figure 12(a) shows the variations in the position of the space-charge region between n- and p-substrate of the n-/p-sub photodiode at reverse biased voltages from 0V to -5V. The width of the space-charge region is increased with the absolute reverse biased voltage, such that two boundaries of the space-charge region extend into n- and p-substrate. Restated, the width of the neutral-region in n- and p-substrate declines as the absolute reverse biased voltage increases. Figure 13(a) depicts the simulated spectral responses of n-, space-charge and p-substrate regions, and the total spectral responses at reverse biased voltages from 0V to -5V when the reflection coefficient is zero. The spectral response in the space-charge region increases with the width of the region, while that in n- and p-substrate falls as the effective charge collection regions decrease in these two layers. In Fig. 13(a), the increase in the spectral response of the space-

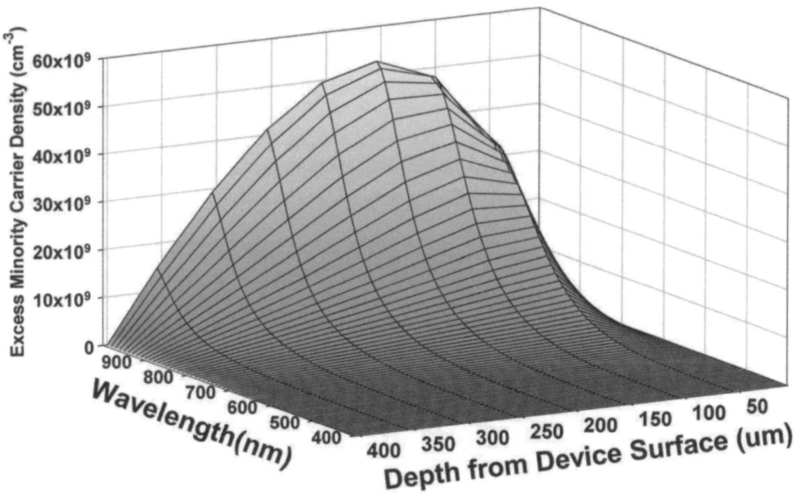

(a)

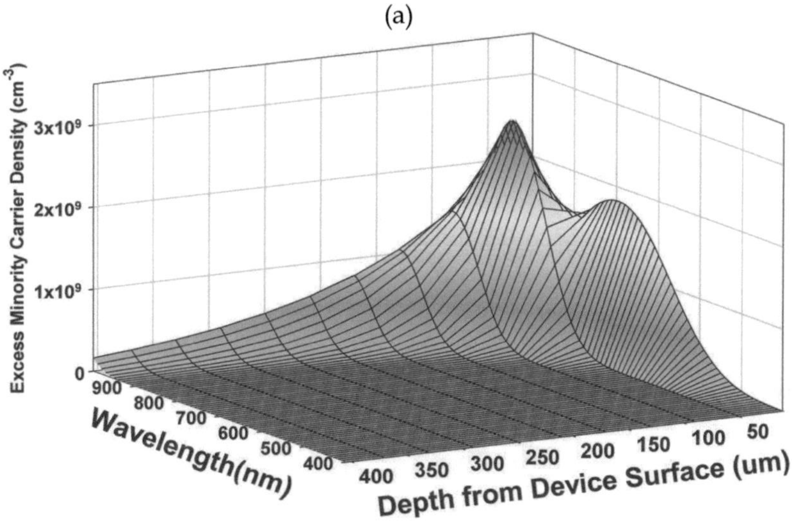

(b)

Fig. 10. Simulated excess minority carrier densities in (a) p-substrate and (b) p-epitaxial/p+substrate.

charge region is nearly compensated for the drop in the spectral response of n- and p-substrate. Hence, the spectral response of this photodiode slightly varies with the reverse biased voltage. The similar phenomenon is at p^+/n-/p-sub and n-/p-epi/p+sub photodiodes as shown in Figs. 11(b) and 11(c). However, the event of the reverse biased voltage affecting the spectral response depends on the recipe and the current density distribution in each layer of a photodiode. Figure 11(d) displays the measured spectral response of p^+/n- at reverse biased voltages of 0V, -1V, -3V and -5V. The spectral response increasing with the reverse biased voltage for this photodiode seems a bit more significantly and regularly than the other three photodiodes. Figure 12(b) depicts the variations in the position of the space-charge region of the p^+/n- photodiode at reverse biased voltages from 0V to -5V. Since the variations of the effective charge collection regions in p^+ and n- are very small, the spectral response of this photodiode varies mainly by that generated in the space-charge region. Because the spectral responses of p+ and n- are decreased rarely little, the increase of the space-charge region dominates the effect of the reverse biased voltage on the

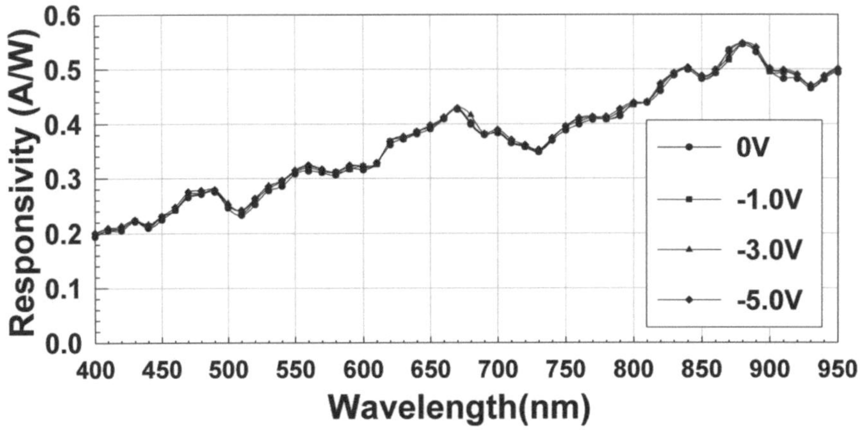

(a)

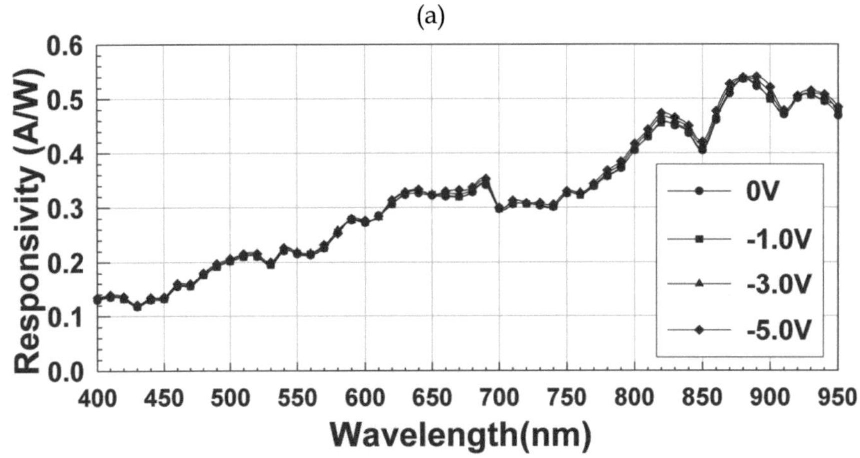

(b)

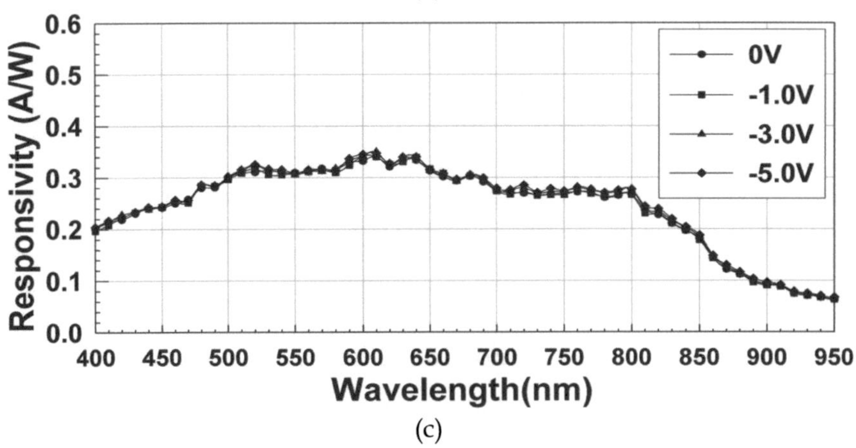

(c)

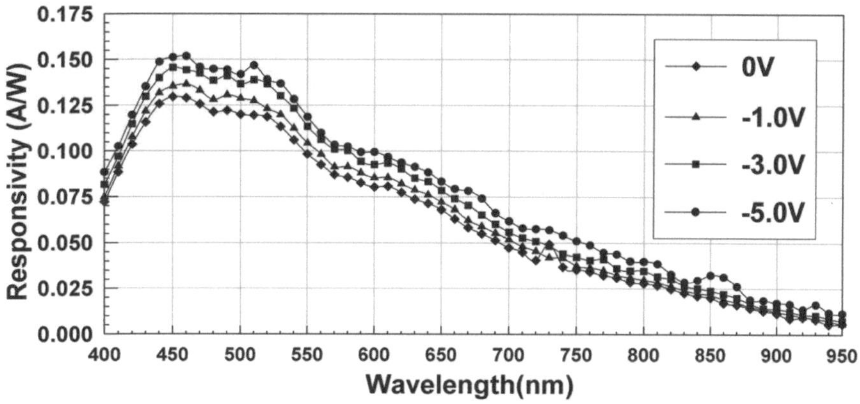

(d)

Fig. 11. Measured spectral responses of photodiodes under different reverse biased voltages in (a) n-/p-sub, (b) p⁺/n-/p-sub,(c) n-/p-epi/p+sub and (d) p⁺/n-.

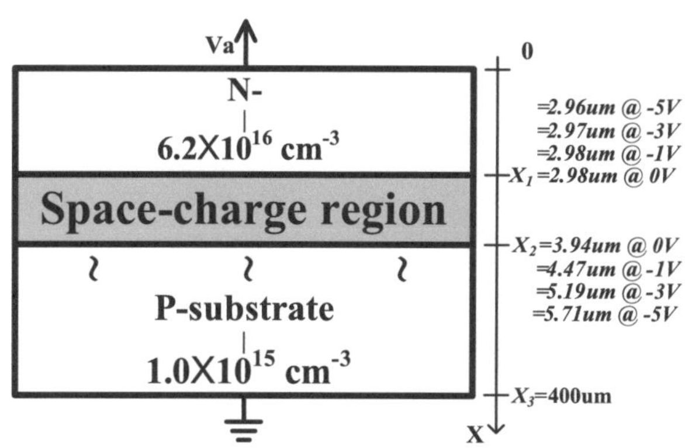

(a)

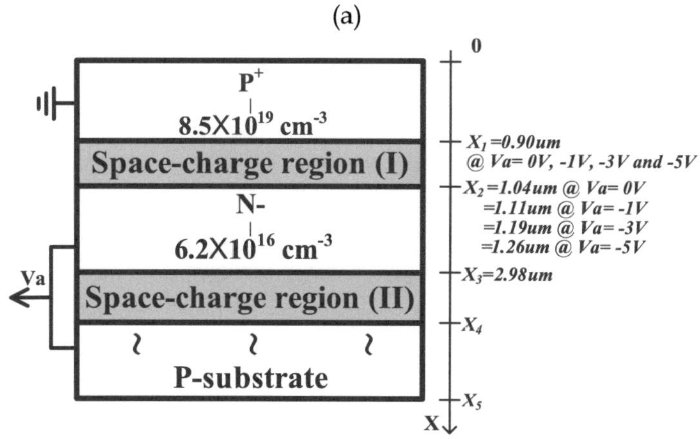

(b)

Fig. 12. Variations in positions of the space-charge regions of (a) n-/p-sub photodiode and (b) p⁺/n- photodiodes, at reverse bias voltages from 0V to -5V (the dimensions of each layer in this structure do not represent actual dimensions).

spectral response. Figure 13(b) shows the simulated spectral responses of n-, space-charge, p-substrate regions, and the total spectral responses at reverse biased voltages from 0V to -5V when the reflection coefficient is zero. The variation of the spectral response for this photodiode increases with the reverse biased voltage more significantly than those in the other three photodiodes.

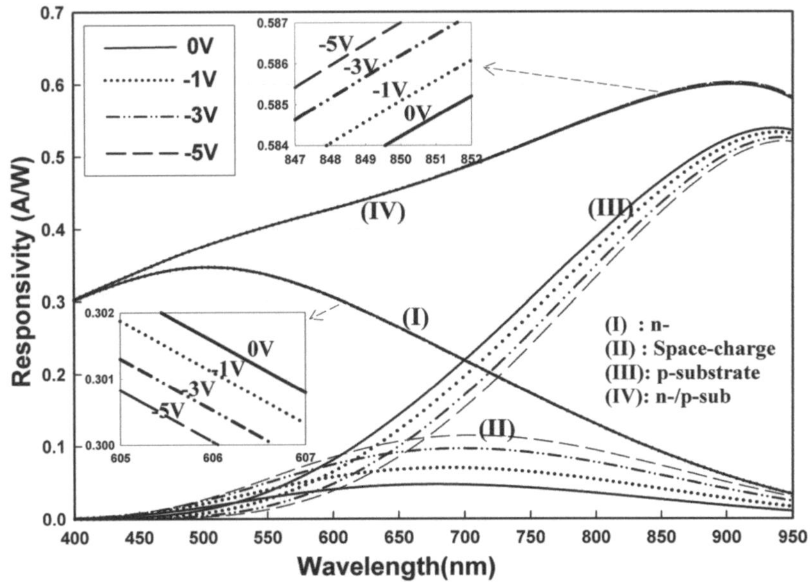

(a)

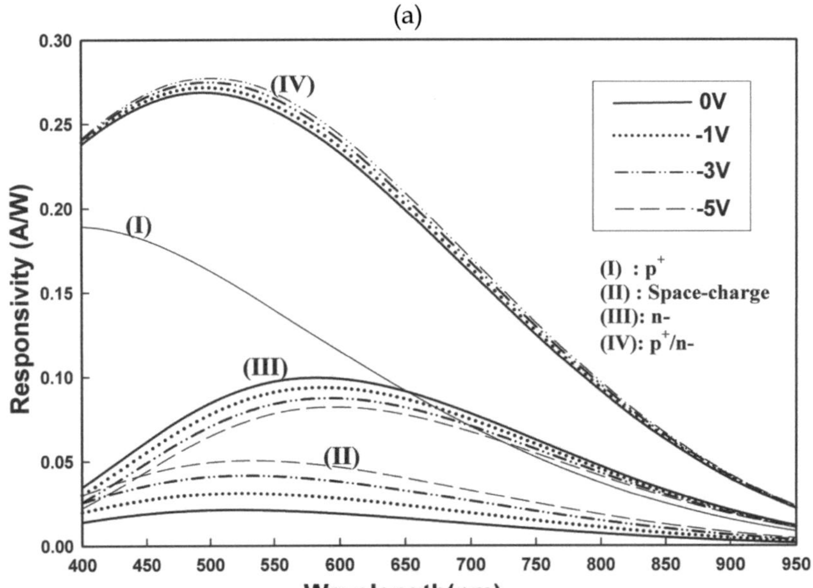

(b)

Fig. 13. Simulated spectral responses in n-type and p-type semiconductors and in space-charge region under different reverse biased voltages ranging from 0V to -5V when the reflection coefficient being zero for (a) n-/p-sub and (b) p⁺/n- photodiodes.

4. Design methodology for color CMOS pixels without color filters

As the abovementioned, we conclude that the color filter technology is still a good choice for color separation presently. In fact, some specific modifications for the semiconductor process or signal processing circuits are applied to color CMOS image sensors without color filters [15]-[17]. In this work, an equation based on the CMOS photodiode model is derived to determine the peak wavelength of the spectral response. The detail of the derivation procedure is illustrated in Appendix. Here, some solutions for obtaining different color spectral responses are briefly sketched. Additionally, the approaches to enhance the capability of separating the color spectral responses are discussed.

1. Reducing the spectral response in the long wavelength region:
   Generally, the thickness of the substrate is as thick as several hundreds of micrometers. Consequently, the spectral response is dominated by the induced photocurrent generated in the substrate region. Since the peak wavelength of the spectral response of substrate is generally located at the infrared region, the peak wavelength of the total spectral response tends to occur at the long wavelength region. There are two approaches to reduce the spectral response in the long wavelength region.
   a. The spectral responses in the long wavelength region can be effectively decreased by shortening the p-n junction in the deep region [16]. The depth of diffusion affects the photodiode to absorb wavelengths of incident light. Referring to the absorption length in Fig. 7, the light with a longer wavelength penetrates to the deeper junction so that the incident light with a longer wavelength can excite electron-hole pairs at the deep region. However, to become photocurrents, the electron-hole pairs should reach to the boundary edges of the space-charge region successfully such that they would be absorbed and transformed to the photocurrent. In other words, the photodiode has a greater response toward the incident light with a longer wavelength at a deeper region whereas for a shallower region it has a better response toward the incident light with a shorter wavelength. Additionally, to prevent CMOS circuits from latch-up, p-substrate is generally connected to the lowest potential in the system. To keep the potential of p-substrate in the lowest level and the photodiode under reverse biased voltages, a connection manner depicted in Fig. 14 is employed to solve the problem of the voltage drop between p and n nodes in the photodiode. Figure 15 shows the simulated results utilizing the recipes in Fig. 14. It clearly reveals that the peak wavelength increases with the depth of the p^+ layer.

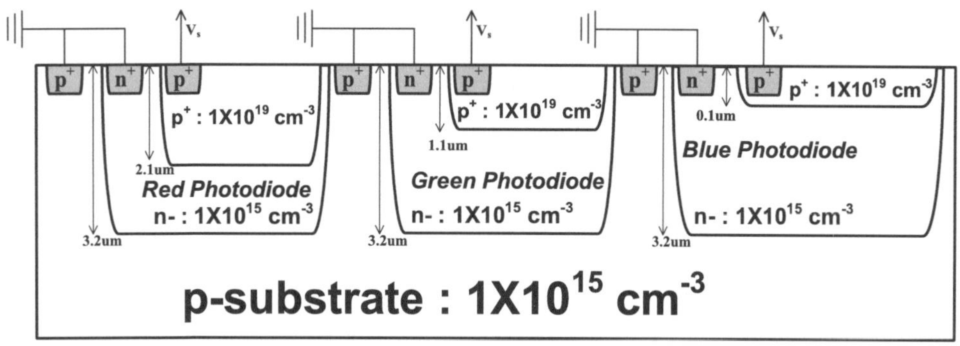

Fig. 14. Connection manner, recipes and structures obtaining three color spectral responses.

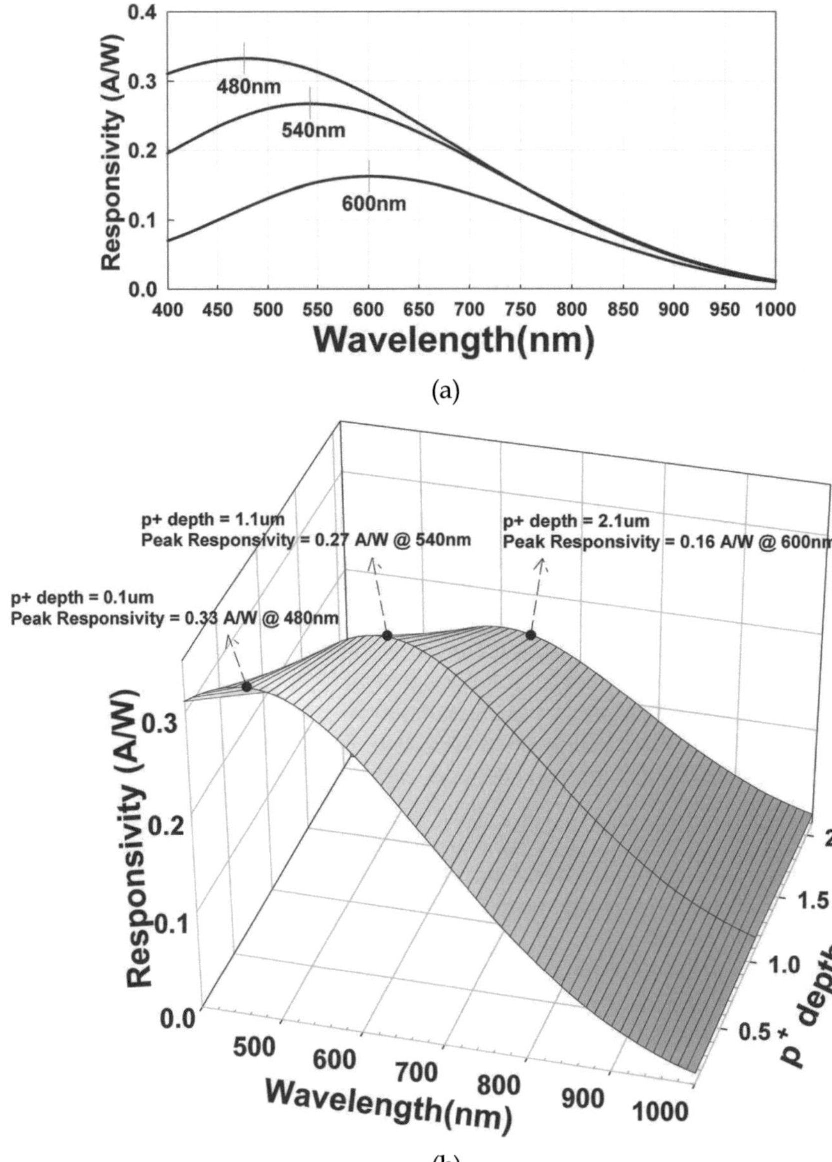

Fig. 15. Structures in Fig. 14 being simulated to yield (a) spectral responses of three recipes for red, green and blue photodiodes and (b) spectral responses of p^+ depth varying from 0.1μm to 2.1μm.

b. The spectral response in the long wavelength region can be also lowered by reducing the thickness of the substrate layer to decrease the region for collecting excess minority carriers. Figure 16 depicts the n-/p-sub photodiode with thin p-substrate of which the thickness is only several micrometers. Figure 17 displays the simulated results by utilizing the corresponding recipes in Fig. 16. It is apparent that the spectral response in the long wavelength region is decayed.

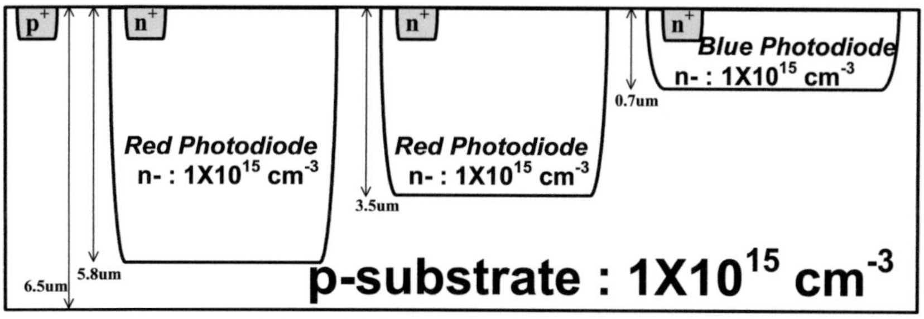

Fig. 16. Structures in Fig. 16 being simulated to yield (a) spectral responses of three recipes for red, green and blue photodiodes and (b) spectral responses of n− depth varying from 0.7μm to 5.8μm.

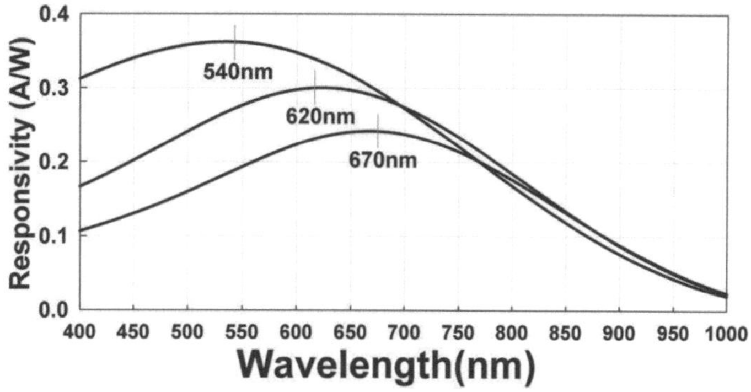

(a)

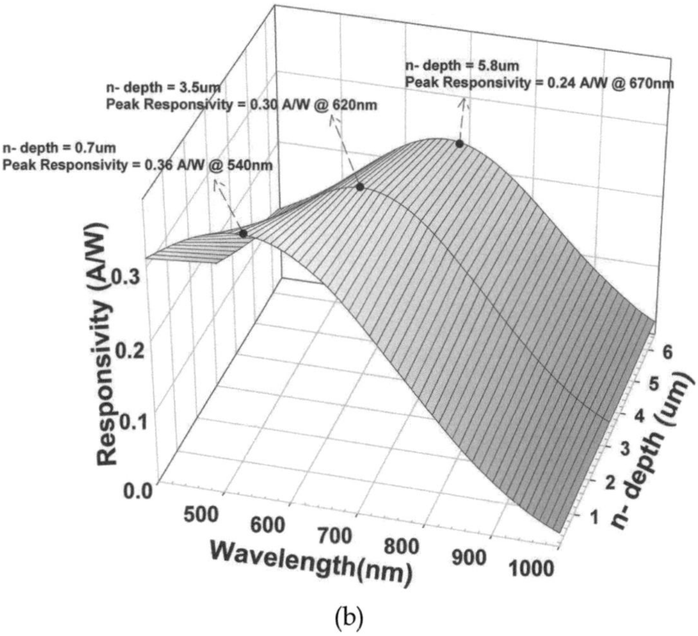

(b)

Fig. 17. Simulated results employing the structures in Fig. 16 under different recipes.

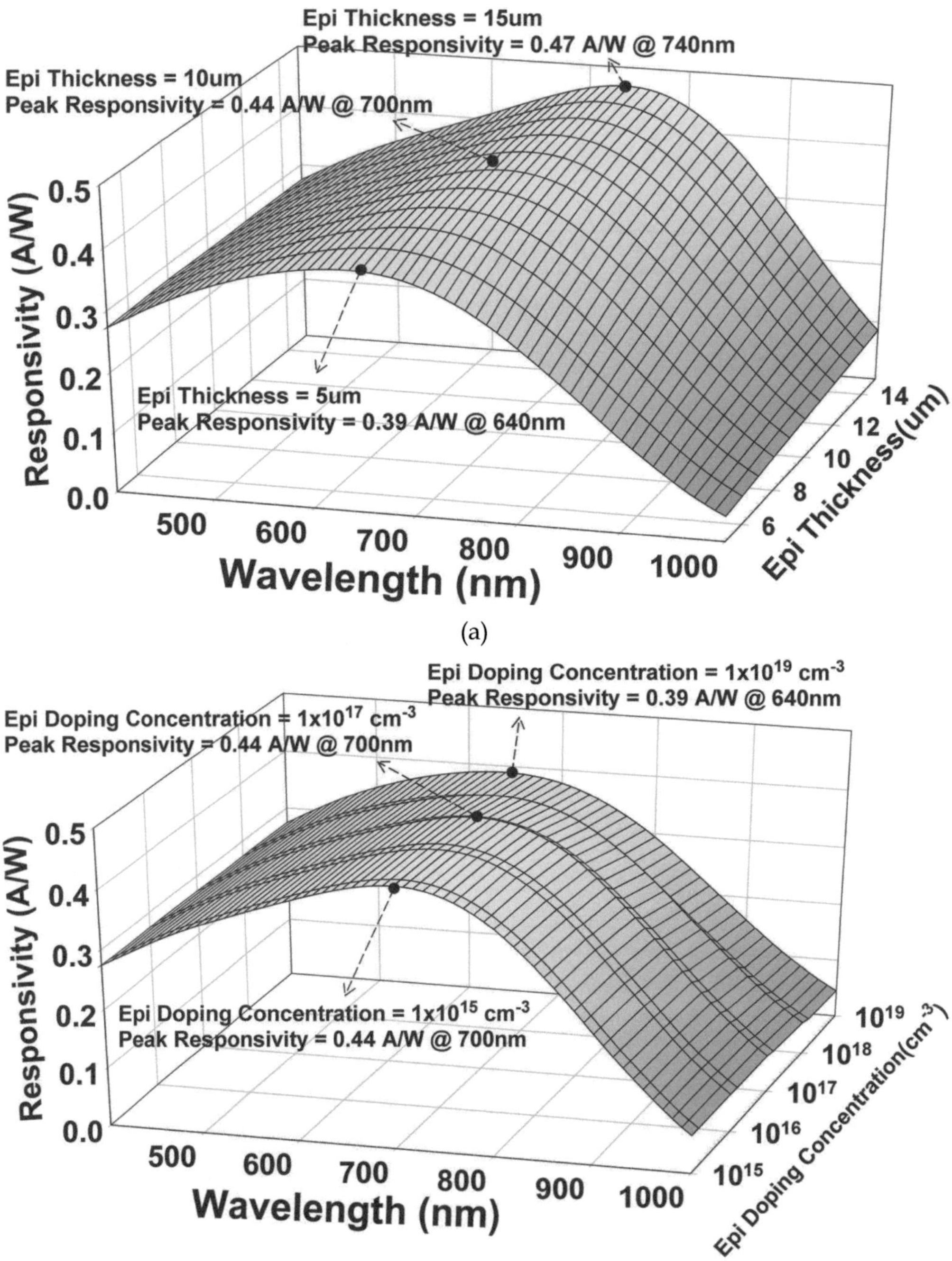

(b)

Fig. 18. Simulated spectral responses of the n-/p-epi/p+sub photodiode in (a) p-epitaxial doping concentration of 1×10^{15} cm^{-3} and p-epitaxial thickness ranging from 5 to 15 *um um* and (b) p-epitaxial doping concentration ranging from 1×10^{15} cm^{-3} to 1×10^{19} cm^{-3} and p-epitaxial thickness of 10 *um* .

2. The spectral response in the long wavelength region can be decreased by heavy doping substrate associated with the p-epitaxial layer. By adjusting the depth of the epitaxial layer, the desired spectral response can be obtained. Figure 18 depicts the simulated spectral responses of the n-/p-epi/p+sub photodiode under different thicknesses and doping concentrations of the epitaxial layer. According to this figure, the thickness and doping concentration of the epitaxial layer apparently affect spectral responses. In practice, some researchers proposed the approach of selective epitaxial growth to obtain various color spectral responses by changing the recipe of the epitaxial layer [20], [21].

5. Conclusion

Adaptive photodiode structures, of which design approach aiming at making the photo-response having a peak value at a specific wavelength, that are realized by the photodiodes with color-selective mechanisms under the condition of without extra color filters is proposed. Moreover, the influences of color filters, photodiode structures, recipes and reverse biased voltages on spectral responses are investigated. Measurement results illustrate that the color filters affect the spectral responses more significantly than the others. The spectral response varies with the reverse biased voltages slightly. The approach of implementing color pixels using the standard CMOS process without color filters is also proposed. This work clearly paves the way for designers to realize color-selective pixels in CMOS image sensors.

Appendix: Derivation for peak wavelength of the spectral response

The n-/p-sub photodiode as shown in Fig. A.1 is employed to illustrate how the proposed model is used to derive the peak wavelength of the spectral response.

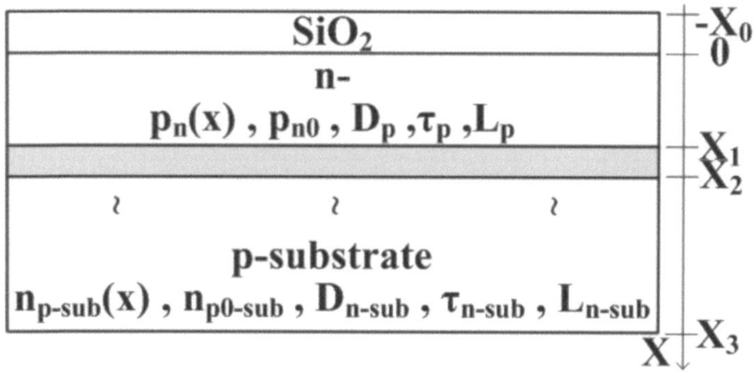

Fig. A.1 n-/p-sub photodiode.

The total current density generated by the n-/p-sub photodiode is

$$J_{total} = J_{n-photo} + J_{p-photo} + J_{drift}$$

$$= qD_p\left(dp_{n-}(x)_{photo}/dx\right)\Big|_{x=x_1} + qD_{n-sub}\left(dn_{p-sub}(x)_{photo}/dx\right)\Big|_{x=x_2} + q\int_{x_1}^{x_2} G_x dx$$

$$= \frac{qD_p\tau_pG_0}{L_p^2\alpha^2-1}\left(\frac{\left(\frac{e^{-\alpha x_1}}{L_p}\left(\frac{D_p}{L_p}sh\left(\frac{x_1}{L_p}\right)+S_pch\left(\frac{x_1}{L_p}\right)\right)-\frac{\left(\alpha D_p+S_p\right)}{L_p}\right)}{\left(S_psh\left(\frac{x_1}{L_p}\right)+\frac{D_p}{L_p}ch\left(\frac{x_1}{L_p}\right)\right)}+\alpha e^{-\alpha x_1}\right) \quad .(A1)$$

$$+\frac{qL_{n-sub}G_0e^{-\alpha(2x_2+x_3)}\left(e^{\alpha(x_2+x_3)}\left(\alpha L_{n-sub}+Coth\left(\frac{x_2-x_3}{L_{n-sub}}\right)\right)-e^{2\alpha x_2}Csch\left(\frac{x_2-x_3}{L_{n-sub}}\right)\right)}{\left(L_{n-sub}^2\alpha^2-1\right)}$$

$$+q\phi_0\left(e^{-\alpha x_2}-e^{-\alpha x_1}\right)$$

The absorption coefficient α can be simplifily represented as a function of the incident light wavelength, i.e. $\alpha = f(\lambda)$, and then Eq. (A1) can be modified to

$$J_{total} = \frac{qD_p\tau_pG_0}{L_p^2f(\lambda)^2-1}\left(\frac{\left(\frac{e^{-f(\lambda)x_1}}{L_p}\left(\frac{D_p}{L_p}sh\left(\frac{x_1}{L_p}\right)+S_pch\left(\frac{x_1}{L_p}\right)\right)-\frac{\left(f(\lambda)D_p+S_p\right)}{L_p}\right)}{\left(S_psh\left(\frac{x_1}{L_p}\right)+\frac{D_p}{L_p}ch\left(\frac{x_1}{L_p}\right)\right)}+f(\lambda)e^{-f(\lambda)x_1}\right)$$

$$+\frac{qL_{n-sub}G_0\left(e^{-x_2f(\lambda)}\left(f(\lambda)L_{n-sub}+Coth\left(\frac{x_2-x_3}{L_{n-sub}}\right)\right)-e^{-x_3f(\lambda)}Csch\left(\frac{x_2-x_3}{L_{n-sub}}\right)\right)}{\left(L_{n-sub}^2f(\lambda)^2-1\right)} \quad . (A2)$$

$$+q\phi_0\left(e^{-f(\lambda)x_2}-e^{-f(\lambda)x_1}\right)$$

In Eq. (A2), the surface generation rate G_0 is

$$G_0 = \frac{\alpha P_{in}\lambda}{Ahc} = \frac{f(\lambda)P_{in}\lambda}{Ahc}. \quad (A3)$$

Additionally, A and P_{in} in Eq. (A3) represent the unit area and unit incident light power, respectively. Hence, Eq. (A2) can be represented as follows.

$$J_{total} = \frac{qD_p\tau_p f(\lambda)\lambda}{hc\left(L_p^2 f(\lambda)^2-1\right)}\left(\frac{\left(\frac{e^{-f(\lambda)x_1}}{L_p}\left(\frac{D_p}{L_p}sh\left(\frac{x_1}{L_p}\right)+S_p ch\left(\frac{x_1}{L_p}\right)\right)-\frac{\left(f(\lambda)D_p+S_p\right)}{L_p}\right)}{\left(S_p sh\left(\frac{x_1}{L_p}\right)+\frac{D_p}{L_p}ch\left(\frac{x_1}{L_p}\right)\right)}+f(\lambda)e^{-f(\lambda)x_1}\right)$$
$$+\frac{qL_{n-sub}f(\lambda)\lambda}{hc\left(L_{n-sub}^2 f(\lambda)^2-1\right)}\left(e^{-x_2 f(\lambda)}\left(f(\lambda)L_{n-sub}+Coth\left(\frac{x_2-x_3}{L_{n-sub}}\right)\right)-e^{-x_3 f(\lambda)}Csch\left(\frac{x_2-x_3}{L_{n-sub}}\right)\right) \quad \text{(A4)}$$
$$+q\phi_0\left(e^{-f(\lambda)x_2}-e^{-f(\lambda)x_1}\right)$$

The peak wavelength of the spectral response can be obtained by taking partial differential of Eq. (A4) by the variable of λ.

$$\frac{\partial J_{total}}{\partial \lambda} =$$
$$=\frac{qL_p}{hc}\left(\begin{aligned}&\frac{e^{-f(\lambda)x_1}L_p\left(f(\lambda)\left(f(\lambda)^2L_p^2-1\right)\left(f(\lambda)D_p+S_p\right)-\lambda f'(\lambda)\left(2f(\lambda)D_p+S_p+f(\lambda)^2L_p^2S_p\right)\right)}{\left(L_p^2f(\lambda)^2-1\right)^2\left(D_pCosh\left(\frac{x_1}{L_p}\right)+L_pS_pSinh\left(\frac{x_1}{L_p}\right)\right)}\\&+\frac{L_pCosh\left(\frac{x_1}{L_p}\right)\left(-f(\lambda)\left(f(\lambda)^2L_p^2-1\right)\left(f(\lambda)D_p+S_p\right)+\lambda f'(\lambda)\left(\left(f(\lambda)D_p\right)\left(2+f(\lambda)x_1\right)\left(f(\lambda)^2L_p^2-1\right)+S_p\left(1-f(\lambda)x_1+f(\lambda)^2\left(1+f(\lambda)x_1\right)\right)\right)\right)}{\left(L_p^2f(\lambda)^2-1\right)^2\left(D_pCosh\left(\frac{x_1}{L_p}\right)+L_pS_pSinh\left(\frac{x_1}{L_p}\right)\right)}\\&+\frac{Sinh\left(\frac{x_1}{L_p}\right)\left(-f(\lambda)\left(f(\lambda)^2L_p^2-1\right)\left(D_p+f(\lambda)L_p^2S_p\right)+\lambda f'(\lambda)\left(D_p\left(1-f(\lambda)x_1+f(\lambda)^2L_p^2\left(1+f(\lambda)x_1\right)\right)+f(\lambda)L_p^2S_p\left(2+f(\lambda)x_1\left(f(\lambda)^2L_p^2-1\right)\right)\right)\right)}{\left(L_p^2f(\lambda)^2-1\right)^2\left(D_pCosh\left(\frac{x_1}{L_p}\right)+L_pS_pSinh\left(\frac{x_1}{L_p}\right)\right)}\end{aligned}\right)$$
$$+\frac{qL_{n-sub}}{hc}\left(\begin{aligned}&\frac{L_{n-sub}f(\lambda)^2e^{-f(\lambda)x_2}+f(\lambda)e^{-f(\lambda)x_2}Coth\left(\frac{x_2-x_3}{L_{n-sub}}\right)-f(\lambda)e^{-f(\lambda)x_3}Csch\left(\frac{x_2-x_3}{L_{n-sub}}\right)}{\left(L_{n-sub}^2f(\lambda)^2-1\right)}\\&+\frac{L_{n-sub}\lambda f(\lambda)f'(\lambda)e^{-f(\lambda)x_2}+\lambda f'(\lambda)e^{-f(\lambda)x_2}Coth\left(\frac{x_2-x_3}{L_{n-sub}}\right)-\lambda f'(\lambda)e^{-f(\lambda)x_3}Csch\left(\frac{x_2-x_3}{L_{n-sub}}\right)}{\left(L_{n-sub}^2f(\lambda)^2-1\right)}\\&+\frac{L_{n-sub}\lambda f(\lambda)f'(\lambda)e^{-f(\lambda)x_2}-L_{n-sub}x_2\lambda f(\lambda)^2f'(\lambda)e^{-f(\lambda)x_2}-x_2\lambda f(\lambda)f'(\lambda)Coth\left(\frac{x_2-x_3}{L_{n-sub}}\right)+x_3\lambda f(\lambda)f'(\lambda)e^{-f(\lambda)x_3}Csch\left(\frac{x_2-x_3}{L_{n-sub}}\right)}{\left(L_{n-sub}^2f(\lambda)^2-1\right)}\\&-\frac{2L_{n-sub}^3\lambda f(\lambda)^3f'(\lambda)e^{-f(\lambda)x_2}+2\lambda f(\lambda)^2f'(\lambda)L_{n-sub}^2e^{-f(\lambda)x_2}Coth\left(\frac{x_2-x_3}{L_{n-sub}}\right)-2\lambda f(\lambda)^2f'(\lambda)L_{n-sub}^2e^{-f(\lambda)x_3}Csch\left(\frac{x_2-x_3}{L_{n-sub}}\right)}{\left(L_{n-sub}^2f(\lambda)^2-1\right)^2}\end{aligned}\right)$$
$$+\frac{q}{hc}\left(e^{-f(\lambda)(x_1+x_2)}\left(e^{f(\lambda)x_2}\left(\lambda f'(\lambda)x_1-1\right)+e^{f(\lambda)x_1}\left(\lambda f'(\lambda)x_2-1\right)\right)\right)$$

(A5)

The calculation result represents the slope of Eq. (A4). When Eq. (A5) equals to 0, the corresponding λ is the peak wavelength of the spectral response.

Equation (A5) is a complex non-exact differential equation. Accordingly, some assumptions are employed to simplify the solution for Eq. (A5). The spectral response induced in the space-charge region is generally too small to be neglected. Additionally, diffusion lengths of the minority carriers in n- and p-substrate are as long as several hundred micrometers owing to low-doped concentrations, and thus wavelengths in the visible region are much smaller than the diffusion lengths. Moreover, there exist the following assumptions

$$L_p^2 f(\lambda)^2 - 1 \cong L_p^2 f(\lambda)^2 , \tag{A6}$$

$$L_{n-sub}^2 f(\lambda)^2 - 1 \cong L_{n-sub}^2 f(\lambda)^2 , \tag{A7}$$

and

$$L_p >> x_1 . \tag{A8}$$

Eq. (A5) can be simplified as follows.

$$\frac{1}{f(\lambda)^4}\left(\begin{array}{l} f(\lambda)^4 + f(\lambda)^4 e^{-f(\lambda)x_1} + f(\lambda)^3 e^{-f(\lambda)x_1}\frac{S_p}{D_p} - 2\lambda f(\lambda) f'(\lambda) e^{-f(\lambda)x_1}\frac{1}{L_p^2} - \lambda f'(\lambda) e^{-f(\lambda)x_1}\frac{S_p}{L_p^2 D_p} \\ -f(\lambda)^2 e^{-f(\lambda)x_1}\lambda f'(\lambda)\frac{S_p}{D_p} + f(\lambda)^3\frac{S_p}{D_p} + 2\lambda f(\lambda)^3 f'(\lambda) + \lambda f(\lambda)^4 f'(\lambda) x_1 + \lambda f'(\lambda)\frac{S_p}{L_p^2 D_p} - \\ \lambda f(\lambda) f'(\lambda) x_1 \frac{S_p}{L_p^2 D_p} + \lambda f(\lambda)^2 f'(\lambda)\frac{S_p}{L_p^2 D_p} + \lambda f(\lambda)^3 f'(\lambda) x_1 \frac{S_p}{L_p^2 D_p} + f(\lambda)^4 e^{-f(\lambda)x_2} \\ + f(\lambda)^4 e^{-f(\lambda)x_2}\frac{1}{L_{n-sub} f(\lambda)} Coth\left(\frac{x_2 - x_3}{L_{n-sub}}\right) - f(\lambda)^3 e^{-f(\lambda)x_3}\frac{1}{L_{n-sub}} Csch\left(\frac{x_2 - x_3}{L_{n-sub}}\right) + \lambda f(\lambda)^3 f'(\lambda) e^{-f(\lambda)x_2} \\ + \lambda f(\lambda)^2 f'(\lambda) e^{-f(\lambda)x_2}\frac{1}{L_{n-sub}} Coth\left(\frac{x_2 - x_3}{L_{n-sub}}\right) - \lambda f(\lambda)^2 f'(\lambda) e^{-f(\lambda)x_3}\frac{1}{L_{n-sub}} Csch\left(\frac{x_2 - x_3}{L_{n-sub}}\right) \\ + \lambda f(\lambda)^3 f'(\lambda) e^{-f(\lambda)x_2} - \lambda f(\lambda)^4 f'(\lambda) e^{-f(\lambda)x_2} x_2 - \lambda f(\lambda)^4 f'(\lambda) x_2 \frac{1}{L_{n-sub} f(\lambda)} Coth\left(\frac{x_2 - x_3}{L_{n-sub}}\right) \\ + \lambda f(\lambda)^4 f'(\lambda) e^{-f(\lambda)x_3} x_3 \frac{1}{L_{n-sub} f(\lambda)} Csch\left(\frac{x_2 - x_3}{L_{n-sub}}\right) - 2\lambda f(\lambda)^3 f'(\lambda) e^{-f(\lambda)x_2} \\ -2\lambda f(\lambda)^2 f'(\lambda) e^{-f(\lambda)x_2}\frac{1}{L_{n-sub}} Coth\left(\frac{x_2 - x_3}{L_{n-sub}}\right) + 2\lambda f(\lambda)^2 f'(\lambda) e^{-f(\lambda)x_3}\frac{1}{L_{n-sub}} Csch\left(\frac{x_2 - x_3}{L_{n-sub}}\right) = 0 \end{array}\right) \tag{A9}$$

6. Acknowledgement

This work was partially supported by National Science Council, Taiwan, under the contract number of NSC 97-2221-E-194-060-MY3.

7. References

[1] B. E. Bayer, "Color Imaging Array," U. S. Patent 3971065, July 1976.

[2] W. J. Liu, Oscal T.-C. Chen, L. K. Dai, P. K. Weng, K. H. Huang and Far-Wen Jih, "A CMOS photodiode model," *Proc. of IEEE International Workshop on Behavioral Modeling and Simulation,* pp. 102-105, 2001.

[3] K. C. Chang, C. Y. Chang, Y. K. Fang and S. C. Jwo, "The amorphous Si/SiC heterojunction color-sensitive phototransistor," *IEEE Electron Device Letters*, vol. 8, no. 2, pp. 64-65, 1987.

[4] H. K. Tsai, S. C. Lee and W. L. Lin, "An amorphous SiC/Si two-color detector," *IEEE Electron Device Letters*, vol. 8, no. 8, pp. 365-367, 1987.

[5] Y. K. Fang, S. B. Hwang, K. H. Chen, C. R. Liu, M. J. Tsai and L. C. Kuo, "An amorphous SiC/Si heterojunction p-i-n diode for low-noise and high-sensitivity UV detector," *IEEE Trans. on Electron Devices*, vol. 39, no. 2, pp. 292-296, 1992.

[6] Y. K. Fang, S. B. Hwang, K. H. Chen, C. R. Liu, M. J. Tsai and L. C. Kuo,"A metal-amorphous silicon-germanium alloy Schottky barrier for infrared optoelectronic IC on glass substrate application," *IEEE Trans. on Electron Devices*, vol. 39, no. 6, pp. 1348-1354, 1992.

[7] G. de Cesare, F. Irrera, F. Lemmi and F. Palma, "Tunable photodetectors based on amorphous Si/SiC heterostructures," *IEEE Trans. on Electron Devices*, vol. 42, no. 5, pp. 835-840, 1995.

[8] K. Eberhardt, T. Neidlinger and M. B. Schubert, "Three-color sensor based on amorphous n-i-p-i-n layers sequence," *IEEE Trans. on Electron Devices*, vol. 42, no. 10, pp. 1763-1768, 1995.

[9] J. Zimmer, D. Knipp, H. Stiebig and H. Wagner, "Amorphous silicon-based unipolar detector for color recognition," *IEEE Trans. on Electron Devices*, vol. 46, no. 5, pp. 884-891, 1999.

[10] M. Topic, H. Stiebig, D. Knipp and F. Smole, "Optimization of a-Si:H-based three-terminal three-color detectors," *IEEE Trans. on Electron Devices*, vol. 46, no. 9, pp. 1839-1845, 1999.

[11] T. Su, P. C. Taylor, G. Ganguly and D. E. Carlson, "Direct role of hydrogen in the Staebler-Wronski effect in hydrogenated amorphous silicon," *Physical Review Letters*, vol. 89, no. 1, pp. 015502-1-015502-4, 2002.

[12] T. Lule, M. Wagner, M. Verhoeven, H. Keller and M. Bohm, "100000-pixel, 120-dB imager in TFA technology," *IEEE Journal of Solid-State Circuits*, vol. 35, no. 5, pp. 732-739, 2000.

[13] Masahiro Kasano, Yuichi Inaba, Mitsuyoshi Mori, Shigetaka Kasuga, Takahiko Murata and Takumi Yamaguchi, "A 2.0-μm pixel pitch CMOS image sensor with 1.5 transistor/pixel and an amorphous Si color filter," *IEEE Trans. on Electron Devices*, vol. 53, no. 4, pp. 611-617, 2006.

[14] Foveon Inc, "Color separation in an active pixel cell imaging array using a triple-well structure," *U.S. Patent 5 965 875*, 12 Oct. 1999.

[15] K. M. Findlater, D. Renshaw, E. D. Hurwitz, R. K. Henderson, M. D. Percell, S. G. Smith and T. E. R. Bailey, "A CMOS image sensor with a double-junction active pixel," *IEEE Trans. on Electron Devices*, vol. 50, no.1, Jan. 2003.

[16] W. J. Liu, O. T.-C. Chen, L. K. Dai, P. K. Weng, K. H. Huang and F. W. Jih, "A color image sensor using adaptive color pixels," *Proc. of IEEE Midwest Symposium on Circuits and Systems*, vol. 2, pp. 441-444, Aug. 2002.

[17] Giacomo Langfelder, Federico Zaraga and Antonio Longoni, "Tunable spectral responses in a color-sensitive CMOS pixel for imaging applications," *IEEE Trans. on Electron Devices*, vol. 56, no. 11, pp. 2563-2569, 2009.

[18] Antonio Longoni, Federico Zaraga, Giacomo Langfelder, and Luca Bombelli, "The transverse field detector (TFD): a novel color-sensitive CMOS device," *IEEE Trans. on Electron Devices*, vol. 29, no. 12, pp. 1306-1308, 2008.

[19] K. De Munck, D. Sabuncuoglu Tezcan, T. Borgers, W. Ruythooren, P. De Moor, S. Sedky, C. Toccafondi, J. Bogaerts and C. Van Hoof, "High performance hybrid and monolithic backside thinned CMOS imagers realized using a new integration process," *Proc. of IEEE IEDM*, pp. 1-4, 2006.

[20] M. Bartek, P. T. J. Gennissen, P. Sarro, P. J. French and R. F. Wolffenbuttel, "An integrated silicon color sensor using selective epitaxial growth," *Sensors and Actuators A: Physical*, vol. 41, no. 1-3, pp. 123-128, 1994.

[21] R. F. Wolffenbuttel, "Color filters integrated with the detector in silicon," *IEEE Electron Device Letters*, vol. 8, no. 1, pp. 13-15, 1987.

[22] Oscal T.-C. Chen, W. J. Liu, L. K. Dai, P. K. Weng, K. H. Huang and Far-Wen Jih, "Extended one-dimensional analysis to effectively derive quantum efficiency of various CMOS photodiodes," *IEEE Trans. on Electron Devices*, vol. 54, no. 10, pp. 2659-2668, 2007.

[23] E. R. Fossum, "CMOS image sensor: electronic camera on a chip," *Tech. Dig. of IEEE IEDM*, pp. 17-25, Dec. 1995.

[24] A. El Gamal and H. Eltoukhy, "CMOS image sensors," *IEEE Circuits and Devices Magazine*, vol. 21, no. 3, pp. 6-20, May/June 2005.

[25] Isao Takayanagi, Miho Shirakawa, Koji Mitani, Masayuki Sugawara, Steinar Iversen, Jørgen Moholt, Junichi Nakamura and Eric R. Fossum, "A 1.25-inch 60-frames/s 8.3-M-pixel digital-output CMOS image sensor," *IEEE Journal of Solid-State Circuits*, vol. 40, no. 1, pp. 2305-2314, 2005.

[26] K. Fife, A. El Gamal and H.-S. P. Wong, "A multi-aperture image sensor with 0.7um pixels in 0.11um CMOS technology," *IEEE Journal of Solid-State Circuits*, vol. 43, no. 12, pp. 2990-3005, 2008.

[27] D. S. Tezcan, S. Eminoglu and T. Akin, "A low-cost uncooled infrared microbolometer detector in standard CMOS technology," *IEEE Trans. on Electron Devices*, vol. 50, no. 2, pp. 494-502, 2003.

[28] C. C. Liu and C. H. Mastrangelo, "A CMOS uncooled heat-balancing infrared imager," *IEEE Trans. on Electron Devices*, vol. 35, no. 4, pp. 527-535, 2000.

[29] C. I. Drowley, M.S. Swenson, L. Arbaugh, R. Hollstein, F. Shapiro, and S. Domer, "A 0.6 *um* CMOS pinned photodiode color imager technology," *Tech. Digest of IEEE IEDM*, pp. 927-929, 1997.

[30] S. M. Sze, *Physics of Semiconductor Devices*, New York: John Wiley and Sons, Inc., 1980.

[31] C. Y. Wu and C. F. Chiu, "A new structure of the 2-D silicon retina," *IEEE Journal of Solid-State Circuits*, vol. 30, no. 8, pp. 890-897, 1995.

[32] Y. Huang and R. I. Hornsey, "Current-mode CMOS image sensor using lateral bipolar phototransistors," *IEEE Trans. on Electron Devices*, vol. 50, no. 12, pp. 2570-2573, 2003.

[33] Xin Liu, Shuxu Guo, Chen Zou, Guotong Du, Yuqi Wang and Yuchun Chang, "Punchthrough enhanced phototransistor fabricated in standard CMOS process," *IEEE Electron Device Letters*, vol. 30, no. 3, pp. 272-274, 2009.

[34] C. Y. Wu, H. D. Sheng and Y. T. Tsai, "The lambda bipolar phototransistor–analysis and applications," *IEEE Journal of Solid-State Circuits*, vol. 20, no. 6, pp. 1227-1234, 1985.

[35] R. R. King, R. A. Sinton and R. M. Swanson, "Studies of diffused phosphorus emitters: saturation current, surface recombination velocity, and quantum efficiency," *IEEE Trans. on Electron Devices*, vol. 37, no. 2, pp. 365-371, 1990.

[36] R. R. King and R. M. Swanson, "Studies of diffused boron emitters: saturation current, bandgap narrowing, and surface recombination velocity," *IEEE Trans. on Electron Devices*, vol. 38, no. 6, pp. 1399-1409, 1991.

[37] D. K. Schroder, *Semiconductor Material and Device Characterization*, New York: John Wiely & Sons, pp. 420-508, 1998.

9

Extrinsic Evolution of the Stacked Gradient Poly-Homojunction Photodiode Genre

Paul V. Jansz and Steven Hinckley
Edith Cowan University
Australia

1. Introduction

The development of fast high-resolution CMOS imaging arrays, for application across a broad spectral range, requires suitable modifications to pixel architecture to improve individual photodiode quantum efficiency and crosstalk suppression (Furumiya et al., 2001; Brouk et al., 2003; Lee et al., 2003; Ghazi 2002). Presented in this chapter are the results of simulation studies that compare the detection efficacy of previous simulated photodiode architectures with the various configurations of the Stacked Gradient Poly-Homojunction (StaG) photodiode genre.

The seed-idea that initiated this line of research, originated from a conference paper demonstrating the benefit of the StaG architecture to near infrared imaging (Dierickx & Bogaerts, 2004). The possibility of controlling photo-carrier direction, led to a radical "out-of-the-box" suggestion of improving the pixel's response characteristics further, by concaving the StaG layers within each pixel, so as to "focus" carrier motion into the pixel's space charge region (SCR). The closest structure to this that was possible to simulate was the first modification to the "flat" StaG architecture: the "U" shaped StaG with interpixel nested ridges (StaG-R). Both this and the concave StaG, having serious fabrication issues, led to further pixel modifications. The result: the evolution of the StaG photodiode genre; driven by the need to improve upon the photodiodes sensitivity and crosstalk suppression for particularly back illuminated pixels, but also for the front illumination mode. This process is "extrinsic" evolution, because the proactive motivations and ideas for device development originated external to the device itself. The present studies have been conducted using 50 μm pitch pixels in order to compare response with previously characterised photodiode architectures. Research into 5 μm pitch StaG pixels is currently under development.

Contemporary research into Camera-on-a-CMOS chip technology has been focused on frontwall-illuminated (FW) architectures, in which the Active Pixel Sensor (APS) and the signal processing circuitry are coplanar-integrated (Shcherback & Yaddid-Pecht, 2003). This architecture is disadvantaged in a number of ways, including the incompatibility of different CCD and CMOS processing technologies and low fill factor. These disadvantages can be overcome by adopting a backwall-illuminated (BW) mode. As well as maximizing the fill factor, back illumination allows the combination of different processing technologies for the two chips. Additionally, it is possible to tailor the spectral response of individual photodiodes, due to the indirect nature of the silicon absorption coefficient, which affects the electron-hole pair photogeneration profile (Hinckley et al., 2000). Back illuminated

CMOS *pin* ultra-thin (75 μm) photodiodes have found application in medical imaging, particularly making x-ray, high quality, real time imaging possible (Goushcha et al., 2007).
However, compared to front illumination, the backwall orientation is disadvantaged in crosstalk, speed and quantum efficiency (QE) due to the distality of the photo generated carrier envelope to the SCR, resulting in diffusion dominated pixels (Jansz Drávetzky, 2003). These problems need to be overcome before back illuminated CMOS photodiode arrays present a serious challenge to the present mature front illuminated active pixel sensor market.
Architectures predicted to reduce these problems for back illuminated sensors :

1. Control the direction of diffusion/drift of the photo-carriers towards the SCR,
2. Bring the SCR closer to the photo-carrier envelope near the pixel backwall by,
   a. Thinning the pixel (Goushcha et al., 2007).
   b. Widening the SCR by,
      i. Increasing the reverse bias to the PN junction, and
      ii. Decreasing the doping on the substrate side of the PN junction, or
      iii. Having no doping (intrinsic Silicon) between the P and N regions, making a *pin* "junction" (Goushcha et al., 2007).
   c. Extending the higher doped well towards the back wall by,
      i. Thinning a single deep well so it is also depleted while at the same time extending the SCR to the pixel backwall, frontwall and side boundaries (2B). This is for small pitch, deep or shallow pixels.
      ii. Using a number of deep thin wells (polywells) across the pixel to extend the SCR to the pixel's backwall, frontwall, side boundaries and between each well (2B). This is for large pitch, deep or shallow pixels.
      iii. Using an inverted "T" shaped well and appropriate doping regimes (2B) that deplete the thin well and the substrate adjacent to the back wall.
3. Incorporate some form of inter-pixel barrier to lateral crosstalk carrier transport by,
   a. Incorporating a single or multiple pixel boundary trench isolation consisting of
      i. Higher doped semiconductor with the same dopant type as the substrate (Jansz-Drávetzky, 2003; Hinckley et al., 2007; Jansz et al., 2008; Jansz, 2003).
      ii. Higher doped semiconductor with opposite dopant type to the substrate
      iii. Insulators such as SiO_2 (Jansz et al., 2008).
   b. Using a guard ring electrode (Hinckley et al., 2004; Jansz, 2003).
   c. Using a guard (double) junction photodiode (Hinckley et al., 2004).

The present interest in the StaG photodiode architectural genre, stems simply from its ability to control the direction of diffusion/drift of photo-carriers. However, StaG incorporation in the photodiode architecture needs to go hand in hand with SCR proximity (2.) and crosstalk barrier incorporation (3.) so that the benefit of the StaG structure in improved speed, crosstalk and sensitivity may be realised.

2. Theory

There are two mechanisms of photo-carrier transport: drift and diffusion. For fast, sensitive and no crosstalk pixels, drift is preferred. Drift is the movement of the majority or minority carriers due to the applied bias field and has a maximum mean thermal velocity of approximately 10^7 cm.s^{-1} in silicon (Streetman et al., 2000). This movement is orders of magnitude faster than diffusion, which depends on carrier concentration gradient.

Transport of photocarriers generated in the SCR is dominated by drift. A wide SCR, drift-dominated, pixel, demonstrates superior carrier capture efficiency as the pixel is swept of carriers faster. Such pixels show far better crosstalk suppression due to the increased efficiency of 'claiming' carriers generated in their borders. Subsequently, they show enhanced sensitivity and lower junction capacitance due to their wider SCR.
The Width of the SCR of a PN junction is dependent mostly on the N or P doping each side of the junction, and the potential bias across the junction,

$$W = \sqrt{\left[\frac{2\varepsilon(V_0 - V)}{q}\left(\frac{N_a + N_d}{N_a N_d}\right)\right]} \qquad (1)$$

where N_a and N_d are the dopent concentrations on the P side and the N side of the PN junction respectively. Also ε, q and V are the permittivity of Silicon (11.8 x 8.85 x 10^{-14} Fcm^{-1}), electronic charge (1.60 x 10^{-19} C) and the external bias voltage, respectively. Due to the concentration gradient of holes and electrons on either side of PN junction, the SCR is generated, having a width W, and an internal equilibrium potential, V_0, across the junction.
The SCR width is more affected by lowering the substrate doping concentration than by increasing the reverse voltage bias. Typical SCR width for 2 volt reverse bias is 6 μm, constrained by a 10^{14} cm^{-3} doping minimum. Lowering the substrate doping to the intrinsic level, 1.5 x 10^{10} cm^{-3}, (using an intrinsic substrate) can expand the SCR to more than 450 μm. For such PIN photodiodes, all photo-carriers are generated within the SCR, and as such are collected quickly and specific to their pixel of origin. Knowledge of the SCR width is needed to determine the best StaG position in the pixel cross section (Jansz & Hinckley, 2010).
The homojunction that is of interest in this chapter, though not as aggressive in carrier collection as a PN homojunction, also relies on an inbuilt potential gradient to capture diffusing carriers and direct their motion towards the SCR. As such, it works in collaboration with the PN junction to better manage pixel carrier capture efficiency. This particular homojunction is characterised by a layering of epitaxially grown epilayers on a substrate of similar doping type (Fig. 1). These epilayers decrease in doping concentration from the substrate towards the pixel well or PN junction at the front of the pixel. As such they represent a poly-homojunction, which is stacked and having a doping concentration gradient: The Stacked Gradient poly-homojunction photodiode – the "StaG".
To explain the StaG dynamics, it is necessary to visualise the cross section of a conventional StaG photodiode pixel in Fig. 1. The epilayer doping concentration decrease towards the front wall, from 10^{18} cm^{-3} in the substrate to 10^{14} cm^{-3} in the uppermost epilayer. This direction of decreasing doping concentration towards the SCR produces a potential gradient that drives the minority carriers vertically towards the SCR. Fig. 2. illustrates this principle using a schematic energy band diagram of the StaG geometry in Fig. 1, developed from Singh (1994).
On average, the direction of reflected carriers is normal to the StaG strata (Hinckley & Jansz, 2007). Carriers diffusing away from the SCR will be reflected back towards the SCR as the StaG structure acts as a minority carrier mirror. This results in increased pixel carrier capture efficiency, reducing crosstalk and increasing pixel sensitivity.
The effects of device geometry on pixel response resolution were measured by the pixel's sensitivity, defined as maximum quantum efficiency (QE) and the electrical crosstalk. The quantum efficiency (η=QE) for an incident wavelength (λ), and radiant intensity (P_{opt}) was calculated using,

$$\eta(\lambda) = \frac{hcI_{\lambda}}{\lambda q P_{opt}} \quad (2)$$

where h is Planck's constant, c is the speed of light, and q is the electronic charge. The simulated electron, hole and total current (I_{λ}) quantum efficiency was calculated.

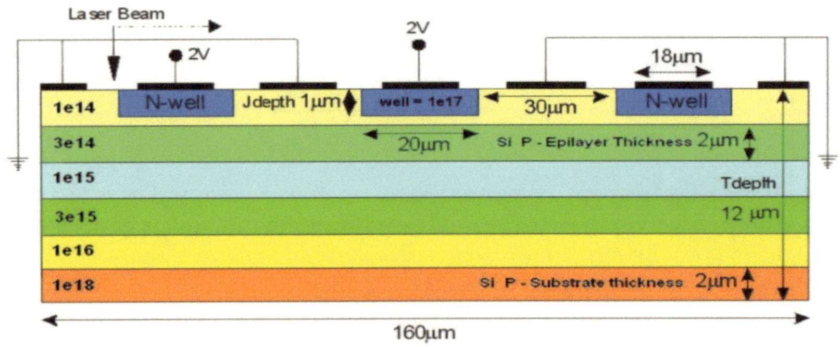

Fig. 1. Cross-section of the simulated front illuminated conventional Stacked Gradient Homojunction (StaG) Photodiode array (Hinckley & Jansz, 2007). The back illuminated array is illuminated upon the bottom surface of the array diagram.

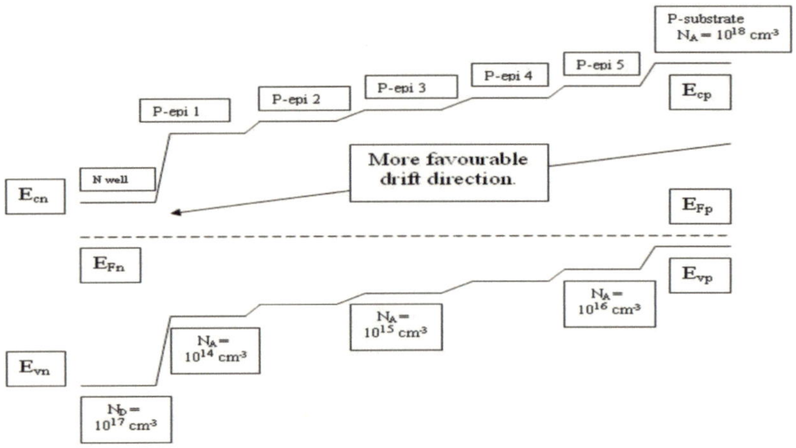

Fig. 2. Energy band diagram schematic of an unbiased five p-epilayer homojunction photodiode, indicating the favourable direction of carrier drift (Hinckley & Jansz, 2007).

3. Method

Imaging arrays consist of repeating light detecting elements called pixels. In these simulation studies, each pixel was configured as a reverse biased vertical p-n junction photodiode. The crosstalk and maximum QE of the central pixel of the three pixel array, 160 µm long and 12 µm deep, having different StaG configurations, were simulated using SEMICAD DEVICE (version 1.2), a two dimensional finite-element simulator. Fig. 3 shows the initial simulated primitive conventional photodiode that began this line of simulation research (Hinckley et al., 2002).

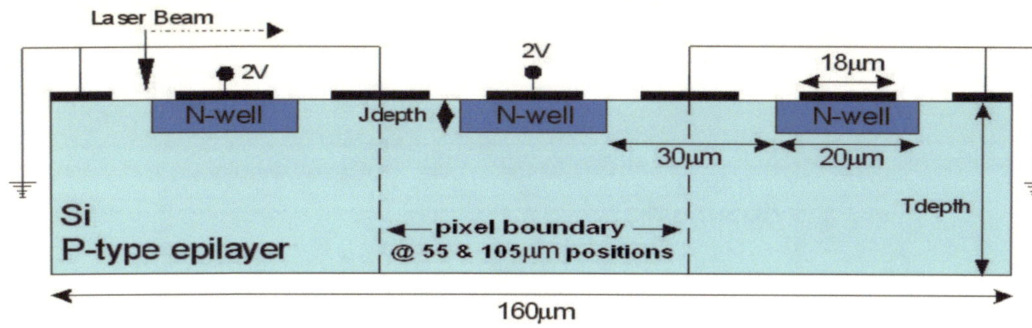

Fig. 3. Cross-section of the simulated front illuminated conventional photodiode array (Hinckley et al, 2002). The back illuminated array is illuminated from underneath.

This photodiode's standard dimensions included a well depth (Jdepth) of 1 µm, and a substrate thickness (Tdepth) of 12 µm. Each photodiode was reverse biased by 2 volts. More recent StaG-polywell hybrid studies (Jansz & Hinckley, 2010; Jansz, Hinckley & Wild, 2010) used 3 volts to compare to previous research (Ghazi et al., 2002). Transparent ohmic contacts were used on the well and substrate surfaces on the front side of the array. The device with identical structure was simulated using back illumination followed by front illumination. The array was scanned at 5 µm intervals along the array, typically using a simulated laser beam of 633 nm wavelength, 5 µm width and 0.1 µW power. The use of 633 nm is for comparison to previous photodiode pixels simulation studies. For the generic StaG and in present StaG-hybrid research, simulation studies have explored pixel response characteristics for ultra-violet to near infrared illumination.

To quantify the electrical crosstalk so that comparison could be made between photodiode configurations, the "relative crosstalk" was calculated. This was defined as the normalized quantum efficiency (NQE) of the photocurrent registering at the central pixel's image (well) electrode for illumination at the 50µm position along the array (Fig. 1). The response resolution of each device was compared using their relative crosstalk and their maximum quantum efficiency (QE). Though pixel speed was not considered, since the simulated source was continuous not modulated, it is clear that there is a relationship between crosstalk suppression and the ability for a pixel to manage its carrier capture efficiency. The latter also impacts on a pixel's speed of photo-carrier capture.

4. The StaG photodiode genre

The following section reports on the characteristic features and performance of each present member of the StaG photodiode genre in chronological order of simulated investigation. The simulated structure, results and discussion are treated separately for each member.

4.1 The Beginning – The "Flat" StaG Photodiode

The "flat" StaG photodiode, designated "StaG" (Fig. 1), QE response, backwall (BW) and frontwall (FW) illuminated, was compared to the QE response of two doping versions of the conventional photodiode (Fig. 3) with the following doping (well/substrate) regimes. Both versions had the same well doping as the flat-StaG, 10^{17} cm-3. One version (17/15) had a substrate doping of 10^{15} cm^{-3} while the other (17/14) had an order of magnitude lower substrate doping of 10^{14} cm^{-3} (Hinckley & Jansz, 2005).

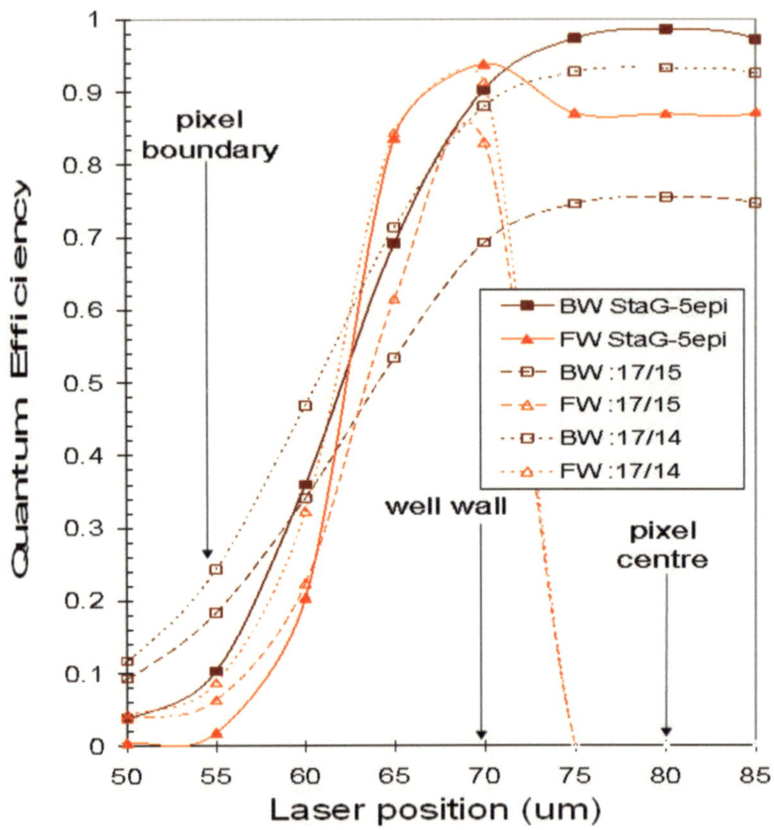

Fig. 4. Comparison of StaG (Fig. 1) and conventional single junction photodiode (Fig. 3) QE, for both back (BW) and front (FW) illuminated cases, as a function of laser position (μm), and 633 nm wavelength (Hinckley & Jansz 2005).

Clearly back and front illumination responses of the flat-StaG architecture is superior in crosstalk suppression and maximum QE (together denoted "response resolution") than either of the standard photodiode configurations. Fig. 4 shows that the response resolution decreases according to the trend: StaG > conventional PD 17/14 > conventional PD 17/15.

4.1.1 StaG relative crosstalk and sensitivity dependence on wavelength

Fig. 5A compares the relative crosstalk (normalized QE for illuminations at the pixel boundary at the 50 μm position allong the array in Fig. 1) dependence on wavelength for the same 12μm thick back and front illuminated StaG (Fig. 1) and conventional photodiodes (PD) (Fig. 3). The PDs have a p-substrate doping of 10^{14} cm^{-3} (17/14) or 10^{15} cm^{-3} (17/15), and an n-well doping of 10^{17} cm^{-3}. Back illuminated relative crosstalk generally decreases with increase in wavelength, because the absorption length increases. This generates more carriers closer to the SCR, resulting in better pixel carrier capture efficiency. The reverse is true for the front illuminated pixels (Hinckley & Jansz, 2005).

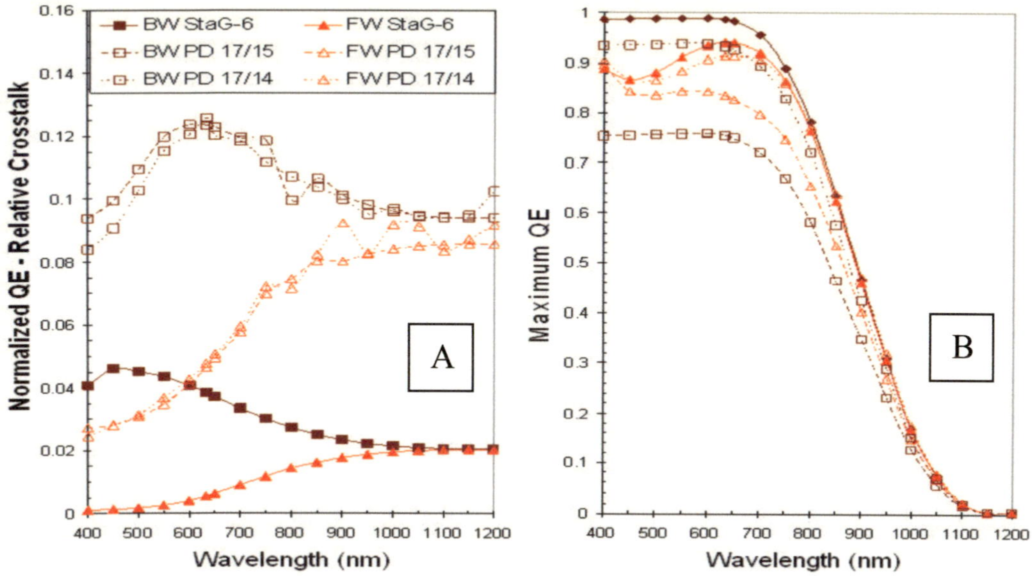

Fig. 5. Relative crosstalk (A) and sensitivity (B) dependence on wavelength for StaG (Fig. 1) and conventional photodiode (PD) (Fig. 3) for 10^{17} cm^{-3} well doping and two p-substrate dopings : 10^{15} cm^{-3} (17/15) and 10^{14} cm^{-3} (17/14) for back (BW) and front (FW) illumination (Hinckley & Jansz, 2005).

Fig. 5B compares the sensitivity - maximum quantum efficiency (QE) - dependence on wavelength for the 12μm thick back and front illuminated StaG and conventional (PD) photodiodes. For both structures, the back (BW) and front (FW) illumination modes have similar maximum QE dependence on wavelength. The StaG shows a higher maximum QE in both modes compared to both conventional photodiodes (PD).
The back illuminated StaG maximum QE is superior to the other geometries, for the depth of well (1 μm). For the shorter absorption length illuminations (λ < 700nm), minority hole generation in the well is significant in front illumination causing significant hole diffusion, suppressing sensitivity. Back illumination is absorbed away from the well so that sensitivity is not suppressed. Note that the lower-doped substrate Naked photodiode (Naked 17/14) enhances carrier capture by increasing the SCR, also enhancing StaG response.

4.1.2 StaG relative crosstalk dependence on epilayer thickness and wavelength

Fig. 6A demonstrates that, though the StaG has a better response resolution than the photodiode without the StaG, even for the StaG, widening the epilayers increases the chance of lateral carrier diffusion, reducing the pixels carrier capture efficency: crosstalk increasing across the given wavelength band. For any given epilayer thickness, front illumination crosstalk increasing while back illumination slightly decreases, and both responses level off at the same wavelengths. The increase or decrease is proportional to the increase in absorption length with wavelength increase. This is due to Silicon being an indirect band gap semiconductor: as the wavelength increases, front and back illumination generates carriers further and closer to the SCR, respectively. For thicker pixels, more of the longer wavelength light is absorbed, thus the larger the wavelength at which the pixel saturates; for any longer wavelengths more light passes though the pixel without being absorbed.

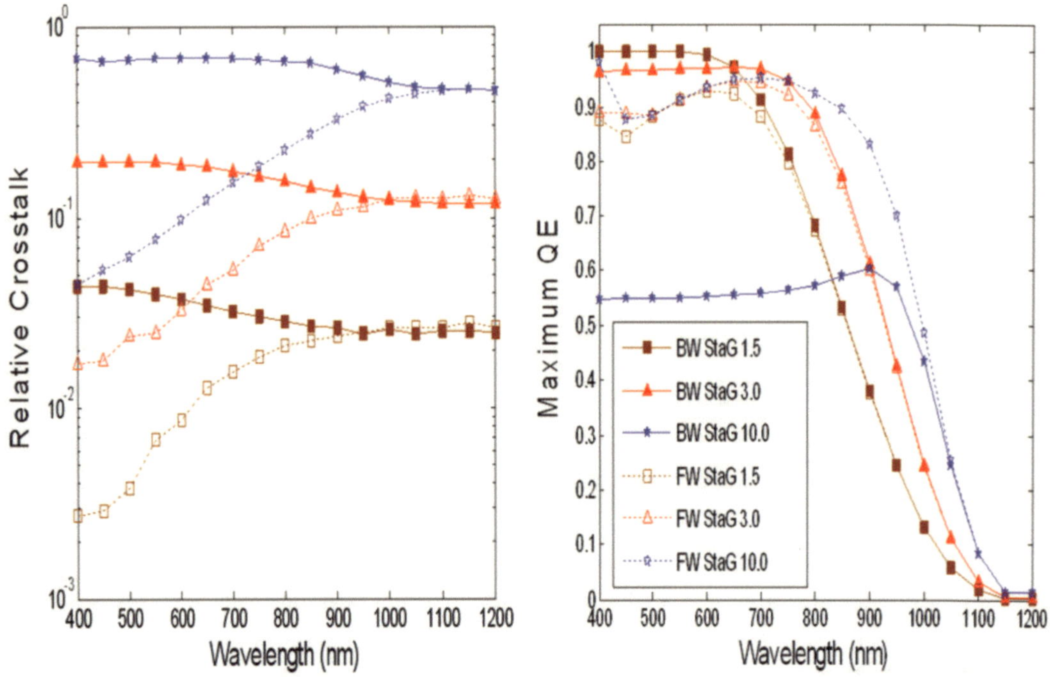

Fig. 6. StaG (Fig. 1) relative crosstalk (A) and sensitivity (B) dependence on wavelength and epilayer thickness of 1.5, 3 and 10 μm. (Hinckley & Jansz, 2005).

Fig. 6B demonstrates that the thinner the epilayers, the better the sensitivity (maximum QE) for back illumination until a wavelength-saturation sensitivity switch-point. There are two switch points: 650 and 900 nm. From 650 upwards, the most sensitive StaG geometry switches from the thinnest pixel (1.5 μm epilayer) to the next thinnest pixel (3 μm). The latter remains the most sensitive until 900nm, when the thickest pixel (10 μm) becomes the most sensitive. For the longer wavelengths and thicker pixels, the light that otherwise would have passed through a thinner pixel, now generates carriers in a larger pixel volume, increasing its carrier capture and so benefiting sensitivity. Below 650 nm, the light absorption length in silicon is less than the depth of the thinnest pixel (1.5 μm epilayers = 9 μm total pixel depth), resulting in all of the illumination being absorbed and generating carriers in close proximity to the SCR. The result: maximum sensitivity for both modes of illumination.

4.1.3 StaG crosstalk and sensitivity score table: comparing photodiodes

Table 1 compares, for illumination at 633nm, the relative crosstalk and maximum QE of the

- StaG photodiode (Fig. 2) (Hinckley & Jansz, 2005).
- Conventional single-junction photodiode (SJPD) (Fig. 4); (Jansz-Drávetzky, 2003)
- The SJPD with 8μm deep boundary trench isolation (BTI);
- The SJPD with guard-ring electrodes (Guard);
- An N^+PN^- guard junction photodiode (DJPD) with well, guard and substrate depth of 1 μm, 2 μm and 12μm respectively; with SJPD pixel pitch (Jansz-Drávetzky 2003).

Photodiode Type	Back Illuminated Crosstalk	Front Illuminated Crosstalk	Back Illuminated Maximum QE	Front Illuminated Maximum QE
StaG	0.105	0.020	0.986	0.940
SJPD	0.260	0.096	**0.933**	0.915
BTI	0.269	0.096	0.952	0.994
Guard	0.069	0.010	0.134	0.436
DJPD	0.001	0.001	0.004	0.543

Table 1. Comparison of crosstalk and maximum QE of the StaG and previously simulated photodiode geometries, for 633 nm illumination (Hinckley & Jansz, 2005).

This embryonic StaG (Fig. 1), for illumination at 633 nm, is already superior in sensitivity to these other back illumination photodiodes. Sensitivity for front illumination is trumped by the SJPD-BTI geometry, while StaG sensitivity is second best.

For back illumination, the carrier envelope falls within the StaG layers, which act as minority carrier mirrors reflecting the carriers towards the SCR. For the SJPD, with or without BTI, the same carrier envelope is not constrained by a StaG lamination or by the BTI that extends only 8 μm into the pixel; 4 μm from the back wall. Carriers are then lost to crosstalk or recombination, reducing sensitivity and increasing crosstalk for SJPD-BTI.

Alternatively the reverse is true for front illumination. For the SJPD-BTI, the carrier envelope is now proximal to the SCR and constrained by the BTI. This results in it's sensitivity being enhance above that of the StaG response.

Considering the relative crosstalk, the StaG is superior to the SJPD with and without BTI. It is inferior to the SJPD with guard-ring-electrode and guard-junction. However the guard configurations work on the basis of selective capture of the outer part of the carrier envelope by the guard electrode and junction. A much reduced envelope is captured, reducing crosstalk, but also reducing sensitivity especially for back illumination. Alternatively, StaG dynamics works on the basis of capturing and focusing towards the SCR as much of the carrier envelope as possible, with benefit to crosstalk and sensitivity (response resolution).

Plots of the electric field strength show that the StaG configuration has greater electric field strength and extent around the pixel well, which improves its carrier capture efficiency, which again translates to improving pixel response resolution (Hinckley & Jansz, 2005)

4.1.4 StaG – the first step

The advantage of the StaG configuration is that carrier diffusion direction is controllable. This vertical directionality is controlled by the doping concentration gradient of the substrate and epilayers. Carriers generated in any epilayer that diffuse towards the back of the pixel will strike a higher doped stratum which will reflect them back into their parent epilayer so that their net displacement will be in the direction of the decrease in doping concentration. Though there will still be lateral diffusion, there will be less recombination of carriers diffusing away from the surface, while pixel capture volume will increase.

In this section, StaG carrier vertical directionality is imposed on the system by the planar epilayers and the direction of epilayer doping gradient. In the next section, this directional control is extrapolated to include an additional StaG structure that gives additional benefit to the pixel's carrier capture efficency.

4.2 StaG with inter-pixel nested ridges

Captalizing on the StaG control of carrier direction, the original seed idea was to concave the StaG epilayers so that the focal point of the epilayers would be within the SCR. It was hypothesised that this would focus additional carriers, primarily lateral crosstalk carriers, towards the SCR, benefiting the pixel's carrier capture efficiency. The closest analogy to this 'StaG-concave' configuration that was able to be defined using the simulation tool, was the StaG with Inter-Pixel Nested Ridges (StaG-R).

Fig. 7 shows the cross section of the simulated StaG-R tri-pixel array. The diagram is squashed laterally making the 1 μm lateral spacing between the vertical nested epilayer ridges appear much closer. This makes each ridge horizontal width, from the highest epilayer ridge down to the substrate ridge, 10, 8, 6, 4 and 2 μm respectively.

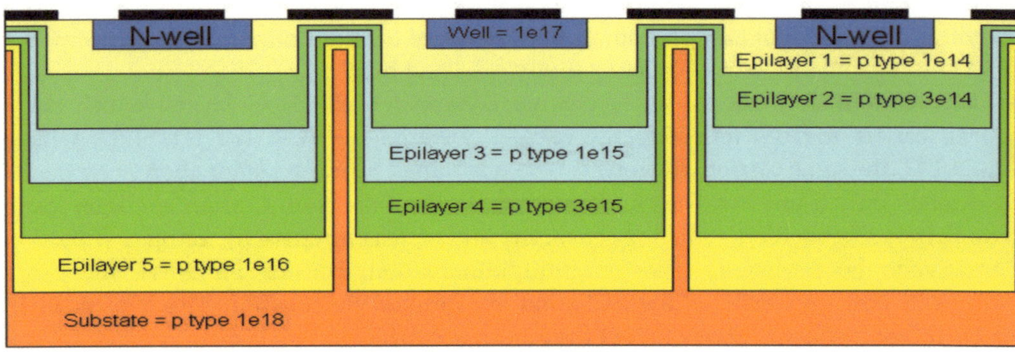

Fig. 7. Cross-section of the simulated Stacked Gradient Homojunction Photodiode array with 5 epilayer inter-pixel nested ridges (Hinckley & Jansz 2007).

Simulations at 633 nm, have shown that it is possible to enhance the StaG PD's response resolution further by including a laterally stacked gradient homojunction in the form of inter-pixel nested ridges. These ridges extend from each epilayer, symetrically about the pixel's lateral boundaries, towards the frontwall of the photodiode: lower ridges nesting into upper ridges. The new hypothesis, an extention of the StaG-concave hypothesis, reasoned that by having both laterally and vertically stacked gradient homojunctions, two dimensional control of photo-carrier transport can be achieved: the vertical stacking reducing diffusion towards the backwall while the lateral stacking reducing lateral carrier diffusion; a primary source of crosstalk. Pixel carrier capture efficiency was enhanced as predicted, benefiting pixel response resolution (Hinckley & Jansz, 2007).

4.2.1 StaG-R relative crosstalk dependence on ridge height.

Fig. 8 shows relative crosstalk dependence on ridge height, or more correctly, dependence on the extent of ridge nesting for 633 nm illumination. Ridge height refers to the height of the lowest ridge which extends upwards from the substrate (Fig. 7). Higher ridges may be of equal or lesser height than the substrate ridge, because of the proximity of the epilayer ridge to the frontwall and the vertical gaps between the tops of ridges being equal for a given ridge height.

The effect of increasing ridge height on relative crosstalk (Fig. 8), for 633 nm back illumination, is to monotonically reduce crosstalk. For front illumination, crosstalk reduces even faster than back illumination, with ridge increase, except for the lower ridges.

Fig. 9 shows a maximum of 80% (back illumination) to 95 % (front illumination) reduction in relative crosstalk. This is significant, demonstrating that the StaG-R configuration fulfills the predicted benefit to crosstalk reduction (Hinckley & Jansz, 2007).

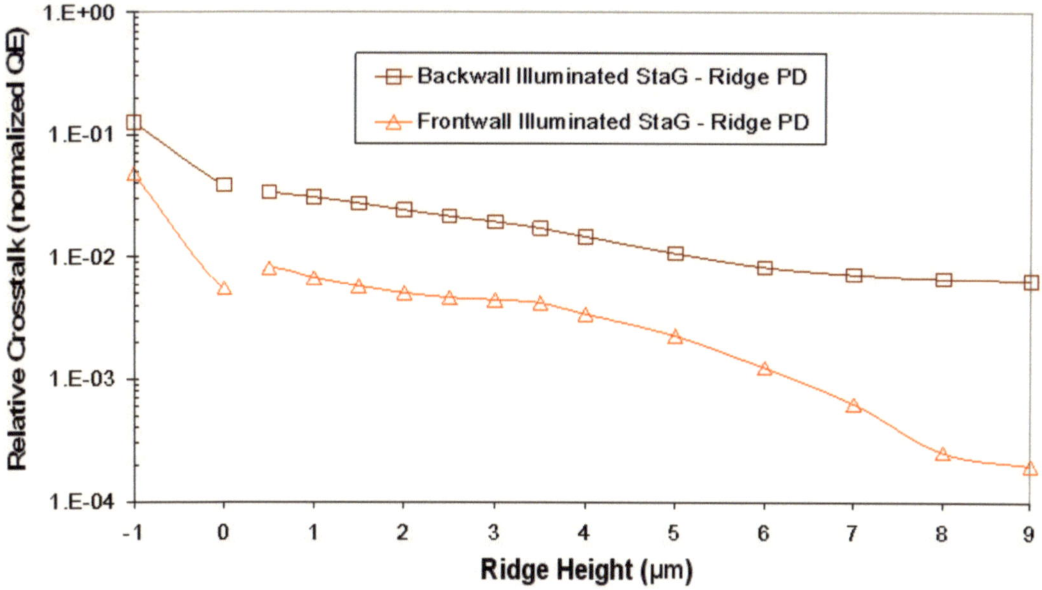

Fig. 8. Relative crosstalk of StaG-R (Fig. 7) compared to the StaG (ridge height = 0) (Fig. 1) and the normal photodiode (ridge height = -1) (Fig. 3) at 633 nm (Hinckley & Jansz 2007).

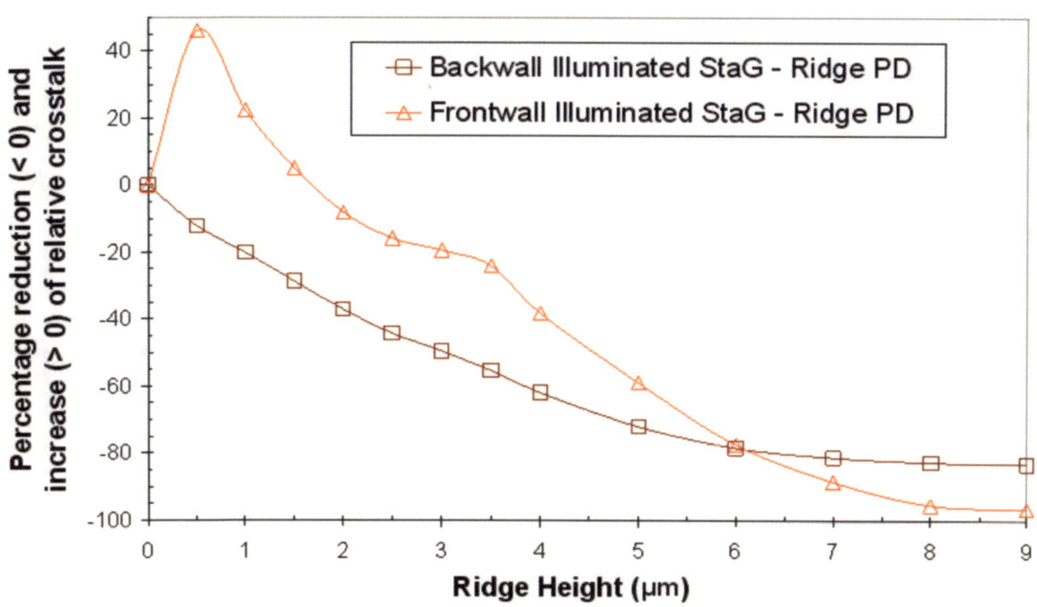

Fig. 9. Percentage reduction of relative crosstalk for StaG-R compared to the StaG PD (ridge height = 0) as a function of ridge height, at 633 nm (Hinckley & Jansz, 2007).

The crosstalk for front illuminated StaG-R is above StaG for ridge heights less than 2μm, because the ridges are broader at the front wall (10 μm) and only one ridge thick, not yet being nested. Front illumination at the 50 μm position generates a carrier envelope in the wider and higher doped ridges towards the front of the pixel. This allows the possibility of lateral crosstalk diffusion. However, if the width of the uppermost ridge was less than 5μm, 2.5 μm either side of the 55 μm position along the array, the 5 μm wide beam front illuminating at the 50 μm position (the defined position for the measure of relative crosstalk), would fall outside the ridges, in the StaG epilayers of the neighbouring pixel. Generated carriers would be reflected off the un-nested ridges, resulting in a reduction in the relative crosstalk compared to the StaG configuration.

Alternatively, for back illumination, the carrier envelope falls outside the thinner shallower, un-nested ridges, which act as doped boundary trench isolation (effectively, bi-layer lateral StaGs) enhancing crosstalk reduction. However, back illumination shows a poorer reduction in crosstalk than front illumination, for the higher ridges, because the generated carrier envelope is now no longer as near the frontwall as for front illumination. It, therefore does not benefiting from the same degree of StaG nesting as front illumination.

4.2.2 StaG-R relative crosstalk dependence on ridge height.

Relative crosstalk was also investigated for dependence on the lateral gap between ridges for 633 nm illumination. Fig. 10 shows the normalized QE of front (FW) and back (BW) illuminated StaG-R dependence on the lateral ridge gap thickness for illumination outside (40μm & 50μm positions) and inside (60μm position) the central pixel (Fig. 7). The relative

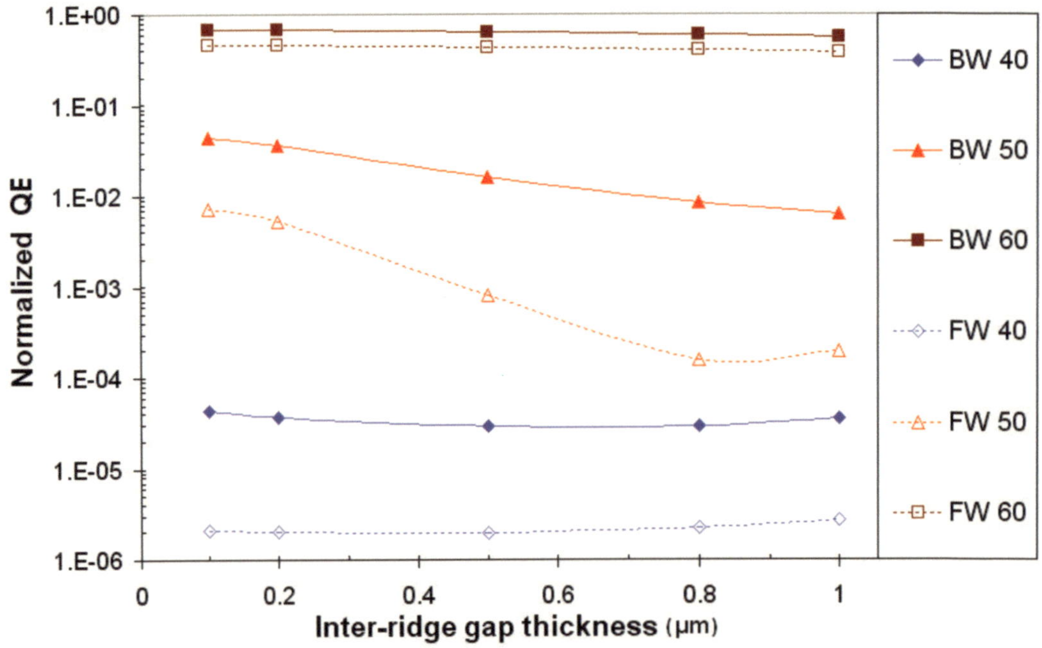

Fig. 10. The normalized QE of Frontwall (FW) and Backwall (BW) illuminated StaG-R dependence on lateral inter-ridge gap thickness for 633 nm illumination outside (40μm & 50μm positions) and inside (60μm position) the central pixel (Hinckley & Jansz, 2007).

crosstalk is represented by the BW50 and FW50 curves. The ridge height (9µm) and the outer ridge width (10µm) were fixed, while the other ridge widths were varied by a constant amount producing a range of inter-ridge gaps from 0.1µm to 1µm. This means that the maximum doped central substrate ridge was the widest for the thinnest gap of 0.1µm, and thinnest for the thickest gap of 1µm.

As the gap between adjacent ridges increased, the relative crosstalk reduced. This was because the central substrate ridge width was decreasing with increasing gap. As the gap increased, the illuminations close to, but outside the central pixel (i.e. BW50 & FW50), fell inside the central ridge or were channeled into the central ridge (BW case) to a lessening extent. Thus, fewer carriers were generated in or channeled into the central ridge. This reduced the relative crosstalk. The further the illumination position was from the pixel boundary (i.e. the 55µm position along the array), the more the pixel response became independent of the gap thickness: illumination at the 40µm (BW40, FW40) and 60µm (BW60, FW60) positions were less affected by the variation in ridge gap size. At these positions the illumination fell outside the nested ridges effectively reflecting carriers away from the pixel (40 µm position) and into the pixel (60 µm position), affecting the QE accordingly (Fig. 10).

4.2.3 StaG-R sensitivity dependence on ridge height

Sensitivity (maximum QE) dependence on ridge dimensions was also investigated for 633 nm illumination. Fig. 11 demonstrates the sensitivity dependence on ridge height for the StaG-R (Fig. 7) compared to the StaG (Fig. 1) and conventional photodiode (Fig. 3).

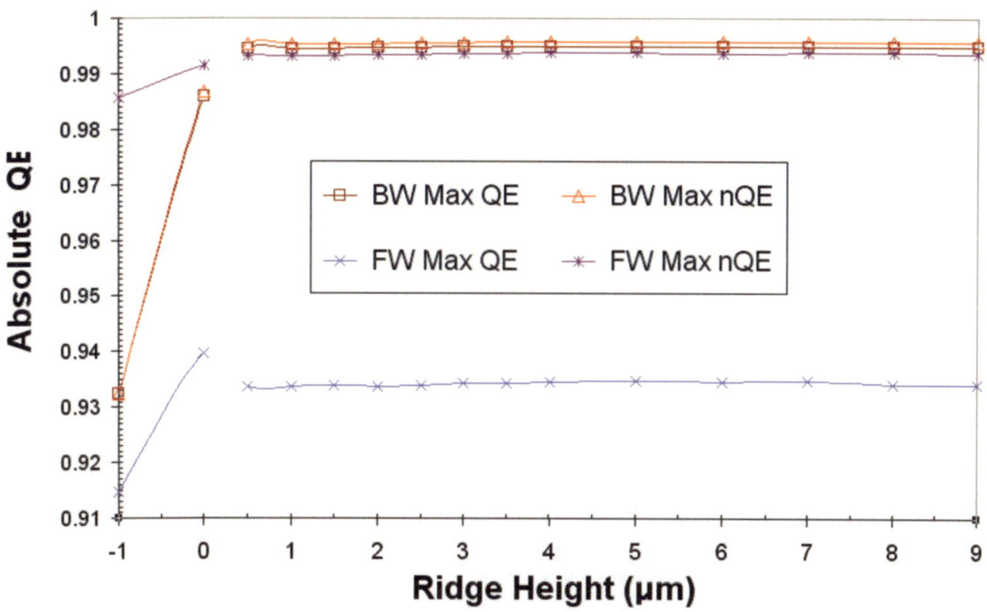

Fig. 11. Maximum electron (nQE) and total Quantum Efficiency (QE) dependence on ridge height, for StaG-R (Fig. 7), compared to the StaG PD (ridge = 0), (Fig. 1) and conventional PD (ridge = -1) (Fig. 3) for backwall (BW) and frontwall (FW) illumination at 633 nm (Hinckley & Jansz, 2007).

Noted is the 0.9 % improvement in sensitivity for the back illuminated StaG-R compared to the StaG. Though, for front illumination, the maximum electron QE (Max nQE) for the StaG-

R was larger than for the StaG, the well minority hole QE (i.e. the difference between the maximum nQE (FW max nQE) and the maximum total QE (FW max QE) in Fig. 11, was greater than for the StaG, resulting in a lower (0.6%) total QE for the StaG-R. The contrast between the back and front illuminated photodiode response was due to the distality of the generated carrier envelope from the photodiode's SCR, for back illumination. Fewer diffusing minority holes were generated inside the well and more drifting electrons in the region below the well, for back illumination.

4.2.4 StaG-R sensitivity dependence on inter-ridge gap size

Fig. 12 shows the sensitivity of the StaG-R for maximum ridge height as a function of lateral inter-ridge gap thickness. This was for back (BW) and front (FW) illumination inside the central pixel, i.e. 60μm position (BW60, FW60), 70μm position (BW70, FW70) and maximum QE (BW Max QE, FW Max QE). The ridge height (9μm) and the outer ridge width (10μm) were fixed, while the other ridge widths were varied from 0.1μm to 1μm as in Fig. 10.

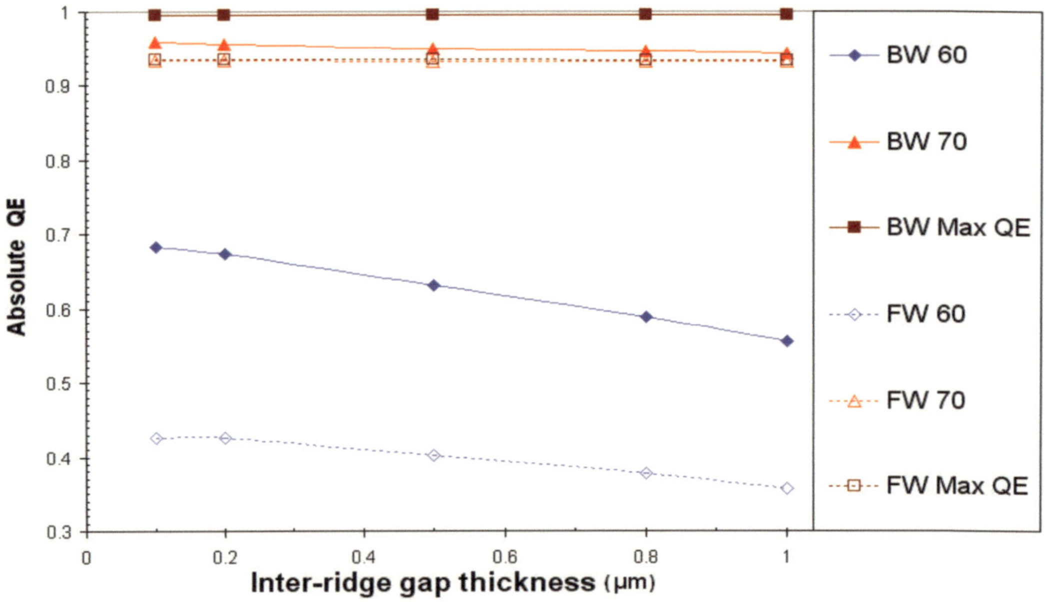

Fig. 12. The absolute QE of Frontwall (FW) and Backwall (BW) illuminated StaG-R dependence on lateral inter-ridge gap thickness for 633 nm illuminations inside (60μm, 70μm and maximum QE positions) the central pixel (Hinckley & Jansz 2007).

Illuminations falling outside the nested ridges (70μm & Max QE) produced absolute QE responses that were affected minimally by a variation in lateral inter-ridge gap thickness. Here the carrier envelope is proximal to the outer layer of the nested ridges only. Any changes within the nested ridges does not connect with the associated carrier envelope. Noted is the decreasing trend for the closest illumination position (60μm) which intersects the nested ridge (Fig. 7). The thinner the gap between nested ridges the larger the potential gradient (Fig. 3) and drift coefficient resulting in more carriers being reflected into the pixel's capture volume, resulting in greater QE for illumination at the 60 μm position.

4.2.5 StaG-R crosstalk and sensitivity score table: comparing photodiodes

Table 2 & 3 compare, for 633nm illumination, the relative crosstalk and maximum QE of the

- StaG photodiode (Fig. 2) (Hinckley & Jansz, 2005);
- Conventional single-junction photodiode (SJPD) (Fig. 4) (Jansz-Drávetzky, 2003);
- The SJPD with 8μm deep double boundary trench isolation (DBTI) (Jansz, 2003);
- The SJPD with guard-ring electrodes (Guard) (Jansz-Drávetzky & Hinckley, 2004);
- The SJPD with guard-ring electrode and 8μm deep DBTI (Guard-DBTI) (Jansz, 2003);
- An N^+PN^- guard junction photodiode (DJPD) with well, guard and substrate depth of 1 μm, 2 μm and 12μm respectively; with SJPD pixel pitch (Jansz-Drávetzky 2003).

Backwall Illumination		Frontwall Illumination	
Photodiode Type	Relative Crosstalk (% QE)		Photodiode Type
DJPD (NPN)	0.0012	1×10^{-5}	DJPD (NPN)
StaG-R	0.65	0.020	StaG-R
Guard-DBTI	3.6	0.55	StaG
StaG	3.8	0.6	Guard-DBTI
Guard	4.1	0.9	Guard
DBTI	14.9	7.3	DBTI
SJPD	21.6	14.5	SJPD

Table 2. Relative crosstalk at 633 nm for StaG-R and other simulated photodiodes.

Back Illuminated		Front Illuminated	
Photodiode Type	Maximum % QE - Sensitivity		Photodiode Type
StaG-R	99.5	99.4	DBTI
StaG	98.6	98.4	SJPD
DBTI	95.2	94.0	StaG
SJPD	92.1	93.5	StaG-R
Guard-DBTI	15.3	54.2	DJPD (N^+PN)
Guard	13.4	45.9	Guard-DBTI
DJPD (N^+PN)	0.444	43.6	Guard

Table 3. Sensitivity at 633 nm for StaG-R and other photodiodes (Hinckley & Jansz, 2007).

Comparing the ***relative crosstalk*** parameter, the StaG-R geometry is second only to the double junction photodiode for both modes of illumination. This confirms the initial hypothesis: By adding an extra StaG dimension to the control of carrier direction, normal to the existing StaG's vertical direction, will capture more lateral carriers -benefiting crosstalk reduction.

Comparing the ***sensitivity***, the StaG-R geometry back illuminated is the best. Again the initial hypothesis is vindicated: A second StaG layer, normal to the first will enhance sensitivity. This is a significant result for back illumination applications. Though front illuminated StaG-R sensitivity is below the StaG sensitivity, the result is not significant and front illuminated StaG-R still shows significant crosstalk suppression upto 95% (Fig. 9). Further characterisation of the superiority of the StaG-R response resolution for illumination with other wavelength, similar to the StaG, 400 - 1200 nm, is required.

4.2.6 StaG-R – the next step.

The results indicate that StaG-R response resolution can be improve further by:

- Increasing the degree of ridge nesting and to approximate a continous concentration gradient by increasing the number of StaG layers.
- Reducing the total and individual thickness of the StaG layers, so increasing the potential gradient and improving the drift coefficient.
- Reducing the nested ridge thickness by reducing the vertical gap between ridges, so again benefiting potential gradient and drift coefficient.

HOWEVER, the most obvious obstical to StaG-R's or StaG-concave's physical application is their complexity of fabrication: the ridge nesting and StaG concaving procedures are non-existant in industry. This hurdle has led to the next step in the evolution of the StaG genre.

4.3 StaG-BTI hybrid

Boundary trench isolation (BTI) and double-BTI in particular, incorporated in single junction photodiode pixels, have shown benefit for response resolution, especially combined with a guard-ring electrode (Table 2 & 3). Similarly, a StaG-BTI hybrid could be beneficial. Replacing the nested ridges with a single highly doped BTI ridge, extending from the substrate to the frontwall, would represent a two layer lateral StaG. The rationale is that the BTI would replicate the benefits of the StaG-R's nested ridge by removing the outer lower doped ridges, leaving behind the central substrate ridge as a BTI, with associated benefit to carrier capture efficiency (Fig. 13) (Jansz & Hinckley, 2006).

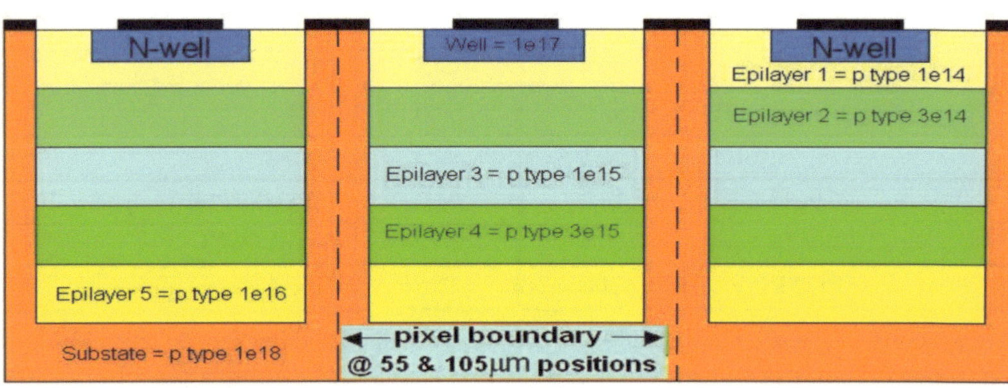

Fig. 13. The simulated Stacked Gradient Homojunction (StaG) Photodiode array with pixel Boundary Trench Isolation (BTI) extending to the frontwall (Jansz & Hinckley, 2006).

4.3.1 StaG-BTI relative crosstalk dependence on BTI width

Fig. 14 compares the relative crosstalk at 633 nm of the StaG-R (Fig. 7) and StaG-BTI (Fig. 13) with that of the StaG (ridge height = 0) (Fig. 1) and normal photodiode (ridge height = -1)

(Fig 3). Here Fig. 8 results for StaG-R are superimposed on the StaG-BTI results for comparison. The horizontal scale is a dual scale for both StaG-R ridge height and BTI width.

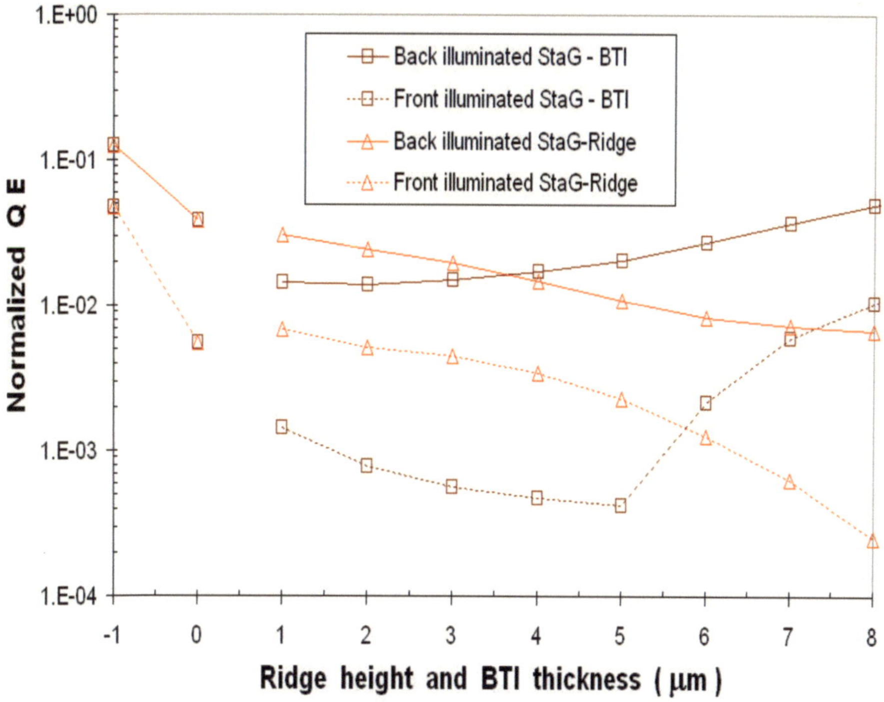

Fig. 14. Relative crosstalk for 633 nm illumination, of StaG-R and StaG-BTI compared to StaG PD (BTI thickness = 0) and conventional photodiode (BTI thickness = -1) as a function of ridge height and BTI thickness (Jansz & Hinckley, 2006).

Comparing the lowest back illumination crosstalk percentage reduction below the StaG response, the StaG-R is significantly superior (80%) to the StaG-BTI (60%). The front illumination results are similar: StaG-R is moderately superior (95%) to the StaG-BTI (90%). The back illuminated StaG-BTI relative crosstalk increases for thicker BTI, because the illumination is proximal to the substrate, allowing carriers to diffuse laterally as well as channel more and more into a widening BTI. Conversly, the front illuminated StaG-BTI relative crosstalk reduces up to 5 μm BTI width, because the illumination generates carriers in the StaG layers close to the frontwall where the higher doped BTI presents a barrier to crosstalk; a barrier that improves with thickness. Larger than 5 μm BTI width, the illumination increasingly intersects the BTI in which generated carriers can increasingly channel as crosstalk as the BTI widens (Jansz & Hinckley, 2006).

4.3.2 StaG-BTI sensitivity dependence on BTI width

Fig. 15 shows that the StaG-BTI is superior in sensitivity to the StaG-R, StaG and conventional photodiodes. Back and front illuminated StaG-BTI have equivalent sensitivities. Noted is the dramatic increase in sensitivity for the front illuminated StaG-BTI, while the FW StaG-R drops in sensitivity below that of the FW StaG.

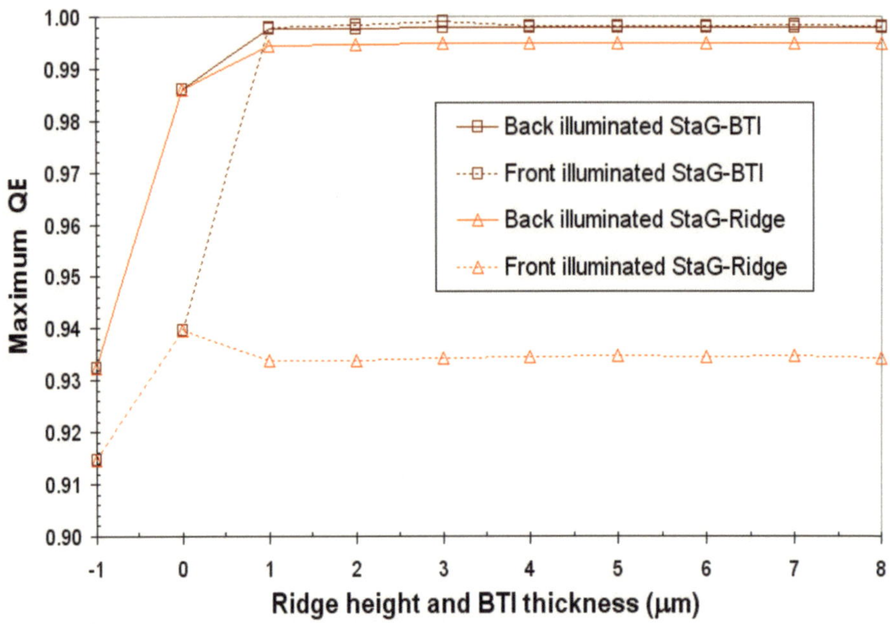

Fig. 15. Maximum quantum efficiency of StaG-R and StaG- BTI compared to StaG PD (ridge height/BTI thickness = 0) as a function of ridge height and BTI thickness (Jansz & Hinckley, 2006).

In Fig. 15, the front illuminated StaG-BTI maximum QE occurs just outside the well in the p-epilayer, while the StaG-R maximum QE occurs on the well wall, resulting in a higher minority hole diffusion current for the latter. Back illuminated StaG-BTI and StaG-R have their maximum sensitivity at the pixel centre (80μm position) with photo-generation of minority hole diffusion current in the well marginally higher for the StaG-R, while their electron currents are similar.

4.3.3 StaG-BTI crosstalk and sensitivity score table: comparing photodiodes

Using the previous Tables 2 and 3, StaG-BTI relative crosstalk and sensitivity can be compared to the other photodiodes. For relative crosstalk, StaG-BTI back and front illuminated is third best, just below StaG-R, in Table 2, at 1.4% and 0.042% QE, respectively. For sensitivity, StaG-BTI back and front illuminated is at to top of Table 3 at 99.8 % and 99.9% QE, respectively. Sensitivity is slightly superior to StaG-R because the substrate doped BTI extends to the frontwall, while the central ridge for the StaG-R is 2 μm shorter.

4.3.4 StaG-BTI – The next step.

Though the nested StaG ridges is still more effective as a minority carrier mirror, the StaG-BTI is significantly less complex for fabrication. Though sensitivity is superior for the StaG-BTI, the primary issue with its elevated crosstalk is the problem of carrier diffusion channelling; a problem that is also present in the StaG-R. The main reason for this is that the nested ridges and the BTI straddle the pixel boundary. Therefore any illumination in a neighbouring pixel, next to the pixel boundary, will always intersect the BTI or nested ridge, resulting in carrier diffusion channelling and its related crosstalk.

The next step makes appropriate changes to the inter-pixel architecture so as to elliminate this pixel boundary straddle problem. Using insulator (SiO2) BTI is also explored.

4.4 StaG-Double-BTI hybrid

Introducing a BTI either side of the pixel boundary removes the problem of channelling, because the boundary illumination now intersects a dead space between the BTI, where carriers are trapped and eventually recombine. Using insulation BTI (SiO_2) can also prevent the problem of channelling for both single and double BTI. The effect of both doped double BTI (DBTI) (Fig. 16) as well as insulated (SiO_2) single BTI and DBTI (Fig. 17) have been characterised using the same device simulator, with device and laser characteristics similar to previous photodiode configurations simulated to allow useful comparisons.

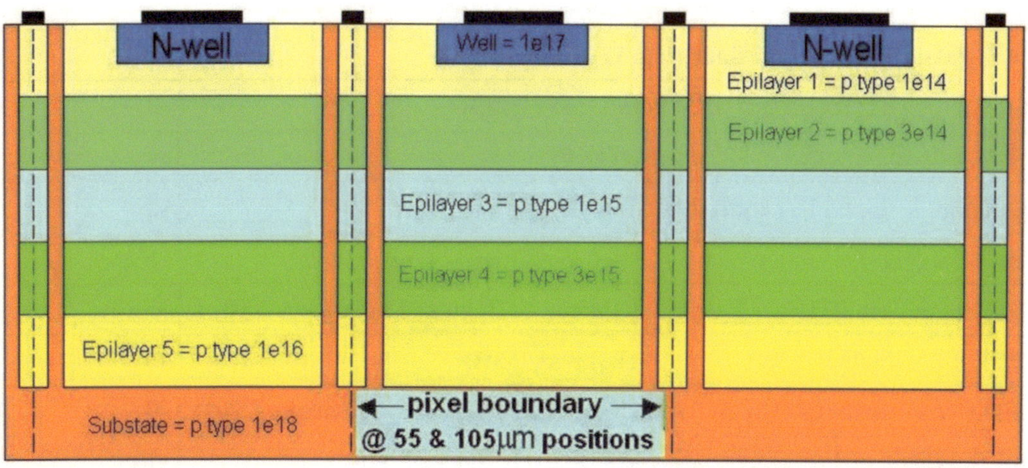

Fig. 16. The StaG photodiode array with inter-pixel Double Boundary Trench Isolation (DBTI) with p+ substrate doping, extending to the frontwall (Jansz & Hinckley, 2008).

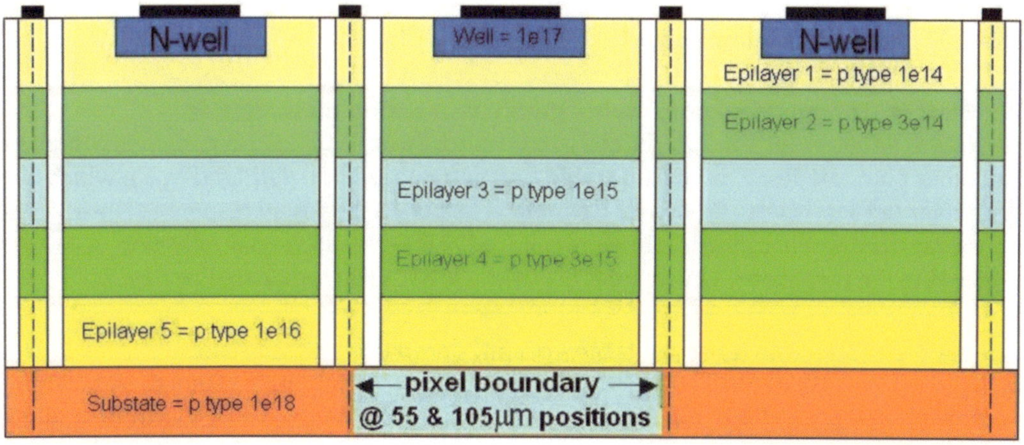

Fig. 17. The StaG Photodiode array with inter-pixel Double Boundary Trench Isolation (DBTI) consisting of SiO_2, extending to the frontwall (Jansz & Hinckley, 2008).

4.4.1 Score table – graph legend: comparing photodiodes

Table 4 contains the horizontal axis legend of the photodiode configurations (negative values) for Fig. 18 and Fig. 19. The positive values on the same axis refer to the doped DBTI widths in microns. "SJPD" refers to "single junction photodiode".

Photodiode Configuration	**Horizontal axis number (Fig 20 & 21)**
BTI width (μm) for StaG Twin BTI 6 μm apart (Fig. 16)	1 - 5
Double Junction photodiode - 12 μm substrate (Jansz-Drávetzky, 2003)	-1
StaG twin BTI SiO_2 1 μm thick (Fig. 17)	-2
StaG single BTI SiO_2 1 μm thick (similar to Fig. 13)	-3
StaG with maximum nested ridges (Fig. 7)	-4
StaG single doped BTI 1 μm thick (Fig. 13)	-5
StaG flat (Fig. 1)	-6
SJPD with twin BTI SiO_2 1 μm thick (Jansz, 2003; Jansz-Drávetzky, 2003)	-7
SJPD with single BTI SiO_2 1 μm thick (Jansz, 2003)	-8
SJPD – convensional (Fig. 3) (Hinckley et al., 2002; Jansz-Drávetzky, 2003)	-9
SJPD with Guard ring electrode and single BTI (Jansz, 2003)	-10
SJPD with Guard ring electrode (Jansz-Drávetzky & Hinckley, 2004; Jansz-Drávetzky, 2003)	-11

Table 4. Horizontal axis number legend for Fig. 18 and Fig. 19.

4.4.2 StaG-DBTI crosstalk score table - graph: comparing photodiodes

Crosstalk is superior for the hybids, SiO_2 and doped Twin BTI StaG photodiodes compared to all other photodiodes, except the Double Junction photodiode (DJPD) (Fig. 2), which also shows retarded sensitivity. Frontwall crosstalk is below the backwall response. The physical mechanism driving the reduction in crosstalk for DBTI StaG is internal reflection of carriers generated in the neighbouring pixel and between the twin BTI (Jansz & Hinckley, 2008).

4.4.3 StaG-DBTI sensitivity score table - graph: comparing photodiodes

Sensitivity (BW/FW) of StaG hybrids (99.8/99.8%) is above non-StaG geometries, including the conventional photodiode (SJPD) (93/91%), the SJPD with guard ring electrode and BTI (15/54%), SJPD and guard ring electrode only (13/46%) and the DJPD (0.004/54%). DJPD sensitivity is reduced, especially for the backwall DJPD, as the majority of carriers are generated outside the outer guard SCR (Jansz-Drávetzky, 2003).

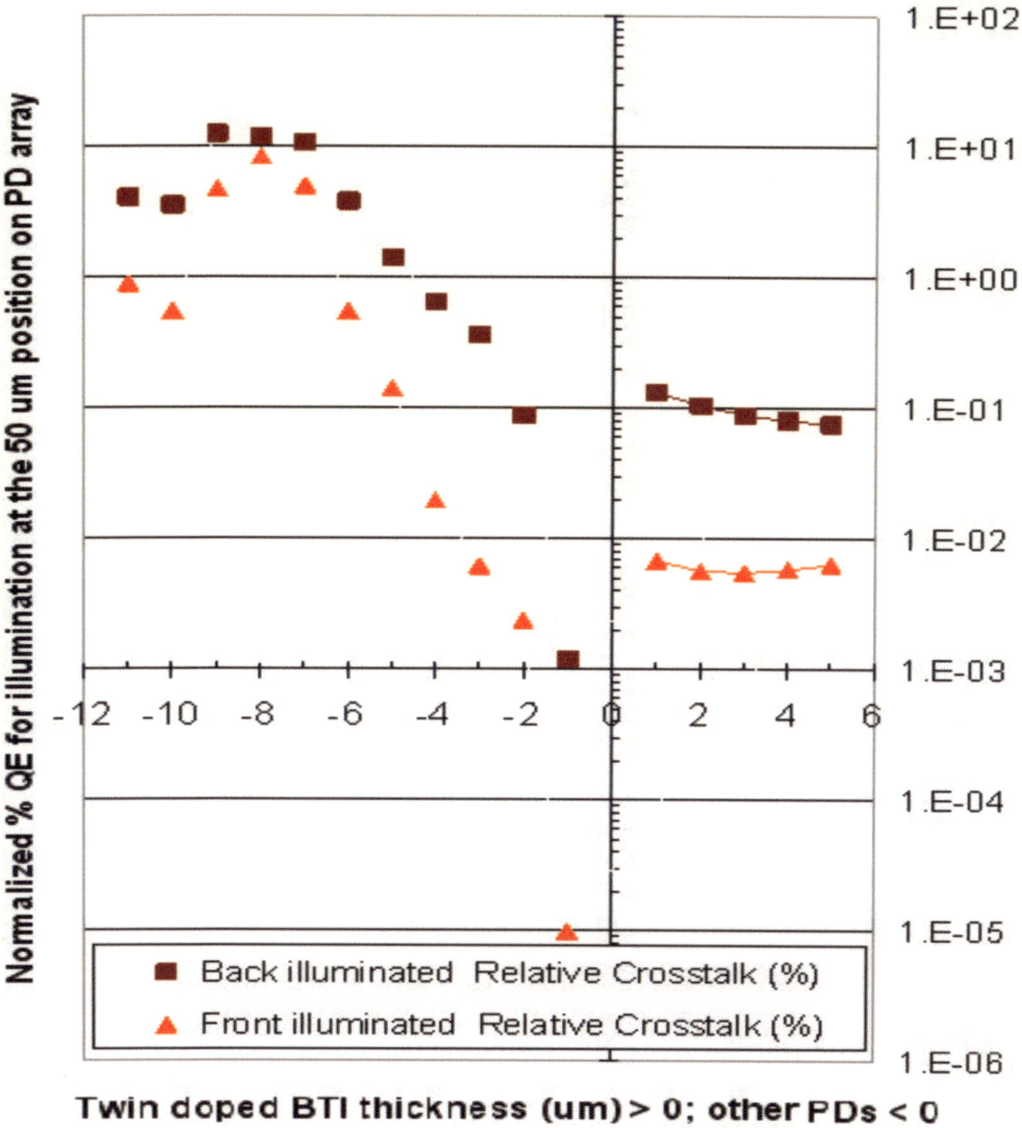

Fig. 18. Relative Crosstalk for Table 4 photodiodes.

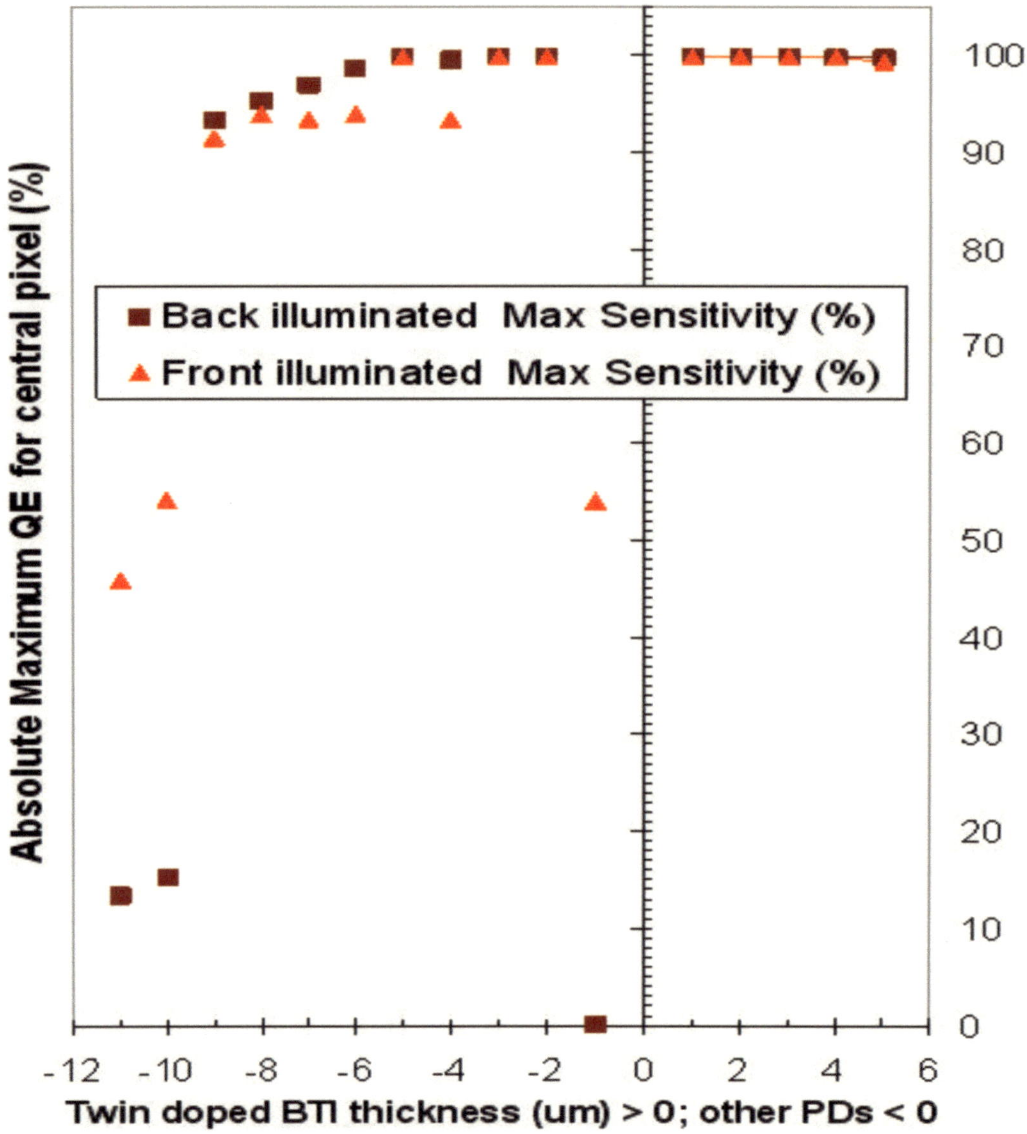

Fig. 19. Maximum QE for Table 4 photodiodes.

5. Future trends for the StaG photodiode genre

One extrinsic evolutionary pressure driving improvement comes from the substrate's minimum doping constraint being only 10^{14} cm^{-3}, resulting in insufficient SCR volume for the primitive SJPD. If substrate doping could be ten times less, at 10^{13} cm^{-3}, each 12 µm thick pixel would be fully depleted with SCR widths of 14 - 21 µm for 1 - 3 volt reverse bias, respectively. The result would be better photodiode response resolution than any of the present doping constrained StaG hybrids. However, the StaG hybrids could also benefit from a lowering of the doping constraint.

Further characterisation of the latter StaG genre in terms of device response resolution for the wavelength range used to characterise the generic StaG photodiode is needed to understand the StaG's response dependence on wavelength between 400 to 1200 nm. Other than 633 nm, other wavelengths are of interest due to niche applications or multi-wavelength specificity. Present research has (Jansz & Hinckley, 2010) and is investigating the suitability of application of the StaG-hybrid configuration to the poly-well geometry to realise back illuminated StaG-polywell photodiodes that have application to ultra-violet/blue sensing.
The consideration at the beginning of this chapter, regarding architectures predicted to benefit back illuminated photodiode response resolution, has opened a number of research directions within the StaG genre as well as within the well-geometry photodiode genre.

6. Conclusion

This StaG genre explosion was sparked by a single idea: exploit the StaG ability to control carrier transport. It was along a path of device extrinsic evolution. This extrinsic pressure was proactive, rather than passive. It resulted in a process that aimed to achieve photodiode architectures that balanced the maximization of response resolution with the minimization of device fabrication complexity. This process has produced a time sequence of individual creations, through simulations, starting with the conventional vertical single junction photodiode (SJPD) with just well and substrate (Fig. 3). From this prototype, various branches have emerged. So far, these branches have form into a penta-dactile tree structure of vertical SJPD genre: Guard ring electrode SJPD, BTI-SJPD, Guard junction SJPD (DJPD), StaG-SJPD and Polywell SJPD.
This development was driven primarily by the need to improve on the backwall illumination CMOS photodiode response, because of its advantages over the frontwall illumination mode. However, most of the improvements also benefit frontwall illuminated CMOS photodiodes across a broad spectrum.
The present results indicate the prospect of obtaining significant crosstalk suppression and sensitivity enhancement in CMOS imaging arrays through achievable modifications to the array structure with the view to producing high-speed high-resolution imaging systems. Research in progress is investigating other StaG hybrids, as well as scaling effects down to 5μm pixel pitch on the benefits of these and other photodiode genre still to be exploited.

7. References

Brouk, I.; Nemirovsky, Y.; Lachowicz, S.; Gluszak, E.A.; Hinckley, S.; Eshraghian, K. (2003). Characterization of crosstalk between CMOS photodiodes. *Solid State Electronics*, 46, 53-6.

Dierickx, B. & Bogaerts, J. (2004). NIR-enhanced image sensor using multiple epitaxial layers. *Proceedings of SPIE – IS&T Electronic Imaging*, 5301, pp. 205–212.

Furumiya, M.; Ohkubo, H.; Muramatsu, Y.; Kurosawa, S.; Okamoto, F.; Fujimoto, Y.; Nakashiba, Y. (2001). High-sensitivity and no-crosstalk pixel technology for embedded CMOS image sensor. *IEEE Transaction on Electron Devices*, 48, 2221 - 6.

Ghazi, A.; Zimmermann, H. & Seegebrecht, P. (2002) CMOS photodiode with enhanced responsivity for the UV/Blue spectral range, *IEEE Trans. Electron Devices*, vol. 49, pp. 1124 – 8.

Goushcha, I.; Tabbert, B.; Popp, A.; Goushcha, A.O. (2007). Photodetectors based on back-illuminated silicon photodiode arrays for x-ray image systems. *IEEE Sensors Applications Symposium (SAS)*, pp1 – 6, San Diego, California USA, 6 – 8 Feb.

Hinckley, S.; Gluszak, E.A.; Eshraghian, K. (2000). Modelling of device structural effects in backside illuminated CMOS compatible photodiodes. *Proc. of Conference on Optoelectronic and Microelectronic Materials and Devices (COMMAD)*, pp 399 - 402, Melbourne. IEEE Press.

Hinckley, S.; Jansz, P.V.; Gluszak, E.A. & Eshraghian, K. (2002). Modelling of device structure effects on electrical crosstalk in back illuminated CMOS compatible photodiodes, *Proc. of Conference on Optoelectronic and Microelectronic Materials and Devices (COMMAD)*, pp 397 – 400, Melbourne. IEEE Press.

Hinckley, S.; Jansz, P. V. & Eshraghian, K. (2004) Pixel structural effects on crosstalk in backwall illuminated CMOS compatible photodiode arrays. *Proc. of DELTA 2004 Conference.* pp 53 - 6, Melbourne, IEEE Press.

Hinckley, S. & Jansz P.V. (2005) "Stacked homojunction effects on crosstalk and response resolution in CMOS compatible photodiode arrays.", Proc. Of IFIP WG 10.5 conference on VLSI-System on a Chip (VLSI-SoC 2005), Perth, IFIP, pp. 383-388.

Hinckley, S. & Jansz, P.V. (2007) The effect of inter-pixel nested ridges incorporated in a stacked gradient homojunction photodiode architecture. *Proc. of SPIE Conference on Smart Structures, Devices and Systems III*, pp 64141T-1 – 12, ISBN: 9780819465221, Adelaide, Dec. 2006, SPIE, Bellingham, Washington State USA.

Jansz P. V. (2003) Pixel boundary trench effects on a CMOS compatible single junction photodiode array, with and without guard-ring electrodes, pp 1 – 4, unpublished.

Jansz-Drávetzky, P.V. & Hinckley, S. (2004). Guard-ring electrode effects on crosstalk in simulated 2D CMOS compatible verticle photodiode pixel arrays. *Proc. of Conference on Optoelectronic and Microelectronic Materials and Devices (COMMAD)*, pp 299 – 302. Melbourne, IEEE press, New Jersey, USA.

Jansz, P.V. & Hinckley, S. (2006) Inter-pixel boundary trench isolation effects on a stacked gradient homojunction single junction photodiode pixel architecture. *Proc. of Conference on Optoelectronic and Microelectronic Materials and Devices (COMMAD2006)*, Melbourne. IEEE Press, New Jersey, USA.

Jansz, P.V. & Hinckley, S. (2008) Double boundary trench isolation effects on a stacked gradient homojunction photodiode array. *Proc. of Conference on Optoelectronic and Microelectronic Materials and Devices (COMMAD)*, pp 156 – 9, ISBN: 9781424427178, Sydney, July 2008, IEEE press, New Jersey, USA.

Jansz, P.V. & Hinckley, S. (2010) Characterisation of a hybrid polywell and stacked gradient poly-homojunction CMOS photodiode. *Proc. of Conference on Optoelectronic and Microelectronic Materials and Devices (COMMAD)*, Canberra, Dec. 2010, in press.

Jansz, P.V., Hinckley, S. & Wild, G. (2010) Effect of a polywell geometry on a CMOS photodiode array. *23rd IEEE International System On a Chip Conference (SOCC)*, Las Vegas, Sept. 2010, in press.

Jansz-Drávetzky, P. V. (2003) *Device structural effects on electrical crosstalk in backwall illuminated CMOS compatible photodiode arrays.* Honours Thesis. Edith Cowan University. Perth, Western Australia.

Lee, J.S.; Jernigan, M.E.; Hornsey, R.I. (2003). Characterization and deblurring of lateral crosstalk in CMOS image sensors, *IEEE Transaction on Electron Devices*, 50, 2361 - 8.

Shcherback, I. & Yaddid-Pecht O. (2003) Photoresponse analysis and pixel shape optimization for CMOS active pixel sensors. *IEEE Transaction on Electron Devices*, 50(1), 12 - 8.

Singh, J. (1994). *Semiconductor Devices: An Introduction.* McGraw-Hill, New York.

Streetman, B. G. & Banerjee, S. (2000) *Solid state electronic devices.* Prentice Hall, ISBN: 0-13-026101-7, New Jersey, 5th edition.

10

Silicon Photodiodes for Low Penetration Depth Beams such as DUV/VUV/EUV Light and Low-Energy Electrons

Lis K. Nanver
Delft Institute of Microsystems and Nanoelectronics (DIMES)
Delft University of Technology
The Netherlands

1. Introduction

When the attenuation lengths of beams in silicon are well below a micron, a high responsivity of silicon photodiodes can only be reached if the photo-sensitive region for detection is close to the surface. The direct way to achieve this is to create ultrashallow, damage-free junctions. This has in fact been one of the challenges that the silicon-based-technology CMOS has been struggling with for the last two decades: such junction depths are specified in the International Technology Roadmap for Semiconductors (ITRS) in order to continue the aggressive downscaling of MOS devices as dictated by Moore's law (ITRS 2009). However, although technologies have been developed for junctions as shallow as 200 nm and below, the resulting diodes are mainly far from being damage-free (Borland et al., 2010). Schottky-type junctions represent the limit in shallowness, but are mainly unattractive due to a high reverse leakage current, surface recombination, reflection and absorption in the front metal, and low surface electric field. Therefore, silicon photodiode research has directed efforts towards increasing the sensitivity near the silicon surface by creating damage-free doped regions with an electric field, also outside the depletion region, that transports the generated carriers to the terminals. An alternative method has also been demonstrated where a Si-SiO_2 interface is used to create an inversion layer with a conductive channel at the interface for transporting the generated carriers. Photodiodes that have found application for the detection of low-penetration beams have been produced by such methods, but they have issues such as poor process control, low yield, and poor radiation hardness (Funsten et al., 2004; Silver et al., 2006; Solt et al., 1996; Tindall et al., 2008).

This chapter reviews a boron-layer silicon photodiode technology that can be used to create an extremely shallow p^+n junction, and therefore no "tricks" are needed to place the photo-sensitive surface within nanometers of the Si surface. An amorphous boron (α-B) layer, which can be down to about 1 nm thick, is formed on the surface of the silicon by chemical vapor deposition (CVD) of pure boron. From this layer an extremely shallow doping of the Si surface is effectuated, forming a p^+n junction that can be readily made in the ~ 1 – 10 nm junction-depth range. This is achieved by applying low processing temperatures from 500 to 700 °C. The properties of the α-B layer, both chemically and electrically, are responsible for achieving exceptional photodiode performance that surpasses that of other existing technologies on points such as internal/external quantum efficiency, dark current, uniformity and degradation of responsivity. At the same time the B-layer process is fully compatible with Si front-end technology, and these photodiodes readily lend themselves to detector integration schemes that allow low parasitic resistance and capacitance as well as on-chip integration with other electronic elements.

These properties have lead to a fast qualification for production of several types of B-layer photodiode detectors for industrial applications. Three examples of such applications are described in Section 4:

- vacuum ultraviolet (VUV) detectors (Shi et al., 2010) for which the attenuation length of the light in Si is as low as 5 nm. This includes the deep ultraviolet (DUV) wavelength of 193 nm used in advanced lithography systems (Sarubbi et al., 2008a);
- extreme ultraviolet (EUV) detectors (Sarubbi et al., 2008b) for detecting light at a wavelength of 13.5 nm. This is essentially soft X-rays that have an attenuation length of 700 nm in Si. This wavelength has been chosen for use in future advanced lithography tools;
- low-energy electrons that for energies around 500 eV have ranges in Si below ~ 10 nm (Šakić et al., 2010b). Particularly the application in Scanning Electron Microscopes (SEMs) is explored here.

2. Nanometer-deep junction formation from α-boron layers

The deposition of α-B layers is performed in a commercially available epitaxial CVD reactor using diborane (B_2H_6) and hydrogen (H_2) as the gas source and carrier gas, respectively. The details of this process, that can be performed either at atmospheric or reduced pressures are given in (Sarubbi et al., 2010a) for deposition temperatures ranging from 500 to 700 °C and various doping gas conditions. The formation of the boron layer is slower the lower the temperature and the diborane partial pressure, but at high gas-flow rates, which provide good conditions for segregation of boron atoms on the Si surface, it is essentially controlled by the exposure time. An example is shown in Fig. 1 for constant temperature, pressure and B_2H_6 flow-rate, where the deposition rate is constant for depositions longer than ~ 1 min.

An example of a B-layer formed after 10 min B_2H_6 exposure at atmospheric pressure for a temperature of 700 °C is seen in the high-resolution transmission electron microscopy (HRTEM) image of Fig. 2, where the segregation of B atoms in an amorphous layer and the reaction with silicon atoms to form a boron-silicide phase at the interface can be discerned. The α-B layer is a conductive semi-metal found to have a high resistivity of ~ 10^4 Ωcm.

To maintain an ultrashallow junction depth that is only determined by thermal diffusion of the boron into the Si at the given processing temperature, it is important that the doping process is free of defects that can cause boron-enhanced or transient-enhanced diffusion (TED) effects. This was evaluated by examining the out-diffusion of epitaxially grown B-doped Si markers after long B-depositions at 700 °C (Sarubbi et al., 2010b). Layers containing more than 10^{17} cm^{-2} boron atoms were deposited, giving about 1 nm of boron silicidation at the interface, but the results gave no indication of TED effects. This result substantiates the conclusion, also drawn from the excellent properties found for B-layer photodiodes, that an effectively damage-free junction is formed.

The c-Si surface is doped up to the solid solubility solely by thermal diffusion. For 500 °C depositions this gives junction depths of ~ 1 nm, and at 700 °C junctions of less than 10 nm

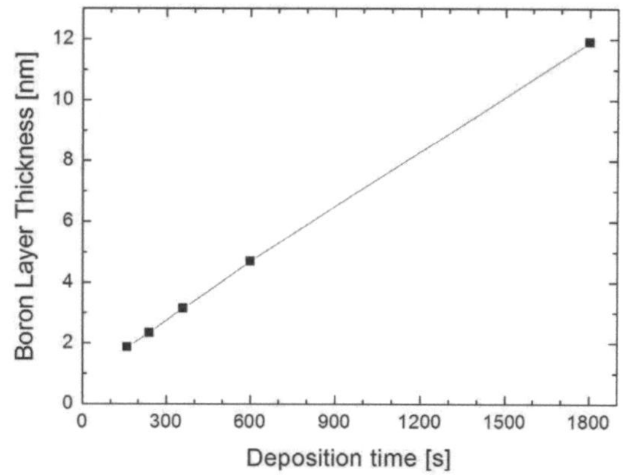

Fig. 1. Thickness of boron layers measured by ellipsometry as a function of time for depositions at a pressure of 760 Torr, a temperature of 700 °C, and a diborane flow-rate of 490 sccm (Šakić et al., 2010a).

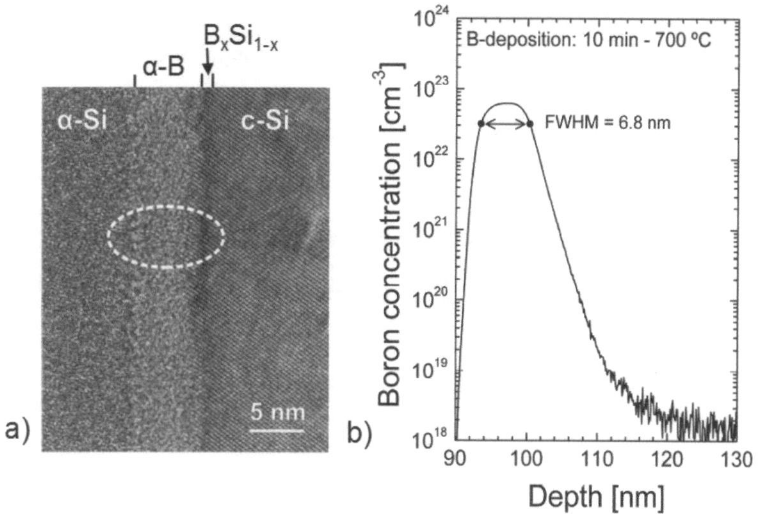

Fig. 2. (a) High-resolution TEM image and (b) SIMS profile (O_2^+ primary ion beam at 1 keV) of an as-deposited B-layer formed on a (100) Si surface at 700 °C after 10 min B_2H_6 exposure (Sarubbi et al., 2010b).

deep are readily formed. In the latter case the doping will be as high as ~ 2×10^{19} cm^{-3} (Vick & White, 1969). For further doping by post-deposition thermal drive-in of the boron, the formation of the α-B layer has two distinct advantages: it acts as an abundant source of dopants and also prevents boron desorption from the Si surface. This is in contrast to the results of other works that also aimed to use depositions from diborane and subsequent thermal annealing to obtain higher dopant activation and deeper junction depths (Inada et al., 1991), (Kim et al., 2000). The difference lies in the fact that in these cases the diborane exposure conditions were designed to avoid or minimize the formation of a distinct layer of boron. To avoid B-desorption during drive-in, an oxide capping layer was proposed, but still the available B will be limited under the given deposition conditions.

To create a p⁺n diode the B-deposition can be performed with high selectivity in a silicon dioxide window on an n-type c-Si surface. This requires that the Si surface is native-oxide free, which can be achieved by HF dip-etching, possibly followed by hydrogen pre-baking such as those seen in Fig. 3. Nevertheless, the α-B layer is continuous and uniform across the Si, and the deposited thickness is independent of the window size. This isotropic boron coverage, selectively on all exposed Si surfaces, considerably enhances the integration potentials of this process among other things for fabricating high-quality diodes.

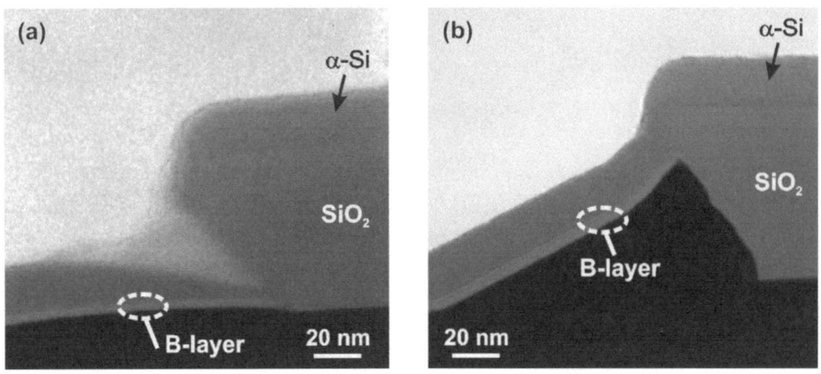

Fig. 3. TEM images of contact windows treated with a 2.5 min B-deposition at 700 °C. The SiO_2 etch geometry has been induced by a low-pressure *in-situ* thermal cleaning at 900 °C before diborane exposure (Sarubbi et al., 2010b).

3. Electrical characteristics of nanometer-deep junctions

Nanometer deep p-n junctions such as the B-layer junctions may exhibit electrical current-voltage characteristics that deviate considerably from those of conventional deep junctions. For the first, the metal acts as a sink for minority carrier injection that hence increases as the junction becomes more and more shallow. Thus the dark current is increased. Moreover, the doping of the junction can become so low that it is completely depleted. This leads to punch-through phenomena that also will increase the current through the diode, and again also increase the dark current, often by decades.

3.1 Theoretical considerations

Consider the case of an n-Si substrate that is exposed to a process for p-doping the surface and that this, upon metallization, may result in anything between a deep metal/p-Si/n-Si (m-p-n) junction and a metal/n-Si (m-n) Schottky diode where effectively no doping is realized. The electron and hole currents in the different situations that can occur in the transition from a deep to an ultrashallow through to a Schottky junction are illustrated by the device simulations shown in Fig. 4. For a detailed analysis of these I-V characteristics, including an analytical model that unifies the standard Schottky and p-n diode formulations, the reader is referred to (Popadić et al., 2009). Three main types of diode behavior can be identified:

(a) an m-n Schottky diode: the diode current is dominated by the injection of the majority carrier (electrons) from the semiconductor into the metal. At the same time, a very small current of holes is injected from the metal into the semiconductor.

(b) an m-p-n diode fully-depleted by a high Schottky barrier height (SBH): an ultrashallow heavily-doped p-type region is created at the surface of the n-type substrate, and the contact to this region is a Schottky contact. Under reverse and small forward bias, the p-region is fully depleted by the metal-semiconductor depletion region and the diode shows the electrical characteristics of an n-Schottky. The current is dominated by electron injection into the metal, but the effective SBH is so high that the total current is much lower than the pure Schottky case and also the hole-current is much higher. At a high enough forward-bias voltage, the p-region can become non-depleted, effectively reducing the electron current to the point where the device behaves as an m-p-n junction diode with the hole-current level approaching that of the electron current. It should be noted that the transition from Schottky-like to pn-junction-like behavior can of course also occur in the reverse voltage region depending on the doping levels.
(c) a non-depleted m-p-n diode: the diode current is dominated by the injection of holes from the p^+ region into the n-substrate and the hole-current is much higher than in cases (a) and (b). For shallow junctions the metal forms a sink for the minority carrier electron injection, and this will give an increase of the otherwise very low electron-current for junction depths below ~ 20 nm. In the example treated in Fig. 4c the electron current becomes about as high as the hole current for a junction depth d = 10 nm. As the junction depth goes to d = 0, the electron current will increase to levels far above the hole current to finally reach Schottky current levels as high as that seen in Fig. 4a.

3.2 Electrical properties of α-B layer diodes

The different diode I-V characteristics described in the previous section have all been observed in the case of B-layer diodes. An example is given in Fig. 5 where the results are displayed for different B-layer deposition times and temperature followed by metal contacting with Al. Even a very short 1 s deposition lowers the current level decades below the pure Al Schottky level, and already for a couple-of-minutes-long deposition the saturation current approaches a value that is typical for conventional deep junctions. If the boron semi-metal layer was functioning as a sink for the electron injection (Sarubbi et al., 2010b, 2010c), this would not be expected for these junction depths of well below 10 nm. In fact, it has been established experimentally that the α-B layer attenuates rather than sinks the electron injection, which also is evidence that it is not pure metallic in this form. This attenuation effect has the attractive consequence that these extremely shallow junctions can be produced with much lower dark currents than would otherwise be possible.
A drawback of the non-metallic nature of the α-B layer is the fact that it has a high resistivity of ~ 10^4 Ωcm. This results in a high series resistance to the contact metal even for nm thin layers, which is clearly seen in Fig. 5 as the large attenuation of the I-V curves at high forward biasing for the longer deposition times. A suitable compromise between series resistance and saturation current level is often found for deposition times of about 1 min where the α-B layer thickness is below the tunneling thickness of ~ 3 nm.

4. Integration of B-layer photodiodes in detectors

The B-layer diode technology is fully compatible with silicon front-end processing, a fact that has been of crucial importance for the realization of the three detector types developed for industrial applications and discussed in the following. Not only have special detector

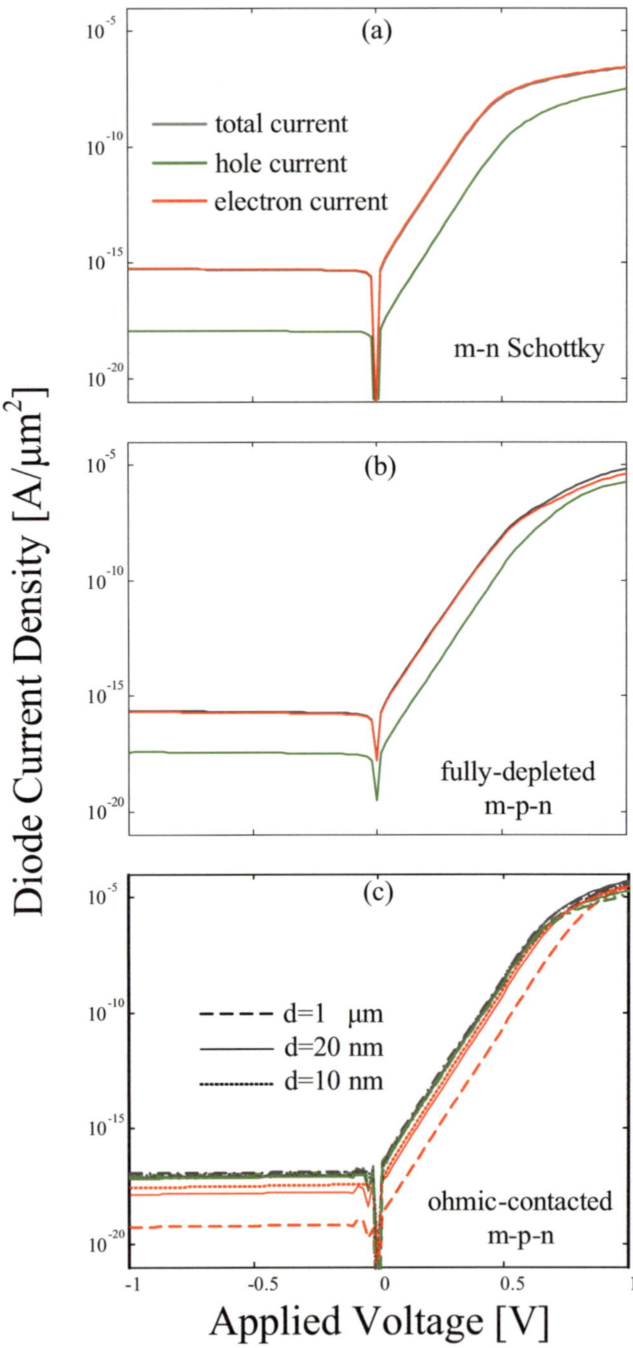

Fig. 4. Simulated output characteristics of (a) an m-n Schottky diode with SBH=0.85 V, (b) a fully-depleted m-p-n diode with SBH=0.75 V and junction depth d=20 nm, and (c) ohmic-contacted m-p-n diodes, both with $N_A=1\times10^{20}$ cm^{-3}. The substrate doping is in all cases $N_D=1\times10^{15}$ cm^{-3}.

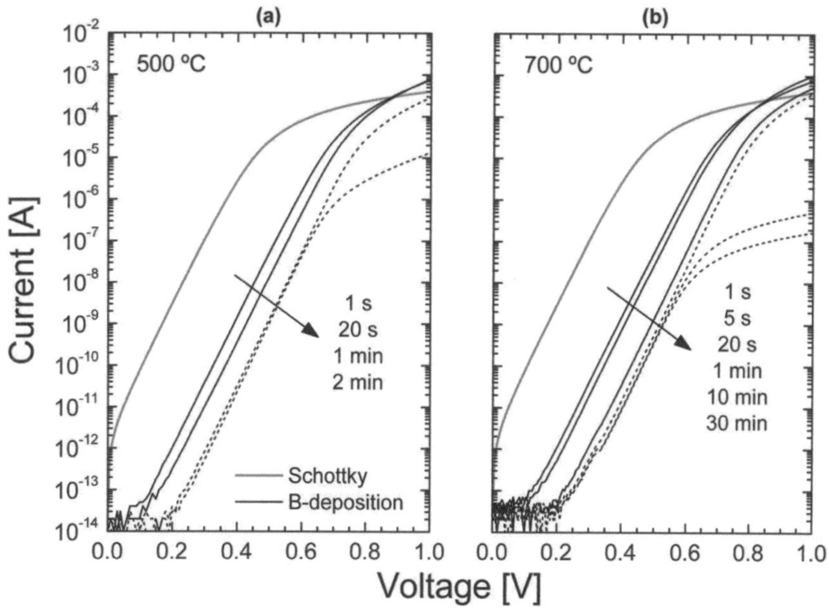

Fig. 5. Diode I-V characteristics for various B-layer deposition times at either (a) 500 °C or (b) 700 °C. The anode area is 2 × 1 μm². For comparison, the I-V curve of a Schottky diode is also included (Sarubbi et al., 2010a).

geometries been realized, but the photodiodes were also integrated with other on-chip active and passive components. For example, bipolar transistors were integrated in EUV detectors as on-chip temperature sensors. The doping regions of such extra devices can often be processed entirely before the anode B-layer deposition since it can be performed below 700 °C, a temperature that will not affect the doping profile of regions already activated at higher temperatures. Likewise, the necessary p-type guard-rings and n-type channel stops that may be essential for keeping the dark current low can be created by implanting and annealing before the anode area of the photodiode is opened for B-deposition. Attention should, however, be paid to the possible autodoping from these pre-fabricated doping regions onto the open anode silicon surface. The position and doping level of extra doping regions must be chosen so that any significant autodoping is avoided if the anode p^+ region is to remain reliably ultrashallow.

Implementing a p-guard-ring can be particularly important for diodes where the n-doping near the metallurgic junction is high, i.e., more than about 10^{16} cm^{-3}, because the very shallow B-layer junctions have a high curvature at the perimeter giving a correspondingly high electric field that can be the cause of early breakdown and high dark current (Sarrubi, 2010c). Moreover, for such extremely shallow junctions the exact topography of the oxide at the edge of the B-layer deposition window, examples of which are shown in Fig. 3, can have an influence on the resulting I-V characteristics. If the post-deposition processing deteriorates the integrity of the oxide at the diode edge, the distance between the n-Si and the contacting metal may become so small that Schottky-like regions with high currents and even shorts to the metal can occur. A p-guard-ring will protect against such effects. An example of the fabrication of p-type guard-rings and n-type channel stops is given in Section 4.3, Fig. 14, where the fabrication of detectors for SEM systems is treated.

The processing after the B-layer deposition is highly facilitated by a number of attractive properties of the layer itself. Pure boron has a very high melt temperature, above 2000 °C, and it is chemically inert with respect to many of the back-end processing materials and etchants applied in silicon IC-technology. In Fig. 6 a typical processing scheme is depicted for the processing of the photodiode contact as well as any extra layer needed to cover the front-entrance window, for example for protection, absorption or filtering purposes.

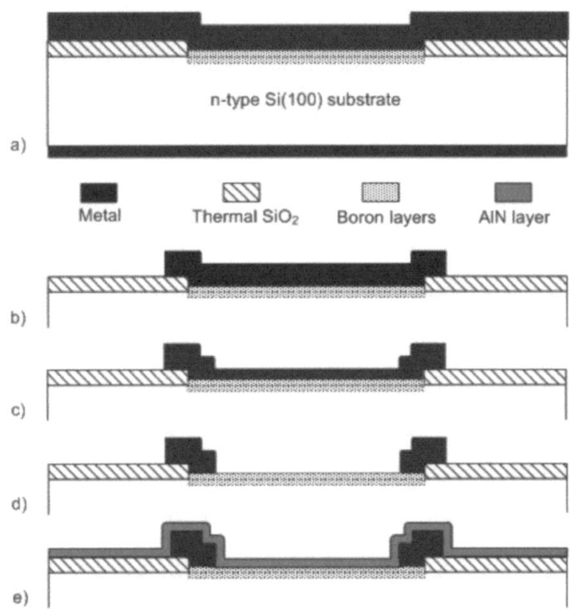

Fig. 6. Schematic of an example of a process flow for anode contact formation when the metal (Al) is deposited directly after the B-layer deposition and removed locally on part of the B-layer region: (a) pure Al deposition, (b) anode contact definition, (c) dry etching to remove most of the Al on the region to be opened, (d) dilute HF etching of the remaining Al using the α-B layer as etch-stop, and (e) deposition of an extra front-entrance window layer (Šakić et al., 2010b).

The high sheet resistance of the B-layer can place a limitation on the series resistance of the photodiode when the as-deposited layer forms the front-entrance window. The series resistance can be significantly lowered by depositing Al directly on the diode surface and patterning it in a grid. This is possible because the Al makes good ohmic contact to the α-B layer and it can be selectively removed by etching in HF. This process is in general applied to contact the p^+ B-doped silicon through the α-B layer, instead of for example contacting through windows in an oxide isolation layer. Due to the high resistivity of the α-B layer the vertical resistance through the layer must also be considered particularly if the contact surface area is small.

The response time of a detector will for millimeter large photodiodes often be dominated by the time constant of the diode RC equivalent circuit formed by the junction capacitance C_j and the series resistance R_S. For the nm-deep B-layer diodes without extra surface layers, the sheet resistance of the p-doped Si can be ~ 10 kΩ/sq or more and this will give the main contribution to the total R_S if photo-generated carriers are collected at a peripheral electrode after flowing through the thin p-doped region (Xia et al., 2008). However, if the application allows it, the conductivity of the active surface layer can be significantly increased by extending the B-deposition cycle *in-situ* with extra depositions, such as p-doped Si, and extra thermal drive-in steps, as described in (Sarubbi et al., 2008b). In addition, conductive films can also be deposited directly onto the entire active area. Such coating layers should be properly optimized for the specific application, since optical absorption will occur at the front-entrance window.

The photodiodes with B-layers deposited at 500 °C offer more flexibility with respect to their integration than the 700 °C diodes. In principle, they could also be fabricated in a back-end processing module, for example on fully-processed CMOS wafers since these will normally be able to tolerate a temperature of 500 °C. However, although the electrical characteristics of the 500 °C diodes have been found to be just as ideal as the 700 °C ones, the reliability of the process is clearly increased by thermally driving the B-dopants further into the Si, thus moving the metallurgic junction away from the surface. At 500 °C the surface doping is extremely limited and a more thorough optical characterization needs to be made to determine whether this has implications for the degradation and radiation hardness of the photodiodes in the different detector applications. All optical characterization reported in the following was performed on diodes fabricated with 700 °C B-layer depositions.

4.1 VUV/DUV radiation detectors

For the VUV spectral range from ~ 100 nm to ~ 200 nm the penetration depth in silicon is extremely small as illustrated in Fig. 7. The name "Vacuum UV" refers to the fact that the light is strongly absorbed by air, and the detectors are mainly operated in vacuum. Particularly high-performance deep-ultra-violet (DUV) photodiodes for 193 nm radiation detection are in high demand due to their application in advanced optical lithography equipment. At this wavelength the penetration depth of the incident radiation in Si is less than 6 nm. However, several materials are transparent at 193 nm, and for example silicon oxides can be used as protection layers that, for the right thickness, may also reduce the reflectivity of the surface thus increasing the responsivity.

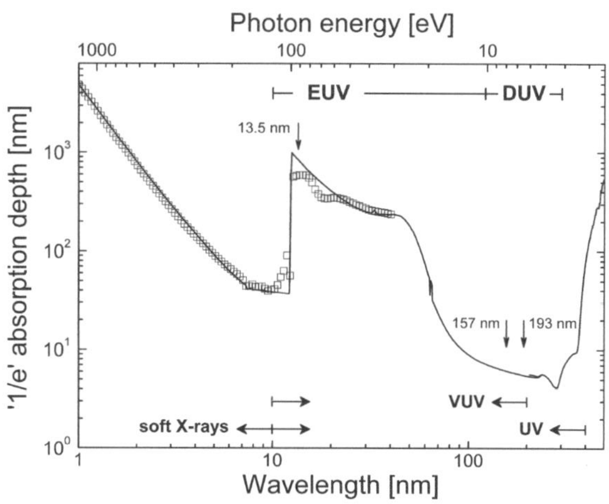

Fig. 7. '1/e' absorption depth in Si as a function of incident radiation wavelength (Palik, 1985; Henke data).

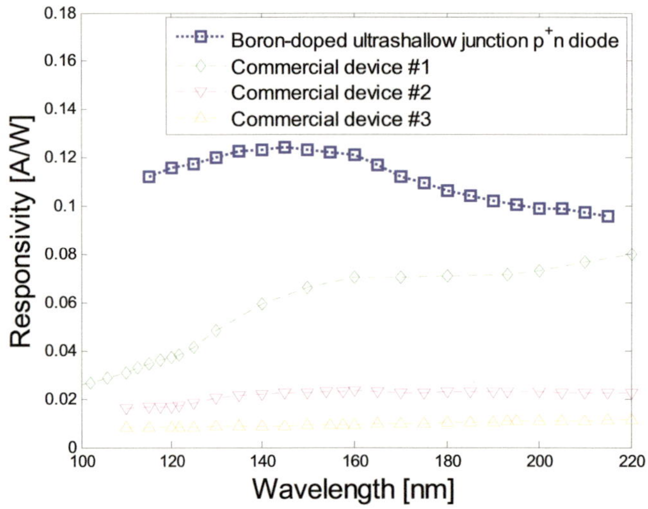

Fig. 8. Responsivity of a B-layer photodiode in the VUV spectral range, compared to three commercially available photodetectors (Shi et al., 2010).

Optical tests were carried out at the synchrotron radiation laboratory of PTB (Physikalisch-Technische Bundesanstalt) in Berlin, Germany (Gottwald et al., 2006, 2010), (Richter et al., 2002). Fig. 8 shows the superior responsivity of the B-layer photodiode compared with other state-of-the-art photodiodes in the VUV spectral range. This excellent optical performance confirms that the B-deposition process can provide ultrashallow and high-quality p^+-doped active surface layers, which can effectively enhance the quantum efficiency by reducing the VUV photon absorption in the front window. The dark current has also been found to be low: on circular photodiodes with an active area of 10.75 mm^2 (a diameter of 3.7 mm) and a reverse biasing of 20 V, the current was only ~ 30 pA as compared to ~ 730 pA for the available commercial diodes of the same area.

The robustness of these diodes was extensively investigated by applying high-dose and/or extended exposures, and it could be concluded that there was no significant responsivity degradation. In all cases, any observed degradation could be related to irradiation-induced charging phenomena. To minimize such effects, the active surface should be made as conductive as possible. This is achieved on diodes with α-B as the top surface layer by avoiding oxidizing post-treatments that create thick native/chemical oxides. In air, the as-deposited α-B layer does not form a native oxide of any significant thickness. To reduce the effective sheet resistance of the anode surface, the aluminum anode contact-metal can be extended to form a grid over the diode surface as, for example, illustrated in Fig. 9. The trade-off is that this additional Al-grid will reduce the effective light-sensitive surface area. In Fig. 10, an example is given of the spectral responsivity during long exposures, in this case for the Al-grid diode monitored for one hour during VUV irradiation at a wavelength of 70 nm. As indicated, the decrease in responsivity stays within 1%. All in all, for VUV radiation detection the B-layer photodiode technology has demonstrated excellent electrical and optical performance in terms of extremely low dark current, outstanding responsivity, and high stability to extended exposures.

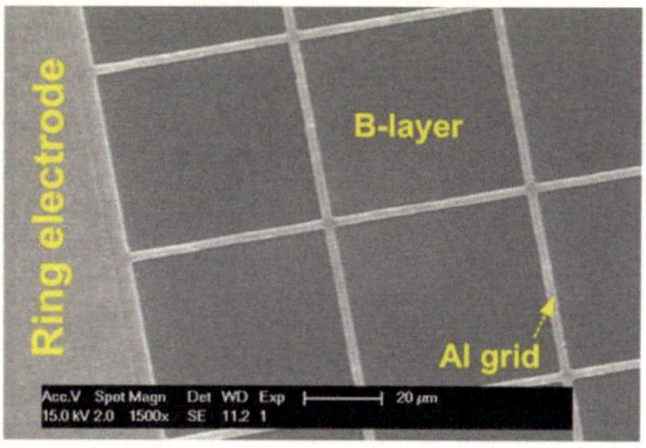

Fig. 9. Photograph of one corner of a photodiode with an Al-grid processed directly on the B-layer anode surface.

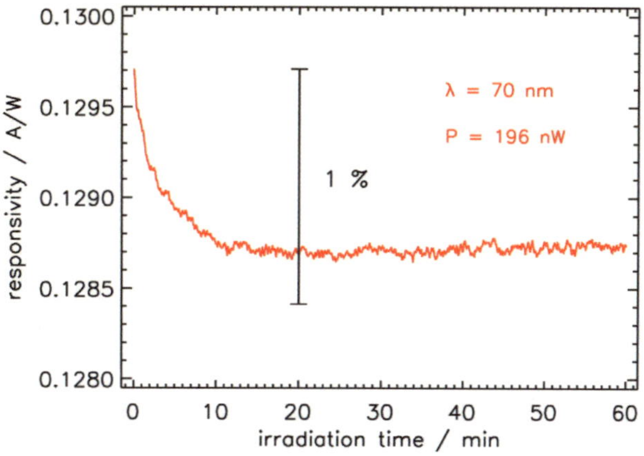

Fig. 10. Degradation of the responsivity of a photodiode with a ~ 3-nm-thick α-B layer and an Al-grid under irradiation at a wavelength of 70 nm as a function of the irradiation time (Shi et al., 2010).

4.2 EUV radiation detectors

The extreme-ultra-violet (EUV) wavelength of 13.5 nm has been selected by ASML Lithography for future EUV lithography equipment, which has been under serious development since 2006 and is forecast to be in high-volume production by 2015 (Benschop et al., 2008). This has spurred interest in photodiode detectors for this wavelength. At the moment the B-layer photodiodes are the only devices qualified for application in the EUV wafersteppers for monitoring the light entering the alignment system from the source (the energy sensor) and for effectuating the actual wafer alignment to the mask (with the transmission imaging sensor and the spot-slit sensor). The penetration depth in Si of this almost soft X-ray light at 13.5 nm is large compared to that of the DUV light, about 700 nm, but the degradation issues are much more severe. Detectors have been fabricated with a number of other on-chip components such as bipolar transistor temperature sensors and optical filter/absorber layers. The latter can be deposited directly on the B-layer surface. For example, zirconium is an attractive filter layer for higher wavelength light (Powell et al., 2010). Moreover, it has the advantage of being metallic with good electrical conductivity, so a surface layer of zirconium can also significantly lower the diode series resistance.

As done for the VUV diodes, the EUV optical characterization was performed at PTB (Klein et al., 2006). A measurement example is shown in Fig. 11 for two B-layer diodes with different surface layers. The one with a ~ 3-nm-thick as-deposited α-B layer as front-entrance window has a responsivity of 0.266 A/W, which is practically that of an ideal lossless system estimated to be 0.273 A/W (Scholze et al., 1998, 2000). This is higher than what was obtained with any of the commercial photodiodes that also were characterized at 13.5 nm.

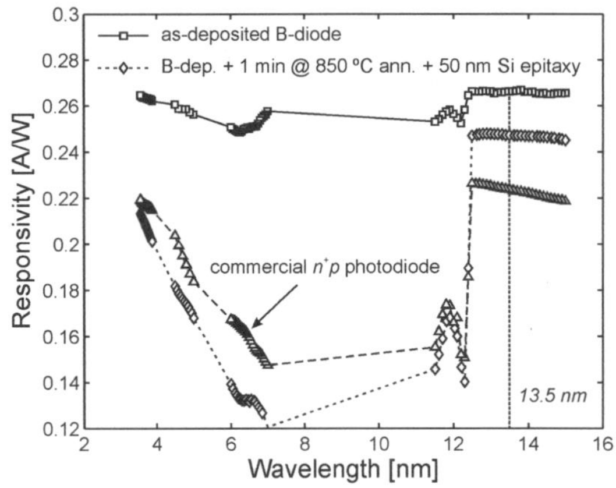

Fig. 11. Measured spectral responsivity (symbols) of B-layer photodiodes with a surface layer of ~ 3 nm thick α-B, with and without an extra in-situ 850 °C thermal anneal and 50 nm epitaxially deposited B-doped Si, compared to a commercial state-of-the-art n^+p photodiode (Sarubbi et al., 2008b).

The degradation of the photodiodes was characterized by prolonged high-dose exposure (0.22 MJ/cm^2 during 24.5 h) to 13.5 nm radiation, and no change in the responsivity due to degradation of the detector itself was observed. As for the corresponding radiation-induced degradation of the electrical behavior, an increase of the photodiode dark current level from below a nA to tens of nA at a reverse bias of 10 V was observed. This effect is common in silicon-based photodiodes, but here the dark current degradation was found to be annealed out by thermal treatment at relatively low temperatures, such as ~ 200 °C, indicating that this is an oxide-related degradation due to charging of the oxide around the diode perimeter. Since for the operating conditions of detectors in future applications the biasing is being scaled down to the mV range, the sensitivity to radiation-induced dark current will be quite low. As for the VUV case, it can be concluded that for EUV applications the B-layer photodiodes exhibit superior responsivity and negligible degradation under high-power exposure. The B-layer photodiodes have superior responsivity and negligible degradation under high-power exposure to EUV light.

4.3 Low-energy electron detectors

In contrast to photodiodes for DUV/EUV detectors, where protective entrance window layers can be found that are fairly transparent to the light, low-energy electron detectors require an entrance window that is essentially free of non-sensitive layers. This is obvious from the projected ranges in matter that have been compiled and are summarized in Fig. 12. At an electron energy of 10 keV the penetration depth is around 1 μm, a depth that is manageable for many conventional silicon photodiodes, but already at 2 keV it drops to about 200 nm and decreases rapidly to the 10 – 40 nm range at 1 keV. At 500 eV it is already below ~ 10 nm, so to have any form of practical responsivity this dictates that all non-sensitive surface layers must be nm-thin. The fact that the B-layer diodes can meet this requirement and are easily integrated in silicon has led to a very fast commercialization of the detectors in scanning electron microscopes.

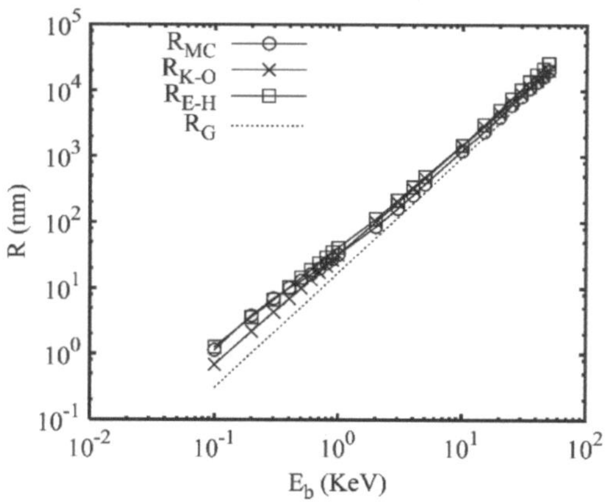

Fig. 12. Electron range R in Si as a function of electron energy E_b (Kurniawan & Ong, 2007).

Low-voltage SEM imaging is widely used for nanometer-scale inspection, among other things in the semiconductor industry. It provides atomic-scale resolution of the specimen surface due to the short range of electrons in matter. Lowering the electron energy gives more surface information and better resolution. Moreover, charging effects become less prominent than for higher voltages where they readily obscure the imaging of the non-conducting materials. The B-layer detectors that have been fabricated for low-energy electron detection are a good example of the versatility of the overall processing capabilities in Si technology. Exceptional imaging capabilities were obtained by the implementation of the following unique (combinations of) processing techniques:

- the B-layer photodiodes themselves, where a ~ 2 nm thin amorphous boron (α-B) layer forms the front-entrance window. The low atomic number of the B is also instrumental in minimizing scattering of the incoming electrons, thus allowing a longer projected range in the detector;
- compact segmented anode layouts, where the photosensitive surface is maximized and low diode capacitance is combined with lateral junction isolation of the segments. This is achieved by epitaxially growing very lightly-doped, tens-of-microns thick n-layers on low-ohmic n-type substrates;
- low photodiode series resistance combined with a large sensitive front-window area by patterning a fine aluminum grid directly on the α-B surface. The low series resistance and capacitance values combine to give low transit times and thus high scanning speeds;
- through-wafer apertures etched close to the anode regions for detectors designed to monitor back-scattered electrons (BSE) in SEM systems such as the one shown in Fig. 13. This is a back-end processing step that would damage the delicate α-B surface layer if it was not protected during the through-wafer hole etching. Therefore, a processing

scheme was developed whereby the final removal of all Al down to the photo-sensitive regions is the very last step of the wafer-scale processing.

In Fig. 14 an illustration is given of the final detector structure with the segmented photodiode layout. A variety of detectors have been fabricated in this way and tested for specific functions in several different SEM systems. An example of the resulting high-resolution images that can be obtained is shown in Fig. 15. By using different read-out sequences of segments in this annular BSE detector an optimal contrast can be attained.

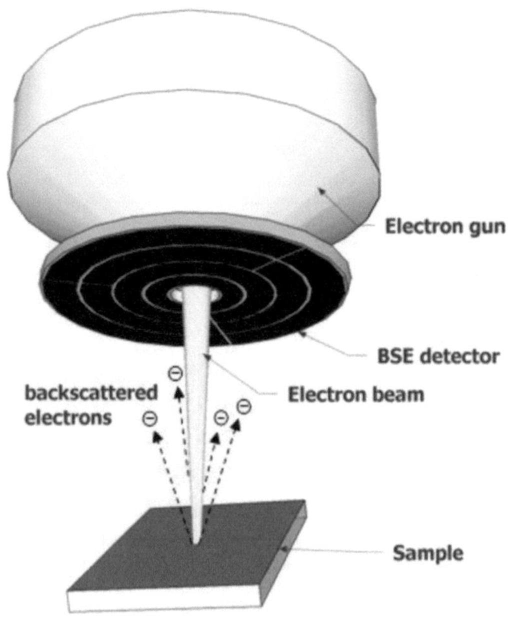

Fig. 13. Conceptual drawing of a SEM system showing the location of an annular BSE detector.

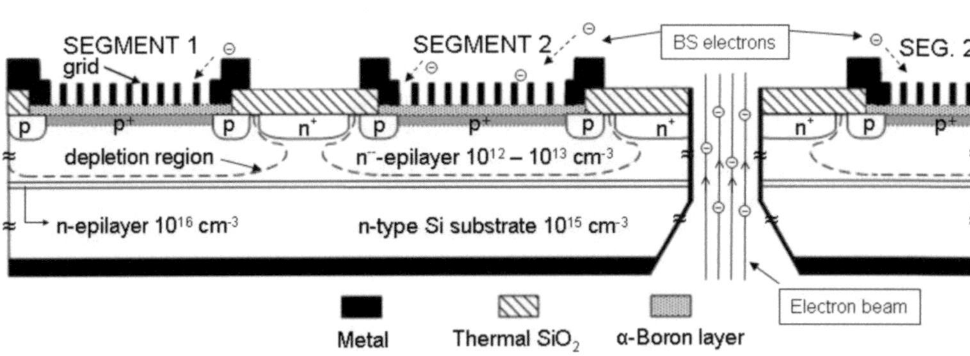

Fig. 14. Schematic cross-section of two neighboring segments of a B-layer detector with a through-wafer hole as aperture for the electron beam. The depletion of the typically 40 μm deep n⁻ epitaxial layer is indicated. Segments are isolated by the n⁺-channel-stop and the undepleted n⁻-layer (Šakić et al., 2010c).

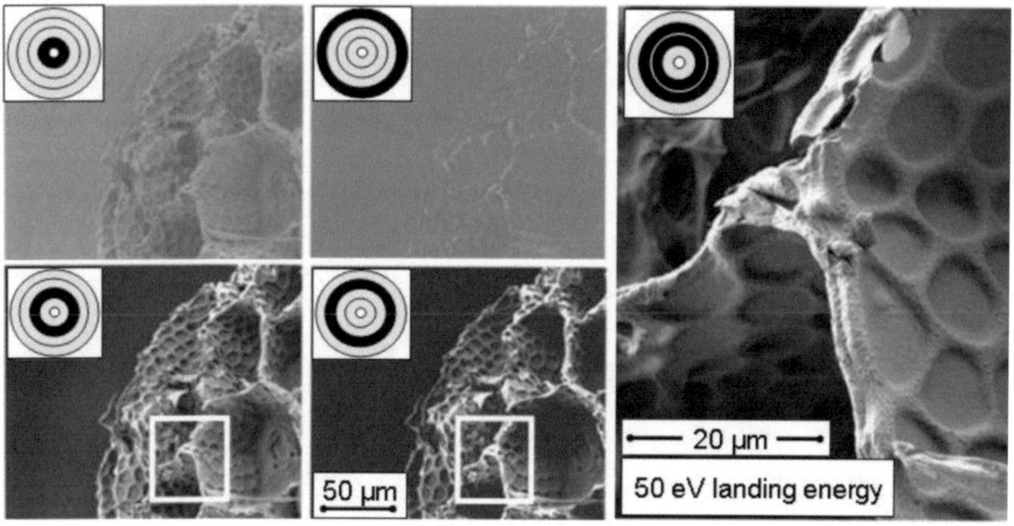

Fig. 15. Imaging of an uncoated pollen sample at 50 eV landing energy. The BSE detector is divided in 8 annular segments that can be operated in various combinations as shown in the inset. The best resolution is obtained with the combination of segments used to produce the image to the right (Šakić et al., 2010c).

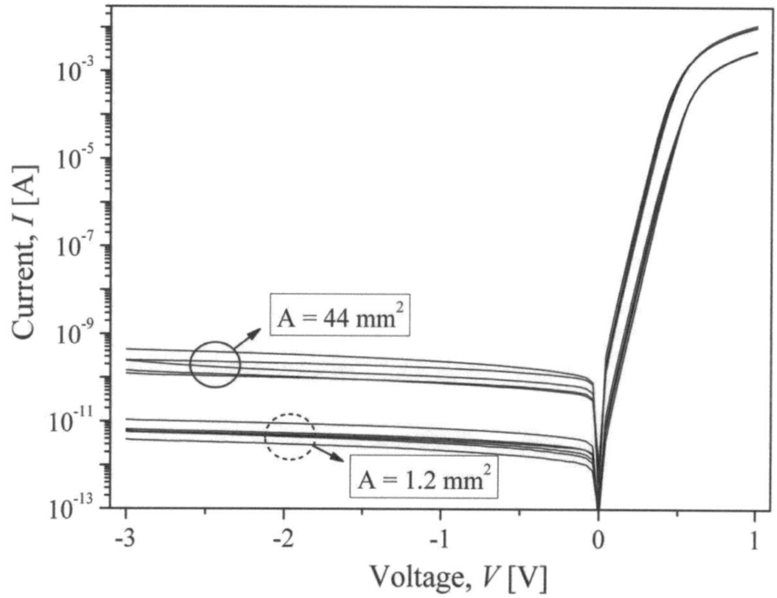

Fig. 16. On-wafer measurements of the I-V characteristics of two types of B-layer diodes with areas 44 mm² and 1.2 mm², respectively. They form 2 out of 8 segments of a BSE detector. For the larger diodes, characteristics are shown for devices with and without a conductive Al-grid on the photosensitive surface (Šakić et al., 2010b).

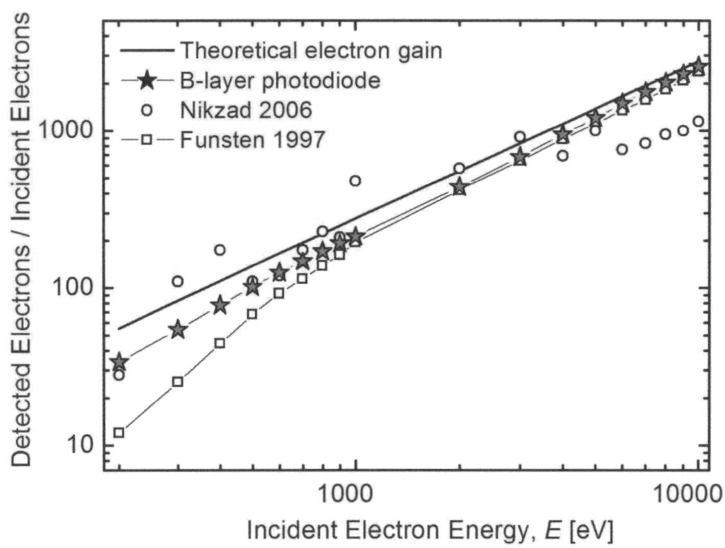

Fig. 17. Measured electron gain of a B-layer photodiode compared to the data reported by (Nikzad et al., 2006) and (Funsten et al., 1997), compared to the theoretical electron gain (Šakić et al., 2010c).

An example of the electrical characteristics of 2 out of 8 photodiode segments of a BSE detector are shown in Fig. 16. The beneficial effect of the Al-grid on the series resistance is seen in the high current forward bias region. Also, a reliably low dark current is seen in the reverse bias region. The electron gain of B-layer photodiodes is shown in Fig. 17. It is measured using the electron beam of a SEM system as electron source, a Faraday cup to measure the reference electron beam current (incident electrons), and a reverse biasing of 3 V is applied to the photodiode during electron detection. Also included are values for other low-energy electron detectors reported in the literature by (Nikzad et al., 2006) and (Funsten et al., 1997). In addition, the theoretical value of the gain is calculated as the incident energy of the electron beam over the creation energy of an electron-hole pair in silicon of 3.61 eV. Although the gain values at low electron energies reported in (Nikzad et al., 2006) are high, they are unreliable due to noise and do not exhibit clear energy dependence. Furthermore, the sensitivity in the region above 4 keV drops off significantly. B-layer photodiodes show the expected energy dependence and operate closer to the theoretical limits than those reported in (Funsten et al., 1997), especially below 1 keV. At 500 eV and 1 keV, they have an electron signal gain at low energies of 60% and 74% of the theoretical gain value, respectively.

5. Conclusion

The three industrial detector applications of the presented B-layer photodiodes, i.e., DUV, EUV and low-energy electron detectors, demonstrate the high performance and integration versatility of this silicon technology. The special doping properties of the low-temperature α-B deposition and the electrical properties of the α-B layer itself are instrumental in securing a damage-free extremely shallow diode that provides superior performance in terms of dark current, optical responsivity, and radiation induced degradation. The DUV/VUV/EUV photodiodes can be implemented to obtain close to 100% internal and external quantum efficiency, and low-energy electron detection with record-high sensitivity has been demonstrated down to 200 eV. Moreover, the silicon fabrication technology is highly reliable and enables flexible configurations of the diodes and other on-chip components, enhancing the detector speed and functionality. This is particularly demonstrated by the low RC constant of the low-energy electron detectors combined with a segmented photodiode design and a through-wafer aperture.

6. Acknowledgements

This chapter summarizes the work of a large group of people to whom the author is greatly indebted. Firstly, thanks go to the original team: Tom Scholtes for the processing experiments, Francesco Sarubbi for measurement/analysis of the experimental material and Koen Buisman for designing the first B-layer diode circuits. Stoyan Nihtianov is acknowledged for securing the ASML DUV/EUV lithography projects and for his work together with Lei Shi producing detector designs and analysing many aspects of the photodiodes in this context. This UV work was performed in fruitful cooperation with PTB, particularly Frank Scholze, Alexander Gottwald and Udo Kroth, who are responsible for the DUV/VUV/EUV optical measurements and interpretations. Thanks are due to Gerard van Veen, Kees Kooijmans and Patrick Vogelsang of FEI Company for two and a half years of intensive and very productive cooperation on the research and development of low-energy electron detectors for scanning electron microscopes. At DIMES the electron detector research was realized mainly by Agata Šakić, Silvana Milosavljević, Negin Golshani, Wim Wien, Caroline Mok and Carel Heerkens, but the success is also a credit to the whole DIMES IC-Processing group, headed by Bert Goudena, for maintaining an outstanding cleanroom. Lastly, the author gratefully acknowledges Gianpaola Lorito, Luigi La Spina, Theodoros Zoumpoulidis and Jaber Derakhshandeh of Iszgro Diodes for bringing this work to a production level, which, particularly for the EUV detectors, has in no way been a trivial task. Beside the industrial support from ASML Lithography and FEI Company, this research has profited from support from and cooperation with the SmartMix project MEMPHIS and Project No. 10024 of The Dutch Technology Foundation STW.

7. References

Canfield, L. R.; Vest, R. E ; Korde, R.; Schmidtke, H. & Desor, R. (1998) Absolute silicon photodiodes for 160 nm to 254 nm photons. *Metrologia*, Vol. 35, No. 4, (1998) pp. 329-334, ISSN 0026-1394

Benschop, J.; Banine, V.; Lok, S. & Loopstra, E. (2009). Extreme ultraviolet lithography: Status and prospects. *Journal of Vacuum Science & Technology B: Microelectronics and Nanometer Structures,* Vol. 26, Issue 6, (November 2008) pp. 2204 - 2207, ISSN 1071-1023

Borland, J.O.; Felch, S.; Wan, Z.; Tanjyo, M.; & Buyuklimanli, T. (2009). Thermal processing issues for 22nm node junction scaling. *Solid State Technology,* Vol. 52, Issue 8, (August 2009) pp. 1-8

Funsten, H. O.; Suszcynsky, D. M.; Ritzau, S. M & Korde, R. (1997). Response of 100% internal quantum efficiency silicon photodiodes to 200 eV–40 keV Electrons. *IEEE Transactions on Nuclear Science,* Vol. 44, Issue 6, (1997) pp. 2561-2565, ISSN 0018-9499

Funsten, H. O.; Ritzau, S. M. & Harper, R. W. (2004). Fundamental limits to detection of low-energy ions using silicon solid-state detectors. *Applied Physics Letters,* Vol. 84, no. 18, (April 2004) pp. 3552-3554, ISSN 0003-6951

Gottwald, A.; Kroth, U.; Krumrey, M.; Richter, M.; Scholze, F. & Ulm, G. (2006). The PTB high-accuracy spectral responsivity scale in the VUV and x-ray range, *Metrologia,* Vol. 43, Issue Newrad 2005, (April 2006) pp. S125-S129, ISSN 0026-1394

Gottwald, A.; Kroth, U.; Richter, M.; Schöppe, H. & Ulm, G. (2010). Ultraviolet and vacuum-ultraviolet detector-based radiometry at the Metrology Light Source, *Measurement Science and Technology,* Vol. 21, No. 12, (December 2010) pp. 1-9, ISSN 0957-0233

Henke data. (no date) Available from <http://henke.lbl.gov/optical_constants>

Inada, T.; Kuranouchi, A.; Hirano, H.; Nakamura, T.; Kiyota, Y. & Onai, T. (1991). Formation of ultrashallow p^+ layers in silicon by thermal diffusion of boron and by subsequent rapid thermal annealing. *Applied Physics Letters,* Vol. 58, No.16, (April 1991) pp. 1748–1750, ISSN 0003-6951

ITRS 2009 Edition, International Technology Roadmap for Semiconductors 2009. Avaibable from <http://www.itrs.net>

Kurniawan, O. & Ong, V. K. S. (2007). Investigation of Range-energy Relationships for Low-energy Electron Beams in Silicon and Gallium Nitride. *Scanning,* Vol. 29, Issue 6, (October 2007) pp. 280–286, , ISSN 0161-0457

Kim, K.-S.; Song, Y.-H.; Park, K.-T.; Kurino, H.; Matsuura, T.; Hane, K. & Koyanagi, M. (2000). A novel doping technology for ultra-shallow junction fabrication: boron diffusion from boron-adsorbed layer by rapid thermal annealing. *Thin Solid Films,* Vol. 369, Issues 1-2, (July 2000) pp. 207–212, ISSN 0040-6090

Klein, R.; Laubis, C.; Müller, R.; Scholze, F. & Ulm, G. (2006). The EUV metrology program of PTB, *Microelectronic Engineering,* Vol. 83, Issue 4-9, (January 2006) pp. 707-709, ISSN 0167-9317

Nikzad, S.; Cunningham, T. J.; Hoenk, M. E.; Ruiz, R. P.; Soules, D. M. & Holland S.E. (2006). Direct detection of 0.1–20 keV electrons with delta doped, fully depleted, high purity silicon *p-i-n* diode arrays. *Applied Physics Letters,* Vol. 89, Issue 18, (October 2006) pp. 182114-1-3, ISSN 0003-6951

Palik, E.D. (Ed.). (1985). *Handbook of Optical Constants of Solids.* Academic Press, Incorperated, Orlando, Florida

Popadic, M.; Lorito, G. & Nanver, L.K. (2009). Analytical carrier transport model for arbitrarily shallow p-n junctions. *IEEE Trans. Electron Devices*, Vol. 56, Issue 1, (January 2009) pp. 116-125, ISSN 0018-9383

Powell, F.R. & Johnson, T.A. (2010). Filter windows for EUV lithography. Available from <http://www.luxel.com/wp/wp-content/uploads/2010/04/EUVL-Paper.pdf>

Richter, M.; Kroth, U.; Gottwald, A.; Gerth, C.; Tiedtke, K.; Saito, T.; Tassy, I. & Vogler, K. (2002). Metrology of pulsed radiation for 157-nm lithography. *Applied Optics*, Vol. 41, Issue 34, (December 2002) pp. 7167-7172, , ISSN 2155-3165

Sarubbi, F. (2010c). *"Characterization of pure boron depositions integrated in silicon diodes for nanometer-deep junction applications"*, Ph.D. dissertation, Delft University of Technology, ISBN 978-90-8570-436-2, Delft, The Netherlands

Šakić, A.; Jovanović, V.; Maleki, P.; Scholtes, T.L.M.; Milosavljević, S. & Nanver, L.K. (2010a). Characterization of amorphous boron layers as diffusion barrier for pure aluminium. *Proceedings of 33rd International Convention MIPRO*, Vol. 1, pp. 52-55, ISBN 978-953-233-051-9, Zagreb, Croatia, May 2010

Šakić, A.; Nanver, L.K.; Heerkens, C.Th.H.; Veen, G. van; Kooijman, K. & Vogelsang, P. (2010b). Silicon photodiodes for high-efficiency low-energy electron detection. *Proceedings of IEEE 40th European Solid-State Device Research Conference (ESSDERC)*, pp. 102-105, ISBN 978-1-4244-6659-7, Sevilla, Spain, September 2010

Šakić, A.; Nanver, L.K.; Veen, G. Van; Kooijman, K.; Vogelsang, P.; Scholtes, T.L.M.; Boer, W. de; Wien, W.H.A.; Milosavljević, S.; Heerkens, C.Th.H.; Knežević, T. & Spee, I. (2010c). Versatile Silicon Photodiode Detector Technology for Scanning Electron Microscopy with High-Efficiency Sub-5 keV Electron Detection. *Proceedings of IEEE International Electron Device Meeting (IEDM)*, pp. 712 -715, ISBN 978-1-4244-7419-6, San Francisco, California, December 2010

Sarubbi, F.; Nanver, L.K.; & Nihtianov, S.N. (2008a). Extremely ultra-shallow p^+-n boron-deposited silicon diodes applied to DUV photodiodes. *Digest IEEE 66th Device Research Conference (DRC)*, pp. 143-144, ISSN 1548-3770, Santa Barbara, CA, USA, June 2008

Sarubbi, F.; Nanver, L.K.; Scholtes, T.L.M.; Nihtianov, S.N. & Scholze, F. (2008b). Pure boron-doped photodiodes: a solution for radiation detection in EUV lithography. *Proceedings of IEEE 38th European Solid-State Device Research Conference (ESSDERC)*, pp. 278–281, ISSN 1930-8876, Edinburgh, Scotland, UK, September 2008

Sarubbi, F.; Scholtes, T.L.M. & Nanver, L.K. (2010a). Chemical vapor deposition of α-boron layers on silicon for controlled nanometer-deep p^+n junction formation. *Journal of Electronic Materials*, Vol. 39, No. 2, (February 2010) pp. 162-173, ISSN 0361-5235

Sarubbi, F.; Nanver, L.K. & Scholtes, T.L.M. (2010b). High effective Gummel number of CVD boron layers in ultrashallow p^+n diode configurations. *IEEE Transactions on Electron Devices*, Vol. 57, Issue. 6, (June 2010) pp. 1269-1278, ISSN 0018-9383

Scholze, F.; Rabus, H. & Ulm, G. (1998). Mean energy required to produce an electron-hole pair in silicon for photons of energies between 50 and 1500 eV. *Journal of Applied Physics*, Vol. 84, No. 5, (September 1998) pp. 2926-2939, ISSN 0021-8979

Scholze, F.; Henneken, H.; Kuschnerus, P.; Rabus, H.; Richter, M. & Ulm, G. (2000). Determination of the electron-hole pair creation energy for semiconductors from the spectral responsivity of photodiodes. *Nuclear Instruments and Methods in Physics Research, section A*, Vol. 439, Issue 2-3, (2000) pp. 208-215, ISSN 0168-9002

Shi, L.; Sarubbi, F.; Nihtianov, S.; Nanver, L.K.; Scholtes, T.L.M. & Scholze, F. (2009), High Performance Silicon-Based Extreme Ultraviolet (EUV) Radiation Detector for Industrial Application, *Proceedings of IEEE Conference on Industrial Electronics (IECON)*, pp. 1891-1896, Porto, Portugal, November 2009

Shi, L.; Nanver, L.K.; Šakić, A.; Nihtianov, S.; Gottwald, A. & Kroth, U. (2010). Optical Stability Investigation of High-Performance Silicon-Based VUV Photodiodes. *Proceedings of IEEE Sensors 2010 Conference*, Waikoloa, USA, November 2010

Silver C. S.; Spallas J.P. & Muray L.P. (2006). Silicon photodiodes for low-voltage electron detection in scanning electron microscopy and electron beam lithography. *Journal of vacuum science & technology B*, Vol. 24, Issue 6, (2006) pp. 2951-2955, ISSN 0734-211X

Solt, K.; Melchior, H.; Kroth, U.; Kuschnerus, P.; Persch, V.; Rabus, H.; Richter, M. & Ulm, G. (1996). PtSi-n-Si Schottky-barrier photodetectors with stable spectral responsivity in the 120-250 nm spectral range. *Applied Physics Letters*, vol. 69, Issue 24, (December 1996) pp. 3662-3664, , ISSN 0003-6951

Sarubbi, F. (2010). *"Characterization of pure boron depositions integrated in silicon diodes for nanometer-deep junction applications"*, Ph.D. dissertation, Delft University of Technology, ISBN 978-90-8570-436-2, Delft, The Netherlands

Tindall, C.S.; Palaio, N.P.; Ludewigt, B.A.; Holland, S.E.; Larson, D. E.; Curtis, D. W.; McBride, S. E.; Moreau, T.; Lin, R. P. & Angelopoulos, V. (2008), Silicon detectors for low energy particle detection. *IEEE Transactions on Nuclear Science*, Vol. 55, Issue 2, (April 2008) pp. 797–801, ISSN 0018-9499

Vick, G.L. & Whittle, K.M. (1969). Solid solubility and diffusion coefficients of boron in Si. *Journal of the Electrochemical Society*, Vol. 116, issue 8, (August 1969) pp. 1142–1144, ISSN 0013-4651

Xia, S.; Sarubbi, F.; Naulaerts, R.; Nihtianov, S.N. and Nanver, L.K. (2008). Response time of silicon photodiodes for DUV/EUV radiation. *Proc. 2008 IEEE International Instrumentation and Measurement Technology Conference (I2MTC)*, pp. 1956-1959, ISSN 1091-5281, Victoria, Vancouver Island, Canada, May 2008

11

Avalanche Photodiodes in Submicron CMOS Technologies for High-Sensitivity Imaging

Gian-Franco Dalla Betta[1], Lucio Pancheri[2], David Stoppa[2], Robert Henderson[3] and Justin Richardson[3,4]
[1]*University of Trento*
[2]*Fondazione Bruno Kessler*
[3]*University of Edinburgh*
[4]*Dialog Semiconductor Ltd., Edinburgh*
[1,2]*Italy*
[3,4]*United Kingdom*

1. Introduction

Vacuum based devices, such as Photo Multiplier Tubes (PMT) and Micro Channel Plates (MCP), have been for many years the sensors of choice for most applications calling for photon counting and timing (Renker, 2004). While providing very good sensitivity, noise and timing characteristics, these photodetectors feature a number of disadvantages: they are bulky, fragile, and sensitive to magnetic fields; they require very high operation voltages, and have large power consumption; in their high-performance models, providing good spatial resolution, they are still very expensive. For high-sensitivity imaging applications, suitable solutions are represented by CCD cameras coupled with either MCP Image Intensifiers (I-CCDs) or Electron Multipliers (EM-CCDs) (Dussault & Hoess, 2004). However, besides being very expensive, their performance is not completely satisfactory in extreme time resolved measurements.

For reasons of cost, miniaturization, ruggedness, reliability, design flexibility, integration density, and signal processing capabilities, a fully solid-state solution (and, particularly, CMOS technology) would be highly desirable. Among the advanced CMOS image sensors so far reported, the most promising ones in terms of high sensitivity and fast timing are those based on Single Photon Avalanche Diodes (SPADs). SPADs are avalanche photodiodes operated in the so-called Geiger mode, i.e., biased above breakdown, so as to be sensitive to single photons (Cova et al., 1996). Although these sensors have been developed for more than 30 years, in particular owing to the group of Prof. Cova at Politecnico di Milano, and single devices have reached outstanding performance (Ghioni et al., 2007), only recently the perspective of making a SPAD-based camera has become feasible. The first SPAD-based pixel arrays in CMOS technology have been demonstrated only a few years ago (Rochas et al., 2003a), but since then further developments rapidly followed, also facilitated by the availability of commercial, High-Voltage CMOS technologies (HV-CMOS) aimed at integrated circuits for power electronics, as well as of specially tailored "imaging" processes, which have been boosted by the huge market of

mobile phone cameras. As a result, also submicron CMOS technologies with ever decreasing minimum feature size, down to 130nm, could be used for SPAD fabrication.
The electro-optical characteristics of CMOS SPADs are not optimized in some respects, but their monolithic integration with readout electronics might offer several advantages for the design of low-cost and high-performance fully integrated imaging systems with time-resolved, single-photon detection capabilities. These devices can be used in many application fields: imaging in extreme low-level light conditions (night, caves, ...), real time imaging of the motion of natural gravity-driven flows (snow avalanches, landslides, ...), ranging and three-dimensional vision, biomedical and molecular biology (single molecule spectroscopy, luminescence microscopy, fluorescence lifetime imaging, etc.), scintillation detection in nuclear medicine (SPECT - Single Photon Emission Compute Tomography; PET - Positron Emission Tomography) and particle/nuclear physics, quantum cryptography, astronomy, adaptive optics, non invasive testing of VLSI circuits, to cite but a few.
While a considerable number of CMOS image sensors based on SPADs have already been presented, CMOS Avalanche Photodiodes (APDs) operating in the linear regime have not evolved to the same extent. Only a few successful implementations have been proposed so far, and large arrays are not available yet. These devices could be interesting for some applications, especially in the detection of blue and UV light, where the noise factor of the device is small, whereas at larger wavelengths the noise is normally higher.
In this Chapter, after reviewing basic operation principles, figures of merit, and state-of-the-art, we report on the design and experimental characterization of both Geiger- and linear-mode avalanche photodiodes devices fabricated in different submicron CMOS technologies. The electro-optical properties are evaluated, underlining the impact of technology scaling on the device characteristics. Moreover, we discuss the main design issues relevant to integrated read-out channels for SPADs to be used in active pixel sensor arrays for high-sensitivity imaging applications.

2. Operation principle

Avalanche photodiodes are p-n junction photodiodes purposely made to be operated at high electric fields in order to achieve an internal gain. In reverse biased photodiodes, the electric field increases with the applied voltage, causing the drift velocity and kinetic energy of charge carriers injected in the depletion region to increase. By doing so, an electron (or a hole) can reach an energy high enough to break a bond when colliding with lattice atoms, thus generating a new electron-hole pair, and losing part of its energy in this process, which is called impact ionization (Sze & Ng, 2007). Both the original carrier (electron or hole) and the secondary electron and hole will be accelerated by the electric field and possibly contribute to the generation of more electron-hole pairs, this resulting in a positive feedback loop which gradually increases the overall number of carriers, hence the term avalanche. Note that this applies both to optically generated carriers and to carriers generated by any other mechanism (e.g., thermally generated carriers). The magnitude of the avalanche phenomenon is governed by two concurrent factors: the carrier ionization rates, which are defined as the number of pairs created by a carrier per unit distance travelled, and the rate at which electrons and holes leave the high-field region and are collected at the device electrodes (Aull et al., 2002). Ionization rates are strongly increasing with the electric field. Although in silicon electrons have a higher ionization rate than holes, on average an electric field value of about 3×10^5 V/cm is required to create one electron-hole pair per 1 μm travelled. For bias voltages below the breakdown voltage, ionization rates are balanced by the extraction rate, so that carrier concentration and output current are increased by a finite multiplication factor or gain, M, normally in the range between a few tens and a few hundreds. This is the case of linear-mode APDs, which provide an output current proportional to the impinging light intensity.

For bias voltages beyond the breakdown voltage, ionization rates are so high that the extraction rate does not keep pace with them, so that the carrier concentration and output current increase to very high values. This is the case of Geiger-mode APDs, also known as SPADs. These concepts can be better appreciated with the aid of Fig. 1(a), which shows a sketch of the quasi-static current-voltage curve and of the corresponding gain-voltage curve. As the voltage reaches V_{APD}, the current starts increasing due to onset of multiplication phenomenon, and then tends to diverge as the voltage exceeds the breakdown voltage V_{BD}. Correspondingly, the gain starts being larger than 1 in linear mode avalanche, and virtually tends to infinite in Geiger mode (practical values can largely exceed 10^6).

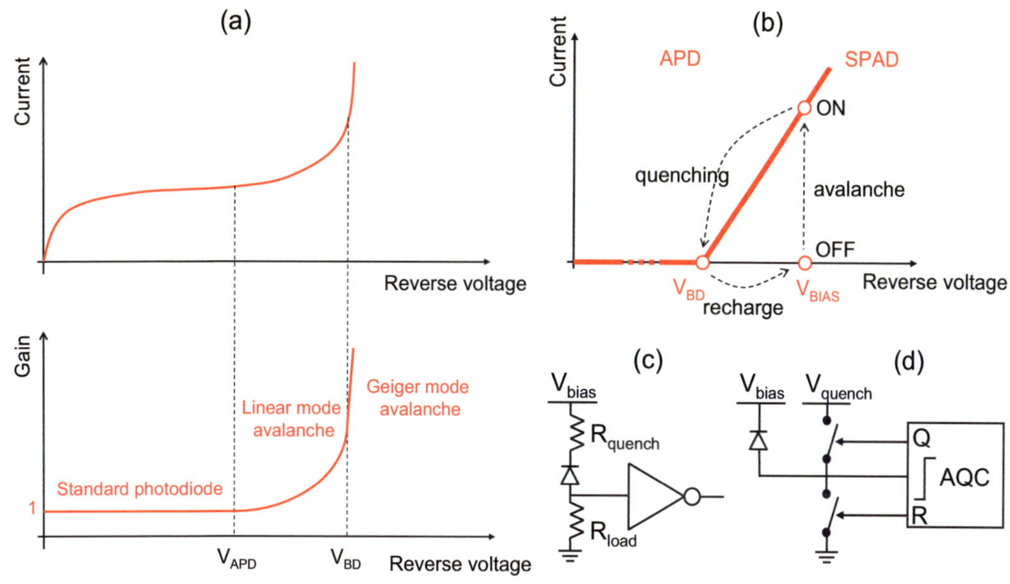

Fig. 1. Basic operation principle of avalanche photodiodes. (a) Sketch of quasi-static current-voltage characteristic and corresponding gain-voltage characteristic; (b) qualitative description of mechanisms involved in the dynamic behaviour of a SPAD; (c) basic circuit of SPAD with passive quenching; (d) basic circuit of SPAD with active quenching.

Fig. 1(a) represents the quasi-static (i.e., averaged on time) behaviour, as could be measured with a semiconductor parameter analyzer. When considering the device operation beyond breakdown, it is necessary to consider the dynamic behaviour, which can be explained with the aid of Fig. 1(b). When a SPAD is biased beyond the breakdown voltage (the difference between V_{BIAS} and V_{BD} is called the excess voltage, V_{ex}), it will stay in an OFF state for a short time, until a carrier (electron or hole) will trigger an avalanche event bringing the device into its ON state. The corresponding current pulse would be self-sustaining at a very large value limited by an intrinsic resistance due to space charge effects. Nevertheless, in order for the SPAD to be useful as a photodetector, the avalanche current must to be turned off by using proper quenching mechanisms, able to reduce the bias voltage down to or below the breakdown point, and to finally restore it to its initial value, so that a new incoming photon can be detected (Cova et al., 1996). In other words, the device is operated in a binary mode: the avalanche can be sensed using a comparator (as simple as a logic inverter in some cases), so that single photons can be detected and counted, but the need for quenching/recharging introduces a dead time between two consecutive events. The simplest way to quench the avalanche is by means of a high ohmic resistor in series with the SPAD (Fig.1(c)), so that the voltage drop caused by the avalanche current lowers the SPAD bias down to the breakdown point. More effective solutions exploit specially designed Active Quenching Circuits (AQC), which are feedback circuits able to sense the avalanche pulse and to control the activation of quenching and recharging blocks, such as switches (Fig.1(d)).

3. Figures of merit

As far as linear-mode APD are concerned, the figures of merit are essentially the same as for standard photodiodes (e.g., quantum efficiency, dark current, bandwidth, etc.), apart from the fact that the gain M, defined as the average number of electron-hole pairs per absorbed photon, is higher than one. However, due to the statistical nature of impact ionization processes, the actual number of electron-hole pairs per photon varies and these fluctuations in the gain result in a multiplication noise that is higher than that simply obtained by considering the shot noise associated to M times the photocurrent. This effect is normally characterized by the excess noise factor, F, which depends on the gain and on the ratio between the ionization coefficients of electrons and holes, k (McIntyre, 1966). Since F is increasing with M, this sets an ultimate limit to the gain value in order for the signal-to-noise ratio not to be degraded by the multiplication noise.

For SPADs, due to the substantially different operation mode, a specific set of performance parameters should be introduced, as reported in the following.

a. *Dark Count Rate* (DCR). A dark count is an avalanche event caused by non photo-generated carriers, which can be originated from four factors (Haitz, 1965): diffusion from neutral regions, thermal generation, band to band tunnelling or by release from a charge trap (see also afterpulsing). The per-second rate at which dark counts occur is the DCR in Hertz. DCR would scale with device area, but in real implementations is generally reported as exhibiting a steeper than exponential profile. DCR increases exponentially with temperature and therefore may be reduced by using cooling methods such as thermoelectric Peltier elements or by forced air-cooling. DCR also varies linearly with the electric field strength because of the increasing avalanche initiation probability.
b. *Afterpulsing*. Charge traps due to unintentional impurities and crystal defects can result in generation-recombination (GR) centres. The high current peak through the junction during an avalanche breakdown introduces a probability that the trap is filled by a carrier which is then later released, initiating a second, follow-on Geiger 'after-pulse' (Haitz, 1965). Trap occupancy has an associated lifetime. Traps located halfway between the valence and conduction bands, called 'deep traps' have longer lifetime and are therefore a major contributor to afterpulsing. For this reason manufacturing processes should be kept as clean as possible (e.g., by means of gettering techniques), and charge flow during an avalanche event should be minimised by proper design of the quenching circuits.

c. *Photon Detection Efficiency* (PDE). The PDE is the percentage of incoming photons that create an output pulse over an incident light bandwidth. The probability of a photon arrival causing an output pulse is reduced by three main factors: reflectance, absorption, and self-quenching. Firstly, an incoming photon may be reflected at the surface of the device or at the interface between the many layers that constitute the optical stack of the detector. An antireflection top coating should ideally be used to maximise photon transmission through the optical stack. Secondly, a photon may be absorbed above the SPAD within the optical stack materials, just at the surface of the active region, or too deep within the silicon substrate in order to initiate an avalanche. Thirdly, an avalanche event may be initiated but stall, becoming self quenched. Such an event may not yield enough potential difference in order to trigger an output pulse. Self-quenching can be minimised by ensuring a high enough electric field is present, so as to increase chance of impact ionization taking place.

d. *Timing resolution* (Jitter). When a SPAD is struck repetitively with a low-jitter, short-pulsed laser, the position in time of the resulting avalanche breakdown pulses has a statistical variation. The timing resolution, or 'jitter' of the detector is the full-width, half-maximum (FWHM) measure of this temporal variation. Among the timing resolution components are the variation caused by the generated carrier transit time from depletion layer to multiplication region, which is dependent on the depth of absorption of the incident photon (as a guideline, the transit time at carrier saturation velocity is 10ps per micron), and, more important, the statistical build up of the avalanche current itself (Ghioni et al., 1988). This is impacted by the electric field strength, and so jitter may be minimised by employing high overall bias conditions. In larger area SPADs, also the timing uncertainty introduced by the avalanche lateral propagation can be non negligible. The shape of the histogram of avalanche events in response to a time accurate photon arrival provides information regarding the location and speed of avalanche build up. A predominantly Gaussian shape indicates that the bulk of photon initiated avalanches occur in the high field active region of the detector, whereas the presence of a long tail indicates that part of the avalanche events are initiated by photon-generated carriers diffusing into the high field region of the detector after a short delay (in this case the timing response at full-width 100th of maximum is often reported). Of course the method employed for detecting the onset of an avalanche event is of high importance, and the readout circuit should be designed to minimize time walk effects.

e. *Dead time* (T_D). The SPAD is not responsive to further incoming photons during the period comprising the avalanche quenching and the reset of the final bias conditions (Haitz, 1964). However, in the case of a passively quenched SPAD this is not strictly the case. As the device is recharged via the quenching resistor (a phase that can last from several tens to a few hundreds of nanoseconds), it becomes increasingly biased beyond its breakdown voltage, so that it is able to detect the next photon arrival prior to being fully reset. This behaviour is coupled with a significant fluctuation in the reset waveform. Clearly the dead time should be kept as small, and as consistent as possible in order to achieve the highest possible dynamic range of incident photon flux and least variation in photon count output to a certain photon arrival rate. In this respect, active quenching circuits offer the best performance with short and well-defined dead times and high counting rates. However, short dead times are often accompanied by enhanced afterpulsing probability due to inadequate trap flushing time.

f. When dealing with arrays of SPADs, other parameters become significant. Among them are *Fill Factor* and *Crosstalk*. The active area of a SPAD is the central photon-sensitive portion of the detector. The electric field strength should be consistent across this part of the structure, so as to yield a homogenous breakdown probability. Zones that exhibit higher field strength will exhibit locally higher photon detection probability and dark count compared with the rest of the active area. Such zones should be avoided by proper design solutions, such as guard rings. The proportion of active region area to total SPAD area is the Field Factor (FF) and is commonly expressed in percent. Crosstalk between adjacent SPADs can occur in two ways. Firstly, a photon absorbed deep in one detector may result in a lateral diffusion of carriers to an adjacent device where an avalanche can be initiated. Secondly, an avalanche event may result in an electro-luminescent emission of photons that are then detected by an adjacent detector (Lacaita et al., 1993a). Electrical and optical crosstalk can be minimised by detector design introducing proper electro-optical isolation structures, at the expense of a reduction in the active area (Sciacca et al., 2006).

4. State of the art

4.1 Geiger-mode APD (SPAD)

SPADs can be traced back to the deep planar/reach through structures created in the 1960's (McIntyre, 1961; Ruegg, 1967; Haitz, 1963). These large, deep junction devices and their subsequent developments required high reverse bias voltages and were stand-alone structures incompatible with other circuit elements. Perkin-Elmer, Rockwell Science Center and Russian research groups have all since contributed to the development of these devices, as well captured in (Cova et al., 2004; Renker, 2006).

Apart from III-V devices and silicon reach-through structures (not suited to arrays), that are not covered in this chapter, the state of the art in modern SPADs may be described in terms of manufacturing process employed and construction details.

Three main process categories can be distinguished:

i. CMOS compatible, full custom processes, optimized to yield the best possible performing single detector element (e.g., Lacaita et al., 1989): among the adopted features are low implant doping concentrations, slow diffusion and annealing steps to minimise silicon lattice damage, gettering phases and embedded constructions to improve DCR and crosstalk. These implementations allow for small size arrays (Zappa et al., 2005; Sciacca et al, 2006), but are not compatible with very large-scale integration (VLSI) of on chip circuitry.

ii. High voltage CMOS processes: both single detectors and arrays have been implemented with this approach, together with quench circuitry (Rochas et al., 2003b; Stoppa et al., 2007) and single channel Time Correlated Single Photon Counting (TCSPC) systems (Tisa et al., 2007), but the scope for large scale integration is limited.

iii. Standard CMOS processes, without any modifications to the layers normally available to the designer. While providing great potential for VLSI integration and low cost, thus enabling new applications, this approach has to cope with the limitations imposed by the shallow implant depths, high doping concentrations, non optimized optical stacks, and design rule restrictions in advanced manufacturing processes. For this SPAD category, examples have been reported featuring: high DCR, even up to 1MHz, most likely due to tunnelling (Niclass et al., 2006; Gersbach et al., 2008) and/or to crystal lattice stress caused by shallow trench isolation process (Finkelstein et al., 2006a); low PDE (Faramarzpour et al., 2008; Marwick et al., 2008), possibly due to non optimized optical stack.

As far as construction details are concerned, the essential features of a SPAD are the method of formation of the guard ring, the overall shape (i.e. circular, square, others), active area diameter and the diode junction itself. Today's nanometer scale processes provide features such as deep well implants and shallow trench isolation (STI) that may be utilised in detector design. Additionally some custom processes provide features such as deep trench isolation, buried implants, and scope for optical stack optimisation.

Even in the very first samples by Haitz and McIntyre, premature edge breakdown was addressed by an implant positioned at the edge of the junction active region, and hence the first SPAD guard ring was created. State of the art SPAD constructions can be grouped according to the method of implementation of the guard ring, as discussed in the following with the aid of Fig.2 (different constructions are discussed with reference to the different cross sections shown in Fig.2 from (a) to (f)).

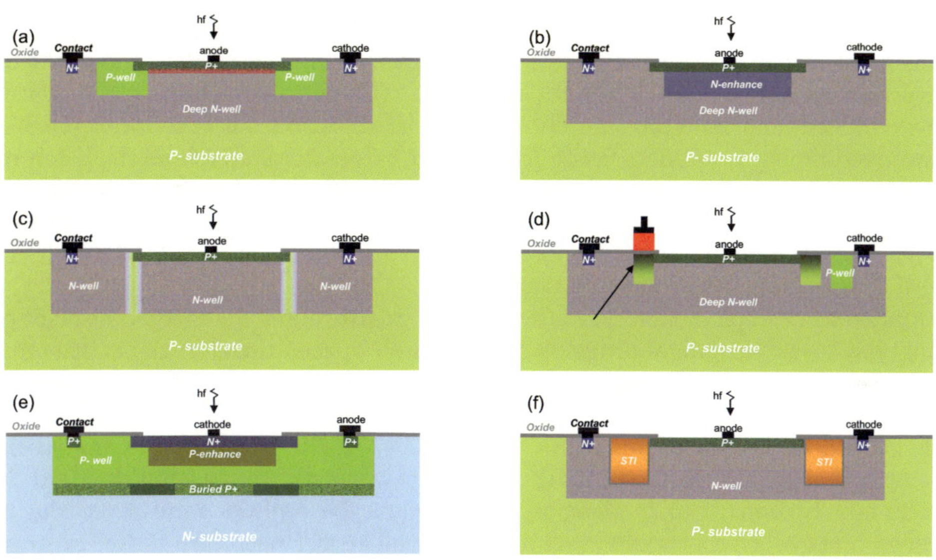

Fig. 2. Cross sections of different SPAD constructions: (a) Diffused Guard Ring; (b) Enhancement Mode; (c) Merged Implant Guard Ring; (d) Gate Bias and Floating Guard Ring; (e) Timing Optimised; (f) Shallow Trench Isolation Guard Ring.

a. First introduced for the purposes of investigating microplasmas in p-n junctions under avalanche conditions (Goetzberger et al., 1963), the diffused guard ring is a lower doped, deeper implant at the device periphery able to reduce the local electric field strength. This construction has since been implemented by several research groups (Cova et al., 1981; Kindt, 1994; Rochas et al., 2002; Niclass et al., 2007). Whilst enabling a low breakdown voltage using implants that are commonly available in most CMOS processes, this structure has several limitations. Firstly, when implemented in a modern CMOS process, the high doping concentration, shallow implants lead to a high electric field structure, resulting in a high DCR due to tunnelling as predicted by (Haitz, 1965), confirmed by (Lacaita et al., 1989) and also recently reported in (Niclass et al., 2007).

Secondly, if long thermal anneal times are employed in relation to the guard ring implant, the resultant field curvature around this key feature creates a non-uniform, dome-shaped electric field profile, peaking at the centre of the device. This in turn implies a breakdown voltage variation across the active region, which strongly affects the homogeneity of the photon detection efficiency (Ghioni et al., 2007). Thirdly, the increase of the quasi-neutral field region at the detector edge promotes late diffusion of minority carriers into the central high-field region, resulting in a long diffusion tail in the timing resolution characteristic, first reported in (Ghioni et al., 1988), and also evident in the 130nm implementation of (Niclass et al., 2007). Fourthly, this structure has a minimum diameter limitation due to merging of the guard ring depletion region as the active region is reduced, illustrated in (Faramarzpour et al., 2008). This limits the scalability of the structure for array implementation purposes.

b. The enhancement mode structure, first introduced in the reach-through device of (Petrillo et al., 1984), was employed in a hybrid diffused guard ring/enhancement structure (Ghioni et al., 1988), and finally used without the diffused guard ring structure (Lacaita et al., 1989), relying on a single central active region enhancement implant (with doping polarities reversed, and embedded in a dual layer P-epitaxial substrate), which is referred to as a 'virtual' guard ring structure. The benefits of this structure are significant. Firstly, the quasi neutral regions surrounding the guard ring are removed, and therefore the minority carrier diffusion tail is reduced resulting in improved timing resolution. Secondly the device does not suffer from depletion region merging when scaling down the active region diameter, easing the prospect of array implementations with fine spatial resolution. Both versions were recently repeated but with dual orientation, range of active areas and active quench circuits in a high voltage CMOS technology (Pancheri & Stoppa, 2007).

c. An alternative implementation to the diffused guard ring is the merged implant guard ring, relying on the lateral diffusion of two closely spaced n-well regions, that creates a localised low-field region, preventing edge breakdown. This technique was demonstrated using only the standard layers available in a CMOS process (Pauchard et al., 2000a). Whilst successful with 50Hz DCR, >20% PDE and 50ps timing resolution being reported (Rochas et al., 2001), this design normally violates the standard design rules and is difficult to implement. Nevertheless the authors were successful in co-integrating quenching circuitry and forming small arrays in a 0.8μm CMOS process.

d. Other guard ring ideas are based either on a metal or polysilicon control 'gate' biased appropriately to control the depth of the depletion region in the zone immediately beneath (Rochas, 2003c), or on a 'floating guard ring' implant inserted near the edge of the active region in order to lower the electric field around the anode periphery (Xiao et al., 2007). The gate bias method is relatively unproven compared to the more common guard ring implementations. This is an interesting solution since polysilicon is normally available to move STI out of the active region. The floating guard ring construction has parallels with the diffused guard ring construction: it lends itself to be employed with certain anode implant depths, but requires careful modelling to determine the optimum layout geometry and can be area inefficient. Implemented in a high voltage technology, it yielded SPADs with both low DCR and good timing resolution, albeit with a high breakdown voltage and an off-chip active quench circuit (Xiao et al., 2007).

e. The work of the research group at Politecnico di Milano has prioritised timing resolution as the key performance metric since early publications utilising the diffused guard ring structure (Cova et al., 1981). This focus was maintained in the progression to devices implemented in a custom epitaxial layer (Ghioni et al., 1988). It was observed that previously published devices exhibited a long diffusion tail in the timing response. This was due to minority carriers generated deep in the quasi-neutral regions beneath the SPAD reaching the depletion layer by diffusion. The single epitaxial layer devices helped to greatly reduce the diffusion tail by drawing away deep photo-generated minority carriers via the secondary epitaxial-substrate diode junction. This technique was taken further in the double-epitaxial structure (Lacaita et al., 1989), and again in the more complex structure of (Lacaita et al., 1993b). The goal of the 'double epitaxial' layer design was to reduce the thickness of quasi-neutral region below the SPAD in order to limit the diffusion tail whilst maintaining a high enough electric field to provide fast response without a high dark count penalty. In the more complex structure of (Lacaita et al., 1993b) the buried p^+ layer was interrupted underneath the active region in order to locally fully deplete the main epitaxial layer by reverse biasing the substrate, for the purpose of eliminating diffusion carriers. Whilst resulting in unprecedented timing performance (35ps FWHM) with DCR of a few hundred Hertz for a 20μm diameter structure, the design required full customisation of the manufacturing process, resulting in limited co-integration capability.

f. The shallow trench guard ring structure was first introduced in 0.18μm CMOS (Finkelstein et al., 2006a). The main goal of this innovation was to increase FF and allow fine spatial resolution. The etched, oxide filled trench, that is a feature of deep submicron processes, is used as a physically blocking guard ring, so containing the high field zone in the active region. This structure was successful in addressing FF and potential pixel pitch, although only single devices were reported. However, the subsequent publication by the same author group (Finkelstein et al., 2006b) revealed a very high DCR of 1MHz for a small diameter 7μm device. This was possibly caused by etching-induced crystal lattice defects and charge trapping associated with STI, as well as band-band tunnelling through the conventional p^+/n-well diode junction. The same author group noted in (Hsu et al., 2009) that, despite the high dark count, the timing resolution characteristic was unspoiled by a diffusion tail due to reduced quasi-neutral field regions associated with implanted guard rings. Further, it was observed that increasing the active region diameter had no effect on the 27ps jitter, suggesting that these structures do not suffer from the lateral avalanche build up uncertainty. Additionally, the lower junction capacitance yields reduced dead time. There are two further related publications associated with the use of STI in SPADs. In (Niclass et al., 2007) the clash of STI with the sensitive active region is avoided by drawing 'dummy' polysilicon to move the etched trench to a safe distance away. This was progressed by (Gersbach et al., 2008), applying a low doped p type implant around the STI interface. However, the DCR was still around 80kHz for an 8μm active region diameter and 1V of V_{EX}. Timing performance remains acceptable for many applications at ~140ps.

4.2 Linear-mode APD

Commercial linear-mode silicon APDs evolved from the same precursors as SPADs and are nowadays a mature technology, with outstanding performances in terms of Quantum Efficiency, Noise Factor and Bandwidth. Several APD products are sold by big companies (APD producers) for applications requiring low-noise and high speed detection such as laser ranging, particle detection, molecule detection, optical communications, etc. Commercial APDs can have a peak quantum efficiency (QE) exceeding 80%, an excess noise factor $F \approx M^{0.3}$ for reach-through devices and as low as $F \approx M^{0.17}$ in the case of Slik™ devices fabricated by EG&G. APDs are commonly used for imaging by mechanical scanning as in the case of laser range finding or confocal microscopy. Only a few small APD arrays are currently present on the market, and the maximum number of devices in a single array is currently limited to 64 (8x8 module by RMD). From year 2000, several linear-mode APDs fabricated in CMOS technologies have appeared in the literature, in various technology nodes. Although their performance is still far from the one obtained with commercial APDs, their low cost and possibility of monolithic integration with readout electronics make them appealing in several application domains. Table 1 lists a selection of CMOS linear-mode APDs so far presented with some of their performance indicators.

Reference	Node [μm]	Type	Guard Ring	V_{APD} [V]	QE [%]	F @ M = 20
Biber 2001	2	p^+/n-well	p-base	42	40 @ 500nm	36000 @ 635nm
Biber 2001	2	n^+/p-sub	n-well	80	75 @ 650nm	1800 @ 635nm
Rochas 2002	0.8	p^+/n-well	p-well	19.5	50 @ 470nm	7 @ 400nm
Stapels 2007	0.8	n^+/p-sub	n-well	n.a.	>60 @ 700nm	5 @ 470nm
Stapels 2007	0.8	p^+/n-well	p-tub	25	50 @ 550nm	50 @ 470nm
Kim 2008	0.7	n^+/p-body	virtual	11	30 @ 650nm	n.a.
Pancheri 2008	0.35	p^+/n-well	p-well	10.8	23 @ 480nm	4.5 @ 380nm 6 @ 560nm

Table 1. Selected characteristics of CMOS avalanche photodiodes.

Although both p^+/n-well and n^+/p-well structures have been presented, the former structure is preferred if an integrated readout is to be fabricated, because both photodiode terminal electrodes are available and low voltage circuits can be implemented. However, an n^+ on p substrate is also feasible, provided it is isolated by means of an n^+ buried layer as in the case of (Kim et al., 2008). From the point of view of readout noise, an n^+/p structure is favored for visible light wavelengths, because the avalanche is electron-initiated. This can be clearly observed comparing the two structures in (Stapels et al., 2007), where p+/n-well APD has a noise 10 times higher than n+/p-sub APD. For wavelength shorter than about 400nm, however, an n^+/p structure can be convenient and has a low noise factor. Passing from old 2μm technologies to 0.8μm and 0.35μm a positive trend is observed. While the breakdown voltage is reduced due to the higher doping levels, the noise also becomes generally lower. One of the reasons is that the ionization coefficient ratio k is closer to unity, so the noise due to hole initiated avalanche is lower, as can be observed comparing the p+/n structures in (Biber et al., 2001) and (Stapels et al., 2007). When the doping levels are even higher, the width of the multiplication region is reduced and standard McIntyre theory is no longer adequate to describe the avalanche process because of the dead space effect (Hayat et al., 1992). In this case, the noise factor becomes lower than the one predicted by standard model both for electron- and hole-initiated avalanche, as observed in (Pancheri et al., 2008). The positive trend in noise as technology is scaled is, however, accompanied by a reduction of quantum efficiency due to the reduced absorption region depth and an increase of dark current due to the contribution of tunneling.

CMOS APDs are appealing for short distance communications because of their large bandwidth, exceeding 1GHz. A few successful examples of CMOS APDs in 0.18μm technology have been reported (Iiyama et al., 2009; Huang et al., 2007; Kang et al., 2007), with bandwidth figures up to 2.6 GHz and good dark currents, in the nA range. Nevertheless, the STI used for guard ring implementation can be inefficient for devices with deeper junctions, resulting in much higher dark currents (Huang et al., 2007).

5. Design and characterization of advanced CMOS avalanche photodiodes

5.1 Geiger-mode APD (SPAD)

(a) High fill factor linear array in a standard 0.35μm CMOS technology

One of the main drawbacks in SPAD arrays presented so far is the low FF, which is in the order of some percent in the best cases. Even if microlens arrays can be used to improve FF, their use is subject to a series of technological constraints. Therefore, a reasonably good FF is important even if the use of optical concentrators is foreseen. There are at least three aspects that limit the FF: the guard ring, the need to reduce optical cross-talk between neighbouring pixels and the size of the readout channel, that needs to be much larger than 3T topology used in standard active pixels.

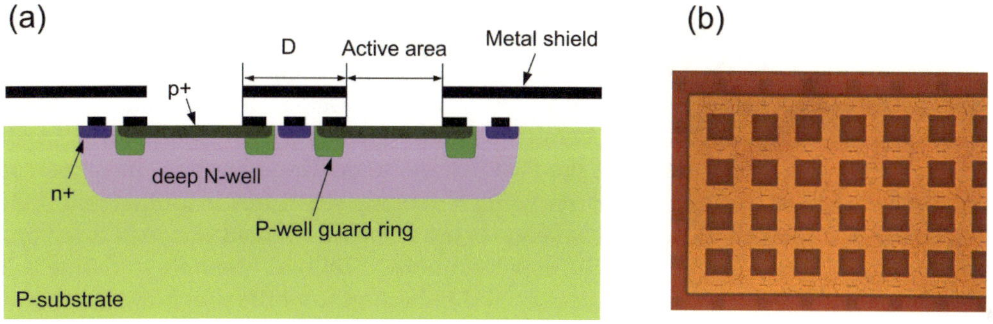

Fig. 3. Schematic cross section (a) and micrograph (b) of the 4 line SPAD array in 0.35μm CMOS technology.

We have started tackling this problem with a 4 line array, fabricated in 0.35μm CMOS technology (Pancheri & Stoppa, 2009). The SPAD array was a 64x4 array, with the 4 devices in a column sharing the same digital readout channel. In order to improve the FF as much as possible, SPADs have a square geometry with rounded corners, and have been implemented in a shared deep n-well. A schematic cross section and a micrograph of the SPAD array are shown in Fig. 3, and a summary of the main characteristics of the SPADs is reported in Table 2. A remarkable 34% FF is obtained, which could easily be doubled with the use of optical concentrators. More than 80% of the SPADs have a dark count rate of approximately 1kHz, while the remaining 20% have increasingly larger dark counts, a small percentage exceeding 100 kHz. In addition to the characteristics of single SPADs, in arrays it is important to consider also PDP non-uniformity and cross talk between pixels. To measure the first one, we have used a uniform incident light onto the SPAD array by using a stabilized lamp and an optical diffuser. Light intensity was adjusted to obtain count rates below 10% of the maximum SPAD count rate, so as to avoid saturation effects. Dark count rate was subtracted from the recorded counts to take into account only the optically generated photons. A non-uniformity lower than 2% was measured at V_{ex} = 4V, which is remarkably good for this kind of device. Cross talk was evaluated by measuring the DCR variation of a low-DCR SPAD (SL) in the neighborhood of a high-DCR SPAD (SH). When SH is enabled, a DCR increase of about 1% of the DCR of SH is observed in SL because of optical cross talk effect. A further effort to increase the FF would increase the cross-talk to higher levels and have a negative impact for the array performance.

SPAD pitch	26 μm
Fill Factor	34%
Breakdown voltage	31V
Dark count rate @ Room Temperature	1kHz typ.@ V_{ex} = 4V
Photon Detection Efficiency	32% @ λ = 450nm
Jitter (FWHM)	< 160ps
Afterpulsing rate	6% @ T_D=200ns
PDP non-uniformity (σ)	< 2% @ V_{ex} = 4V
Cross-talk	1%

Table 2. Summary of the main characteristics of SPAD array in 0.35μm CMOS technology.

(b) Low noise SPADs in a 0.13μm imaging CMOS technology

Advanced CMOS technologies have been optimised for high performance transistors, somewhat contrary to the requirements for low-noise Geiger mode CMOS avalanche photodiodes. The active region is generally defined between the p^+ source-drain implant and n-well used to define the bulk of the PMOS transistors. These implants have increased in doping density and the breakdown mechanism in many structures in 0.18μm and 0.13μm has switched to tunneling. A further challenge is the presence of shallow trench isolation for isolation of NMOS transistors from the substrate noise. The first attempts to define a low DCR SPAD with these technologies, summarized in Section 4, whilst providing good timing characteristics, still suffered from poor DCR performance.

A low-DCR SPAD structure is proposed here which largely attenuates the tunneling problem of the p^+/n-well junction by constructing the deep anode from p-well (contacted by p^+) in conjunction with the buried n-well. The resulting SPAD represents an alternative to that reported in (Richardson et al., 2009a), showing outstanding noise performance. A further novelty in this structure is that the cathode and a new guard ring structure are formed simultaneously by the use of buried n-well without n-well. This latter technique requires a drawn p-well blocking layer to inhibit the automatic generation of p-well as the negative of n-well by the CAD mask Boolean operation. The result is a progressively graded doping profile in the guard ring region, reducing in concentration near the substrate surface, as indicated by the shading of the buried n-well zone in Fig. 4(a). This lowers the electric field at the periphery in comparison to the main p-n junction, as shown in the simulated plot of Fig. 4(b). The guard ring zone can be kept free of STI by using a poly ring around the periphery defining a thin oxide region. STI formation is known to introduce defects and crystal lattice stresses which cause high DCR so it is normally important to move the trench away from the main diode p-n junction. Finally, the connection to the buried n-well cathode is implemented by contacting to drawn n^+ and n-well at the outer edge. An attractive property of this p-well SPAD structure is that it can be implemented in any standard digital triple-well CMOS technology. SPADs implemented at older process nodes have required low-doped wells only found in more costly high voltage technologies.

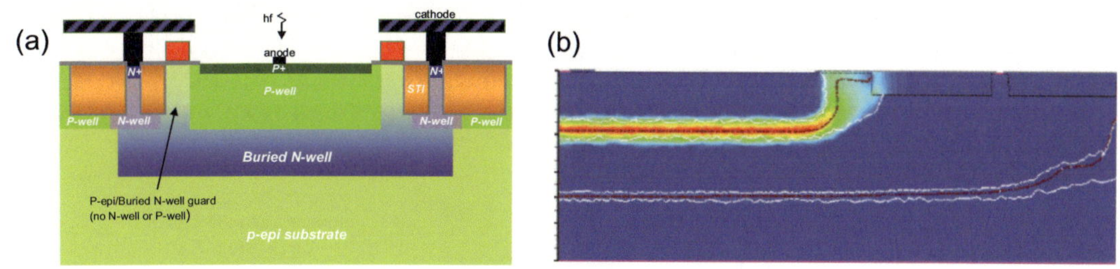

Fig. 4. (a) Cross section of SPAD with retrograde buried n-well cathode and p-well anode; (b) TCAD simulated electric field distribution (arbitrary scale) showing low field at the surface increasing at depth with the grading profile of the buried n-well.

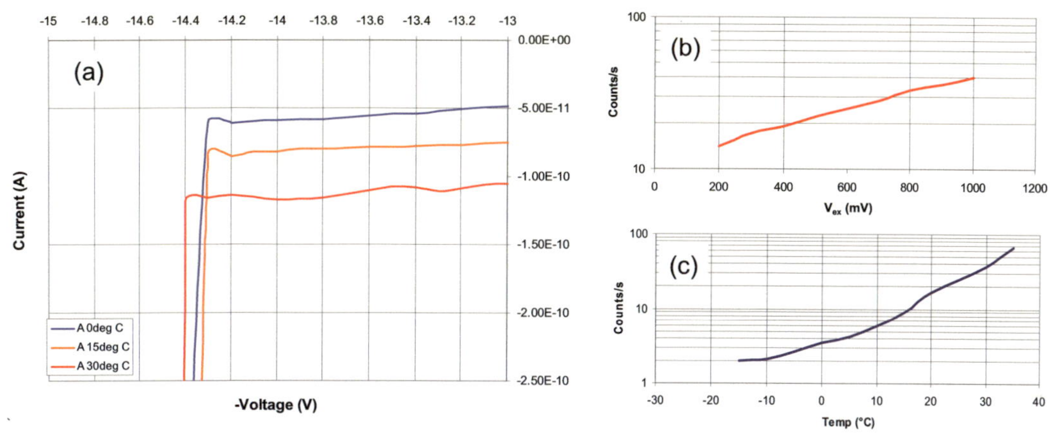

Fig. 5. Dark characteristics of the SPAD shown in Fig. 4: (a) Current-Voltage curves; (b) dark-count rate variation with the excess bias voltage at room temperature; (c) dark count rate variation with temperature at 0.6V of excess bias voltage.

The characterisation results for an 8μm active diameter SPAD implemented in a 0.13μm CMOS image sensor process according to the previous construction are now discussed. Figure 5(a) shows the current-voltage characteristics at three different temperatures: the breakdown knee occurs at 14.3V, in very close agreement with TCAD simulation (14.4V), and with 83pA of dark current at breakdown. The variation of the breakdown voltage over temperature is +3.3mV/°C, indicating that the breakdown mechanism is avalanche (Sze & Ng, 2007). Dark count rate is reported in Figs. 5(b) and 5(c): the first graph set shows that this detector has a very low dark count rate with the expected exponential relationship between DCR and excess bias; the second graph set shows the dark count rate is dominated by thermal carrier generation at temperatures larger than 5°, the DCR values doubling every 7-8 degrees, whereas at lower temperatures the reduced slope indicates that tunnelling starts to be non negligible. Also afterpulsing (not shown) is very low, in the order of 0.02%. This is in line with the low junction capacitance and photoelectric gain of the considered SPAD.

Furthermore, the implementation of this detector design in a 32x32 array as part of the MEGAFRAME Project (MEGAFRAME) provided the opportunity for analysis of DCR population distribution for 1024 elements, as shown in Fig. 6. The detector had the same structure as those above but a slightly smaller active region of 7μm diameter. It can be seen that the population splits roughly into two groups: those with low DCR well below 100Hz, and those with higher DCR up to 10 kHz. The split ratio is ~80:20. All 1024 SPADs were functional. The impact on DCR by increased excess bias is evident from the two traces in Fig. 6.

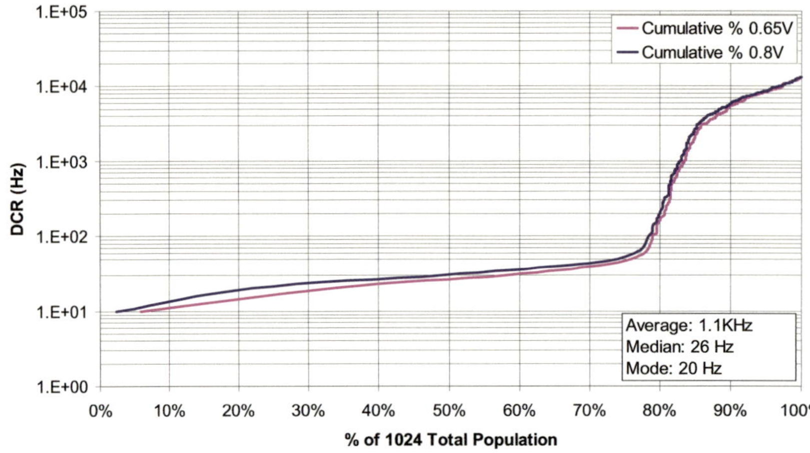

Fig. 6. Distribution of dark count rate measured at room temperature in 1024 p-well buried n-well SPADs of 7μm active region diameter, at two different excess bias voltages.

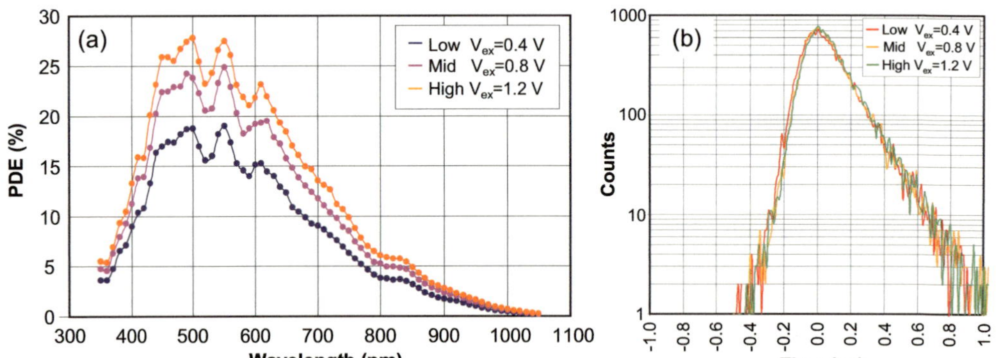

Fig. 7. Electro-optical characteristics of p-well buried n-well SPAD: (a) photon detection efficiency as a function of wavelength at three different excess bias voltages; (b) time jitter measured at 470nm wavelength at the same three excess bias voltages.

Fig. 7(a) shows the photon detection efficiency as a function of the wavelength. As expected, PDE improves with the excess bias voltage at all wavelengths. The deeper p-well junction of this SPAD results into a peak at around 500nm (green) rather than 450nm of more conventional p^+/n-well SPADs. The peak value at 1.2 V excess bias voltage is about 28%. PDE curves extend beyond 800nm where values ~6% are still observed at higher excess bias.

The increased response in the near infra-red is useful for several applications, among them time-of-flight 3D range sensing applications. The perturbations in the response are due to constructive/destructive interference patterns caused by the dielectric stack above the detector, which consists of several different materials with varying refractive indices.
Fig. 7(b) shows the timing histograms measured at 470nm wavelength and at the same excess bias voltages as in Fig. 7(a). Timing resolution is measured as the FWHM of the distributions. Its values are 199 ps, 192 ps, and 184 ps, at low, mid, and high excess bias voltage, respectively. A distinct, exponential tail extends for around 1ns after the peak, evidence of a diffusion related timing delay.
Table 3 shows a performance comparison for SPADs fabricated in 130nm CMOS technologies. As can be seen, a dramatic improvement in DCR can be achieved by adopting proper implant layers which lower the electric field, at the expense of a degradation in the jitter, that however remains good enough for most applications.

Reference	This work	Gersbach 2009	Niclass 2007
Active area	50.3 μm^2	58 μm^2	53 μm^2
Type	p$^+$/p-well/buried n-well	p$^+$/n-well	p$^+$/n-well
Guard ring	buried n-well with poly gate	STI with p-type passivation	p-well/STI with poly gate
V_{BD}	14.3V	9.4V	9.7V
DCR	40Hz@V_{ex}=1V	670kHz@V_{ex}=2V	100kHz@V_{ex}=1.7V
PDE	28%@ 500nm	30%@480nm	34%@450nm
Jitter	184ps FWHM	125ps FWHM	144 ps FWHM
Afterpulsing	0.02%@T_D=100ns	<1%@T_D=180ns	n.a.

Table 3. Performance comparison of SPADs made in 130nm CMOS technologies.

5.2 Linear-mode APD

In order to implement APD-based pixels with integrated low-voltage readout channels, both terminals of the photodiode must be available: one to apply a high voltage and the other to connect the readout channel. This restricts the choice of the structure to the p+/n-well type (although also n$^+$/floating p-well would be feasible). With technology scaling, the well doping levels are steadily increasing to allow the integration of smaller-size MOSFETs. Therefore, APDs fabricated in standard wells would have decreasing breakdown voltages and increasing tunneling dark currents when migrating the design to more advanced technology nodes. However, in High Voltage processes, a larger choice of layers with different doping levels are available, and can be exploited in the realization of APDs, both to obtain an efficient guard ring and a suitable doping for the multiplication region. The High Voltage option can then enable the fabrication of APDs also in deep sub-micron technologies. One key aspect to understand is how the noise factor scales with technology. Unfortunately, there are only a few examples of noise-characterized CMOS APDs in the literature, so that only some basic guidelines can be drawn.
Fig. 8 shows the measured noise factor for two p$^+$/n-well APDs fabricated in two different technologies. The APD in Fig.8(a), fabricated in a 0.7μm High Voltage technology, has a breakdown voltage of 20.8V, and exhibits a low noise factor at low wavelengths because of electron-initiated avalanche. At higher wavelengths, however, noise increases considerably due to the hole-initiated avalanche noise. The measurements are in good agreement with

McIntyre noise theory, considering an ionization ratio k=0.2. The device of Fig. 8(b), fabricated in 0.35μm CMOS technology, exhibits low noise also in the case of hole-initiated avalanche. The measured noise does not correspond to the one predicted by McIntyre theory because of the dead-space effect. This device is therefore more suitable for imaging applications in the visible spectral region. A further improvement in the noise factor could be given by n^+ on floating p-well, which would combine the low noise of electron initiated avalanche with the requirement to access both device terminals. However, for this device structure, there is no experimental noise characterization available so far.

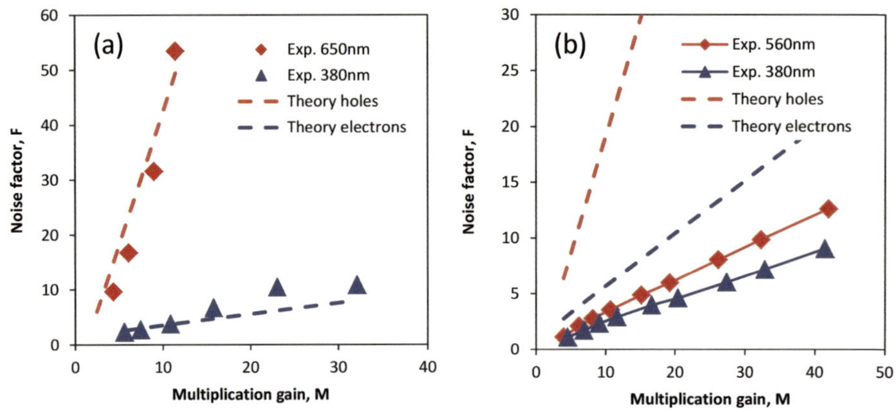

Fig. 8. Noise factor as a function of multiplication gain at two illumination wavelengths in two APDs made from: (a) 0.7μm High Voltage and (b) 0.35μm standard CMOS technology.

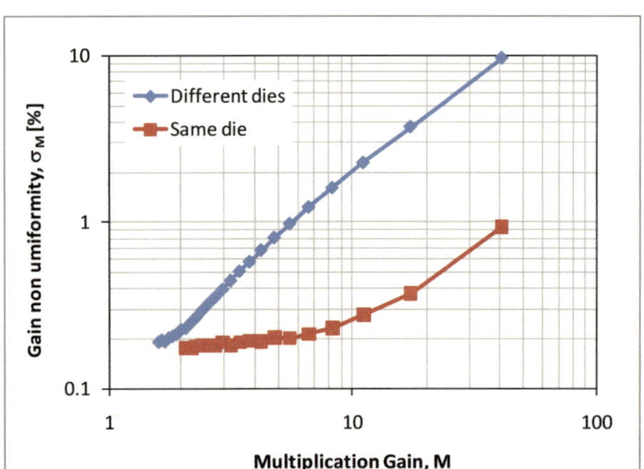

Fig. 9. Gain uniformity between 0.35μm CMOS APDs fabricated on the same die and on different dies in the same production batch.

In the implementation of avalanche photodiode arrays, an important factor to take into account is gain uniformity. As multiplication gain increases, the voltage range allowing a tolerable gain variation progressively reduces. Therefore, if small non-uniformities between different APDs can be corrected via pixel-by-pixel calibration, large non-uniformities are detrimental to the correct operation of the array. In order to evaluate the uniformity of APDs made on the same die, a small set of identical APDs, whose characteristics were reported in

(Pancheri et al., 2008), were fabricated in a 0.35μm CMOS technology. For every APD in the die and for several dies, the gain was measured as a function of voltage. Gain non-uniformity, expressed as standard deviation, is reported as a function of average gain in Fig. 9. It can be observed that the gain non-uniformity of APDs fabricated on different dies (in the same production batch) is 10 times higher than the non-uniformity of gains for APDs belonging to the same die. In this last case, a non-uniformity better than 1% is achieved for a gain M=40. This demonstrates the feasibility of linear-mode APD arrays in that specific technology.

6. Integrated read-out channels for SPADs

Two basic circuit elements are required to read-out a SPAD: i) a quench circuit to detect and stop the avalanche event, followed by restoration of bias conditions; ii) a buffer to isolate the SPAD from the capacitance of external processing electronics. The integration of more complex processing and support circuits such as time-to-digital conversion, gated ripple counters and charge pumps is the subject of much recent research (Niclass et al., 2008; Richardson et al., 2009b; Guerrieri et al., 2010). However, as the main properties of the optical system are set by the choice of quench circuit and associated biasing arrangement for the SPAD we will focus on these aspects in more detail.

(a) SPAD Orientation Options

Since a SPAD is a two terminal diode, operated at an excess bias potential beyond its breakdown voltage, it may be oriented in two ways; either with a negative potential on the anode, or positive on the cathode. As previously discussed, it is desirable to minimise detector dead time and charge flow quantity during an avalanche event. The optimal connectivity to permit this depends on the parasitic capacitances which are specific to the diode construction which has been implemented.

To illustrate this point, two possible passive quench configurations are shown in Fig. 10 for a SPAD implemented within a deep n-well, p-substrate technology such as that of (Pauchard et al., 2000b) Fig. 10(a) shows a large negative voltage applied to the anode, with a much smaller positive excess bias applied via the passive quench PMOS to the cathode. Fig. 10(b) employs an NMOS quench component with source connected to ground potential, and a large positive bias potential V_{bias_total} applied to the SPAD cathode. In both cases the key moving 'sense' node of the circuit is at the buffer/comparator input terminal. The time domain output signal is determined by the threshold voltage of the buffer, the SPAD excess bias voltage and recharge time constant.

In configuration (a) the cathode is the moving node and therefore the additional capacitance of the n-well to p-substrate parasitic diode (C_{nwell}) must also be charged and discharged during the detector operating cycle. As well as contributing to the lengthening of detector dead time due to the increased RC load, this adds to the volume of charge flow through the detector, so increasing probability of afterpulsing. Configuration (a) also places a limitation on the maximum negative $V_{breakdown}$ that can be applied without forward biasing the n-well p-substrate junction. The maximum voltage V_{ex} that can be applied is determined by the transistor gate-oxide breakdown voltages, typically 3.3V for thick oxide devices. Thus voltages beyond $V_{breakdown} + V_{ex} + V_{diode}$ across the SPAD will induce a latch-up behaviour where the buffer input voltage will transit below $-V_{diode}$ drawing current from the substrate ground potential through the forward biased parasitic diode.

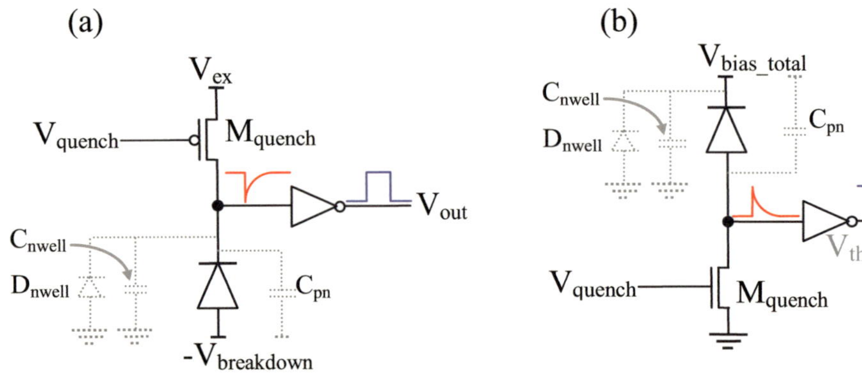

Fig. 10. SPADs biased with (a) PMOS quench (b) NMOS quench

In the case of the 'flipped' configuration (b) the anode of the detector is the moving node, and therefore there is only the charging of the SPAD junction capacitance to consider. Importantly, this permits the sharing of n-well regions by multiple elements within an array implementation leading to reduced pixel pitches and improved fill factor. However, it should be noted that it is common for the mobility of an NMOS transistor channel to be higher than PMOS in deep submicron processes (the intrinsic conductivity K_p can be typically ¼ of K_n), which leads to a larger equivalent area used when compared to a PMOS implementation for the same target quench resistance.

Configuration (b) will only operate correctly as a SPAD provided the voltage applied to the SPAD active region $V_{bias\_total} = V_{breakdown} + V_{ex}$ is lower than the breakdown voltage of the parasitic n-well/p-substrate or n-well/p-well peripheral diodes. Indeed the n-well of these parasitic diodes can act as the base of a parasitic bipolar transistor inducing a latch-up condition and so the resistivity of the buried n-well is of concern.

Regardless of orientation, for reliability reasons it is vital not to exceed the maximum gate oxide potential of the output buffer otherwise permanent damage can occur. For this reason it is common to employ thick gate oxide transistors for this circuit if available in the target technology. A clamping diode may be used in configuration (b) to ensure safe operating region of the buffer transistors are respected.

Both orientations are prevalent in the published literature, for example (Ghioni et al, 1996) implementing configuration (b), and (Pancheri & Stoppa, 2007), and (Niclass, 2008). Yet more bias options are shown in Fig.11(a) and (b) where the SPAD is capacitively coupled to the output buffer and quenched by a passive resistor. The resistors can be conveniently made from high ohmic polysilicon and the capacitors by metal-metal finger (MOM) capacitors paying attention to the dielectric reliability at the high voltages. Such circuits decouple the d.c. bias conditions of the SPAD from those at the input of the digital buffer (fixed by M_{keep}). Only the pulse of the SPAD is transferred to the output. The capacitor C_{ls} must be chosen such that capacitive division by the input capacitance of the buffer does not attenuate the SPAD pulse below the minimum required by the buffer trigger threshold. In practice, C_{ls} can easily be as small as a few femto Farads. The circuit of Fig. 11(a) allows the p-well of the SPAD to be biased at ground potential (in common with all NMOS p-wells). The circuit of Fig. 11(b) allows the moving node of the SPAD to be the lower capacitance p-well while the n-well is biased at a low voltage avoiding breakdown issues with n-well/p-substrate junctions. These floating circuits also permit a differential bias arrangement whereby V_{bias_pos} and V_{bias_neg} are set to $+V_{bias\_total}/2$ and $-V_{bias\_total}/2$ respectively. This minimises the stress on the parasitic wells and eases power supply generation.

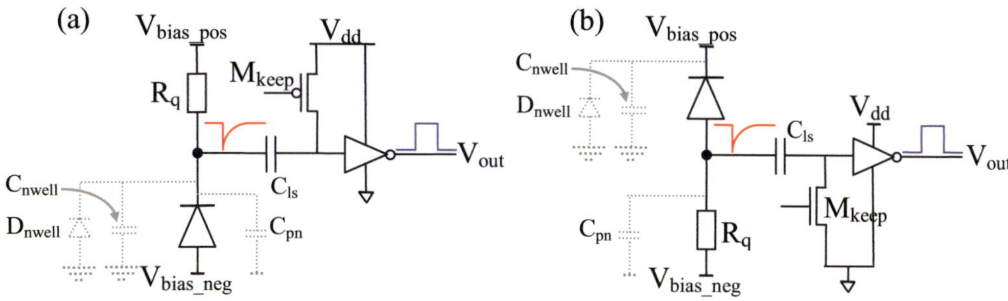

Fig. 11. Floating SPAD orientation options.

(b) Quench Resistor Optimisation

Incorrectly sized passive quench circuitry can result in undesirable circuit operation, such as lengthy and inconsistent dead times. This introduces noise, imposes severe dynamic range limitations, particularly impacting photon counting applications and those employing gated counters such as ranging, 3D cameras and fluorescence lifetime imaging microscopy (FLIM). The p-well to buried n-well diode and parasitic n-well to p-substrate diodes are often modelled in foundry technology information allowing extraction of junction capacitance. Parasitic capacitive elements of the quench circuit can be predicted using layout-vs-schematics (LVS) extraction tools. This means that the passive quench MOS element can be sized appropriately for linear mode operation, and accurate simulations performed.

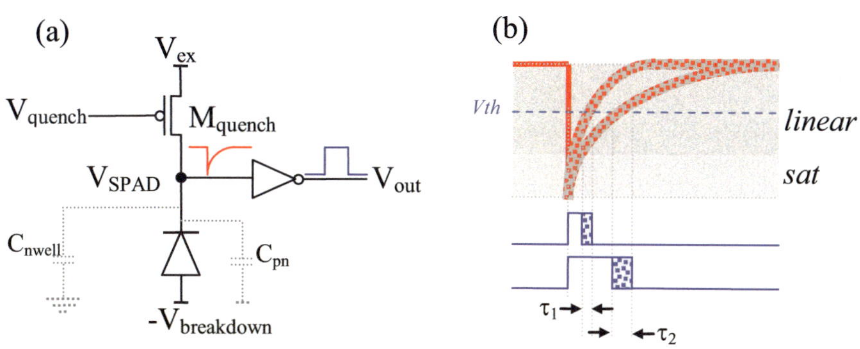

Fig. 12. Passive Quench Optimisation

This optimisation process is illustrated in Fig. 12. Fig. 12(a) shows the basic passive quench circuit, parasitic elements and waveforms. Fig. 12(b) shows how noise can cause increased dead time variation in a non-optimised system. In an optimised configuration the crossing of the inverter threshold (V_{th}) is performed at a steep gradient. Noise on the moving node results in variation of dead time τ_1 at the inverter output. If the passive quench element W/L ratio is too small, the extended RC recharge time degrades noise immunity and results in a higher dead time variation τ_2. In the most basic configuration the gate of the PMOS passive quench element is simply grounded. Under the aforementioned non-optimal circumstances a small negative voltage can be applied to the gate of the PMOS transistor but this is considered an undesirable complication. Similarly, in the case of an undersized PMOS element, a positive voltage must be applied to its gate to ensure an output pulse is generated. In a low voltage process, with comparatively low overall excess bias conditions, for the bulk of the quenching and reset cycle the PMOS element is mainly operating in linear mode, with a brief entry into saturation (see Fig. 12(b)). Thus, equations (1), (2), and (3) can be applied to optimise the passive quench element aspect ratio:

$$V_{SPAD} = V_{ex}\left(1 - e^{t / R_{quench} \cdot C_{tot}}\right) \tag{1}$$

$$R_{quench} = V_{ex} / (2 \cdot Id_{Mquench}) \tag{2}$$

$$Id_{Mquench} = K_p \frac{W}{L}\left[\left(V_{gs} - V_t\right) - \frac{V_{ds}}{2}\right] \cdot V_{ds} \tag{3}$$

For a parasitic capacitance C_{tot} = 50fF, V_{ex}=1.2V and inverter threshold V_{th}=0.6V, for a dead time of 20ns and a PMOS passive quench element with K_p= 30 μA/V^2, equations yield: R_{quench}≈400kΩ, Id=0.6μA, and (W/L)=1/8. Therefore for a 0.35μm width, grounded gate PMOS transistor, length L should be ~2.7μm to meet the above specification.

7. Conclusion

We have reported on CMOS avalanche photodiodes, reviewing the most significant devices so far proposed in the literature and discussing selected results from our research activities and relevant to both Geiger-mode and linear-mode APDs. In spite of the technological constraints and design rule restrictions in advanced manufacturing processes, smart design solutions for SPADs and the read-out electronics exist that allow very for good performance even in the 130nm CMOS technological node, thus paving the way to new application fields for high resolution, SPAD based image sensors. As far as linear-mode CMOS APD are concerned, although these devices have not been developed to the same extent as SPADs, our results demonstrate that acceptable excess noise figures can be achieved at moderate gain even at relatively large wavelengths. Moreover, the spatial uniformity of the gain is found to be reasonably good in devices from the same die, so that the design of APD based pixel arrays for high-sensitivity imaging can be envisioned.

8. References

APD producers: Advanced Photonix (www.advancedphotonix.com), EG&G (www.urscorp.com), Hamamatsu Photonics (www.hamamatsu.com), Perkin Elmer (www.PerkinElmer.com), RMD (www.rmdinc.com)

Aull, B.F.; Loomis, A.H.; Young, D.J.; Heinrichs, R.M.; Felton, B.J.; Daniels, P.J., Landers, D.J. (2002) Geiger-Mode Avalanche Photodiodes for Three-Dimensional Imaging. *Lincoln Laboratory Journal*, Vol. 13, No. 2 (2002), 335-350, ISSN 0896-4130

Biber, A.; Seitz, P. & Jäckel, H. (2001). Avalanche Photodiode Image Sensor in Standard Silicon BiCMOS Technology. *Sensors and Actuators A: Physical*, Vol. 90, No. 1-2 (May 2001), 82-88, ISSN 0924-4247

Cova, S.; Longoni, A.; Andreoni, A. (1981) Toward Picosecond Resolution with Single-Photon Avalanche Diodes, *Review of Scientific Instruments*, Vol. 52, No. 3 (March 1981), 408-412, ISSN 0034-6748

Cova, S.; Ghioni, M.; Lacaita, A.; Samori, C.; Zappa, F. (1996) Avalanche Photodiodes and Quenching Circuits For Single-Photon Detection. *Applied Optics*, Vol. 35, No. 12 (April 1996), 1956-1976, ISSN 1943-8206

Cova, S.; Ghioni, M.; Lotito, A.; Rech, I.; Zappa, F. (2004) Evolution and Prospects for Single-Photon Avalanche Diodes and Quenching Circuits. *Journal of Modern Optics*, Vol. 51, No. 9-10 (June-July 2004), 1267-1288, ISSN 0950-0340

Dussault, D.; Hoess, P. (2004) Noise Performance Comparison of ICCD with CCD and EMCCD Cameras, *Proceedings of SPIE, Vol. 5563,* 195-204, ISBN 0-8194-5501-6, Denver, August 2004, SPIE, Bellingham, WA, USA

Faramarzpour, N.; Deen, M.j.; Shirani, S.; Fang, Q. (2008) Fully Integrated Single Photon Avalanche Diode Detector in Standard CMOS 0.18μm Technology. *IEEE Transactions on Electron Devices*, Vol. 55, No. 3 (March 2008), 760–767, ISSN 0018-9383

Finkelstein, H.; Hsu, M.J.; Esener, S.C. (2006a) An Ultrafast Geiger-Mode Single-Photon Avalanche Diode in a 0.18μm CMOS Technology. *Proceedings of SPIE, Vol. 6372,* 63720W.1-63720W.10, ISBN 0-8194-6470-8, Boston, USA, Oct 2006, SPIE, Bellingham, WA, USA

Finkelstein, H.; Hsu, M.J.; Esener, S.C. (2006b) STI-Bounded Single-Photon Avalanche Diode in a Deep-Submicrometer CMOS Technology, *IEEE Electron Device Letters*, Vol. 27, No. 11 (November 2006), 887-889, ISSN 0741-3106

Gersbach, M. ; Niclass, C.; Charbon, E.; Richardson, J.; Henderson, R.; Grant, L. (2008) A Single Photon Detector Implemented in a 130nm CMOS Imaging Process, *Proceedings of the 38th European Solid-State Device Research Conference (ESSDERC 2008)*, pp. 270-273, ISBN 978-1-4244-2363-7, Edinburgh (UK), September 2008, IEEE, Piscataway, NJ, USA

Gersbach, M. ; Richardson, J.; Mazaleyrat, E.; Hardillier, S.; Niclass, C.; Henderson, R.; Grant, L. Charbon, E. (2009) A Low-Noise Single-Photon Detector Implemented in a 130nm CMOS Imaging Process, , *Solid-State Electronics*, Vol. 53, No. 7 (July 2009), 803-808, ISSN 0038-1101

Ghioni, M.; Cova, S.; Lacaita, A.; Ripamonti, G. (1988) New Silicon Epitaxial Avalanche Diode for Single-Photon Timing at Room Temperature. *Electronics Letters*, Vol. 24, No. 4 (November 1988), 1476-1477, ISSN 0013-5194

Ghioni, M.; Cova, S.; Zappa, F.; Samori, C. (1996) Compact Active Quenching Circuit for Fast Photon Counting with Avalanche Photodiodes. *Review of Scientific Instruments*, Vol. 67, No. 10 (October 1996), 3440-3448, ISSN 0034-6748

Ghioni, M.; Gulinatti, A.; Rech, I.; Zappa, F.; Cova, S. (2007) Progress in Silicon Single-Photon Avalanche Diodes. *IEEE Journal of Selected Topics in Quantum Electronics*, Vol. 13, No. 4 (July/August 2007), 852-862, ISSN 1077-260X

Goetzberger, A.; Mc Donald, B.; Haitz, R.H., Scarlett, R. M. (1963) Avalanche Effects in Silicon p---n Junctions. II. Structurally Perfect Junctions. *Journal of Applied Physics*, Vol. 34, No. 6 (June 1963), 1591-1600, ISSN 0021-8979

Guerrieri, F.; Tisa, S.; Tosi, A.; Zappa, F. (2010); Single-Photon Camera for High-Sensitivity High-Speed Applications, *Proceedings of SPIE, Vol. 7536,* 753605.1-753605.10, ISBN 0-8194-7929-2, San Jose, USA, January 2010, SPIE, Bellingham, WA, USA

Haitz, R.H., Goetzberger, A.; Scarlett, R. M. ; Shockley, W. (1963) Avalanche Effects in Silicon p---n Junctions. I. Localized Photomultiplication Studies on Microplasmas. *Journal of Applied Physics,* Vol. 34, No. 6 (June 1963), 1581-1590, ISSN 0021-8979

Haitz, R.H. (1964) Model for the Electrical Behavior of a Microplasma. *Journal of Applied Physics,* Vol. 35, No. 5 (May 1964), 1370-1376, ISSN 0021-8979

Haitz, R.H. (1965) Mechanisms Contributing to the Noise Pulse Rate of Avalanche Diodes. *Journal of Applied Physics,* Vol. 36, No. 10 (October 1965), 3123-3131, ISSN 0021-8979

Hayat, M. M.; Sargeant, W. L. & Saleh, B. E. A. (1992). Effect of Dead Space on Gain and Noise in Si and GaAs Avalanche Photodiodes. *IEEE Journal of Quantum Electronics,* Vol. 28, No. 5 (May 1992), 1360 – 1365, ISSN 0018-9197

Huang, W.-K.; Liu, Y.-C. & Hsin, Y.-M. (2007). A High-Speed and High-Responsivity Photodiode in Standard CMOS Technology. *IEEE Photonics Technology Letters,* Vol. 19, No. 4 (February 2007), 197 – 199, ISSN 1041-1135

Hsu, M.J.; Esener, S.C.; Finkelstein, H. (2009) A CMOS STI-Bounded Single-Photon Avalanche Diode with 27-ps Timing Resolution and a Reduced Diffusion Tail, *IEEE Electron Device Letters,* Vol. 30, No. 6 (June 2009), 641-643, ISSN 0741-3106

Iiyama, K.; Takamatsu, H. & Maruyama, T. (2009). Silicon Lateral Avalanche Photodiodes Fabricated by Standard 0.18 µm CMOS Process. *Proceedings of the 35th European Conference on Optical Communication (ECOC '09),* P2.06, ISBN 978-1-4244-5096-1, Vienna, Austria, September 2009, VDE VERLAG GMBH, Berlin, Germany

Kang, H.-S.; Lee, M.-J. & Choi, W.-Y. (2007). Si Avalanche Photodetectors Fabricated in Standard Complementary Metal-Oxide-Semiconductor Process. *Applied Physics Letters,* Vol. 90, No. 15 (April 2007), 151118, ISSN 0003-6951

Kim, Y. S.; Jun, I. S. & Kim, K. H. (2008). Design and Characterization of CMOS Avalanche Photodiode With Charge Sensitive Preamplifier. *IEEE Transactions on Nuclear Science,* Vol. 55, No. 3 (June 2008), 1376-1380, ISSN 0018-9499

Kindt, W.J. (1994) A Novel Avalanche Photodiode Array, *Conference Record of the IEEE Nuclear Science Symposium (NSS 1994),* pp. 164-167, ISBN 0-7803-2544-3, Norfolk (USA), October/November 1994, IEEE, Piscataway, NJ, USA

Lacaita, A. ; Ghioni, M.; Cova, S. (1989) Double Epitaxy Improves Single-Photon Avalanche Diode Performance. *Electronics Letters,* Vol. 25, No. 13 (June 1989), 841–843, ISSN 0013-5194

Lacaita, A. ; Zappa, F.; Bigliardi, S.; Manfredi, M. (1993a) On the Bremsstrahlung Origin of Hot-Carrier-Induced Photons in Silicon Devices. *IEEE Transactions on Electron Devices,* Vol. 40, No. 3 (March 1993), 577–582, ISSN 0018-9383

Lacaita, A. ; Zappa, F.; Bigliardi, S.; Manfredi, M. (1993b) Single-Photon Avalanche Diode with Ultrafast Pulse Response Free from Slow Tails. *IEEE Electron Device Letters,* Vol. 14, No. 7 (July 1993), 360-362, ISSN 0741-3106

Marwick, M.A.; Andreou, A.G. (2008) Single Photon Avalanche Photodetector with Integrated Quenching Fabricated in TSMC 0.18 µm 1.8 V CMOS process. *Electronics Letters,* Vol. 44, No. 10 (May 2008), 643-644, ISSN 0013-5194

McIntyre, R.J. (1961) Theory of Microplasma Instability in Silicon. *Journal of Applied Physics,* Vol. 32, No. 6 (June 1961), 983-995, ISSN 0021-8979

McIntyre, R.J. (1966) Multiplication Noise in Uniform Avalanche Diodes. *IEEE Transactions on Electron Devices*, Vol. 13, No. 1 (January 1966), 164–168, ISSN 0018-9383

MEGAFRAME: Million Frame Per Second, Time-Correlated Single Photon Camera, IST FP6 FET Open Project 029217, http://www.megaframe.eu

Niclass, C.; M. Sergio, M.; Charbon, E. (2006) A Single Photon Avalanche Diode Array Fabricated in 0.35μm CMOS and Based on an Event Driven Readout for TCSPC Experiments. *Proceedings of SPIE, Vol. 6372*, 63720S.1-63720S.12, ISBN 0-8194-6470-8, Boston, USA, Oct 2006, SPIE, Bellingham, WA, USA

Niclass, C.; Gersbach, M.; Henderson, R.; Grant, L.; Charbon, E. (2007) A Single Photon Avalanche Diode implemented in 130-nm CMOS Technology. *IEEE Journal on Selected Topics in Quantum Electronics*, Vol. 13, No. 4 (July/August 2007), 863-869, ISSN 1067-260X

Niclass, C. (2008) Single-Photon Image Sensors in CMOS : Picosecond Resolution for Three-Dimensional Imaging, PhD Thesis EPFL, no. 4161 (2008), DOI: 10.5075/epfl-thesis-4161

Pancheri, L. & Stoppa, D. (2007) Low-Noise CMOS Single-Photon Avalanche Diodes with 32 ns Dead Time. *Proceedings of the 37th European Solid-State Device Research Conference (ESSDERC 2007)*, 362-365, ISBN 978-1-4244-1123-8, Muenchen, Germany, September 2007, IEEE, Piscataway, NJ, USA

Pancheri, L.; Scandiuzzo, M.; Stoppa, D; Dalla Betta, G.-F. (2008). Low-Noise Avalanche Photodiode in Standard 0.35-μm CMOS Technology. *IEEE Transactions on Electron Devices*, Vol. 55, No. 1 (January 2008), 457-461, ISSN 0018-9383

Pancheri, L. & Stoppa, D. (2009). A SPAD-based Pixel Linear Array for High-Speed Time-Gated Fluorescence Lifetime Imaging. *Proceedings of the European Solid-State Circuits Conference (ESSCIRC 2009)*, 428-431, ISBN 978-1-4244-4354-3, Athens, Greece, September 2009, ERGO Publications, Athens, Greece

Pauchard, A.; Rochas, A.; Ranjelovic, Z.; Besse, P.A.; Popovic, R.S. (2000a) Ultraviolet Avalanche Photodiode in CMOS Technology, *Technical Digest of the IEEE International Electron Device Meeting (IEDM 2000)*, 30.6.1-30.6.4, ISBN 0-7803-6438-4, San Francisco (USA), December 2000, IEEE, Piscataway NJ, USA.

Pauchard, A.; Besse, P.A.; Martek, M.; Wolffenbuttel, R.F.; Popovic, R.S. (2000b) Ultraviolet-Selective Avalanche Photodiode. *Sensors and Actuators A: Physical*, Vol. 82, No. 1-3 (May 2000), 128-134, ISSN 0924-4247

Petrillo, G.A.; McIntyre, R. J.; Lecomte, R.; Lamoureux, G.; Schmitt, D. (1984) Scintillation Detection with Large-Area Reach-Through Avalanche Photodiodes. *IEEE Transactions on Nuclear Science*, Vol. 31, No. 1 (January 1984), 417–423, ISSN 0018-9499

Renker, D. (2004) Photosensors. *Nuclear Instruments and Methods in Physics Research A*, Vol. 527, No. 1-2, (July 2004), 15-20, ISSN 0168-9002

Renker, D. (2006) Geiger-Mode Avalanche Photodiodes, History, Properties and Problems. *Nuclear Instruments and Methods in Physics Research A*, Vol. 567, No. 1 (November 2006), 48-56, ISSN 0168-9002

Richardson, J.A.; Grant, L.A.; Henderson, R.K. (2009a) Low Dark Count Single-Photon Avalanche Diode Structure Compatible with Standard Nanometer Scale CMOS Technology. *IEEE Photonics Technology Letters*, Vol. 21, No. 14 (July 2009), 1020-1022, ISSN 1041-1135

Richardson, J.; Walker, R.; Grant, L.; Stoppa, D.; Borghetti, F.; Charbon, E.; Gersbach, M.; Henderson R.K. (2009b) A 32x32 50ps Resolution 10 bit Time to Digital Converter Array in 130nm CMOS for time Correlated Imaging, *Proceedings of the IEEE Custom Integrated Circuits Conference (CICC 2009)*, 77-80, ISBN 978-1-4244-4071-9, San Jose, USA, September 2009, IEEE, Piscataway, NJ, USA.

Rochas, A.; Besse, P.A.; Popovic, R.S. (2001) A Geiger Mode Avalanche Photodiode Fabricated in a Conventional CMOS Technology. *Proceedings of the 31st European Solid-State Device Research Conference (ESSDERC 2001)*, 483-486, ISBN 2-914601-01-8, Nurenberg, Germany, September 2001, IEEE, Piscataway, NJ, USA.

Rochas, A.; Gösch, M.; Serov, A.; Besse, P.A.; Popovic, R.S.; Lasser, T.; Rigler., R. (2002) Low-Noise Silicon Avalanche Photodiodes Fabricated in Conventional CMOS Technologies. *IEEE Transactions on Electron Devices*, Vol. 49, No. 3 (March 2002), 387-394, ISSN 0018-9383

Rochas, A.; Gösch, M.; Serov, A.; Besse, P.A.; Popovic, R.S.; Lasser, T.; Rigler., R. (2003a) First Fully Integrated 2-D Array of Single-Photon Detectors in Standard CMOS Technology. *IEEE Photonics Technology Letters*, Vol. 15, No. 7 (July 2003), 963-965, ISSN 1041-1135

Rochas, A., Gani, M.; Furrer, B.; Besse, P. A.; Popovic, R.S.; Ribordy, G.; Gisin, N. (2003b) Single photon detector fabricated in a complementary metal--oxide--semiconductor high-voltage technology. *Review of Scientific Instruments,* Vol. 74, No. 7 (July 2003), 3263-3270, ISSN 0034-6748

Rochas, A. (2003c) Single Photon Avalanche Diodes in CMOS Technology, PhD Thesis EPFL, no. 2814 (2003), DOI: 10.5075/epfl-thesis-2814

Ruegg, H.W. (1967) An Optimized Avalanche Photodiode. *IEEE Transactions on Electron Devices*, Vol. 14, No. 5 (May 1967), 239–251, ISSN 0018-9383

Sciacca, E.; Lombardo, S.; Mazzillo, M.; Condorelli, G.; Sanfilippo, D.; Contissa, A.; et al. (2006) Arrays of Geiger Mode Avalanche Photodiodes. *IEEE Photonics Technology Letters*, Vol. 18, No. 15 (August 2006), 1633-1635, ISSN 1041-1135

Stapels, C. J.; Squillante, M. R.; Lawrence, W. G.; Augustine, F. L. & Christian, J. F. (2007). CMOS-Based Avalanche Photodiodes for Direct Particle Detection. *Nuclear Instruments and Methods in Physics Research A*, Vol. 579, No. 1, 94-98, ISSN 0168-9002

Stoppa, D.; Pancheri, L.; Scandiuzzo, M.; Gonzo, L.; Dalla Betta, G.F.; Simoni, A. (2007) A CMOS 3-D Imager Based on Single Photon Avalanche Diode. *IEEE Transactions on Circuits and Systems - I*, Vol. 54, No. 1 (January 2007), 4–12, ISSN 1057-7122

Tisa, S.; Tosi, A.; Zappa, F. (2007) Fully Integrated CMOS Single Photon Counter. *OSA Optics Express*, Vol. 15, No. 6 (March 2007), 2873-2887, ISSN 1094-4087

Sze, S.M.; Ng., K.K. (2007) *Physics of Semiconductor Devices*, 3rd Ed., John Wiley & Sons, Inc., ISBN 978-0-471-14323-9, Hoboken, NJ, USA

Xiao, Z.D.; Pantic. D.; Popovic, R.S. (2007); A New Single Photon Avalanche Diode in CMOS High-Voltage Technology, *Proceedings of the 14th International Conference on Solid-State Sensors, Actuators and Microsystems (TRANSDUCERS 2007),* 1365-1368, ISBN 1-4244-0842-3, Lyon, France, June 2007, IEEE, Piscataway, NJ, USA

Zappa, F.; Gulinatti, A.; Maccagnani, P.; Tisa, S.; Cova, S. (2005) SPADA: Single-Photon Avalanche Diode Arrays. *IEEE Photonics Technology Letters,* Vol. 17, No. 3 (March 2005), 657-659, ISSN 1041-1135

12

The Use of Avalanche Photodiodes in High Energy Electromagnetic Calorimetry

Paola La Rocca[1,2] and Francesco Riggi[2]
[1]*Museo Storico della Fisica e Centro Studi e Ricerche "E.Fermi"*
[2]*Department of Physics and Astronomy, University of Catania*
Italy

1. Introduction

Avalanche Photodiodes (APD) are now widely used for the detection of weak optical signals. They find applications in a large number of fields of science and technology, from physics to medicine and environmental sciences. The request for sensitive detectors, capable to respond to weak radiations emitted from scintillation materials, has produced over the last decades an increasing number of studies on avalanche photodiodes and their applications as photo-sensors in particle detection. The first APD prototypes were developed more than 40 years ago. The initial size of such devices was however very small (below 1 mm^2) and their spectral response confined to the near-infrared region. As a result, although available since several years, they did not receive much attention, also because of their initial high cost and low gain. However, large progresses have been made since then, and it has been possible to design and produce, at a reasonable cost, devices which have now a much larger area (tens of squared millimetres), with a high spectral sensitivity in the blue and near ultra-violet wavelength region. For such reasons, avalanche photodiodes are now widely used as sensitive light detectors in the construction of particle detectors in high energy physics. One of such examples is the impulse received by the design and construction of large scale electromagnetic calorimeters for the high energy experiments currently running in the world largest Laboratories. At present, APDs exhibit excellent quantum efficiency, with values around 80% in the near ultra-violet range, dropping to about 40% in the blue region, which is to be compared to typical values of 5-8 % in the blue for standard photomultipliers. Additional advantages which make them preferable over photomultipliers are discussed more specifically in the Chapter.

The overall set of problems and solutions related to the use of Avalanche Photodiodes in the design, construction, test and operation of large electromagnetic calorimeters in nuclear and particle physics experiments, is described in this Chapter, as observed within a Collaboration at the CERN Large Hadron Collider. Section 2 briefly recalls the principles on which electromagnetic calorimetry for particle physics experiments is based. Relative merits of Avalanche Photodiodes in comparison to traditional devices, mostly photomultipliers, are discussed in Sect.3, in connection with the light collection from scintillation detectors and the readout and front-end electronics. A review of the large detectors which have employed in the recent past or are currently employing such devices as photo-sensors is given in Section 4. Sect.5 describes the overall set of procedures carried out to characterize a large number of such devices when installing a complex detector. Section 6 discusses also the problems which may be encountered in the digital treatment of the signal and presents a comparison between traditional and alternative approaches in the analysis procedures.

2. High energy electromagnetic calorimetry in nuclear and particle physics

The use of avalanche photodiodes in nuclear and particle physics has largely increased in the last decades especially in connection with the growing impact of calorimetry techniques on accelerator-based physics experimentation being taken in the world largest Laboratories.
The term "calorimetry" comes from the Latin word *calor* (= heat) and indicates the basic detection principle which calorimeters are based on: the incident particles to be measured are fully absorbed in a block of instrumented material and their energy is converted into a measurable quantity (usually charge or light). In the process of absorption showers of secondary particles are generated, causing a progressive degradation in energy and producing some signal which can be detected to gain information on the original energy of the particle.
In order to match the physics potential at the major particle accelerator facilities, a wide variety of possible solutions for calorimeters is today available. Apart from the broad distinction between electromagnetic and hadronic calorimeters, they can be further classified according to the various types of technology employed, sampling calorimeters and homogeneous calorimeters being the most commonly used. This Chapter will focus on the electromagnetic calorimetry, describing the working principle and the practical realizations of electromagnetic calorimeters, as well as the reasons that make such detectors so attractive in the field of nuclear and particle physics. The interested reader is referred to textbooks (Wigmans, 2000) or review papers (Fabjan, 2003) for a more detailed discussion on calorimeters.

2.1 Working principle of an electromagnetic calorimeter

The various interaction mechanisms by which particles of different nature lose their energy in the medium underlie the broad distinction between hadronic and electromagnetic calorimeters: whereas hadronic calorimeters are built in order to exploit mostly the *strong* interactions experienced by hadrons (particles containing quarks, such as protons and neutrons) traversing matter, electromagnetic calorimeters detect light particles (electrons and photons) through their *electromagnetic* interactions with the medium's constituents.
Unlike hadronic showers, which are the result of a number of complex hadronic and nuclear processes, the physics of the electromagnetic showers is quite well-understood since it is based on few elementary processes, depending on the nature and energy of the incident particles. More precisely, *bremsstrahlung* and electron *pair production* are the dominant processes for high-energy electrons and photons: above 100 MeV electrons and positrons radiates photons (process called bremsstrahlung) as a result of the interaction with the nuclear Coulomb field; on the other hand, in the same energy range, photon interactions produce mainly electron-positron pairs. As a consequence, electrons and photons of sufficient high energy incident on a block of material create secondary photons and electron-positron pairs, which may in turn produce other particles through the same mechanisms. The result is a shower that may consist of thousands of different particles with progressively degraded energies. A diagram of an electromagnetic shower initiated by an electron is shown schematically in Fig 1.

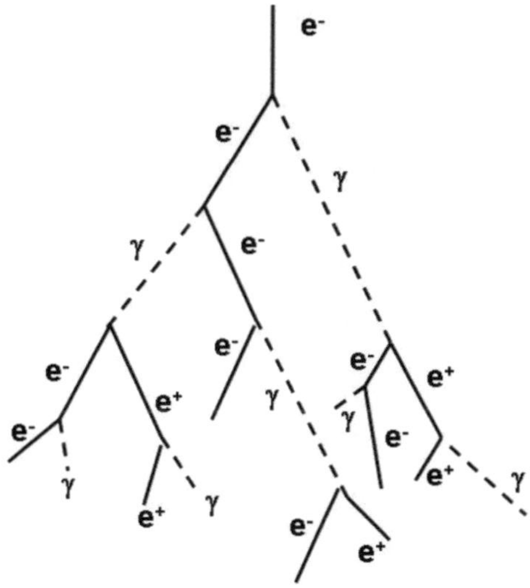

Fig. 1. Schematic diagram of an electron initiated electromagnetic shower.

This multiplication process is arrested when the energy of the secondary electrons produced in the electromagnetic cascade falls below a critical energy ε, which may be defined as the energy at which the average energy losses from bremsstrahlung equal those from ionization. At this energy the electrons and positrons lose their energy through collisions with atoms and molecules of the absorber medium, causing ionizations and thermal excitation, while photons are more likely to lose their energy through Compton and photoelectric interactions. When the critical energy is reached, the shower contains the maximum number of particles; the depth at which this occurs is called *shower maximum*.
Since calorimeters have to measure the energy lost by particles that go through them, they are usually designed to entirely stop or absorb the incident particles, forcing them to deposit most of their energy within the detector. Depending on the particular construction technique, the energy lost by the incident particles is collected in the form of light or charge, producing a physical signal proportional to the amount of energy deposited.
All the energy loss mechanisms that characterize an electromagnetic shower can be calculated with a high degree of accuracy through Quantum Electrodynamics (QED) calculations. For this reason the main features of electromagnetic showers are well known and both the longitudinal and lateral profiles of the showers may be parameterized with simple empirical functions in terms of two parameters, the radiation length X_0 and the Molière radius ρ_M. The first is defined as the mean distance over which a high-energy electron loses all but $1/e$ of its energy by bremsstrahlung or, in case of photons, as 7/9 of the mean free path for pair production. The radiation length only depends on the characteristics of the traversed material and it is used to describe the longitudinal development of the shower in a material-independent way. On the other hand the second parameter, the Molière radius, is a measurement of the transverse size of a shower. More precisely it is the average lateral deflection of electrons at the critical energy after traversing one radiation length. The theoretical background needed for understanding the principle which electromagnetic calorimeters are based on is of great help for the evaluation of the performance characteristics of a real electromagnetic calorimeter, especially during the designing phase. The size, the construction principles and the materials used for an

electromagnetic calorimeter may be accurately chosen according to the environment in which it has to operate and the tasks it has to fulfil. Calorimetry is the art of compromising between conflicting requirements, such as the energy and spatial resolution, the triggering capabilities, the radiation hardness of the materials used, the dynamic range and so on. Some more details about these aspects will be given in the following sections, together with some practical details of building and operating these detectors.

2.2 Homogeneous and sampling calorimeters

From the construction point of view, the possible solutions for calorimeters is very wide and quite ingenious calorimeter systems have been designed to cope with more and more demanding physics goals and requirements.
Here we will divide calorimeters only in two broad categories, sampling and homogeneous calorimeters.

Sampling Calorimeters

Sampling calorimeters consist of layers of *passive* or *absorber* high–density material (lead for instance) interleaved with layers of *active* medium such as solid lead-glass or liquid argon. Absorber layers are used to enhance photon conversions, while active layers to sample energy loss.
The main drawback of these devices is their limited energy resolution which rises from the large fluctuations, caused by the absorbers, in the energy deposited in the active layers. On the other hand, their excellent space resolution (i.e. their capability to reconstruct the impact position of incident particles) and their satisfactory particle identification are the result of the laterally and longitudinally segmentation, that is relatively easy to implement.
Many types of sampling calorimeter exist, which differ one another in the type of materials used. The most common absorber materials are lead, copper and iron, while the active medium can be solid (scintillator and semiconductor), liquid or gaseous.
In sampling calorimeters the energy deposited by showering particles can be collected both in the form of light, as in case of scintillation calorimeters, and in the form of electric charge, as happens in gas, solid-state and liquid calorimeters. Some details about the techniques used to collect the light signal are given in Section 3.1.

Homogeneous Calorimeters

In homogeneous calorimeters the same medium is used both to cause the shower development and to detect the produced particles. As discussed later, the main advantage of these devices is their excellent energy resolution, since the intrinsic fluctuations that occur in the development of the showers are small with respect to sampling calorimeters.
Usually homogeneous calorimeters are difficult to be segmented and this reduces their capabilities to give information about the position of the incident particle and its identification.
Since the materials used to build homogeneous calorimeters are characterized by large interaction lengths, such devices are almost exclusively used for electromagnetic calorimetry. Homogeneous calorimeters can be classified, according to the type of active material, into semiconductor calorimeters, noble-liquid calorimeters, Cherenkov calorimeters and scintillator calorimeters. Whereas the first two types of devices are based on the charge measurement, in scintillator and Cherenkov calorimeters the signal is collected in the form of light: the photons produced are converted into electrons and the electric signal is amplified to reduce the

electronic noise level. This is usually performed by photo-sensitive devices, such as photomultipliers, which are able to reach multiplication gains of the order of 10^6. However, the use of magnetic fields in modern particle experiments prevents the use of traditional photomultipliers: in these cases avalanche photodiodes are a valid alternative; however they provide a moderate gain, causing a non negligible noise term.
The choice of the typology and of the detector parameters for a specific application depends on several factors, such as the physics goals, the energy range that has to be considered, the accelerator characteristics and the available budget as well. Often the choice is supported by results from tests of small prototypes with particles beams, as well as by detailed simulation studies of the calorimeter performance.

2.3 The energy response of a calorimeter

In an electromagnetic calorimeter, the energy dissipated by the charged particles of a shower in the detector material is converted into a detectable signal which provides information on the original energy of the incident particle. This is verified both in case of homogeneous and sampling calorimeters: in the former, the whole energy of an incident particle is deposited in the active material, so that the entire detection volume may contribute to the signals that the particle generates; in the latter, the presence of passive absorber layers reduces the energy lost in the active layers to a fixed fraction (called sampling fraction) of the original energy.
However, the energy response of calorimeters depends on the nature of the incident particle: whereas showering electrons and photons produce a signal that is proportional to the original energy, the same is most certainly not valid for hadrons because part of their energy is used to dissociate the atomic nuclei and does not contribute to the calorimeter signals. Hence, the hadronic signals from electromagnetic calorimeters are non-linear and are not constant as a function of energy. In the following we will focus exclusively on the response of electromagnetic calorimeters to electron and photons.
In order to provide a reliable measurement of the energy, a first requirement that has to be fulfilled concerns the containment of the shower inside the detector volume. As the energy of the incident particle increases, the detector size needed to contain the showers increases as well. For electromagnetic showers, a simple formula that gives approximately the location of the shower depth expressed in radiation lengths is the following:

$$x_{max} \approx \ln(E/\varepsilon) + x_0 \quad (1)$$

where E is the energy of the incident particle, and $x_0 = 0.5$ for photons and -0.5 for electrons. This expression shows that the longitudinal size of a shower increases only logarithmically with energy, allowing to design a compact detector even for electromagnetic showers depositing a large energy (hundred of GeV) into the sensitive volume. On the other hand, the thickness containing 95% of the total shower energy is approximately located at $x_{95\%} \approx x_{max} + 0.08Z + 9.6$, indicating that the realization of compact calorimeters requires the use of high-Z materials.
Therefore, even at the particle energies reached at the Large Hadron Collider, electromagnetic calorimeters are very compact devices and energetic showers lose only a small percentage of their energy beyond the end of the active calorimeter volume.
Besides being intrinsically linear, electromagnetic calorimeters should also give a precise measurement of the energy. The precision with which the unknown energy of a given

particle is measured is called *energy resolution* and represents one of the most important performance characteristics of a calorimeter.
The actual energy resolution of a realistic electromagnetic calorimeter can be in general parameterized as follows:

$$\sigma/E = a/\sqrt{E} \oplus b/E \oplus c \quad (2)$$

where σ is the standard deviation in the energy measurement, the constants *a*, *b* and *c* depend on the detector characteristics and the symbol $\oplus$ indicates a quadratic sum.
From the left-hand side, the three terms are known as *stochastic term*, *noise term* and *constant term* respectively; the importance of each term strongly depends on the energy deposited in the calorimeter.
The stochastic term arises from the shower intrinsic fluctuations that characterize electromagnetic cascade developments event by event. Unlike homogeneous calorimeters, where these fluctuations are moderate, the stochastic term is quite important especially in sampling calorimeters, due to the presence of the absorber layers.
The noise term arises from the electronic noise of the readout chain. Usually it does not weight much upon the total energy resolution, especially for calorimeters in which the signal is collected in the form of light because, in that case, the first step of the electronic chain is a photo-sensor (like a phototube or a Silicon photomultipliers) which amplifies the original signal with almost no noise.
However, for energetic particles as those produced in the new-generation accelerators, the dominant contribution to the energy resolution is the constant term, that does not depend on the energy of the particle and arises from systematic effects (such as detector non-uniformities, shower leakage, mis-calibration,...). Therefore, modern calorimeters are built imposing severe construction tolerances in order to reduce possible instrumental imperfections that may give rise to response non-uniformities.
Additional contributions can make the energy resolution of a calorimeter worse, such as lateral and longitudinal leakages of the energy shower outside the active calorimeter volume, or fluctuations due to energy losses of electrons and photons in inactive materials (mechanical structures and cables) before reaching the calorimeter. These effects mainly affect the energy resolution of calorimeters integrated in complex high-energy physics experiments where the calorimeter is only one component that has to satisfy several mechanical constraints.

2.4 The tasks of an em-calorimeter

Calorimeters were originally conceived as devices used to measure only the energy of the incident particles. Today they have different applications and often their tasks are made more effective when the information they provide is combined with that coming from other sub-detectors.
The reasons that make calorimeters so attractive are various. First of all they provide a precise measurement of the energy in a wide range, since the energy resolution varies with energy as $1/\sqrt{E}$. This characteristic is crucial when looking for narrow resonances characterized by a poor signal/background ratio, as in the case of the search for Higgs bosons decaying into $\gamma\gamma$. Moreover, the capability to entirely absorb high energetic showers in compact distances make electromagnetic calorimeters the suitable instrument to measure electrons and photons over an unprecedented energy range, as requested by the experiments running at recent accelerator facilities like the LHC.

Electromagnetic calorimeters can also provide a fairly good particle identification, being able to distinguish, for instance, electrons and photons from muons and pions on the basis of their different energy deposit profiles. Moreover, calorimeters are sensitive to all kind of particles, including neutral ones. This feature is particularly exploited in particle physics experiments for several applications: for example, when measuring jets of particles (i.e. the characteristic collimated sprays of hadrons coming from hard collisions between energetic partons), electromagnetic calorimeters can give a measurement of the neutral component of the jets, completing the information provided by other kinds of detectors sensitive only to charged particles. Moreover, in modern experiments, calorimeters are built in order to have an angular coverage as large as possible, achieving often hermeticities in excess of 90%; in this way they can provide an indirect measurement of weakly interacting particles, as neutrinos, that can be detected only by observing geometric imbalances (missing energy) in the total transverse energy.

The possibility to segment electromagnetic calorimeters into several identical cells allows to gain information also on the impact position of the incident particles. The shower position is determined by reconstructing the centre of gravity of the energy deposited in the various detector cells that contribute to the signal. The precision with which the position of a given particle is measured mainly depends on the granularity of the detector and can achieve accurate values. The granularity of an electromagnetic calorimeter plays also an important role for the minimization of the *pileup* phenomena, i.e. the overlap of signals coming from different particles hitting the same cell. This problem may occur in calorimeters running in high-density particle environments characteristic of ultra-relativistic heavy-ion collisions. Moreover, the high luminosities achieved in modern hadron colliders may cause another type of pileup due to the overlap of signals coming from the preceding or following bunch crossing. To avoid event overlaps the calorimeter response in time is usually fast enough. This important feature is extensively used to select or trigger on interesting events, especially in those experiments in which the event rate is orders of magnitude beyond the rate at which events may be collected by the data acquisition system. Signals from calorimeters are available at very short time after the particle impact and are easy to process and interpret: the characteristics derived from the calorimeter data allows physicists to select only interesting (and often rare) events, such as those containing jets of particles or very energetic electrons and photons.

Thanks to all these features, calorimeters became key components of particle detectors and today almost every experiment in particle physics relies heavily on calorimetry.

3. Readout systems based on Avalanche Photodiodes: problems and solutions

3.1 Light collection in electromagnetic calorimeters

In order to collect all the energy deposited in each calorimeter cell, a proper readout system must be designed, to convert the light produced in the active material into an electronic signal which can be handled by the analog-to-digital converters and/or used by the trigger logic. Any readout system must be designed in such a way not to disturb the working condition of the calorimeter. For instance, no additional effect has to be introduced as far as the energy, position and time resolution are concerned. Moreover, the hermeticity of the calorimeter needs not be degraded by the readout part of the detector, as well as the overall radiation hardness.

In case of homogeneous calorimeters, which usually consist of individual crystals, scintillation or Cerenkov light is generated by the passage of particles. Each crystal is read out from the back, by the use of a photomultiplier tube (PMT) or by a different light sensor. While traditional readout devices employed photomultipliers in the past, solid-state devices started to be used more than 15 years ago. Silicon photodiodes (Barlow et al., 1999), Hybrid Photon Detectors (Anzivino et al., 1995) or APDs (Lorenz et al., 1994) are among the oldest examples of such devices.

Sampling calorimeters have been used since the very beginning in particle physics. In these structures, metal plates are interleaved with active scintillation materials. The geometry of the first generation calorimeters was very simple, with the scintillation light being extracted by each scintillation portion of the detector, coupled to one or two photomultipliers. In recent years however, the geometry of sampling calorimeters has changed considerably, mainly to design calorimeters with hermeticity properties. The development of wavelength shifting WLS fibers has allowed to design calorimeters with a fine segmentation of the active material, without enlarging the number of readout channels. In sampling calorimeters, the WLS fibers run along the length of each cell or perpendicular to it, and all the portions of the scintillation material are optically coupled to a photon detector located behind the calorimeter. This has the advantage that all the active parts of a cell are read out by the same photo-sensor. Some disadvantages inherent to this technique are related however to the worsening of the signal timing, which is much slower than that of the scintillation process, and to the partial non-hermeticity introduced by the volume itself of the WLS fibers. The latter may be solved with the use of scintillating and wavelength shifting fibers.

As far as the time structure of the signal is concerned, the absorption of the electromagnetic shower induced in a calorimeter by a high energy particle or radiation takes a time in the order of a few nanoseconds, due to the typical size of each individual module. However, the detection of neutrons has a much longer time scale, since they may be scattered from the surrounding materials. As a consequence, a long tail may be observed, which also requires some intervention on the associated electronics, such as to modify the shaping time of the signal in order to cut a large fraction of this tail.

In addition to the late arrival of neutrons, other instrumental effects may also have their influence on the time structure of the signals. One of these effects is due to the nature of the particular scintillation material used as active element in the calorimeter, since the fluorescent processes by which light is produced inside a scintillation material have characteristic decay times which range from nanoseconds to milliseconds. While organic scintillators have decay times of a few nanoseconds, considerably longer times (300 ns) are typical of BGO material.

An important aspect to be considered in understanding the time structure of the signal - especially for large volume calorimeters - is the fact that relativistic particles which produce the light inside the scintillation detector cover the distance between the front face of the detector and the photo-detector in a time which is smaller than the time required to the light photons produced by their passage, due to the refraction index n of the material. Such difference between c and c/n may amount to a few nanoseconds, which may be relevant in case of fast scintillators.

3.2 Avalanche Photodiodes as photo-sensors in calorimeters

Silicon Avalanche Photodiodes are now considered good candidates to replace traditional photomultipliers to detect the light produced in scintillation materials, especially in specific situations, where their advantages become more apparent. APDs have several features which make them attractive for scintillation detection. The discussion of the working principles of such devices and their detailed structure is beyond the scope of the present Chapter. Here only follows a list of different aspects which are relevant for the use of APDs as photo-sensors in electromagnetic calorimeters, and which need to be evaluated when choosing the proper solution:

Size:

Photomultipliers have different sizes, which can be adapted in principle to any specific situation. However, the size of traditional photomultipliers is always much bigger than their sensitive area, which can be a problem for large detector arrays, as it is the case with large particle detectors. On the contrary, APDs have limited sizes (in the order of a 1-50 mm^2) and the required front-end electronics may also be very compact. For such reasons, their use is suitable to the need of a large granularity crystal array (in case of homogeneous or sampling calorimeters). On the other side, their small size may require sometimes to employ more than one device per crystal, in order not to reduce too much the amount of collected light.

Internal gain

APDs are components which possess an intrinsic internal gain. The applied bias voltage produces a region where a high electric field (in the order of 150 kV/cm) is obtained. Such field is able to generate avalanches of secondary particles, and hence to amplify the signal. The exact value of the internal gain coefficient depends on the applied bias voltage and also on the temperature. In comparison to the possibility of "external" gain, the intrinsic gain has the advantage of using all the signal generated by the light development inside the scintillation crystal, whereas the signals generated within the sensitive volume of the diode are not amplified.

Insensitivity to magnetic field

Due to the need to bend charged particles emitted in a nuclear collision and hence measure their momentum according to the curvature of the track, a large fraction of a modern particle detector is usually contained inside a magnetic field. Large magnetic fields, up to several Tesla, may be in order, with the aim to bend very energetic (GeV or more) particles. Such fields however do not allow the use of traditional photomultipliers, since electron trajectories would be highly distorted by the field, and even robust shielding could be ineffective to protect them. Moreover, the use of a thick shielding would cause a large number of secondary interactions in the material, which could be of disturb for the detection of particles and radiations in the active detectors. The use of traditional PMTs would then require to transport the light through optical fibers for a long distance outside the magnetic field. APDs on their side are not sensitive to magnetic fields, which makes them the device of choice for particle detectors embedded in large magnets.

Spectral response

While the spectral response of an Avalanche Photodiode is nearly the same as for a normal photodiode when no bias is applied, this changes as a result of the application of the bias voltage, since the penetration depth of the light inside the device depends on the wavelength. Devices which have an enhanced sensitivity in the near-infrared region or at smaller wavelengths (300 nm) are available nowadays, to allow the user to select the appropriate device.

Quantum efficiency

The overall efficiency of a photo-sensor strongly depends on how good is the matching between the emission spectrum of the scintillation material being used and the spectral response of the photosensor. Avalanche photodiodes usually exhibit a very high quantum efficiency (QE), relatively constant over a wide wavelength interval. As an example, the APDs employed in the ALICE and CMS calorimeters exhibit a quantum efficiency of the order of 85% from 500 to 800 nm, dropping down to 50% at 350 and 950 nm. These values are significantly higher than the average values for standard photomultipliers, which are in the order of 20%.

Gain stability

The gain of an Avalanche Photodiode depends both on the applied voltage and on the temperature, so it is important to measure, for any individual APD, the relevant coefficients, in order to allow for possible variations in these quantities. The stability of the bias voltage depends on the overall quality of the power supply and (to some extent) on the environmental conditions. While it is not so critical to control the bias voltage, a relevant problem is that the gain of each APD needs to be individually adjusted to match a common value of the gain, which requires a software-controlled power supply, able to distribute to each APD channel the appropriate voltage. Concerning the temperature dependence, this is an important factor to maintain the gain stable over time, since the ambient temperature may be subjected to non negligible variations over long time operational periods. Moreover, due to the large number of channels usually involved, sensible temperature differences are experienced by devices which are located even meters apart, so that a monitoring and correction of the bias voltage to overcome such temperature-dependent gain variations is mandatory to equalize the gain in different channels.

Measurements have been also performed (Chartrchyan et al. 2008) on the long term stability of such devices, maintaining them under bias voltage for long periods, up to 250 days, and checking for possible failures through the monitoring of the dark current. Values of the MTTF (mean time to failure) of the order of 10^7 hours have been reported.

Negligible nuclear counter effect

The nuclear counter effect is related to the amount of extra signal produced inside the photodiode by a charged particle traversing it, which adds to the charge produced by the scintillation light in the crystal. It can be quantified by the thickness of a Si PIN diode required to produce the same signal. The signal produced in the APD would be proportional to such equivalent thickness. From this point of view, the effective thickness of the device should be minimized as much as possible in order to reduce the influence of the nuclear counter effect. However, reducing the thickness has also the effect of increasing the APD capacitance, so that a compromise needs to be reached.

Small excess noise

Avalanche photodiodes generate excess noise, due to the statistical nature of the avalanche process. An estimate of the statistical fluctuations of the APD gain is given by the excess noise factor F, where $\sqrt{F}$ is the factor by which the statistical noise on the APD current exceeds that expected from a noiseless multiplier. The excess noise has its origin in the intrinsic statistical nature of the internal charge carrier multiplication inside the device, which depends on the inhomogeneities in the avalanche region and in hole multiplication. If k is the ratio of the ionization coefficients for electrons to holes, at a given gain M, the excess noise factor is given by:

$$F = k \times M + (2 - 1/M) \times (1-k) \quad (3)$$

The result is an additional contribution to the energy resolution, and clearly a small value of the excess noise factor is preferable to optimize the overall resolution. This factor increases with the gain, reaching for instance a value of about 1.9 at M=30 for the APD employed in the ALICE and CMS calorimeters. Large area APDs which have been subsequently developed for the PANDA calorimeter, exhibit smaller values of F (1.38 at M=50).

High resistance to radiations

The use of Avalanche Photodiodes in hostile environments, as far as the radiation level is concerned, is a critical point for large particle physics experiments, where the flux of charged and neutral particles produced in high energy collisions over long operational periods may be very high. The dose absorbed by the detectors and associated electronics is usually evaluated by detailed GEANT simulations which take into account the description of the complex geometry and materials of the detector. Depending on the physics program (proton-proton or heavy-ion collisions, low or high beam luminosity, allocated beam time,...) and on the location of such devices inside the detector, a particular care must be devised to understand whether the photo-sensors will be able to survive during the envisaged period of operation. For such reason, a detailed R&D program has been undertaken within the High Energy Collaborations to expose the devices of interest to different sources of radiations, and measure their performance before and after irradiations.
There are basically two damage mechanisms: a bulk damage, due to the displacement of lattice atoms, and a surface damage, related to the creation of defects in the surface layer. The amount of damage depends on the absorbed dose and neutron fluence.
Whereas experiments like ALICE, which will run with low luminosity proton and heavy ion beams at LHC, do not suffer of big problems with the radiation dose in the electromagnetic calorimeter, the CMS detector, which runs at a much larger luminosity, will have a very large dose in the photo-sensors. As an example, in ten years LHC operation, the planned dose in the CMS barrel is in the order of 300 Gy, with a neutron fluence of 2×10^{13} n/cm^2 (1 MeV-equivalent). This has lead to an extensive set of measurements with different probes (protons, photons and neutrons), an to the successful development of APDs capable to survive to these conditions.

3.3 Front-end electronics

Once the light produced in the active material has been collected by the photosensor, an important step towards the extraction of the signal is the associated front-end electronics. Such electronics has to be used to process the signal charge delivered by the photo-sensors and extract as much information as possible concerning the time and amplitude of the signal. Several aspects are important to understand the requirements which are demanded to front-end electronics.

Dynamic range

In high energy experiments, for instance in the experiments running at LHC, the dynamic range required to a calorimeter is very high. Signals of interest go from the very small amplitudes associated to MIP particles (for instance, cosmic muons used for the calibration, which typically deposit an energy of a few hundred MeV in an individual cell) to highly energetic showers (in the TeV region) produced by hadrons or jets. The dynamic range required may then easily cover 4 orders of magnitude, which requires a corresponding resolution in the digitization electronics (ADC with 15-16 bits). An alternative approach is the use of two separate high-gain and low-gain channels, which requires ADCs with a smaller number of bits, at the expense of doubling the number of channels.

Time information

The extraction of timing information from the individual signals originating from each module in a segmented calorimeter is an important goal for the front-end electronics. Time information may be important in itself, also for calibration and monitoring purposes, and it is mandatory when the information from a calorimeter must be used to provide trigger decisions. The timing performance of the overall readout system also depends on the rest of the electronics, as well as on the algorithms being used to extract such information (See Sect.6).

Number of independent channels

Due to the large granularity usually employed in segmented calorimeters, the number of independent channels is very high, in the order 10^4-10^5. This requirement demands a corresponding high number of front-end preamplifiers and a high level of integration for the associated electronics, which needs to be compacted in a reasonable space.

3.4 Monitoring systems

A common aspect to all kind of detectors which are used to transform the light, produced in the active part of the calorimeter, into an electric signal, is the fact that their exact response (gain) is intrinsically unstable, depending on a number of factors which may vary according to the experimental conditions. Temperature and voltage variations are particularly important in this respect, as discussed before, since the gain of Avalanche Photodiodes is very sensitive to such parameters. Such aspects require usually a careful study of the devices being used, under the specific working conditions, in order to characterize their response as a function of these parameters (see Sect.5). Moreover, a monitoring system is in order, to take into account the variation of the working parameters, and sometimes even to correct the gain by a proper feedback. A LED monitoring system is usually employed in large calorimeters, with the aim to send periodically a reference signal to all readout cells and to check the response uniformity.

4. A review of large APD-based electromagnetic calorimeters

Most of the large experiments devoted to high energy physics make use of calorimeters, to detect hadronic and electromagnetic showers originating from energetic particles and radiations. Electromagnetic calorimeters in particular are employed since several decades, making use in the past of traditional photo-sensors (photomultipliers) and, more recently, of solid-state devices such as photodiodes, APD and silicon photomultipliers. Here a brief review is given of several experiments in high-energy physics which have an electromagnetic calorimeter as an important part of the detection setup.

4.1 Calorimeters based on traditional photo-sensors

Several high-energy experiments installed in the largest nuclear and particle physics Laboratories have employed in the past electromagnetic calorimeters of various configurations and design, with traditional photomultipliers or photodiodes as photon sensitive devices. As an example, Table 1 shows a (non-exhaustive) list of detectors which include an electromagnetic calorimeter, together with some basic information on the organization and design of the detector. As it can be seen, the largest installations have a number of channels in the order of 10^4, which is remarkable for traditional readout systems based on photomultipliers.

Experiment	Laboratory	Type	No.of channels
E731	FNAL	Lead Glass	802
CDF	FNAL	Lead/Scint	956
FOCUS	FNAL	Lead/Scint	1136
SELEX (E781)	FNAL	Lead Glass	1672
BABAR	SLAC	CsI (photodiode)	6580
L3	CERN /LEP	BGO Crystals (photodiode)	10734
OPAL	CERN /LEP	Lead Glass	9440
HERMES	DESY /HERA	Lead Glass	840
HERA-B	DESY/HERA	Pb(W-Ni-Fe)/Scint Shashlik-type	2352
H1	DESY/HERA	Lead-scintillating fibre	1192
ZEUS	DESY/HERA	Depleted uranium-Scint calorimeter, WLS	13500
WA98	CERN /SPS	Lead Glass	10080
KLOE	LNF	Lead-scintillating fibre	4880
STAR	RHIC	Pb/Scint Sampling calorimeter, WLS	5520
PHENIX	RHIC	Pb/scint shashlik-type	15552
PHENIX	RHIC	Pb glass	9216
LHCb	CERN /LHC	Lead/Scint shashlik-type, WLS	5952

Table 1. Summary of detector installations which make use of an electromagnetic calorimeter with traditional readout devices.

4.2 Calorimeters making use of Avalanche Photodiodes

Only in the last years Avalanche Photodiodes have been routinely employed as photo-sensors for large electromagnetic calorimeter installations. Here we want to briefly summarize a few examples of recent detectors which have been installed and commissioned or in the stage of being constructed.

The electromagnetic calorimeter of the CMS experiment at LHC

CMS (Compact Muon Solenoid) is one of the large experiments running at the CERN Large Hadron Collider (LHC). A general description of the CMS detector is reported in (Chartrchyan et al. 2008). A large electromagnetic calorimeter, based on lead tungstate crystals with APD readout, is included in the design of the CMS detector.

The barrel part of the CMS electromagnetic calorimeter covers roughly the pseudo-rapidity range $-1.5 < \eta < 1.5$, with a granularity of 360-fold in φ and 2x85-fold in η, resulting in a number of crystals of 61200. Additional end-caps calorimeters cover the forward pseudo-rapidity range, up to $\eta=3$, and are segmented into 4 x 3662 crystals, which however employ phototriodes as sensitive devices.

The use of lead tungstate crystals with its inherent low light yield and the high level of ionizing radiations at the back of the crystals has precluded in this case to employ conventional silicon PIN photodiodes. In collaboration with Hamamatsu Photonics, an intensive R&D work has led the CMS Collaboration to the development of Si APDs particularly suited to such application (Musienko, 2002). As a result of this work, a compact device (5x5 mm^2 sensitive area, 2 mm overall thickness) has been produced, which is now used also by other experiments. The performances of such device are its fast rise time (about 2 ns) and the high quantum efficiency (70-80 %), at a reasonable cost for large quantities. To overcome the inherent limitations of a reduced gain at wavelength smaller than 500 nm, and a high sensitivity to ionizing radiation, an inverse structure for such devices was implemented. In these APDs the light enters through the p^{++} layer and is absorbed in the p^+ layer. The electrons generated in such layer via the electron-hole generation mechanism drift toward the pn junction, amplified and then drift to the n^{++} electrode, which collects the charge. The APD gain is largest for the wavelengths which are completely absorbed in the p^+ layer, which is only a few micron thick; as a result, the gain starts to drop above 550 nm. Moreover, with this reverse structure, the response to ionizing radiation is much smaller than a standard PIN photodiode.

An important issue for the APD installed in the CMS detector is the effect of radiation on the working properties of the device, due to high luminosity at which this experiment is expected to run for most of its operational time. In ten years of LHC running, the neutron fluence (1 MeV equivalent) in the barrel region is expected in the order of 10^{13} n/cm^2, with a dose of about 300 Gy. The extensive irradiation tests performed in the context of this Collaboration have provided evidence that the devices are able to survive the long operational period envisaged at LHC.

Due to the large area of the crystals employed in the CMS calorimeter, compared with the sensitive area of the APD devices, two individual Avalanche Photodiodes are used to detect the scintillation light from each crystal.

The electromagnetic calorimeter of the ALICE experiment at LHC

The ALICE detector (Aamodt et al., 2008) is another large installation at LHC, mainly devoted to the heavy ion physics program. It is equipped with electromagnetic calorimeters of two different types: the PHOS (PHOton Spectrometer), a lead tungstate photon spectrometer, and the EMCAL, a sampling lead-scintillator calorimeter. These two detectors are able to measure electromagnetic showers in a wide kinematic range, as well as to allow reconstruction of neutral mesons decaying into photons.

The PHOS spectrometer is a high resolution electromagnetic calorimeter covering a limited acceptance domain in the central rapidity region. It is divided into 5 modules, for a total number of 17920 individual Lead tungstate (PWO) crystals. Each PHOS module is segmented into 56 x 64=3584 detection cells, each of size 22 x 22 x 180 mm, coupled to a 5 x 5 mm^2 APD.

An additional electromagnetic calorimeter (EMCal) was added to the original design of ALICE, to improve jet and high-pt particle reconstruction. This is based on the shashlik technology, currently employed also in other detectors. The individual detection cell is a 6 x 6 cm^2 tower, made by a (77+77) layers sandwich of Pb and scintillator, with longitudinal wavelength shifting fiber light collection. The total number of towers is 12288 for the 10 super-modules originally planned (which cover an azimuth range of 110°). Recently a new addition of similar modules started, to enlarge the electromagnetic calorimeter (DCAL), providing back-to-back coverage for di-jet measurements. This will roughly double the number of channels.

The active readout element of the PHOS and EMCal detectors are radiation-hard 5 x 5 mm^2 active area Avalanche Photodiodes of the same type as employed in the CMS electromagnetic calorimeter. These devices are currently operated at a nominal gain of M=30, with a different shaping time in the associated charge-sensitive preamplifier.

The electromagnetic calorimeter of the PANDA experiment at FAIR

PANDA is a new generation hadron physics detector (Erni et al., 2008), to be operated at the future Facility for Antiproton and Ion Research (FAIR). High precision electromagnetic calorimetry is required as an important part of the detection setup, over a large energy region, spanning from a few MeV to several GeV. Lead-tungstate has been chosen as active material, due to the good energy resolution, fast response and high density. To reach an energy threshold as low as possible, the light yield from such crystals was maximized improving the crystal specifications, operating them at -25 °C and employing large area photo-sensors. The largest part of such detector is the barrel calorimeter, with its 11360 crystals (200 mm length). End-cap calorimeters will have 592 modules in the backward direction and 3600 modules in the forward direction. The crystal calorimeter is complemented by an additional shashlyk-type sampling calorimeter in the forward spectrometer, with 1404 modules of 55 x 55 mm^2 size.

The low energy threshold required of a few MeV and the employed magnetic field of 2 T precludes the use of standard photomultipliers. At the same time, PIN photodiodes would suffer from a too high signal, due to ionization processes in the device caused by traversing charged particles. In order to maximize the light signal, new prototypes of large area (10 x 10 mm^2 or 14 x 6.8 mm^2), APDs were studied, devoting particular care to the radiation tolerance of these devices.

In the forward and backward end-caps, due to the high expected rate and other requirements, vacuum phototriodes (VPT) were the choice. Such devices, which have one dynode, exhibit only weak field dependence, and have high rate capabilities, absence of nuclear counter effect and radiation hardness.

5. Characterization of Avalanche Photodiodes for large detectors: procedures and results

As discussed in the previous Sections, the construction of a large electromagnetic calorimeter based on Avalanche Photodiodes as readout devices may require a large number (in the order of 10^3-10^5) of individual APDs to be tested and characterized, after the

R&D phase has successfully contributed to produce a device compliant with the specifications required by the experiment. Not only the devices have to be checked for their possible malfunctioning, but to minimize the energy resolution for high energy electromagnetic showers, it is important to obtain and assure a relative energy calibration between the different modules into which the calorimeter is segmented. The uncertainty in the inter-module calibration contributes to the constant term in the overall energy resolution, which becomes most significant at high energy. An additional motivation to have a good module-to-module calibration comes from the possibility to implement on-line trigger capabilities, especially for high energy and jet events. In such case, it is mandatory to adjust the individual gains of the various channels within a few percent.

For all such reasons, a massive work is usually required to choose the optimal APD bias for each individual device. Such massive production tests allow also to check the functionality of the device under test and the associated preamplifier, prior to mounting them in the detector. Mass production tests carried out in the lab prior to installation usually consist of measurements of the gain versus voltage dependence of each APD at fixed and controlled temperature, and in the determination of the required voltage to reach a uniform gain for all the devices.

Several properties may be measured during this screening operation, depending on the amount of information required, the desired precision and the amount of time at disposal to carry out all the required operations in a reasonable time schedule. If the device under consideration originates from a stable production chain at the manufacturer's site, as it is usually for APDs which have been in use for several applications, a complete set of characterization procedures may be carried out only for limited samples of devices. These may include the evaluation of the quantum efficiency, of the excess noise factor, of the capacitance, dark current and gain uniformity over the APD surface, as well as the temperature dependence of the gain curve in a wide range of temperatures (Karar, 1999). Massive tests, to be carried out on each individual APD, at least require the measurement of the gain-bias voltage curve at one or more temperatures, close to the operational one, and (possibly) the measurement of the dark current at different gain values. From the measured data one can extract the bias voltage required to match a fixed value of the gain, and the voltage coefficient.

The basic equipment to carry out such tests includes a system to maintain and measure the APD temperature while performing the measurements (usually within 0.1 °C), a pulsed light source (for instance a pulsed LED in the appropriate wavelength region), the front-end electronics and some acquisition system to store the data for further analysis. Due to the large number of devices usually under test, a suitable procedure must be designed, which tries to minimize as much as possible the time required to carry out a complete scan. As an example, the test of several APDs (8-32) at the same time may be planned with a proper choice of the readout system. Moreover, bias voltage may be software controlled together with acquisition, thus allowing to carry out automatic measurements in controlled steps of bias voltage.

Fig.2 shows an example of a typical gain curve obtained during the characterization of a large number of Hamamatsu S8148 APDs within the ALICE Collaboration (Badalà, 2008).

The output signal was measured for different values of the bias voltage, from 50 V (where a plateau is expected, corresponding to unitary gain) to about 400 V. The data were fitted by the function:

$$M(V) = p_0 + p_1 \exp(-p_2 V) \tag{4}$$

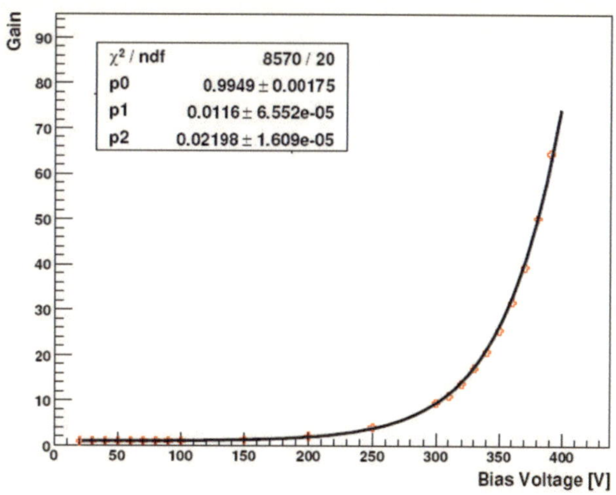

Fig. 2. Gain curve as a function of the APD bias voltage, for one of the Hamamatsu S8148 employed in the ALICE electromagnetic calorimeter. A common gain of 30 is usually set for all the modules.

in order to extract the coefficients p_0, p_1, p_2 and thus determine the voltage V_{30} at which the gain equals M=30, which is the required value in the ALICE EMCal.

The relative change in the gain with the bias voltage is an important parameter to extract from such measurements, especially in the region where the APD will work. Fig.3 reports one of such results, showing a value of 2.3 %/V at M=30.

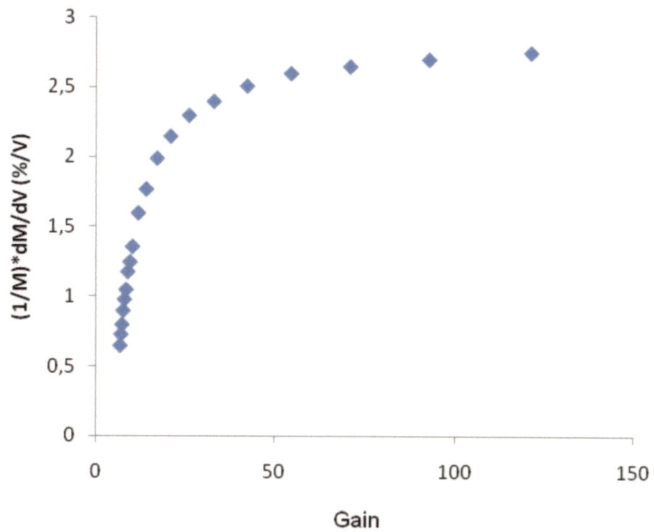

Fig. 3. The relative change in the APD gain is here reported at different values of the gain.

Due to the strong dependence of the APD gain from the temperature, the investigation of the gain versus temperature is an important issue of the characterization phase, at least for subsamples of the complete set of devices. Gain curves have to be measured for different values of the temperature - spanning the region of interest - in order to extract a temperature coefficient. Fig.4 shows an example of a set of different gain curves measured in the range 21 to 29 °C, for the Hamamatsu S8148 APDs.

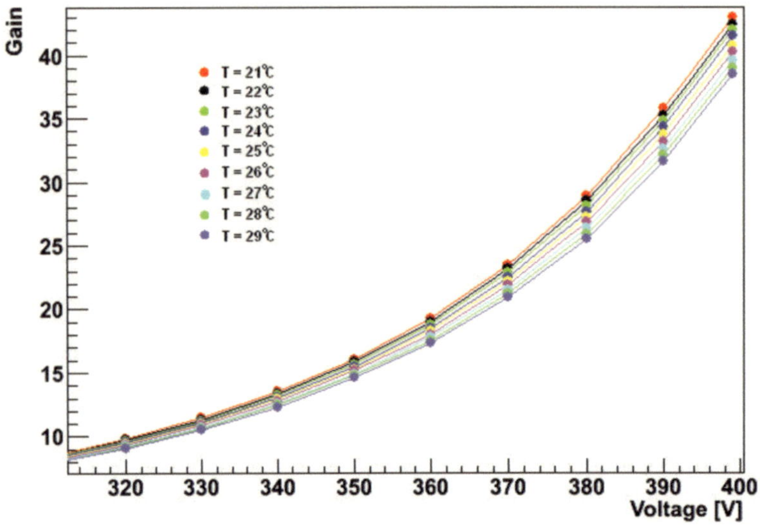

Fig. 4. Gain curves measured at different temperatures.

This or similar sets of measurements allow to extract the gain versus temperature dependence (Fig.5) and finally a value of the temperature coefficient, which decreases with the temperature, as shown in Fig.6.

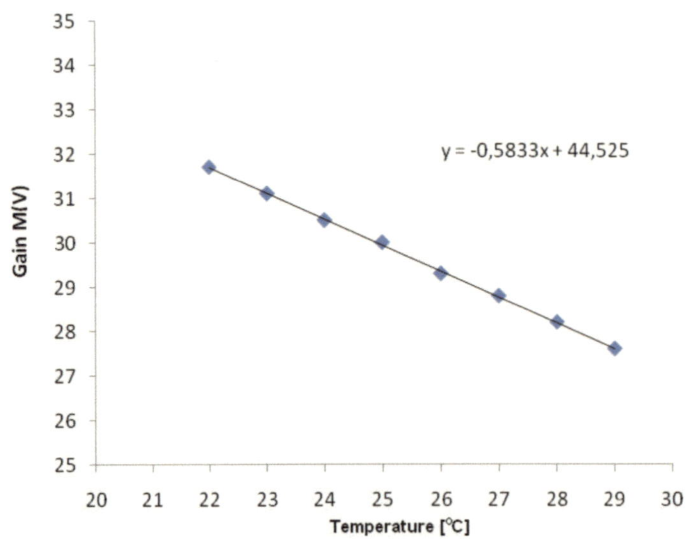

Fig. 5. APD gain as a function of the temperature.

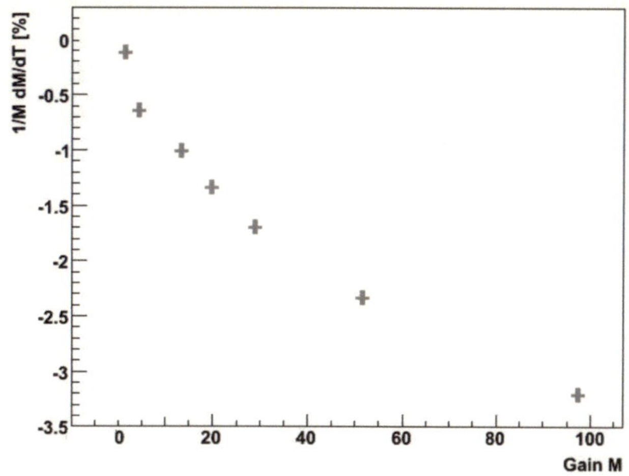

Fig. 6. Temperature coefficient of the APD gain, reported as a function of the APD gain.

All these procedures allow to classify the individual devices into different categories (for instance according to the voltage required to match a given gain, or to the temperature coefficient) for the sake of response uniformity, and to reject APDs with inadequate performance. Carrying out systematic characterization of a large number of individual devices permits to investigate statistical distribution of several quantities of interest, and establish classification criteria, to be used for the next samples. As an example, Fig.7 shows the distribution of the bias voltages required to have a common gain (M=30) in a set of 1196 APDs which were used in one of the super-module of the ALICE electromagnetic calorimeter.

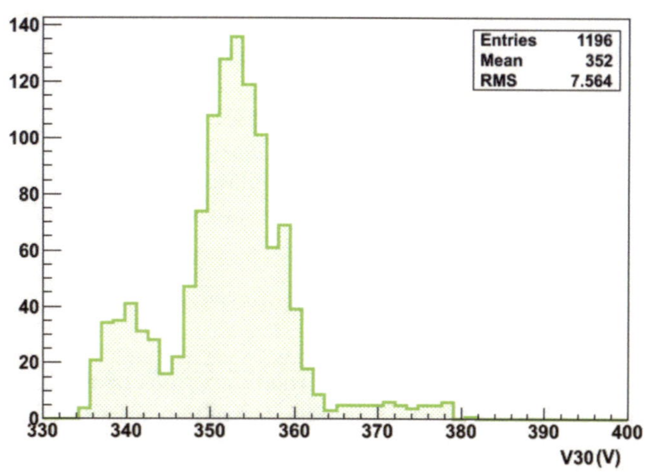

Fig. 7. Statistical distribution of the APD bias voltages required to match a common gain M=30, for a set of 1196 devices employed in one of the super-modules of the ALICE calorimeter.

While the distribution shows clearly the presence of two populations (due to different production lots), all devices showed a bias voltage smaller than 400 V, which was the limit set by the electronic circuitry to power the APD with a sufficient resolution. Fig.8 shows also the distribution, for the same set, of the voltage coefficient, which has an average value of 2.3%/V, with an RMS in the order of 0.08 %/V.

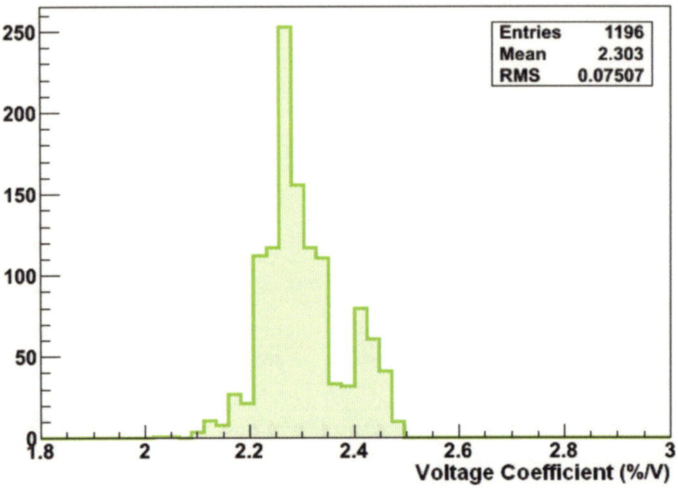

Fig. 8. Statistical distribution of the voltage coefficients, for the same set of 1196 APDs.

6. Extraction of amplitude and time information: traditional methods and alternative approaches

The output signal from Avalanche Photodiodes needs to be analyzed to extract as much as possible the information contained. Particularly relevant are of course the amplitude information, related to the amount of energy deposited in the individual module, and the timing information associated to it. The procedures to extract such information are not trivial, especially when analyzing events which span a large dynamical range, as it is the case for electromagnetic calorimeters in high energy experiments. In such a case, various algorithms have been developed and used, whose relative merits may be compared according to the precision and CPU time required. Even methods based on neural network topologies may be implemented and applied to simulated and real data.
With reference to Fig.9, which shows a typical signal, as sampled by a flash ADC, the shape of the signal may be fitted by a Gamma function

$$ADC\,(t) = Pedestal + A^{-n}\, x^{n}\, e^{\,n(1-x)} \quad , \quad x = (t-t_0)/\tau \qquad (5)$$

where $\tau = n\,\tau_0$, τ_0 being the shaper constant, and n ~2.
Such fit procedure is certainly able to provide reliable values of the amplitude A and time information t_0 in case of large-amplitude signals, for which the number of time samples is relatively high (larger than 5-7). However, there are two main drawbacks inherent to this method: the algorithm is relatively slow, if one considers that it has to be applied to a large number of individual modules on an event-by-event basis, which is dramatic especially for on-line triggering. Secondly, in case of signals with very low amplitudes, the fit quite often provides unreliable values, since the signal shape is no longer similar to a Gamma function.
For such reasons, alternative approaches have been tested and compared to the standard fitting procedure: fast fitting methods, peak analysis and so on. Here we want to show an example based on a neural network approach, which was recently tested on a sample of

LED calibration data obtained for a large number (a few thousands) of channels in the ALICE electromagnetic calorimeter.

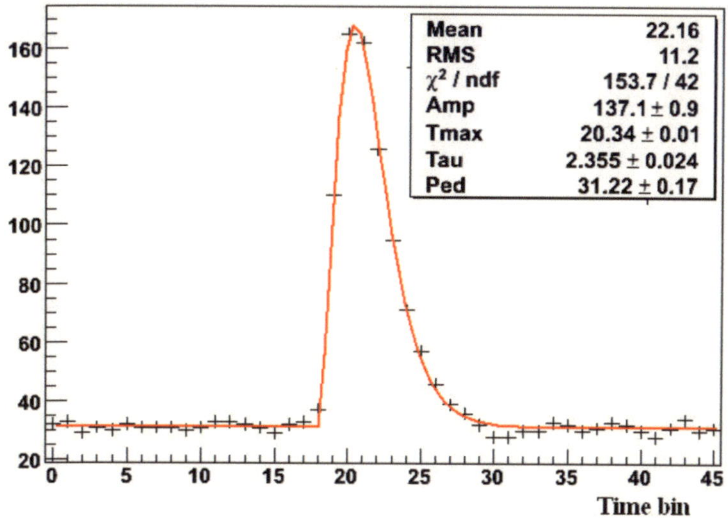

Fig. 9. Shape of the signal, as extracted from a sampling ADC.

In order to prepare a data sample which exhibits its maximum at different times, as it could happen for real data, the LED signal was shifted in time every 100 events.
Also the amplitude distribution is very broad, in order to span a region as large as possible, similarly to real data. This was due to the inevitable difference in the distribution of the light signal to the different modules. As a result, Figs.10 and 11 show two examples of a high amplitude (number of time samples = 12) and a low amplitude signal (number of time samples = 6). All the data were processed with the standard fit algorithm, which provided the reference for the learning phase in the neural network approach.

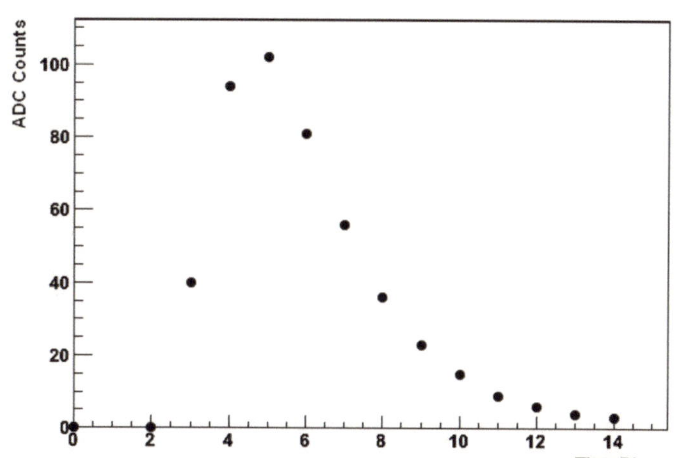

Fig. 10. An example of a high amplitude signal, including 12 time samples.

A feedforward multilayered neural network (Bishop, 1995) consists of a set of input neurons, one or more hidden layers of neurons, a set of output neurons, and synapses connecting each layer to the subsequent layer. The synapses connect each neuron in the first layer to each neuron in the hidden layer and each neuron of the hidden layer to the output (Fig.12). Several topologies may be chosen, as far as the number of input neurons and hidden layers are concerned.

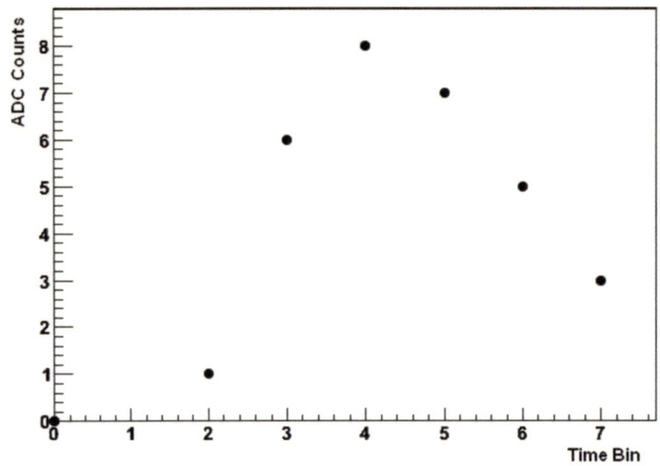

Fig. 11. An example of a low amplitude signal, including only 6 time samples.

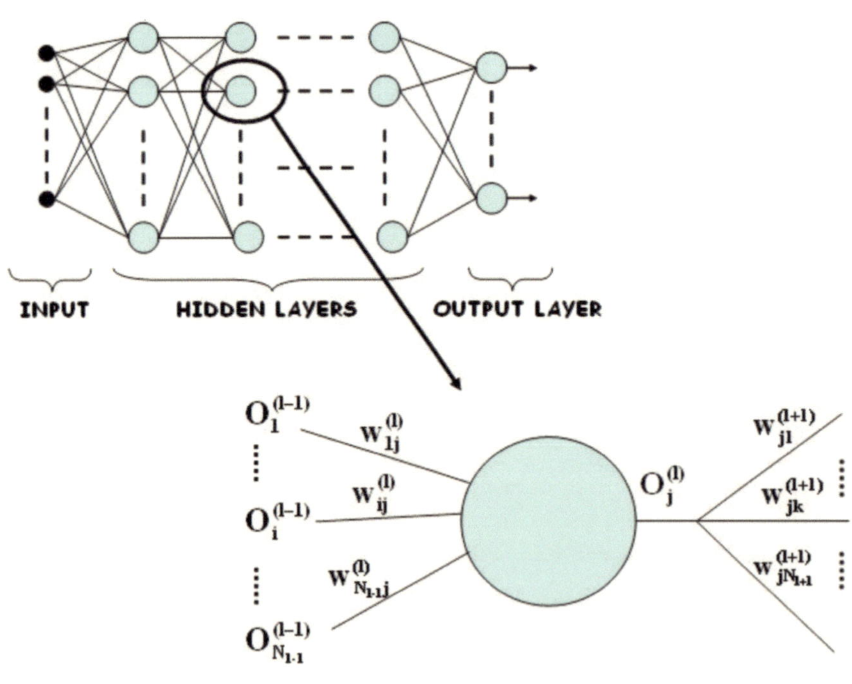

Fig. 12. Schematic layout of a neural network.

The signal provided by the *j*-th neuron of the *l*-th layer is given by the linear combination of the neuron input values, where the w's are the weights:

$$a_j = \sum_{i=1}^{N_{l-1}} w_{ij}^{(l)} O_i^{(l-1)}$$

A backpropagation algorithm was used in the learning phase, in order to modify the initial values of the weights and minimize the error function:

$$E = \frac{1}{2} \sum_{p=1}^{N_p} \sum_{j=0}^{N_O} [y_{jp} - O_{jp}^{N}]^2$$

Best results were obtained in this case with 5 input neurons (the 5 values of the signal amplitude closest to the maximum), 10 hidden neurons and 2 output neurons (the amplitude and the time of the signal peak). Fig.13 shows the minimization of the error function with the number of epochs employed in the learning and testing phases.
Figs.14 and 15 show the distributions of the differences between the reference values (provided by the Gamma-fit) and the output values from the neural network, both for the amplitude and the time. An RMS of 0.26 ADC channel was obtained for the signal amplitude, while a value of 0.007 channel bin (corresponding to 700 ps) was obtained for the time.
Such performance was compared to more traditional methods, based on fast fitting procedures or peak analysis methods, and it was shown that after a proper training phase, comparable results may be in principle obtained by a neural network, with a reduced CPU time.

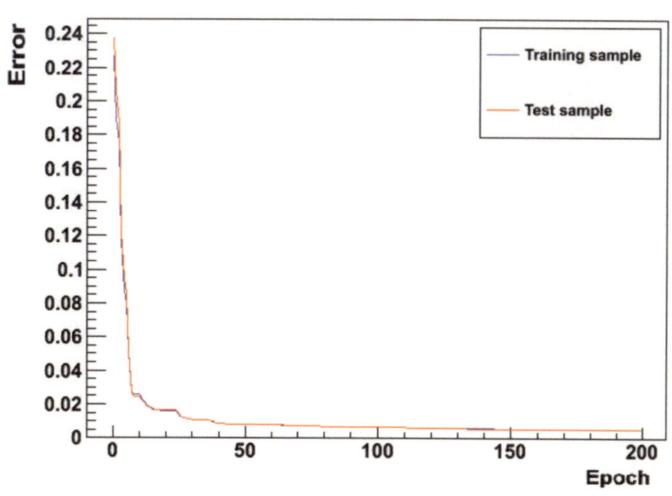

Fig. 13. Minimization of the error function with a neural network.

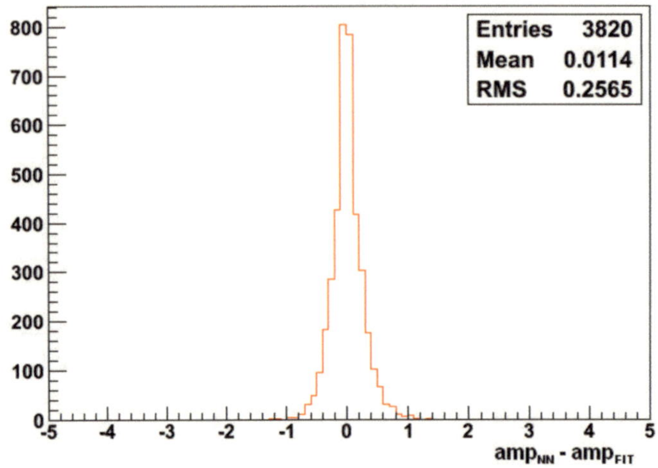

Fig. 14. Distribution of the differences between the "true" value (provided by the fit with a Gamma-function) and the value provided by the neural network, in case of the signal amplitude.

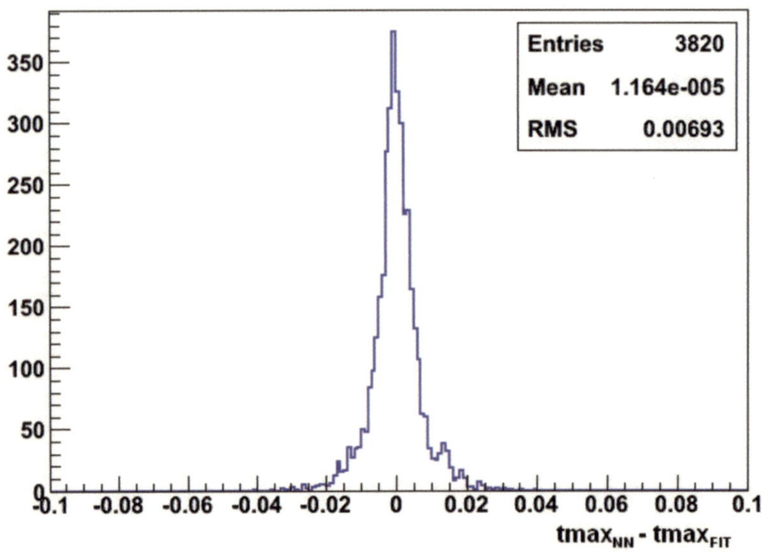

Fig. 15. As for fig.14, for the time information.

7. Conclusion

After several years of R&D work, Avalanche Photodiodes have proved to be a mature technology to be routinely employed in the design and construction of large high-energy calorimeters for the readout of the scintillation light produced in the individual calorimeter cells. The use of APDs in high energy electromagnetic calorimetry has required large efforts from both physics Laboratories and Industries in order to improve several aspects allowing an efficient usage of these devices in particle detectors. As a result of these combined efforts, several devices have been developed which have a reasonable sensitive area, a suitable spectral sensitivity and a good resistance to radiations. Different experiments incorporating one or more electromagnetic calorimeters in their setup make now use of a large number (in the order of 10^5) of these devices with good results, and additional projects are looking forward to this solution. Several progresses are however possible along different directions. One aspect is certainly related to the increase in the sensitive area of the individual devices, without loosing any advantage originating from their intrinsic properties. This will allow a more efficient coupling of APDs to the scintillation crystals. Optimization of the spectral response in connection with the choice of the scintillation material is certainly another direction where some development could be expected in the next future. Additional improvements could come from the monitoring and control of such devices, in order to optimize and stabilize their gain as a function of the bias voltage and of the operating temperature.

8. References

Aamodt, K. et al., The ALICE Collaboration (2008). The ALICE detector at LHC, *Journal of Instrumentation* 3, S08002

Anzivino, G. et al. (1995). *Review of the hybrid photo diode tube (HPD) an advanced light detector for physics*, Nuclear Instruments and Methods A365, 76-82

Badalà, A. et al.(2008). *Characterization of Avalanche Photodiodes for the electromagnetic calorimeter in the ALICE experiment*, Nuclear Instruments and Methods A596, 122-125

Badalà, A. et al.(2009). *Prototype and mass production tests of avalanche photodiodes for the electromagnetic calorimeter in the ALICE experiment at LHC*, Nuclear Instruments and Methods A610, 200-203

Barlow, R.J. et al.(1999). *Results from the BABAR electromagnetic calorimeter beam test*, Nuclear Instruments and Methods A420, 162-180

Bishop, C.M. (1995). *Neural Networks for Pattern Recognition*, Clarendon, Oxford

Chartrchyan, S. et al., The CMS Collaboration (2008). The CMS detector at LHC, *Journal of Instrumentation* 3, S08004

Erni, W. et al., The PANDA Collaboration (2008). *Technical Design Report*, arXiv :0810.1216v1

Fabjan, C.W. & Gianotti, F. (2003), Review of Modern Physics 75,1243-1286

Lorenz, E. et al. (1994), *Fast readout of plastic and crystal scintillators by avalanche photodiodes*, Nuclear Instruments and Methods A344, 64-72

Karar, A.; Musienko, Yu. & Vanel, J.Ch. (1999), *Characterization of Avalanche Photodiodes for calorimetry applications*, Nuclear Instruments and Methods A428,413-431

Musienko, Yu. (1992), The CMS electromagnetic calorimeter, Nuclear Instruments and Methods A 494, 308-312

Wigmans, R. (2000), *Calorimetry: Energy Measurements in Particle Physics*, University Press, Oxford

13

Low-Energy Photon Detection with PWO-II Scintillators and Avalanche Photodiodes in Application to High-Energy Gamma-Ray Calorimetry

Dmytro Melnychuk and Boguslaw Zwieglinski
The Andrzej Soltan Institute for Nuclear Studies, Hoza 69, PL-00681 Warsaw Poland

1. Introduction

APDs (Avalanche Photodiodes) referred to in this Chapter differ by their construction and characteristics from those commonly used in long-distance optical communication. Common to both applications is the usage of an internal gain mechanism that functions by applying an adequate reverse voltage. In the optical communication industry one is mainly interested in small diameter devices to be coupled to optical fibres in near infrared domain. In nuclear physics they are used to convert light pulses, induced by particles and photons in scintillating crystals, into electronic signals. These emit at shorter wavelengths, moreover, it is advantageous to cover up to several cm^2 of scintillator exit face with the sensing element to maximize the detector signal for low energies deposited in the scintillator. Therefore, progress in large area APDs in short-wavelength domain has been mainly driven by nuclear physics applications. A vast amount of research and development work invested by the joint CERN + Hamamatsu Photonics team resulted in a 5x5 mm^2 device (S8664-55), which paved the way to larger area APDs. A notable feature of S8664-55 is its outstanding radiation hardness, so that by its application, the CMS-ECAL expects 10 years of failureless operation in a hostile radiation environment of the CERN-LHC. Let numbers illustrate the volume of APD usage: the barrel part of CMS-ECAL has 122400 pieces and the ALICE-PHOS 35840 of them.
The meaning of the term *High-Energy Gamma-Ray* used in the title and meant in the rest of this chapter is related roughly to the maximum antiproton energy of 14 GeV from the HESR accumulator at the future FAIR facility in Darmstadt. The electromagnetic calorimeter (EMC) (Erni et al., 2008) of the PANDA detector, to be used in studies of hadron physics in antiproton-proton annihillations, will thus deal with photons in the energy range extending nearly from zero up to the maximum energy of the order of 10 GeV. *Low-Energy Photon* does not have its common meaning of a photon emitted from a radioactive source but rather is related to the practically achievable EMC low-energy detection threshold of several MeV. The results reported in Sect. 7 refer to a study intended to investigate detector resolution with 4 - 20 MeV photons, relevant for the latter energy range. Low-energy proton capture reactions are used to generate these gamma-rays.
A reader interested in medical application of APDs, for example in Positron Emission Tomography (PET), which uses photons emitted in annihilation of positrons from radioactive

sources, will find useful references in (Phelps, 2006). For additional data a review paper of (Moszynski et al., 2002) is recommended.

2. High-energy gamma-ray detection with inorganic scintillators

The choice of material of an individual EMC detection cell is related to the nature of high-energy photon interaction with matter. This initiates an electromagnetic cascade (shower) as e^+/e^- pair production and bremsstrahlung induced by them in the medium generate more electrons and photons with lower energy. Electron energies eventually fall below the critical energy, and then dissipate their energy by ionization and excitation, rather than by generation of more shower particles. In this way the cascade terminates. The spacial distribution of a shower is determined by radiation length, X_0, in longitudinal direction, and Molière radius, R_M, in transverse direction relative to the photon propagation direction. A summary of X_0, R_M and other important parameters typifying scintillators

Parameter		CeF_3	LSO/LYSO:Ce	BGO	PWO	PWO-II
ρ	g/cm^3	6.16	7.40	7.13	8.3	
X_0	cm	1.77	1.14	1.12	0.89	
R_M	cm	2.60	2.07	2.23	2.00	
τ_{decay}	ns	30	40	300	$30^s/10^f$	$30.4^s/6.5^f$
λ_{max}	nm	330	402	480	$425^s/420^f$	
n at λ_{max}		1.63	1.82	2.15	2.20	2.17
relative LY	%(LY NaI)	5	83	21	$0.083^s/0.29^f$ at RT 0.8 at -25°C	0.6 at RT 2.5 at -25°C
hygroscopic		no	no	no	no	
dLY/dT	%/°C	0.1	-0.2	-0.9	- 2.7 at RT	-3.0 at RT
dE/dx (MIP)	MeV/cm	6.2	9.6	9.0	10.1	

Table 1. Properties of a few scintillators used or planned to be used in high-energy gamma-ray calorimetry. f = fast component, s = slow component. The light yield of NaI(Tl), taken here as a reference, is 40000±2000 photons/MeV (Moszynski et al., 2002).

already used (BGO, PWO) or planned to be used (PWO-II, LSO/LYSO:Ce) in high-energy photon calorimetry is collected in Table 1. One may note an inverse correlation of X_0, R_M with, ρ, the material density. $PbWO_4$ (standard abbreviation PWO), posessing the highest density among those listed, thus offers the most compact calorimeter design, a valuable feature in view of the cost increasing as cube of the scintillator length. Economy considerations plus a short scintillation decay time, $\tau_{decay} = 6.5$ ns, motivated the PANDA Collaboration to choose PWO as the EMC material. The latter is important in order to assure short response time, necessary at high counting rates, expected to occur at small angles relative to antiproton beam direction. As the shape of an individual scintillator a truncated pyramid was decided with the front face of 20.2×20.2 mm and 200 mm (roughly $22 \cdot X_0$) length, which guarantees a tolerable energy loss due to longitudinal leakage of the shower in the forseen photon energy range. Further perfection of the PWO technology during the last decade resulted in the development of PWO-II (Novotny et al., 2005), (Borisevich et al., 2005) with doubled light yield at room temperature relative to that reported by CMS, e.g. in (Annenkov et al., 2002), and listed in Table 1. The improvement was reached by the producer - Bogoroditsk Plant of Technochemical Products (BTCP) in Russia by growing the crystals from melt with precise tuning of the stoichiometry and co-doping with Y and La with a total concentration

up to 40 ppm. To achieve further increase in light yield PANDA-EMC will be operated at a reduced temperature of T=-25°C extending electromagnetic calorimetry with sufficient energy resolution down to photon energies of a few MeV.
Lutetium oxyorthosylicate (Lu_2SiO_5:Ce or LSO) and lutetium-yttrium oxyorthosylicate ($Lu_{2(1-x)}Y_{2x}SiO_5$:Ce or LYSO) by their large light output, significantly larger than PWO, and fast decay time are the scintillators of the future, see e.g. (Ren-yuan Zhu et al., 2007). However, the present high cost associated with fabrication of scintillators with sufficient length was an obstacle to plan their usage in PANDA. The situation may improve with time, hence it is not excluded that the calorimeters of the planned SLHC and ILC will see application of LSO or LYSO.
Besides EMC the PANDA detector will contain other subsystems used for charged-particle identification and tracking goals. For the latter purpose the central part of the detector will work inside magnetic field, up to 2.0 T, of a superconducting solenoid. This precludes application of standard photomultiplier tubes as sensors for PWO-II readout. Led by the successful application of avalanche photodiodes (APDs) in PWO calorimetry by the CMS (Deiters et al., 2000) and ALICE (Aleksandrov et al., 2005) Collaborations at CERN, the PANDA Collaboration initiated a collaborative effort with the Hamamatsu Photonics K.K. (Japan) in order to develop an APD with a significantly larger sensitive area than 5×5 mm^2 posessed by S8664-55 used in those detectors. This was required by significantly lower energies of photons, as indicated above, to be encountered with PANDA in comparison with multi-GeV deposits in the search for Higgs boson at the CERN-LHC. An APD S8664-1010 with the sensitive area 10×10 mm^2 was developed to meet the needs of PANDA Collaboration for the initial R&D stage of the EMC. Ultimately, an application of two 20×10 mm^2 Hamamatsu APDs is forseen completely covering the exit face of a scintillator.

3. Principle of operation of an APD

The principle of operation of an Si APD, used in conjunction with PWO, or any other scintillator emitting short wavelength light, which is strongly absorbed in Si, is illustrated with the aid of Fig. 1. The basic elements of an APD are contained between the cathode and anode contacts reversed biased with the positive potential on the cathode so that the wafer is fully depleted. The electic field, reaching values as high as $2.5 \cdot 10^5$ V/cm at the P-N junction, as a function of depth is indicated schematically. The surface, through which the detected light enters, is protected with an antireflective coating, Si_3N_4 or SiO_2, which improves the quantum efficiency by reducing reflection losses from the surface of the Si wafer. The P-type material in front of the amplification region, which forms the P-N junction (*buried junction*) is made less than 7 μm thick to reduce sensitivity to the *nuclear counter-effect*. For e-h pairs generated within the first few microns of the depletion layer, the electron is collected and undergoes full multiplication, whereas for a pair generated within the wide drift region behind the multiplying region only the hole enters the multiplying region, where it undergoes a much reduced gain. This additional layer of N-type material is introduced to decrease the APD capacitance and to improve stability with respect to changes in bias voltage. A notable feature is the groove cut around the p - n structure and slightly into the drift space (Deiters et al., 2000) (not shown in Fig. 1) to limit the surface currents.
The *nuclear counter-effect* refers to the situation when the charged component (e^+/e^-) of an electromagnetic shower leaks out through the rear face of the scintillator and causes ionization in the attached light sensor. This is an undesired effect, since it superimposes the signal produced by scintillation light causing a decrease in its resolution. Comparing an APD with

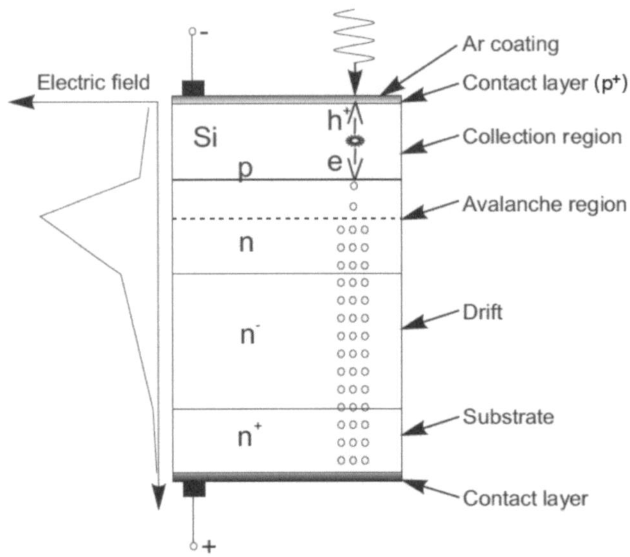

Fig. 1. The concept of an Si APD intended for short-wavelength light detection is demonstrated with a section of the Si wafer. The electric field distribution, *E(x)*, is plotted schematically to the left of the structure. Broken line marks the position of a P-N junction. Note that relative thicknesses of the different regions are not shown in scale. Ar coating is an abbreviation for Anti-reflective coating.

a PIN diode of the same thickness, one concludes that the *nuclear counter-effect* is very much reduced in the former because of a narrow width of the collection region (see above).

4. Energy resolution of an APD in scintillation light detection

The performance of an APD depends on the number of primary electron-hole (e-h) pairs produced by scintillation light, N_{eh}, APD excess noise factor, F, and dark noise level of the device-preamplifier system. The quantum efficiency, $\epsilon_Q(\lambda)$, of the APD and spectral distribution of the light emitted by a scintillator, $N_{phot}(\lambda)$, define the number of (e-h) pairs. The excess noise is due to the statistical nature of the multiplication process, which causes additional fluctuation of the measured signal. The excess noise factor depends on the ratio, $k = \beta(E)/\alpha(E)$, of ionization coefficients for electrons, $\alpha(E)$, and holes, $\beta(E)$, both functions of the electric field, E. F is a function of the internal structure of the diode, profile of the electric field and the device operating gain. The statistical variance of the APD signal, σ_N^2, is expressed as:

$$\sigma_N^2 = M^2\sigma_n^2 + N_{eh}\sigma_A^2, \tag{1}$$

where, σ_N^2, is the variance of the output signal, expressed in the number of electrons, M, is the APD gain, σ_n^2 is the variance of the number of primary electrons and σ_A^2 is the variance of single electron gain. Dividing both sides of Eq. 1 with M^2 and taking into account the definition of F:

$$F = 1 + \sigma_A^2/M^2, \tag{2}$$

we get the statistical variance of the signal from an APD, σ_{st}^2:

$$\sigma_{st}^2 = \sigma_n^2 + N_{eh}(F-1). \tag{3}$$

The first term in Eq. 3 corresponds to the statistical error of the detected signal, while the second one to the contribution of the avalanche gain of an APD. Taking into account that σ_n^2 is governed by Poisson statistics:

$$\sigma_n^2 = N_{eh}, \tag{4}$$

we find that the variance of APD signal is given by:

$$\sigma_{st}^2 = N_{eh}F. \tag{5}$$

Assuming that the detected peak is Gaussian and taking into account the relation between its variance and full width at half maximum (FWHM), we may write that the FWHM resolution is:

$$(\Delta E)^2 = (2.36)^2(N_{eh}F + \delta_{noise}^2), \tag{6}$$

where the dark noise contribution δ_{noise}, expressed in rms electrons has been explicitely included. One may rewrite Eq. 6 in energy units by taking into account that ε=3.6 eV is required for one e-h pair creation in Si:

$$(\Delta E)^2 = (2.36)^2(FE\varepsilon + \Delta_{noise}^2), \tag{7}$$

where Δ_{noise} is the dark noise contribution of the diode-preamplifier system (FWHM in energy units). The relative energy resolution (in %) is:

$$\Delta E/E = 2.36(F/N_{eh} + \delta_{noise}^2/N_{eh}^2)^{1/2}. \tag{8}$$

One may conclude from Eq. 8 that the relative energy resolution of the light signal is a decreasing function of both the number of primary e-h pairs and the signal-to-noise ratio. The high light output of a scintillator and high quantum efficiency of the employed APD are of primary importance to reduce $\Delta E/E$.

4.1 Dark noise contribution to energy resolution

The sources of dark noise in Eq. 6 are parallel and series noise of an APD. The parallel noise originates from the surface and bulk dark currents of the device. The series noise is the effect of shot noise of a preamplifier; it is proportional to the sum of APD capacitance and input capacitance of preamplifier. Following Ref. (Lorenz et al., 1994) we may write δ_{noise}^2 in Eq. 6 as:

$$\delta_{noise}^2 = 2q\left(\frac{I_{ds}}{M^2} + I_{db}\cdot F\right)\tau + 4kTR_s\frac{C_{tot}^2}{M^2}\frac{1}{\tau}, \tag{9}$$

where q is the electron charge, I_{ds} the surface leakage current, I_{db} the bulk current, τ the shaping time constant of the amplifier (assumed $\tau = \tau_{diff} = \tau_{int}$), k the Stephan-Boltzmann constant, T the absolute temperature, R_s the preamplifier series noise resistance, C_{tot} the parallel capacitance (APD plus preamplifier). We will show in Sect. 5 that APD capacitance decreases rapidly with the reverse bias voltage.

Fig. 2 illustrates the dependence of the noise contribution vs. gain for different shaping time constants of the amplifier. The measurements were performed in Ref. (Moszynski et al., 1997). The lowest noise contribution is observed close to the maximum attained gain of 160 and the shortest shaping time constant of 50 ns. Eq. 9 reflects the measured trends. The initial decrease of noise with APD gain is related to a simultaneous action of several factors: the attenuation of an unamplified noise component, related to I_{ds} in Eq. 9, decreasing capacitance of the APD with increasing bias voltage (see Fig. 3D) and the preamplifier noise. The curves have slightly increasing tendency at high gain, passed the minimum, because of dark current excess noise and fluctuations in avalanche gain, both of which increase with the excess noise factor,F.

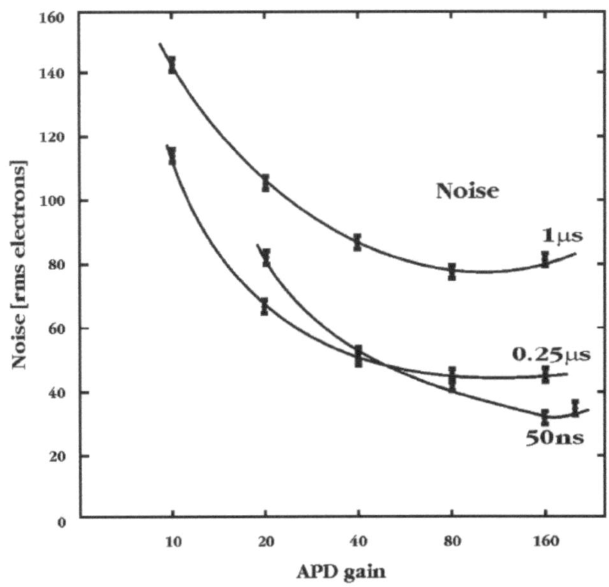

Fig. 2. Noise (in rms electrons) as a function of the gain for different shaping time constants measured in (Moszynski et al., 1997) [*reproduced from (Moszynski et al., 2002) with permission of Elsevier Ltd*].

5. Main characteristics of Hamamatsu silicon APDs

The quantum efficiency of Hamamatsu APDs S8664-55/S8664-1010 as a function of light wavelength is presented in Fig. 3A. One may see that at the wavelength of 420 nm, corresponding to the maximum of PWO blue emission (see Table 1), the efficiency is above 70% and shows a broad plateau thereafter with an efficiency of about 85%. This should be compared with about 18% of a typical photomultiplier with a bialkali photocathode in the same spectral range. One may conclude that these APDs are very well matched for detection of light from PWO. Light absorbed behind the P-N junction in Fig. 1 produces electrons and holes, but only holes go to the avalanche region and multiply, while electrons drift towards the back contact and are collected without multiplication. The multiplication factor for holes is much smaller than that for electrons, as a result the gain for light with long wavelengths is smaller than for short ones, which is reflected also as a drop with wavelength in Fig. 3A in the quantum efficiency.

Fig. 3B presents plots of gain vs. reverse voltage at the different working temperatures. With increasing voltage these curves approach the breakdown voltage, which is characterized by an uncontrollable growth of the dark current (see Fig. 3C). On the other hand, the asymptotic value of gain at low voltages is unity (not reached in Fig. 3B), at which carriers created in the collection region are transferred though the p-n junction without multiplication. A typical operating gain used with the indicated APDs is M=50 at room temperature, +20 °C. One may note that decreasing the temperature down to -20 °C, which is close to the forseen operating temperature of PANDA at -25 °C, will bring an increase of about a factor of 3 - 3.5 in gain. This increase in gain is ascribed to decreasing excitation of lattice phonons, which permits carriers to acquire higher energies used in avalanche multiplication. One may conclude that both these operating parameters, voltage and temperature, should be carefully stabilized for

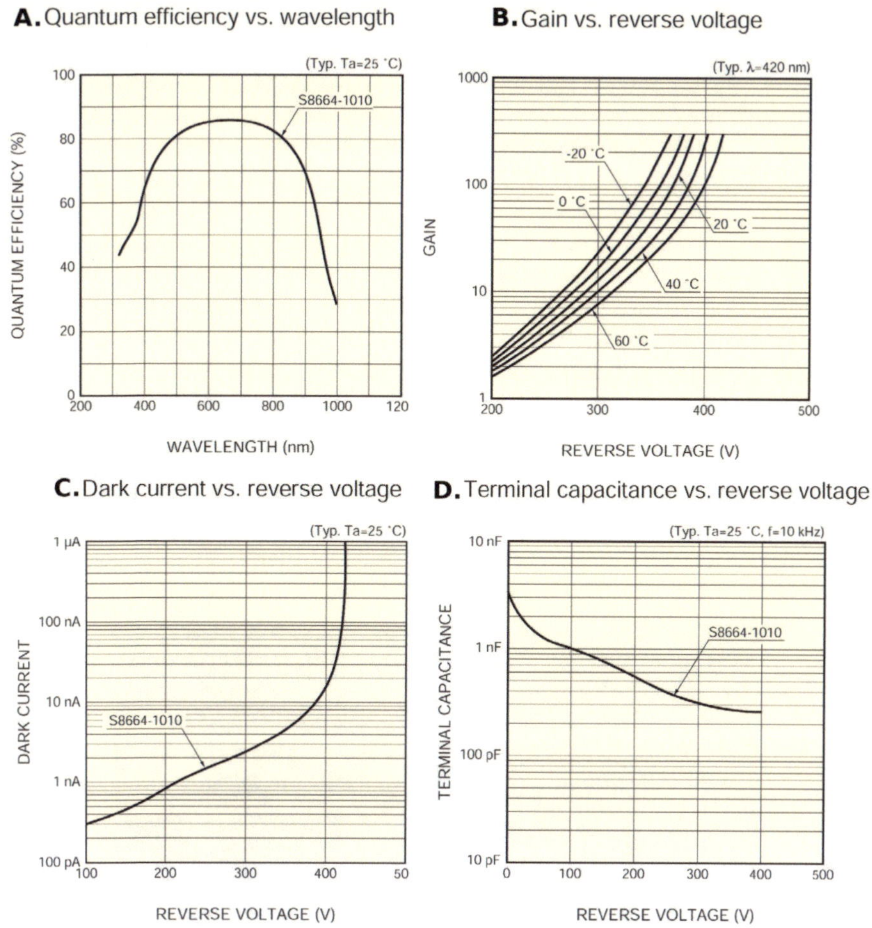

Fig. 3. Selected characteristics of an APD S8664-1010 of Hamamatsu Photonics K.K. [*by courtesy of Hamamatsu Photonics K.K.*].

a stable operation. The bias at room temperature cannot be chosen too high, because cooling that follows, may end-up in break-down. M=50 is a safe initial choice, as proved in practice, with the corresponding voltage and reverse current values at room temperature provided for each APD by the producer. A further fine tuning of gain is accomplished by distributing a reference light signal through a system of optical fibers in contact with an exit face of each individual scintillator.

Comparing Hamamatsu Si-APDs with other producers, one notes several additional features making the Japanese products more convenient in large-scale application, as in the PANDA-EMC. The silicon wafer of S8664-1010 is installed on a thin ceramic plate, only slightly exceeding in size the sensitive area $10 \times 10\ \text{mm}^2$. Moreover, the surface through which light enters is covered with a transparent plastics. This prevents from damaging the APD upon exerting stress when a contact with the scintillator is done using an optical grease. Also, much lower bias voltage at the same gain deserves stressing as a factor in favor of Hamamatsu in large-scale applications.

6. Measurements of $\Delta E/E$ using high-energy tagged photon beams and detector matrices

A comprehensive information on the performance of a PWO+APD combination is obtained from an experiment in which the PWO scintillator is irradiated along its axis with a narrow beam of photons and the scintillation light converted into an electronic signal with the aid of an APD. It has been stressed in Sect. 2 that high-energy photons create an electron-positron shower, which propagates both along and perpendicular to the scintillator axis. The lateral dimensions of scintillators of about 20×20 mm^2 are determined by the required angular resolution [*granularity*] in the forseen experiments (see (Erni et al., 2009)). With the Molière radius (see Table 1) of 2.0 cm for PWO, one needs a matrix of at least nine closely packed scintillators in order to intercept with the scintillating material and convert into light the shower originating from the central one. An experiment using high-energy tagged photons is illustrated in Fig. 4. Photons are products of bremsstrahlung of a high energy electron beam from the MAMI-B microtron facility at Mainz in a thin carbon foil. There is a unique relation between the energy of a photon and the momentum of an electron that it radiated, so that the highest energy photons are accompanied with low-momentum electrons and *vice-versa*. The post-radiation electron is bent in the magnetic field of a magnetic spectrometer and detected with a position-sensitive detector along its focal plane. This illustrates the method of *photon-tagging* with the aid of coincident electron detection in a certain range of positions

Fig. 4. Section of the magnetic spectrometer of the MAMI-B tagging facility along its central plane. The iron flux return yoke is shown hatched. The dotted line is the trajectory of electrons with the incident momentum. The solid lines are trajectories of electrons that suffered bremsstrahlung with emission of photons with progressively increasing energy. The matrix of nine PWO-II scintillators is located behind an iron collimator (not shown) with its central scintillator axis along the photon beam [*reproduced from (Anthony et al., 1991) with permission of Elsevier Ltd*].

in the focal plane (Anthony et al., 1991; Hall et al., 1996). In the experiment that we refer to (Novotny et al., 2008) sixteen photon energies in the range 40.9 - 674.5 MeV were selected, and

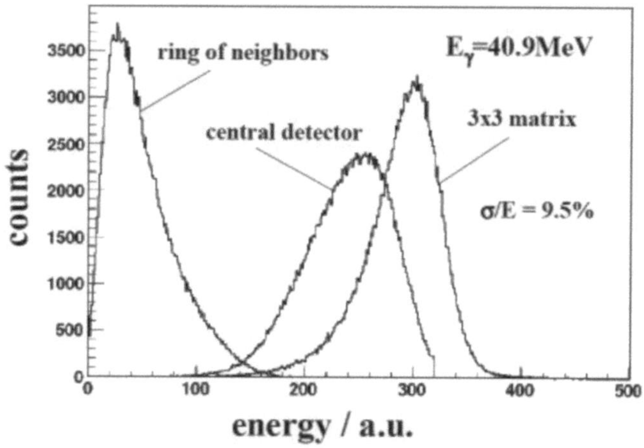

Fig. 5. Amplitude spectra illustrating the response of the central scintillator and the entire 3x3 PWO-II matrix, taken eventwise, to photons with E_γ= 40.9 MeV selected with the MAMI tagger. The matrix was kept at 0°C [*reproduced from (Novotny et al., 2008a)*].

the primary electron beam energy was 840 MeV. Typical energy width per tagging channel varied between 2.3 MeV at 50 MeV to 1.5 MeV at 500 MeV photon energy, respectively. The matrix of 3x3 PWO-II scintillators was installed in a thermally isolated container, in which it was cooled down to 0°C. The container could be moved remotely in the plane perpendicular to the photon beam, so that each of the nine scintillators could be inserted into the beam and its amplitude spectra calibrated in energy at the sixteen points. After the calibration runs, the central scintillator was inserted and the spectra in all of them simultaneously measured. By summing eventwise the energy deposits, the energy response to a photon shower for the entire matrix was determined. Fig. 5 compares the spectrum in the central scintillator with the result of summing individual scintillator responses eventwise for the incident photon energy E_γ= 40.9 MeV. The summed spectrum is almost Gaussian, with only a slight indication of the tailing seen. The reduced Gaussian widths, σ/E, of the summed peaks, are plotted in Fig. 6 as a function of the photon energy. The solid line is a fit to the measured points with a formula:

$$\frac{\sigma}{E} = \frac{1.86\%}{\sqrt{E[GeV]}} + 0.65\%, \tag{10}$$

where the energy is expressed in GeV. One recognizes here Eq. 8 with the first statistical, and the second constant term. A comparison with the GEANT4 simulations (lower curve in Fig. 6), which considers just the pure energy deposition into the scintillator material, demonstrates the contribution due to photon statistics and the effect of experimental thresholds. The resolution approaches asymptotically the simulation reflecting the decreasing relative importance of the latter two factors. The value obtained by extrapolating Eq. 10 to 1.0 GeV is 2.5% - best resolution ever measured at this energy for PWO with APD readout. One should stress that energies lower than 40.9 MeV could not be reached with the MAMI photon tagger, having the maximum photon energy set at 674.5 MeV, because of limitations imposed by the focal plane detector.

7. Measurements of $\Delta E/E$ using low-energy photons produced in proton capture reactions

Identification of π^0 and η mesons by detecting both γ-rays from their decay is a prerequisite for suppressing undesired background in studies of photon transitions between the states of charmonium in the physics program of PANDA (Erni et al., 2009). The detection threshold of the PANDA - EMC should be as low as possible to achieve this goal. Thus the question arises, whether the relative resolution, $\sigma(E)/E$, will or will not follow Eq. 10 at energies lower than 40.9 MeV, where it is expected to deteriorate rapidly because of $\sqrt{E}$ term in the denominator. The authors of Ref. (Melnychuk et al., 2009) decided to answer this question using gamma-rays produced in proton-induced reactions with light nuclei. The reactions ^{11}B(p,γ)^{12}C and T(p,γ)^{4}He are of primary concern, because they proceed to well-separated, narrow states in the final nuclei, so that the photon energy spread is determined by the target thickness and beam energy spread. These two factors gave negligible contribution to σ/E in the reported experiment. In order to make a direct comparison with the results of (Novotny et al., 2008) and Eq. 10, we used a PWO-II scintillator of the same size $20\times20\times200$ mm, as used in the scintillator matrix of Sect. 6, the same type of reflector, an APD S8664-1010 of Hamamatsu Photonics and the same preamp type LNP as used in that work. What differed our experiments was the working temperature. We aimed at -25°C, forseen as the working temperature for PANDA, however reached only -21.6 °C, because of insufficient cooling power of our JULABO F32-ME refrigerated circulator and thermal losses from the cooled volume to the ambient.

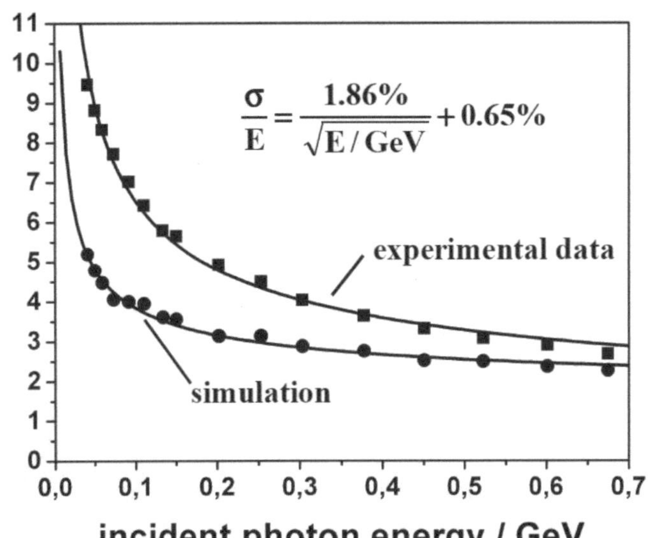

Fig. 6. Experimental reduced energy resolutions, σ/E, (upper set of points) and the results of GEANT4 simulations (lower set) for the 3x3 PWO-II matrix in the energy range 40.9 - 674.5 MeV. The simulation incorporates only the pure energy deposit. Solid lines are the results of fits, with the upper line given by the indicated formula [*reproduced from (Novotny et al., 2008a)*].

An isolation from the ambient is provided in our system by vacuum common with the Van-de-Graaff accelerator, which was 3 - 7×10^{-6} mbar in the performed measurements. Our

experimental system is intended for PWO response measurements in the energy range 4.4 - 20 MeV, which is accessible with the aid of the above mentioned reactions. In this energy range pair production is the main interaction process of photons with heavy elements as occuring in $PbWO_4$. Consequently, we used a cylinder of the plastic scintillator EJ-200, cut along the vertical axis, which covered the PWO-II over the length of 13.4 cm. Both halves were readout independently via light-guides with the 3 inch diam. photomultipliers XP4312.
A detailed presentation of the experimental system is contained in Ref. (Melnychuk et al., 2009). The PWO-II received photons along its axis through a lead collimator from the beam spot on a ^{11}B target. The events in which an electron-positron pair was generated within its volume, and both 0.511 MeV positron annihilation quanta absorbed therein, could be separated with the aid of the two plastics from the events in which one or both of the annihilation quanta escaped from the PWO-II and ended up to be detected in the plastics. The spectra from these three channels were registered with a 14-bit 100 MHz sampling ADC N1728B of CAEN. Fig. 7 is the full-energy spectrum in PWO sorted-out using the above mentioned criterion. The solid line is a sum of the three Gaussians with indicated energies and reduced resolutions, σ/E, fitted to the spectrum. The two highest energy peaks correspond to gamma-ray transitions from the capturing state in ^{12}C to its ground and first excited states, respectively. An intense peak with E_γ=6.13 MeV originates from the 6.13 MeV state in ^{16}O, excited in ^{19}F(p,α)^{16}O with a fluorine contamination of the ^{11}B target, because of its strong affinity to ^{19}F. A small contribution of the 6.92 and 7.12 MeV lines in ^{16}O is also taken into

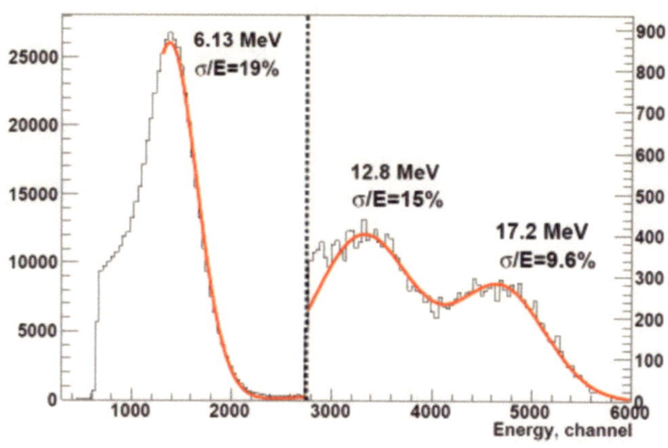

Fig. 7. Full-energy spectrum of γ-rays excited with the 1390 keV proton beam in a thin ^{11}B target, detected with the PWO-II scintillator cooled down to -21.6°C. The solid line is a fit with the three Gaussians. The relative dispersions, resulting from the fit, σ/E, are indicated [*reproduced from (Melnychuk et al., 2009) with permission of Elsevier Ltd*].

account. These assignments of the produced gamma-rays to proton induced reactions were obtained with a large NaI(Tl) detector offering a much better resolution than PWO with the same boron target. In order to compare our dispersions with the smooth trend established in Eq. 10, one needs to take into account the increase of light yield upon lowering the working temperature from 0°C in (Novotny et al., 2008) to -21.6 °C in (Melnychuk et al., 2009).
Taking into account the dependence of PWO-II light yield on the temperature, quoted in Ref. (Erni et al., 2008) one comes to a formula which should approximately govern σ/E, if both

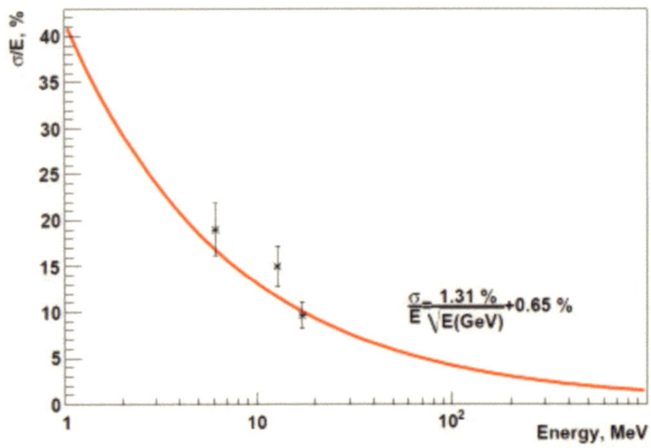

Fig. 8. Comparison of the relative dispersions, σ/E, measured in Ref. (Melnychuk et al., 2009) (points with error bars) with the smooth trend established in Ref. (Novotny et al., 2008) in the 40.9 - 674.5 MeV energy range at MAMI. The curve is corrected for the difference in working temperatures between these two experiments [*reproduced from (Melnychuk et al., 2009) with permission of Elsevier Ltd*].

experiments used -21.6°C:

$$\frac{\sigma}{E} = \frac{1.31\%}{\sqrt{E[GeV]}} + 0.65\%. \tag{11}$$

We have tacitely assumed in Eq. 11 that contributions to resolution beyond photon statistics in both experiments were equal. Eq. 11 is plotted with the solid line in Fig. 8. One may conclude from this figure that a smooth trend established in Ref. (Novotny et al., 2008) in the photon energy range 40.9 - 674.5 MeV gives a valid extrapolation of resolution for cooled PWO-II scintillators with APD readout to energies as low as 6 MeV. It should be stressed that with a tritium target and the $T(p,\gamma)^4He$ reaction, which yields monoenergetic gammas with about 20 MeV energy at $E_p \approx 1$ MeV, a much simpler pattern than in Fig. 7 is expected. We plan to perform such an experiment in order to get a clean access to the line-shape to learn on its possible deviation from a pure Gaussian as implied in Fig. 5.

8. Summary

This Chapter is intended to give the interested reader an insight into application of APDs as sensors for the readout of scintillation light in multielement electromagnetic calorimeters. This application promotes high-energy and medium-energy nuclear physics to major users of large-area APDs in the blue domain of light spectrum, consequently nominates them to the title of driving force of large-area APD development. Sect. 1 introduces the main actors on the scene: CMS and ALICE Collaborations at the CERN-LHC at high-energies and the PANDA Collaboration at FAIR-Darmstadt in the medium-energy domain. The CMS-ECAL was the pioneer by its development together with Hamamatsu Photonics K.K. of an APD S8664-55, which fitted well the large light signals from over hundred-GeV photon detection in PWO. The PANDA Collaboration considered other options for a scintillator besides PWO, however, its fast scintillation decay time and good radiation hardness were the major factors to stay also with this scintillator choice. A significant R&D effort together with the Bogoroditsk

Plant of Technochemical Products (BTCP), the scintillator producer, has been invested into perfectioning the properties of PWO, which resulted in an improved version PWO-II, with doubled light output in comparison with the average used in the CMS-ECAL (Sect. 2). To achieve further increase in light yield it was decided to operate PANDA-EMC at a reduced temperature of -25°C.

A parallel research program conducted together with Hamamatsu Photonics K.K. resulted in an APD S8664-1010, having the sensitive area increased fourfold and posessing a superior radiation hardness of its predecessor. Sect. 3 is intended for a qualitative presentation of the essential features of S8664-1010 and explaining how its narrow *collection region* helps to minimize an adverse *nuclear counter effect*. Sect. 4 summarizes the factors determining energy resolution, which is manifested as a full-width at half maximum, ΔE, of an output signal of the sensor under excitation of the scintillator with monoenergetic photons. We use also an equivalent σ parameter of a Gaussian fitted to the line shape to characterize energy resolution. Sect. 4.1 extends the discussion to the dark-noise contribution to energy resolution and illustrates how this component is influenced by APD gain and shaping time-constant of the amplifier. Sect. 5 presents several characteristics of an APD S8664-1010 and relates them to the discussion in Sect. 4. Sections 6 and 7 are devoted to a detailed presentation of the methods used to measure the relative resolution, σ/E an important property of the scintillator plus APD combination, in a wide range of photon energies, i.e. light signals. The electron-photon avalanche at high photon energies requires application of the detector matrices (Sect. 6), whereas at low energies, where the first stage of a shower, the pair formation dominates, a simpler system might prove sufficient (Sect. 7). It is demonstrated that a smooth functional dependence of σ/E on the photon energy, E, established in (Novotny et al., 2008) (Sect. 6), when corrected for the difference in applied temperatures, accounts well also for the rapid increase of σ/E in the low-energy domain.

9. Acknowledgments

We gratefully acknowledge the effort of the staff and technical personnel of the SINS and JLUG Institutes in performing the measurements reported in this Chapter. The authors are grateful to Professor Rainer W. Novotny for the critical reading of the manuscript and constructive remarks, which greatly improved its quality. This work was supported in part by the European Community Framework Programme FP6 under Contracts #RII3-CT-2003-506078; Hadron Physics and #515873; DIRACsecondary-Beams (PANDA4). This work received also support on behalf of the Polish Ministry for Science and Education and the International Post-Graduate Studies of the Polish Academy of Sciences.

10. References

Aleksandrov, D.V. et al. (2005) *Nuclear Instr. and Meth. in Phys. Res. A*, Vol. 550, No. 1-2, September 2005, 169 - 184.

Annenkov, A.A. et al. (2002) *Nuclear Instr. and Meth. in Phys. Res. A*, Vol. 490, No. 1-2, Sept. 1 2002, 30 - 50.

Anthony, I. et al. (1991) *Nuclear Instr. and Meth. in Phys. Res. A*, Vol. 301, No. 2, March 1 1991, 230 - 240.

Borisevich, A. et al. (2005) *Nuclear Instr. and Meth. in Phys. Res. A*, Vol. 537, No. 1-2, Jan. 21 2005, 101 - 104.

J. Chen, L. Zhang and Ren-yuan Zhu, (2007) *Nuclear Instr. and Meth. in Phys. Res. A*, Vol. 572, No 1, March 1, 2007, 218 - 224.

Deiters, K. et al. (2000) *Nuclear Instr. and Meth. in Phys. Res. A*, Vol. 442, No. 1-3, March 2000, 193 - 197.

Erni, W. et al. (2008). *Technical Design Report for PANDA Electromagnetic Calorimeter (EMC)*, Heinsius F.-H., Kopf B., Lewandowski B., Löhner H., Novotny R., Peters K., Rosier Ph., Schmitt L., Vasiliev A., (Eds.), October 2008, 1 - 199, arXiv:0810.1216v1

Erni, W. et al. (2009). *Physics Performance Report for: P̄ANDA Strong Interaction Studies with Antiprotons*, Bettoni D., Bussa M.P., Düren M., Feliciello A., Gillitzer A., Iazzi F., Johansson T., Kopf B., Lehrach A., Lutz M.F.M., Mass F., Maggiora M., Negrini M., Peters K., Pochodzalla J., Schmitt L., Scholten O., Stancari G., Timmermans R., (Eds.), March 2009, 1 - 204, arXiv:0903.3905v1

Hall, S.J. et al. (1996) *Nuclear Instr. and Meth. in Phys. Res. A*, Vol. 368, No. 3, Jan. 11, 1996, 698 - 708.

Lorenz, E. et al. (1994) *Nuclear Instr. and Meth. in Phys. Res. A*, Vol. 344, No 1, April 21, 1994, 64 - 72.

Melnychuk, D. et al. (2009) *Nuclear Instr. and Meth. in Phys. Res. A*, Vol. 607, No. 3, Aug. 21, 2009, 600 - 606.

Moszynski, M. et al. (1997) *IEEE Trans. Nucl. Sci.*, Vol. NS-44, No 3, June 1997, 436 - 442.

Moszynski, M. et al. (2002) *Nuclear Instr. and Meth. in Phys. Res. A*, Vol. 485, No 3, June 11, 2002, 504 - 521.

Novotny, R.W., Döring W.M., Hjelm F., Melnychuk D., Makonyi K., Reiter A., Salz Ch., Steinacher M., Thiel M., Zwieglinski B. (2005) *IEEE Nucl. Sci. Symp. Conf. Record*, Oct. 23 - 29, 2005, 244 - 248, ISSN 1082 - 3654.

Novotny, R.W. et al. (2008) *IEEE Trans. Nucl. Sci.*, Vol. NS-55, No 3, June 2008, 1295 - 1298.

Novotny, R.W. et al. (2008a) *GSI Scientific Report 2007*, GSI Report 2008 - 1, (2008), 29.

Phelps, M.E. (2006). *PET - Physics, Instrumentation and Scanners*, Springer, ISBN-10: 0-387-32302-3, Berlin, Heidelberg, New York.

Part 3

Emerging Technologies

14

High-Power RF Uni-Traveling-Carrier Photodiodes (UTC-PDs) and Their Applications

Tadao Nagatsuma[1] and Hiroshi Ito[2]
[1]Osaka University,
[2]Kitasato University
Japan

1. Introduction

Research on exploring millimeter waves and/or terahertz (THz) waves, which cover the frequency range from 30 GHz to 10 THz, has lately increased since the nature of these electromagnetic waves is suited to spectroscopic sensing as well as to ultra-broadband wireless communications. One of the obstacles to developing applications of millimeter waves and terahertz waves is a lack of solid-state signal sources.

For the generation of millimeter waves and terahertz waves, photonic techniques are considered to be superior to conventional techniques based on electronic devices with respect to wide frequency bandwidth, tunability, and stability. Moreover, the use of optical fiber cables enables us to distribute high-frequency RF signals over long distances. In this scheme, optical-to-electrical (O-E) converters, or **"photodiodes"**, which operate at long optical wavelengths (1.3-1.55 μm), play a key role, and high output current is required in addition to large bandwidth for practical applications.

Among various types of long-wavelength photodiode technologies, a **uni-traveling-carrier photodiode (UTC-PD) and its derivatives** have exhibited the highest output powers at frequencies from 100 GHz to 1 THz, with improvement in layer and device structures since its debut in 1997 (Ishibashi et al., 1997).

This paper describes recent progress in the high-power RF UTC-PDs, which operate at millimeter-wave and terahertz-wave frequencies. In Section 2, we discuss how its operation mode differs from that of the conventional photodiode. Next, some typical characteristics of the UTC-PD, such as output RF powers and their frequency characteristics, as well as some techniques for enhancing the output powers from the UTC-PD, such as resonant design, antenna integration, and packaging, are presented. In Section 3 we show recent analog applications of the UTC-PD such as wireless communications, spectroscopy, and imaging, which have not been realized with conventional electronics or photodiode technologies.

2. Uni-traveling-carrier photodiode (UTC-PD)

2.1 Concept and operations

Uni-traveling-carrier photodiode (UTC-PD) (Ishibashi et al., 1997, Ito et al., 2004a, 2008) is a novel photodiode that utilizes only electrons as active carriers. This feature is the key for realizing its superior performance. Band diagram of the UTC-PD is shown in Fig. 1(a) together with that of conventional pin-PD in Fig. 1(b). The active part of the UTC-PD consists of a p-type doped InGaAs light absorption layer and an undoped (or a lightly n-type doped) InP carrier-collection layer. In the UTC-PD, the photo-generated minority electrons in the neutral absorption layer diffuse into the depleted collection layer and then drift to the cathode layer. On the other hand, because the absorption layer is quasi-neutral, photo-generated majority holes respond very fast within the dielectric relaxation time by their collective motion. Therefore, the photoresponse of a UTC-PD is determined only by the electron transport. This is an essential difference from the conventional pin-PD (Fig. 1(b)), in which both electrons and holes contribute to the device response and the low-velocity hole-transport determines the total performance (Furuta et al., 2000). This qualitatively different operation mode of the UTC-PD provides three important advantages.

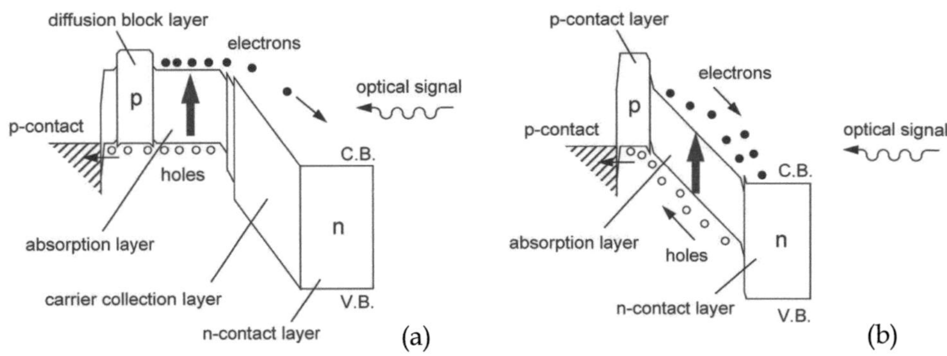

Fig. 1. Band diagram of a UTC-PD (a) and a pin-PD (b).

One is higher operation speed due to an order of magnitude higher electron velocity (Ishibashi et al., 1994) in the depletion layer compared to hole velocity (Hill et al., 1987). Here, we are also benefitted from the velocity overshoot of electrons (Ishibashi et al., 1994) in the depletion layer. Thus, with an appropriate UTC-PD design, electron diffusion time in the absorption layer mainly determines the device operation speed. In general, diffusion velocity is considered to be smaller than the drift velocity. However, in the absorption layer of the UTC-PD, the electron diffusion velocity can be very large for a relatively thin absorption layer due to the large minority mobility of electrons in p-InGaAs (Harmon et al., 1993). In addition, we can independently design the depletion layer and the absorption layer thicknesses in the UTC-PD structure. Thus, a very thin absorption layer can be used to attain an extremely high 3-dB bandwidth (f_{3dB}) without increasing the CR charging time. This is also an important advantage over the pin-PD, in which the CR charging time becomes significantly larger when the absorption layer thickness is excessively reduced (Kato et al., 1993).

The second advantage is the higher output saturation current due to much less space charge effect in the depletion layer, which also results from the high electron velocity in the depletion layer. In the pin-PD, the band profile is modified under a high-excitation condition (Willams et al., 1993) because of photo-generated carriers accumulated in the absorption layer as shown in Fig. 2(b). Thus, the decreased electric field reduces the carrier velocity, enhances the charge storage, and finally results in output current saturation. Although the situation is similar in the UTC-PD (Fig. 2(a)), the space charge consists of only electrons whose velocity is much higher than that of holes even for the decreased electric field. Therefore, the output does not saturate until the current density becomes an order of magnitude higher than that for the pin-PD. Further, the velocity reduction of holes starts to occur even when the current density is relatively low. This is in contrast to the UTC-PD, where the prominent velocity reduction of electrons does not occur until the electric field becomes much smaller (Ishibashi et al., 1994). Therefore, the linearity range of the UTC-PD is much wider than that of the pin-PD.

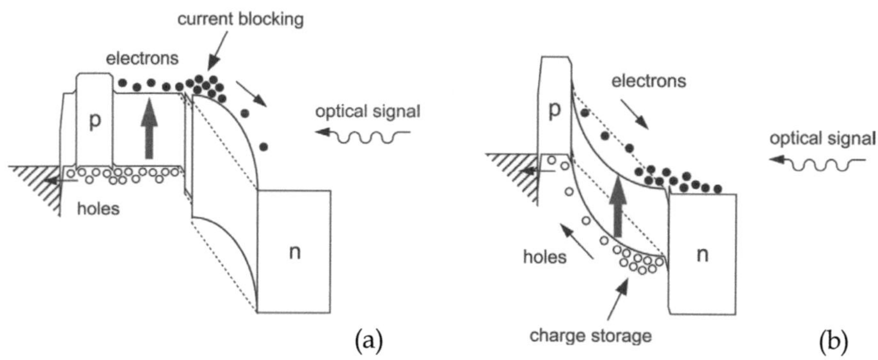

Fig. 2. Modified band diagram of a UTC-PD (a) and a pin-PD (b) at high optical input.

The third advantage is the capability of low-voltage operation. The high speed with high saturation output of the UTC-PD is maintained at a low (or even at zero) bias voltage (Ito et al., 2000a), because the high electron velocity in the depletion layer can be maintained at a relatively low electric field or even with the built-in field of the pn junction. This is in contrast to the large electric field required for holes to maintain relatively large drift velocity (Hill et al., 1987). A smaller operation voltage reduces power consumption, simplifies heat sinking, reduces biasing circuit cost, and improves reliability.

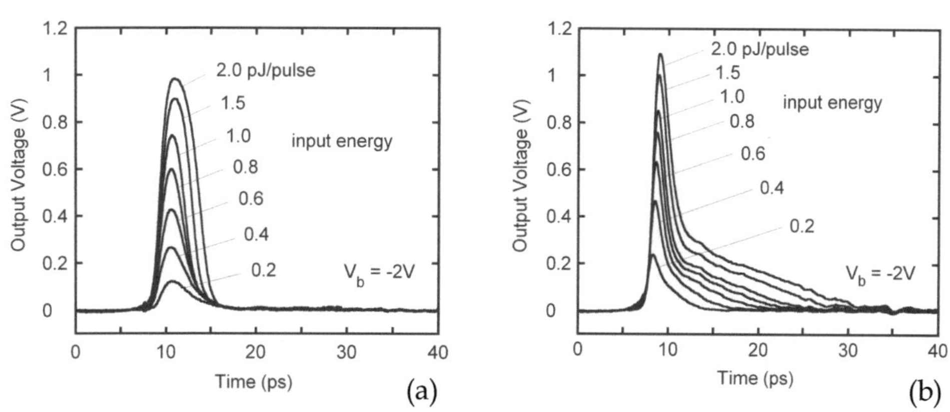

Fig. 3. Typical pulse photoresponses of a UTC-PD (a) and a pin-PD (b).

2.2 Basic characteristics

Figures 3(a) and (b) show typical output waveforms of the pulse photoresponse of a UTC-PD and a pin-PD (Furuta et al., 2000) measured using the electro-optic sampling technique.

In the UTC-PD, the output peak current increases linearly with increasing input energy, and the waveform does not significantly change until it reaches the saturation point. After the saturation occurs, the pulse width increases gradually. However, even at this stage, the fall time of the waveform does not prominently increase. The fast fall time even in the saturation region is attributed to the fast response of electrons accumulated in the depletion layer. On the other hand, the waveform of the pin-PD is quite different and it consists of two current components. The initial fast component is attributed to electron transport, and the slow tail is caused by hole transport. Especially, the full width of the hole-response increases drastically as the input energy increases. This is due to the stronger space charge effect of slow holes in the absorption layer as mentioned previously. The broadening of the waveform in turn results in severe degradation of the bandwidth and linearity of the pin-PD in a high excitation condition.

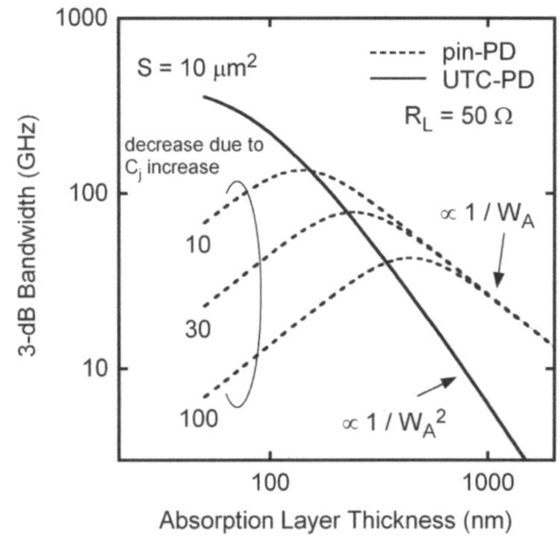

Fig. 4. Calculated relationships between f_{3dB} and W_A for UTC-PD and pin-PDs.

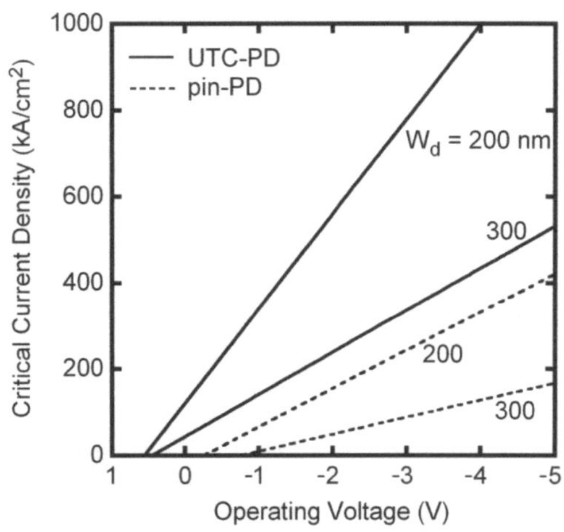

Fig. 5. Calculated critical current densities against operating voltage for UTC-PDs and pin-PDs.

Figure 4 shows calculated f_{3dB} for the UTC-PD and pin-PD as a function of absorption layer thickness (W_A) (Ito et al., 2008). The f_{3dB} is basically proportional to $1/W_A^2$ for the UTC-PD and to $1/W_A$ for the pin-PD for thicker W_A. The former is due to the diffusive electron transport in the neutral absorption layer in the UTC-PD, and the latter is due to the drift motion of both carriers in the depletion layer of the pin-PD. If we try to increase the f_{3dB} of a pin-PD, we have to decrease W_A. However, as W_A is decreased, the bandwidth becomes smaller again at a certain point due to the junction capacitance (C_j) increase as shown by broken curves in Fig. 4. On the other hand, in the UTC-PD structure, the absorption layer and the collection layer are separated. Thus, f_{3dB} increases monotonically up to very thin W_A without C_j increase (solid curve in Fig. 4). Therefore, the prospective maximum f_{3dB} is much larger than that for the pin-PD. The maximum f_{3dB} reported for a UTC-PD is as high as 310 GHz (Ito et al., 2000b), which is twice as large as that reported for a pin-PD (Bach et al., 2007) operating at 1.55 μm.

As explained in the previous sub-section, the high output current of the UTC-PD is an important benefit over the conventional pin-PD. If we assume a simple velocity saturation model for both carriers (Ishibashi et al., 2000), the output current saturation starts to occur when the current density reaches a critical current density (J_c) at a critical electric filed (Tebbenham et al., 1975, Hill et al., 1987). Because of the larger saturation velocity and smaller critical electric field for electrons, J_c can be much larger for the UTC-PD than that for a pin-PD. Figure 5 shows the calculated J_c as a function of the operating voltage (V_b) for UTC-PDs and pin-PDs. Here, W_d is the depletion layer thickness. Although the ratio of J_c for the UTC-PD and the pin-PD depends on V_b, it lies in between about three and ten, and the ratio further increases when the W_d becomes larger. In an actual device, the charge distribution in the depletion layer is not uniform so that the output current saturation starts to occur at lower current density than in this calculation, especially in pin-PDs, resulting in a larger ratio of J_c. In addition, the pin-PD requires a certain negative bias voltage to work properly with the carrier velocity saturated, while the UTC-PD can be operated at zero or even positive bias voltages. A smaller W_d is further effective for expanding the operation area of the UTC-PD towards the positive bias voltage region.

For improving the UTC-PD performance, various modifications, including the structures proposed for pin-PDs and base-collector junctions of heterojunction bipolar transistors (Kroemer, 1982, Bowers et al., 1987, Ishibashi et al., 1988, Hafizi et al., 1993) are possible, and many of them have been applied to the UTC-PD structure. To decrease the electron traveling time in the absorption layer, the introduction of a quasi-field into the absorption layer by means of the band-gap grading and/or doping grading are effective (Ishibashi et al., 1997, Shimizu et al., 1998). For decreasing the electron traveling time in the collection layer, p-type doping or cliff-like structure (Shi et al., 2005) are suitable for optimizing the electric field profile. On the other hand, to increase the saturation current level, increased n-type doping in the collection layer (Shimizu et al., 1998, Li et al., 2004) is preferable. A combination of neutral and depleted absorption layers (Li et al., 2003, Muramoto et al., 2003) will increase the responsivity without considerably sacrificing the saturation current and operation speed. The use of dual-depletion region (Jun et al., 2006) is also effective in this modification to keep the high operation speed. Another way to improve the responsivity with the high operation speed maintained is to employ an external structure to guide the input light to the absorption layer at different angle, such as waveguide structure (Muramoto et al., 1998), evanescently coupled structure (Achouche et al., 2004), velocity-matched distributed structure (Hirota et al., 2001), refracting facet structure (Fukano et al., 1999), and total reflection structure (Ito et al., 2000c).

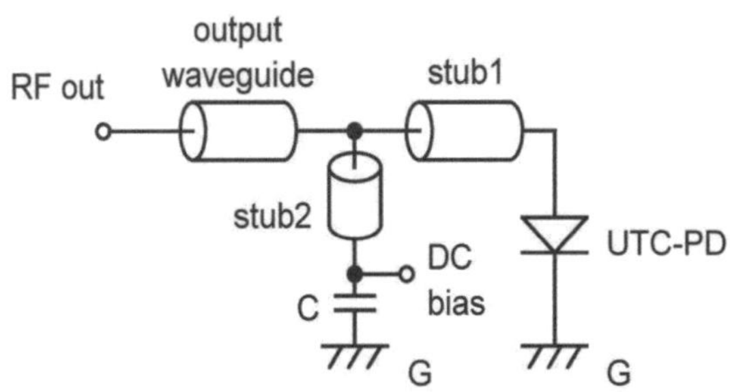

Fig. 6. A short-stub matching circuit integrated with a UTC-PD.

2.3 Circuit and package design

For efficiently utilizing the output power from the UTC-PD, the external circuit as well as the packaging is important. A resonant matching circuit is an effective element that can increase the maximum output power of a PD at desired frequencies. A typical example is a short-stub circuit (Ito et al., 2003a) containing two waveguides and a capacitor (Fig. 6). This circuit is a good solution for its simplicity, ease of integration with the UTC-PD, and limited numbers of optimizing parameters. It can change the effective output impedance, the resonant peak frequency, the peak output power, and the bandwidth. It also acts as a bias-tee circuit. All the elements fit within a standard-size chip, and can be fabricated without modifying the standard UTC-PD fabrication processes.

For a practical use, the device should be assembled in a module that is suitable for handling very high frequency signals. There are two possible approaches for transmitting very-high-frequency signals from a PD. One is to use a rectangular-waveguide-output module, and the other is to use a quasi-optic one. The waveguide-output approach is effective for transmitting the confined output signal to the objective. In addition, we can easily control the radiation pattern of the signal by using an appropriate external antenna. The drawback of this approach is that the bandwidth of the fundamental mode signal is essentially restricted by the waveguide nature. On the other hand, the key advantage of the quasi-optic module is its extremely-large bandwidth, and thus the full bandwidth can be covered with only one device. Drawbacks of this approach are that it requires a quasi-optic collimation system, the integration of a proper planar antenna with a PD, and proper tailoring of the radiation pattern for specific applications.

For the first approach, compact waveguide output UTC-PD modules have been developed. Figure 7 shows the one for operation in the J-band (220-325 GHz) (Ito et al. 2006). The output electrode of a matching-circuit-integrated UTC-PD is connected to an MSL-based waveguide fabricated on a quartz substrate, and it leads the output signal to a rectangular waveguide with a coupler. The module size is 12.7 mm x 30 mm x 10 mm, excluding the optical fiber and the bias connector. The modules for operation in the W-band (75 - 110 GHz) (Ito et al. 2002), F-band (90-140 GHz) (Ito et al. 2003b), and D-band (110 - 170 GHz) (Furuta et al. 2005) have also been developed. This package has an additional advantage for integrating other components for higher functionality (Ito et al., 2005a, 2010). Another possibility is that it can be operated at much higher frequency than its fundamental frequency range. For example, the F-band module could be operated up to 813 GHz (Ito et al., 2004b), and the J-band module up to 650 GHz (Wakatsuki et al., 2008).

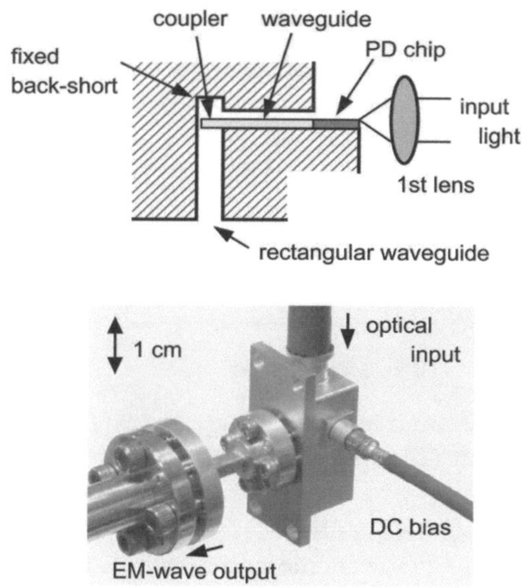

Fig. 7. Cross-sectional diagram and micrograph of a waveguide-output UTC-PD module.

For the second approach, a quasi-optic UTC-PD module (Ito et al., 2005b) shown in Fig. 8 has been developed. Here, the UTC-PD was integrated with a planar antenna, and the device was mounted on a hyper-hemispherical lens made of high-resistivity silicon to collimate the output EM wave.

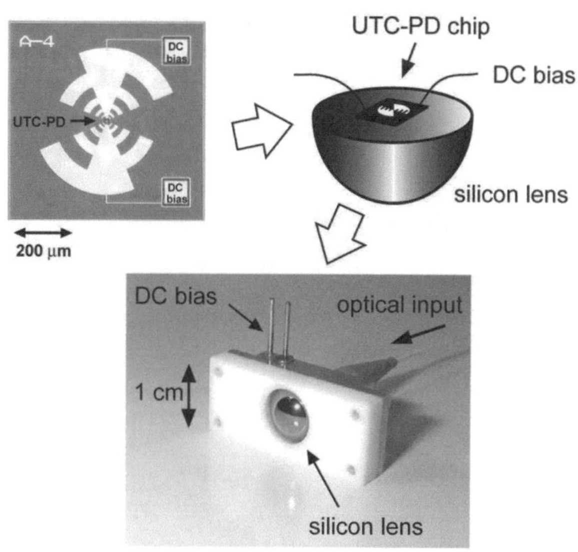

Fig. 8. Assembly of the quasi-optical module employing a UTC-PD integrated with a log-periodic antenna.

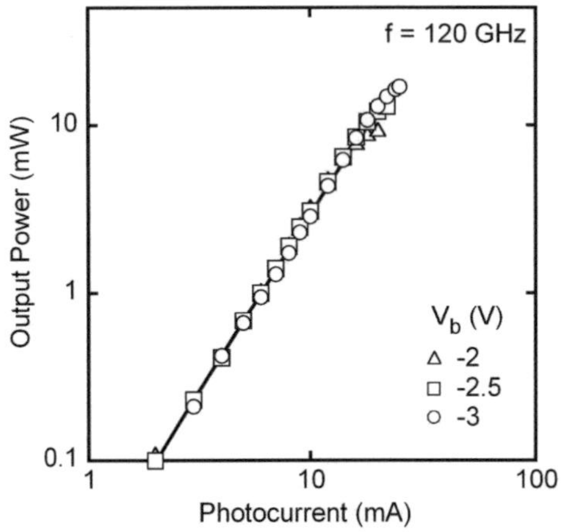

Fig. 9. Relationship between measured mm-wave output power and the diode photocurrent for an F-band UTC-PD module at 120 GHz.

2.4 Generation of millimeter and THz waves

Figure 9 shows the relationship between measured millimeter-wave output power and photocurrent for the F-band UTC-PD module at 120 GHz (Ito et al. 2003b). A wide linearity range is maintained up to a very high mm-wave output power of over 10 mW. The saturation point of the output power increased with increasing bias voltage, and the maximum output power of 17 mW was obtained at a bias voltage of -3 V with a photocurrent of 25 mA.

Figure 10 shows the output power against frequency for a J-band UTC-PD module at a photocurrent of 20 mA (Wakatsuki et al., 2008). The output 3-dB down bandwidth was about 120 GHz, and the 10-dB bandwidth was about 300 GHz. Here, most of the frequency variation is that of the integrated matching circuit. The steep decrease of the output power

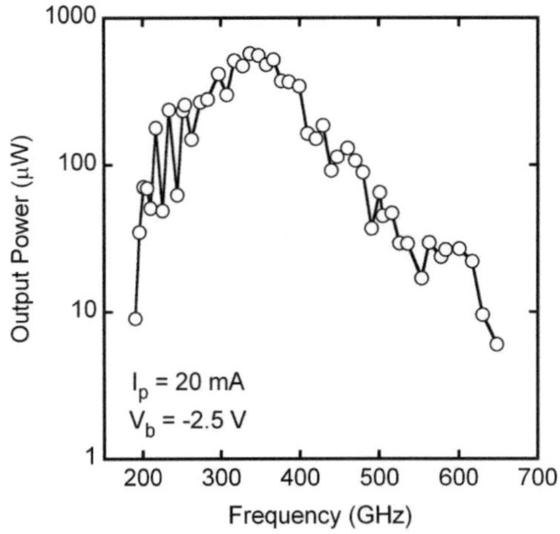

Fig. 10. Output powers against frequency for a J-band UTC-PD module designed for an over-mode operation.

at the low-frequency side is due to the cut-off characteristics of the WR-3 waveguide (at 173 GHz). In the high-frequency region, on the other hand, the output power decrease with increasing frequency is more gradual. This is because there is no cut-off frequency for the fundamental mode signal and the higher-order-mode output is also possible in the frequency range above 347 GHz.

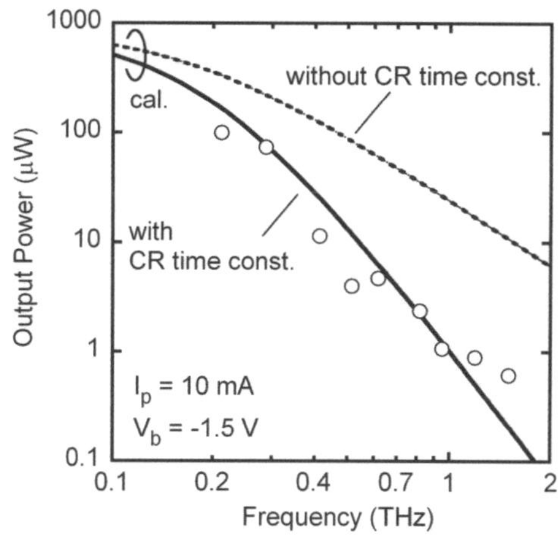

Fig. 11. Frequency dependence of the output power from a quasi-optic UTC-PD module.

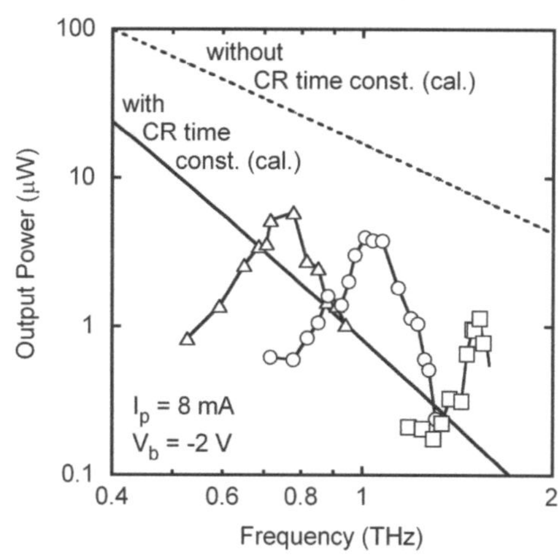

Fig. 12. Output powers against frequency for UTC-PDs integrated with resonant twin-dipole antennas.

As an example for the quasi optical approach, the UTC-PD was integrated with a wideband self-complementary log-periodic antenna (Ito et al., 2005b) as shown in Fig. 8. Figure 11 shows the frequency characteristics of the output power from the fabricated quasi-optical module at a photocurrent of 10 mA. The output power decreased gradually with increasing frequency, and the sub-mm waves were detected at frequencies of up to 1.5 THz. The solid curve in the figure is a calculation, which only takes into account the CR-limited and transit-time-limited bandwidth of the UTC-PD with a constant loss. The experimental result agrees well with the calculation, indicating that only the device parameters of the PD determine the basic frequency dependence. Thus, if we compensate for the influence of the CR time constant by employing a resonant matching circuit, it will be possible to increase the output power toward the level indicated by the broken curve in the figure, where only the transit-time-limited bandwidth is considered.

For this purpose, UTC-PDs integrating resonant twin-dipole planar antennas were also fabricated (Nakajima et al., 2004). Figure 12 shows the frequency characteristics of the fabricated devices for a photocurrent of 8 mA. The calculations for the wideband UTC-PD integrating a log-periodic antenna at the same photocurrent are also shown for comparison. The output power for each resonant device exhibits peaking behavior at different frequencies, namely, at 0.78, 1.04, and 1.53 THz. The peak output values are considerably higher than the output power from the wideband device for the same photocurrent, indicating the effectiveness of the resonant design. The peak output powers lie in between the two curves, one is with CR time constant and the other is without. This is because there is a tradeoff between the bandwidth and peak output power even in the resonant design so that the influence of the CR time constant is only partially compensated by the resonant circuit. Figure 13 shows the relationship between the detected output power and the photocurrent at 1.04 THz for the device having a peak at the same frequency with a bias voltage of -2 V. The inset is a micrograph of a UTC-PD with a resonant twin-dipole antenna. The output power increased linearly in proportion to the square of the photocurrent, and the maximum output power obtained was 10.9 μW at a photocurrent of 14 mA.

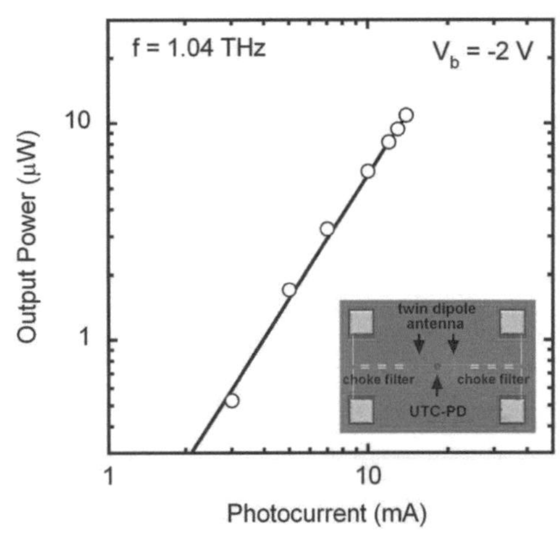

Fig. 13. Output powers against photocurrent for a UTC-PD integrated with a twin-dipole antenna having a resonant peak at 1.04 THz.

Figure 14 compares reported RF output powers for UTC-PDs (Ito et al., 2005b), pin-PDs (Huggard et al., 2002, Stöhr et at., 2003, Malcoci et al., 2004) and LT-GaAs photomixers (Duffy et al., 2001) with wideband designs. The results for narrowband UTC-PDs (Ito et al., 2003b, Nakajima et al., 2004, Wakatsuki et al., 2008, Rouvalis et al., 2010) are also shown for comparison. Although the output power for each type of device decreases inversely proportional to the fourth power of the frequency, the results for UTC-PDs are about two orders of magnitude higher than those for pin-PDs. This difference is mainly due to the high saturation current levels of the UTC-PD. The output power of the UTC-PD is even higher than those reported for LT-GaAs photomixers. These results clearly indicate that the UTC-PD is a promising device for generating a continuous terahertz wave with a high output power. The output powers of narrowband UTC-PDs are even higher than those with wideband design. However, they also tend to decrease with increasing frequency along the results for devices with wideband design. This indicates that, even for a device with a resonant design, it is essential to improve the intrinsic device performance for further increasing the output power of the UTC-PD.

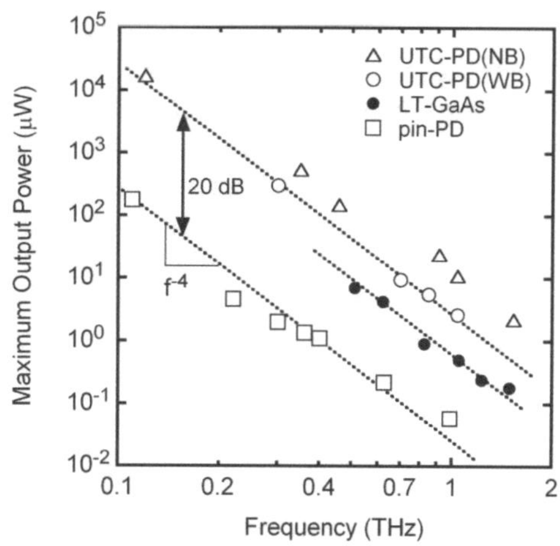

Fig. 14.Comparison of reported mm-/sub-mm-wave output powers against the operation frequency for UTC-PDs, pin-PDs, and LT-GaAs photomixers.

Application	Generator	Detector
Communications	**Photonic emitter using UTC-PD**	Schottky-barrier diode detector
Spectroscopy Active I	**Photonic emitter using UTC-PD**	Schottky-barrier diode detector
Active II	**Photonic emitter using UTC-PD**	SIS mixer with **photonic LO**
Active III	**Photonic emitter using UTC-PD**	**Photonic mixer using UTC-PD**
Passive	NA	SIS mixer with **photonic LO**
Imaging	**Photonic noise source using UTC-PD**	Schottky-barrier diode detector

Table 1. Summary of several applications using UTC-PDs.

3. Applications of the UTC-PD

Table 1 summarizes applications of photonically generated millimeter and terahertz waves in wireless communications, spectroscopic measurement systems and imaging systems. In wireless communications, UTC-PDs are used to generate millimeter and sub-millimeter waves in transmitters. In spectroscopy applications, UTC-PDs are employed as mixers in detectors as well as in generators. Photonically generated noise signals can be used as illuminators in the active imaging systems. In the following, these applications are described in more detail.

3.1 Wireless communications

Here, we describe wireless link systems using two bands; one is 120-GHz band and the other is 300-GHz band. One of common concerns when we use >100-GHz radio waves for wireless communications is a large propagation loss in the air. From 100 to 300 GHz, there are three valleys, where the attenuation is a local minimum; 75-100 GHz, 120-160 GHz, and 220-320 GHz. Our initial choice is the 120-GHz band centered at 125 GHz.

A block diagram of a 120-GHz-band wireless link system with 10-Gbit/s transmission capability is shown in Fig. 15 (Hirata et al. 2003, 2006, 2008, Suzuki et al., 2005). A photonic millimeter-wave generator is used in the transmitter. An optical millimeter-wave source generates optical subcarrier signals whose intensity is modulated at 125 GHz. An optical intensity (ASK) modulator (MZM) modulates the optical subcarrier signal using data signals. The modulated subcarrier signal is amplified by an optical amplifier (EDFA) and input to the high-power UTC-PD. The UTC-PD converts the optical signals into millimeter-wave signals, which are amplified and radiated toward the receiver via an antenna. The received millimeter-wave signals are amplified and demodulated by a simple envelope detection scheme, for example. The millimeter-wave receiver is composed of all-electronic devices such as pre-amplifier and Schottky-barrier diode (SBD) detector using InP-HEMT technology.

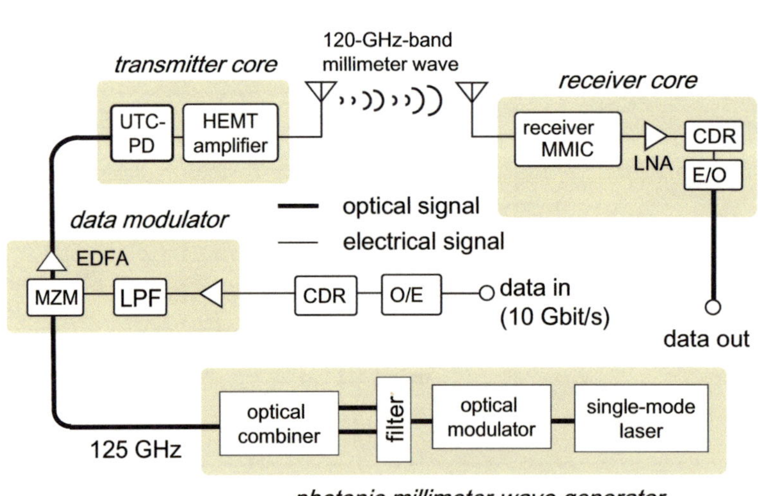

Fig. 15. Block diagram of 120-GHz band wireless link system using the UTC-PD in the transmitter.

The promising application of the above 10-Gbit/s wireless link is found in the broadcasting industry. A wireless link system that can transmit "uncompressed" high-definition television (HDTV) signals has been strongly desired, because TV program production based on the HDTV standard is spreading rapidly in TV stations due to the launch of digital TV broadcasting all over the world. An uncompressed HDTV signal (HD-SDI: high definition serial digital interface) requires a data rate of 1.5 Gbit/s per channel. Commercial wireless links using 60-GHz-band MMWs have a data rate of over 1.5 Gbit/s and thus a capability of transmitting one channel of uncompressed HD-SDI signals. However, large-scale live relay broadcasts, such as golf tournaments and music concerts, requires multiple channels of uncompressed HD-SDI signals. The 120-GHz-band system allows up to 6 channels of uncompressed HDTV material to be sent over a wireless link with no latency.
For such a purpose, this link uses a high-gain (~50 dBi) Cassegrain antenna, and can support the optical network standards of both 10 GbE (10.3 Gbit/s) and OC-192 (9.95 Gbit/s) with a bit error rate of 10^{-12}.
To increase the bit rate to 20-40 Gb/s or more, use of higher carrier frequencies of over 200 GHz is promising. In particular, the use of terahertz waves at frequencies above 275 GHz has attracted a great deal of interest for wireless communications. This is mainly because that these frequency spectra have not yet been allocated to specific applications and thus we can possibly make use of extreme bandwidth for high-speed communications.

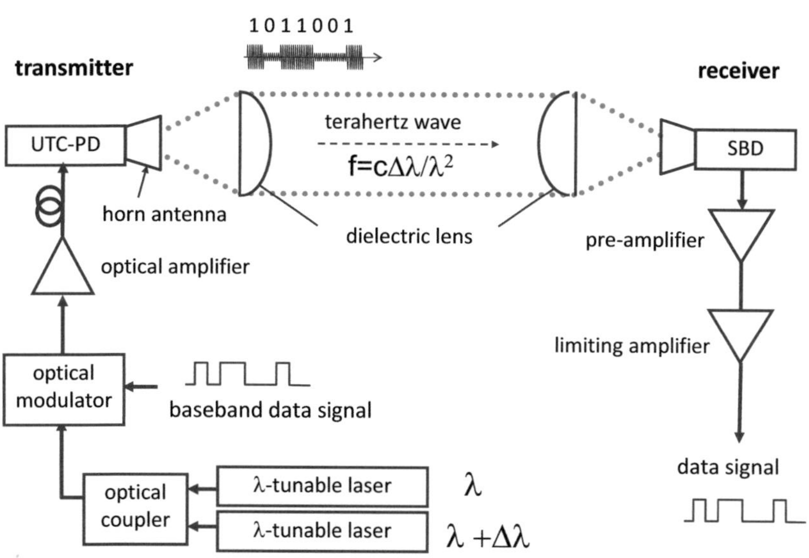

Fig. 16. Block diagram of 300-GHz band wireless link system using the UTC-PD in the transmitter.

Figure 16 shows a block diagram of 300-400 GHz band system using a photonics-based transmitter (Nagatsuma et al., 2009a, 2010a). This system is intended for use in short-distance (<10 m) applications, since there are no amplifiers. An optical RF signal is generated by heterodyning the two wavelengths of light from the wavelength-tunable light sources. The optical signal is digitally modulated by the optical intensity modulator driven by the pulse pattern generator (PPG). Finally, the optical signal is converted to an electrical signal by the UTC-PD. The terahertz wave is emitted to the free space via a horn antenna with a gain of 25 dBi, and it is well collimated by a 2-inch-diameter Teflon lens. The receiver consists of a Schottky barrier diode (SBD) and an IF filter followed by a low-noise pre-amplifier and a limiting amplifier.

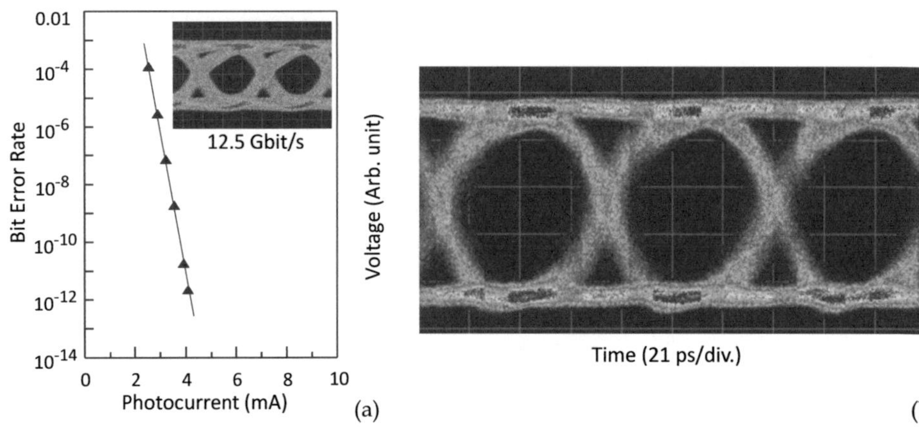

Fig. 17. (a) BER characteristics of 300-GHz wireless link. (b) 14-Gbit/s eye diagram.

Figure 17(a) shows BER characteristics at 12.5 Gbit/s with a carrier frequency of 300 GHz. Horizontal axis is a photocurrent of the transmitter. An error-free transmission at 12.5 Gbit/s has been achieved with 4-mA current, which corresponds to the transmitter output of around 10 μW. Currently, the upper limits in the bit rate of PPG and BER tester are 14 Gbit/s and 12.5 Gbit/s, respectively. Figure 17(b) shows the eye diagram at 14 Gbit/s. Although the BER could not be measured, an error-free transmission was confirmed from the clear eye opening. Based on the design, >20Gbit/s will be feasible.

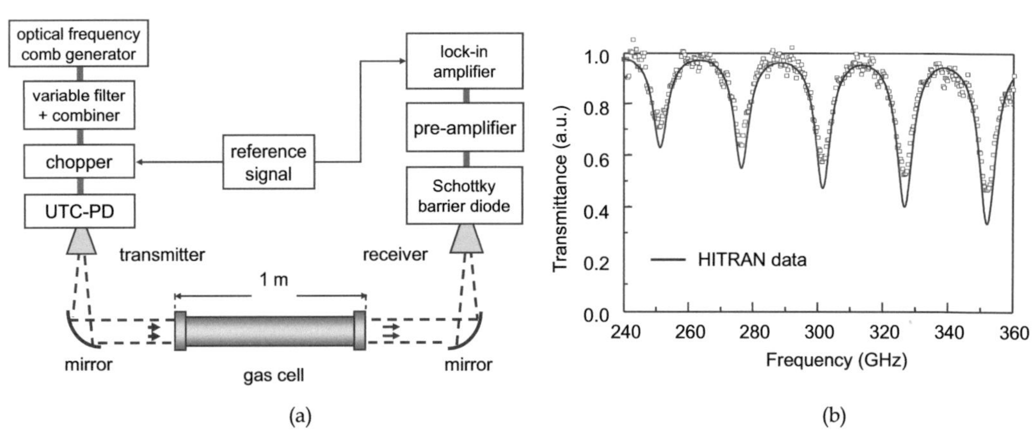

Fig. 18. Block diagram of the spectroscopy system using the UTC-PD and the Schottky barrier diode in the transmitter and receiver, respectively. (b) Experimental results.

3.2 Spectroscopy

In the following, we describe measurement results obtained by spectroscopy systems using photonic millimeter-/terahertz-wave emitters in combination with the different detection techniques. First, using the photonic millimeter-/terahertz-wave generator and the direct detection scheme, simple spectroscopic measurements were demonstrated in the frequency range between 240 and 360 GHz, as shown in Fig. 18 (Song et al., 2008a). The sample under the test was a mixture of N_2O and N_2 in the ratio of 3:1 (75 %), and it was filled in a 1-m long gas cell under atmospheric pressure. The experimental setup is shown in Fig. 18(a). The millimeter-/terahertz-wave signal generator was computer controlled to sweep the frequency and the optical millimeter-/terahertz-wave signal before the UTC-PD was intensity modulated at a frequency of 10 kHz. Then the generated signal was radiated and collimated with a diagonal horn antenna and a gold-coated off-axis parabolic mirror, respectively. The signal transmitted through the gas cell was received with a Schottky barrier diode detector followed by a lock-in amplifier tuned to 10 kHz.

The measured transmittance for the gas is plotted in Fig. 18(b) with the simulated results based on the HITRAN (**hi**gh-resolution **tran**smission molecular absorption) database (HITRAN homepage). As can be seen, the positions and tendency of the magnitude and the shape of absorption peaks from the measurement coincide well with those of HITRAN. The measurement bandwidth can be extended by using a photomixer and a receiver, which are integrated with broadband antennas (Hirata et al., 2002).

The ultralow-noise characteristics of the photonically generated millimeter-/terahertz-wave signal were verified through their application to the local oscillator (LO) for the superconducting mixers in heterodyne receivers, which operate at 4.2 K. Figure 19(a) shows an experimental setup of the cryogenic receiver pumped by the photonic LO signal (Kohjiro et al., 2008). Measured noise temperatures of the receiver with a photonic LO and a conventional Gunn LO are plotted in the frequency range from 180 GHz to 500 GHz. Three Gunn oscillators were required to cover this frequency range even in part, while it was fully covered by the single photonic LO. As shown in Fig. 19(b), the noise temperature was as low as 300 K, which corresponds to better than pico-watt sensitivity over an octave frequency range.

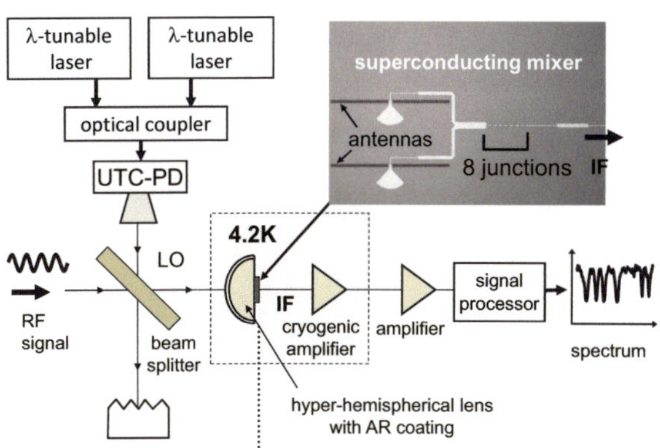

(a)

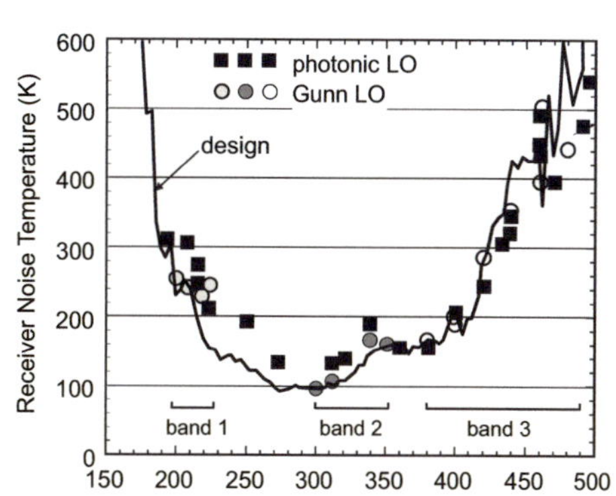

(b)

Fig. 19. Experimental setup of the cryogenic receiver pumped by photonic local oscillator. (b) Frequency characteristic of the noise temperature.

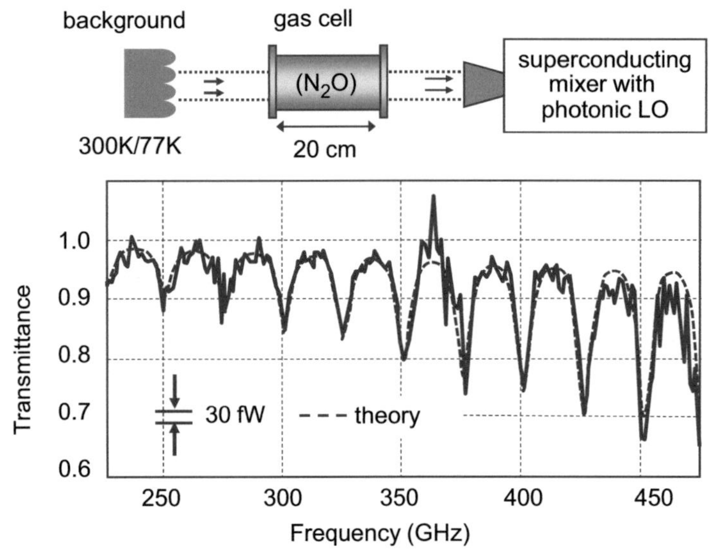

Fig. 20. Passive spectroscopic measurement of gas.

This receiver was successfully applied to the passive spectroscopy of gasses, as shown in Fig. 20. By using black-body radiation sources at 77 K and 300 K, the absorption can be calibrated accurately. It takes 1 - 2 minutes to scan the frequency range from 200 GHz to 450 GHz.

By combining the above ultra-sensitive SIS receiver pumped by the photonic LO with the photonic millimeter-wave sweeper, stand-off detection of toxic and/or dangerous gasses such as CO, CO_2, HCN, HCl, SO_x, NO_x, was successfully performed (Oh et al., 2009).

In addition to the generator, the UTC-PD can be used as a detector. Figure 21 shows the operation principle of the photodiodes as terahertz-wave detectors. There are two operation modes with different voltage-bias conditions; one is a square-law detector under the forward bias condition, and the other is a down-converter under the reverse bias. In case of down-conversion (Fig. 21(b)), the origin of the nonlinearity of the UTC-PD can be explained by the dynamic capacitance associated with charge storage in the photo-absorption layer (Fushimi et al., 2004). Mixing between the input terahertz wave, f_{RF}, and the local oscillator (LO) signal, f_{LO}, photonically generated in the photodiode leads to the intermediate frequency, f_{IF}.

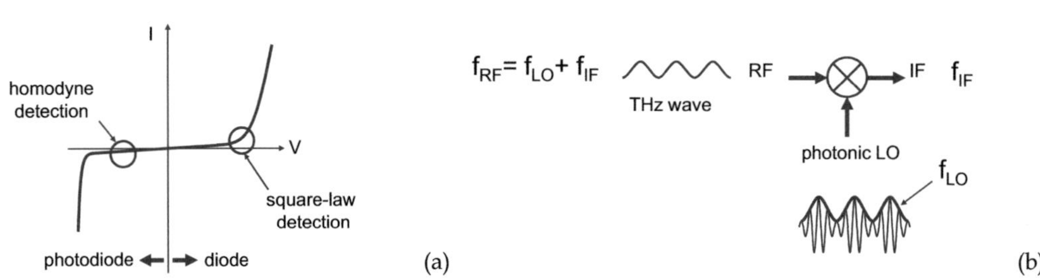

Fig. 21. Operation principle of the UTC-PD for receivers as a down-converter and a square-law detector.

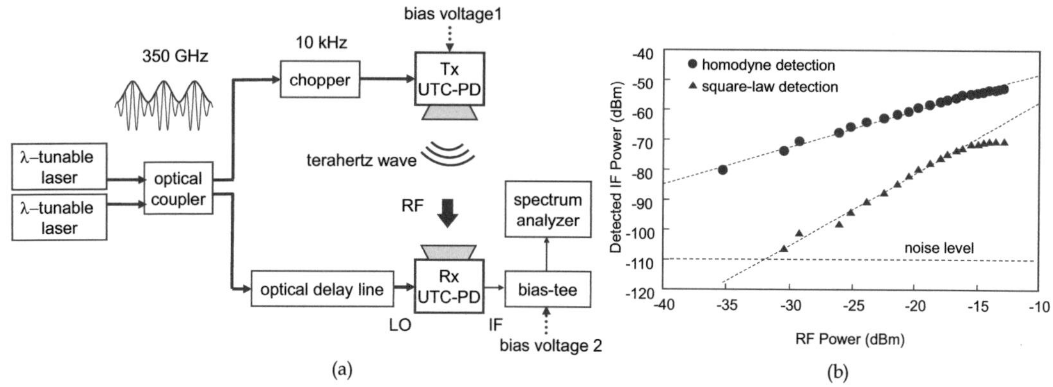

Fig. 22. (a) Block diagram of the spectroscopy system using the UTC-PDs both in the transmitter and receiver. (b) Detection characteristics of the UTC-PD as a down-converter and a square-law detector.

Figure 22(a) shows a block diagram of the CW THz spectrometer consisting of the transmitter (Tx) and the receiver (Rx) based on the down-conversion (Nagatsuma et al., 2010b). This coherent system is often referred to as "homodyne" detection system (Ducournau et al., 2009). Optical delay line was used to maximize the intermediate frequency signal at 10 kHz. Dependence of the IF power on the input RF power at 350 GHz was compared between the homodyne detection and the square-law detection as shown in Fig. 22(b). For the homodyne detection, the optical LO power was set to -15.5 dBm, which is an optimum condition experimentally confirmed. Maximum S/N (ratio of IF power to noise level) is 39 dB for the square-law detection, while it is 58 dB for the homodyne detection. This corresponds to the difference in the maximum conversion efficiency is 19 dB. Since the available output power from the UTC-PD is more than -4 dBm, about 20-dB loss in the transmission between transmitter and receiver and as well as in the object under test is still allowable for the actual spectroscopy. In addition, since the slope in the relationship between the RF power and the IF power is smaller for the homodyne detection, loss caused in the object and transmission has less effect, which may be a merit of the homodyne detection scheme.

3.3 Imaging

Millimeter-wave and terahertz-wave imaging has been extensively studied and deployed in numerous areas such as material inspection, non-destructive testing, and security applications (Chan et al., 2007). Between continuous-wave (CW) and pulsed-wave imaging systems, the CW imaging can be faster, more compact, cost-effective and simpler to operate. However, mono-chromatic waves cause interference effect due to reflection of waves at any boundary in the materials under inspection and components used such as lenses and mirrors, which leads to significant degradation in image quality and effective spatial resolution (Redo-Sanchez et al., 2006).

Here, we present a new approach to achieving high-quality image with use of photonically-generated incoherent millimeter-wave noise signals (Nagatsuma et al., 2009b). Figure 23(a) shows a block diagram of transmission-type millimeter-wave imaging system using incoherent illumination source. The amplified spontaneous emission (ASE) noise from Er-doped fiber amplifier is delivered to the UTC-PD. The UTC-PD converts the optical noise to the electrical noise (Fig. 23(b)). Then, the noise is radiated into free space via a horn antenna.

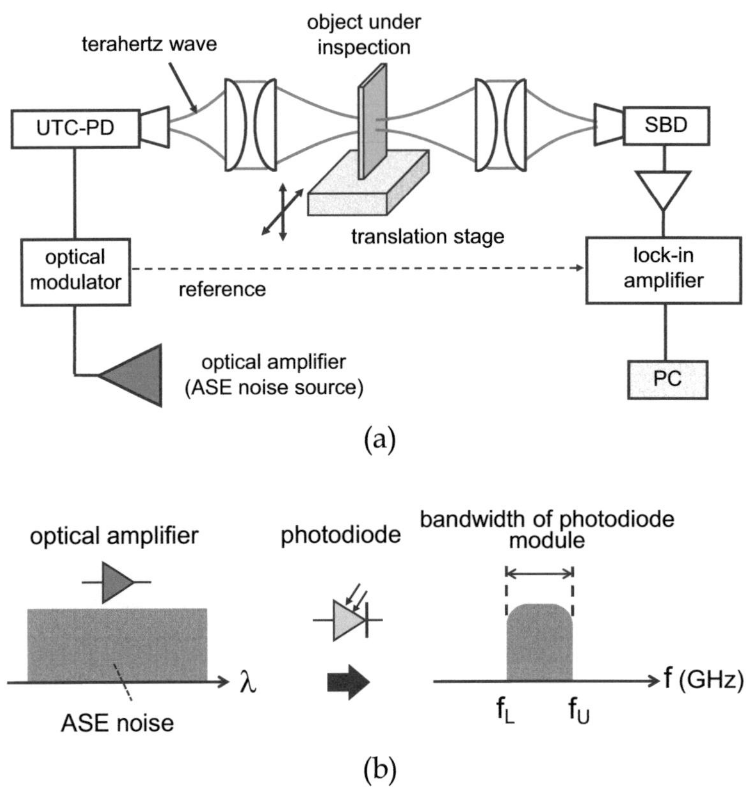

Fig. 23. (a) Block diagram of the imaging system using an incoherent noise source.
(b) Mechanism of the noise generation using the photodiode.

In order to verify the effectiveness of the proposed technique, the imaging of a clip hidden in the paper envelope was conducted as shown in Fig. 24(a). For comparison, the millimeter-wave images obtained by a monochromatic 320-GHz signal are also shown in Figs. 24(b). The 320-GHz signal was generated by injecting two optical signals with a frequency difference of 320 GHz from two wavelength-tunable lasers into the UTC-PD. The linewidth of the generated millimeter-wave signal is about 10 MHz. Even in the case of the thin envelope, interference effect caused artifact near the clip. As clearly shown in Fig. 24, the interference effect has been dramatically reduced by the incoherent source without scarifying the spatial resolution.

Song et al. also demonstrated the effectiveness of noise in the spectroscopic measurement (Song et al., 2008b).

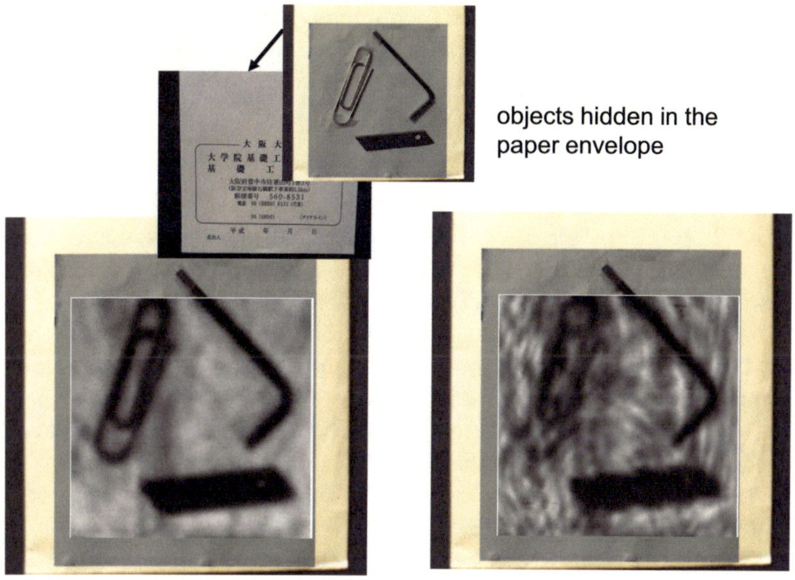

(a) (b)

Fig. 24. Comparison of images obtained by noise source (a) and coherent source at 320 GHz (b).

4. Conclusions

The combination of radio-wave and photonics technologies is an effective way to explore electromagnetic waves in the undeveloped millimeter-/terahertz-wave frequency regions. In this article, we have described photonic generation of continuous millimeter-/terahertz-wave signals using UTC-PDs and their applications to communications, spectroscopic measurement and imaging. Currently, the maximum available output power at 1 THz is around 10 μW. Improvement of photodiode performance in terms of saturation current and thermal management as well as efficient RF coupling between the photodiode and the antenna front-end will allow a single device to emit powers on a level of 100 μW at 1 THz. The power combining technique using an array of photodiodes promises to increase the output power up to the 1-mW level. Thus, photodiode-based CW millimeter-/terahertz-wave technologies are expected to be at the forefront of a new generation of more compact, versatile and cost-effective millimeter-and terahertz-wave systems.

5. Acknowledgments

The authors wish to thank Dr. T. Ishibashi from NTT Electronics, Drs. Y. Kado, N. Kukutsu, A. Hirata, R. Yamaguchi, H. Takahashi, N. Shimizu, H.-J. Song, K. Ajito, T. Furuta, A. Wakatsuki, F. Nakajima, Y. Muramoto, K. Yoshino, T. Yoshimatsu, S. Kodama, T. Kosugi, K. Murata, N. Shigekawa, and K. Iwatsuki all from NTT Laboratories, and Dr. H. Yamamoto from Kitasato University for their contribution and support. Part of this work was supported by the "Research and Development Project for the Expansion of Radio Spectrum Resources" made available by the Ministry of Internal Affairs and Communications, Japan, the National Institute of Communications Technology, Japan, and the Ministry of Education, Science, Sports and Culture, Grant-in-Aid for Scientific Research (A) 20246062, 2008, and (C) 21560360, 2009.

6. References

Achouche, M.; Magnin, V.; Harari, J.; Lelarge, F.; Derouin, E.; Jany, C.; Carpentier, D.; Blache, F. & Decoster, D. (2004). High performance evanescent edge coupled waveguide unitraveling-carrier photodiodes for >40-Gb/s optical receivers, *IEEE Photonics Technol. Lett.*, Vol. 16, No. 2, pp. 584-586.

Bach, H.-G. (2007). Ultrafast waveguide-integrated pin-photodiodes and photonic mixers from GHz to THz range, *Proc. 33rd European Conf. and Exhibition on Optical Communication*, Vol. 5, pp. 247-250.

Bowers, J. E. & Burrus, Jr. C. A. (1987). Ultrawide-band long-wavelength p-i-n photodetectors, *IEEE J. Lightwave Tech.*, Vol. 5, No. 10, pp. 1339-1350.

Chan, W. L.; Deibel, J. & Mittleman, D. M. (2007). Imaging with terahertz radiation", *Rep. Prog. Phys.*, Vol. 70, pp. 1325-1379.

Ducournau, G., Beck, A., Blary, K., Peytavit, E., Zaknoune, M., Akalin, T., Lampin, J.-F., Martin, M., & Mangeney, J. (2009). All-Fiber continuous wave coherent homodyne terahertz spectrometer operating at 1.55 μm wavelengths, *Proc. 2009 Infrared, Millimeter and Terahertz Waves (IRMMW-THz 2009)*, T4A02.0362.

Duffy, S. M.; Verghese, S.; McIntosh, K. A.; Jackson, A.; Gossard, A. C. & Matsuura, S. (2001). Accurate modeling of dual dipole and slot elements used with photomixers for coherent terahertz output power," *IEEE Trans. Microwave Theory and Tech.*, Vol. 49, pp. 1032-1038.

Fukano, H.; Muramoto, Y.; Takahata, K. & Matsuoka, Y. (1999) High-efficiency edge-illuminated uni-traveling-carrier-structure refracting-facet photodiode, *Electron. Lett.*, Vol. 35, pp. 1664-1665.

Furuta, F.; Ito, H. & Ishibashi, T. (2000). Photocurrent dynamics of uni-traveling-carrier and conventional pin-photodiodes, *Inst. Phys. Conf. Ser.*, No. 166, pp. 419-422.

Furuta, T.; Ito, T.; Muramoto, Y.; Ito, H.; Tokumitsu, M. & Ishibashi, T. (2005). "D-band rectangular-waveguide-output uni-traveling-carrier photodiode modules, *Electron. Lett.*, Vol. 41, pp. 715-716.

Fushimi, H.; Furuta, T.; Ishibashi, T. & Ito, H. (2004). Photoresponse nonlinearity of a uni-traveling-carrier photodiode and its application to optoelectronic millimeter-wave mixing in 60 GHz band, *Jap. J. Appl. Phys.*, vol. 43, no. 7B, pp. L966-968.

Hafizi, H.; Liu, T.; Rensch, D. B. & Stanchina, W. E. (1993). Effects of collector doping on DC and RF performance of AlInAs/GaInAs/InP double heterojunction bipolar transistors, *IEEE Trans. Electron Dev.*, Vol. 40, No. 11., pp. 2122-2123.

Harmon, E. S.; Lovejoy, M. L.; Melloch, M. R.; Lundstrom, M. S.; Ritter, D. & Hamm, R. A. (1993). Minority-carrier mobility enhancement in p+ InGaAs lattice matched to InP, *Appl. Phys. Lett.*, Vol. 63, pp. 636-638.

Hill, P.; Schlafer, J.; Powazinik, W.; Urban, M.; Eichen, E. & Olshansky, R. (1987). Measurement of hole velocity in n-type InGaAs, *Appl. Phys. Lett.*, Vol. 50, pp. 1260-1262.

Hirata, A.; Nagatsuma, T.; Yano, R.; Ito, H.; Furuta, T.; Hirota, Y.; Ishibashi, T; Matsuo, H.; Ueda, A.; Noguchi, T.; Sekimoto, Y. ; Ishiguro, M. & Matsuura, S. (2002). Output power measurement of photonic millimeter-wave and sub-millimeter-wave emitter at 100-800 GHz, *Electron. Lett.*, Vol. 38, No. 15, pp. 989-800.

Hirata, A.; Harada, M. & Nagatsuma, T. (2003). 120-GHz wireless link using photonic techniques for generation, modulation, and emission of millimeter-wave signals," *IEEE J. Lightwave Technology*, Vol. 21, No. 10, pp. 2145-2153.

Hirata, A.; Kosugi, T.; Takahashi, H.; Yamaguchi, R.; Nakajima, F.; Furuta, T.; Ito, H.; Sugahara, H.; Sato, Y. & Nagatsuma, T. (2006). 120-GHz-band millimeter-wave photonic wireless link for 10-Gb/s data transmission," *IEEE Trans. Microwave Theory Tech.*, vol. 54, no. 5, pp. 1937-1944.

Hirata, A.; Takahashi, H.; Yamaguchi, R.; Kosugi, T.; Murata, K.; Nagatsuma, T.; Kukutsu, N. & Kado, Y. (2008). Transmission characteristics of 120-GHz-band wireless link using radio-on-fiber technologies, *J. Lightwave Technol.*, Vol. 26, No. 15, pp. 2338-2344.

Hirota, Y.; Ishibashi, T. & Ito, H. (2001). 1.55-μm wavelength periodic traveling-wave photodetector fabricated using unitraveling-carrier photodiode structures, *J. Lightwave Technol.*, Vol. 19, No. 11, pp. 1751-1758.

HITRAN homepage, http://cfa-www.harvard.edu/hitran/welcometop.html

Huggard, P. G.; Ellison, B. N.; Shen, P.; Gomes, N. J.; Davis, P. A.; Shillue, W. P.; Vaccari, A. & Payne, J. M. (2002). Generation of millimetre and sub-millimetre waves by photomixing in 1.55 μm wavelength photodiode, *Electron. Lett.*, Vol. 38, pp. 327-328.

Ishibashi T. & Yamauchi, Y. (1988). A possible near-ballistic collection in an AlGaAs/GaAs HBT with a modified collector structure, *IEEE Trans. Electron Dev.*, Vol. 35, No. 4, pp. 401-404.

Ishibashi, T. (1994). High speed heterostructure devices, In: *Semiconductors and Semimetals*, Vol. 41, Chap. 5, p. 333, Academic Press, San Diego.

Ishibashi, T.; Shimizu, N.; Kodama, S.; Ito, H.; Nagatsuma, T. & Furuta, T. (1997). Uni-traveling-carrier photodiodes, *Tech. Dig. Ultrafast Electronics and Optoelectronics (1997 OSA Spring Topical Meeting)*, pp. 166-168, Incline Village, Nevada.

Ishibashi, T.; Furuta, T.; Fushimi, H.; Kodama, S.; Ito, H.; Nagatsuma, T.; Shimizu, N. & Miyamoto, Y. (2000). InP/InGaAs uni-traveling-carrier photodiodes, *IEICE Trans. Electron.*, Vol. E83-C, No. 6, pp. 938-949.

Ito, H.; Furuta, T.; Kodama S. & Ishibashi, T. (2000a). Zero-bias high-speed and high-output-voltage operation of cascade-twin uni-travelling-carrier photodiode, *Electron. Lett.*, Vol. 36, pp. 2034-2036.

Ito, H.; Furuta, T.; Kodama, S. & Ishibashi, T. (2000b). InP/InGaAs uni-travelling-carrier photodiode with a 310 GHz bandwidth, *Electron. Lett.*, Vol. 36, pp. 1809-1810.

Ito, H.; Furuta, T.; Kodama, S. & Ishibashi, T. (2000c). High-efficiency unitraveling-carrier photodiode with an integrated total-reflection mirror, *IEEE J. Lightwave Technol.*, Vol. 18, No. 3, pp. 384-387.

Ito, H.; Furuta, T. ; Ito, T.; Muramoto, Y.; Tsuzuki, K.; Yoshino, K. & Ishibashi, T. (2002). W-band uni-travelling-carrier photodiode module for high-power photonic millimetre-wave generation, *Electron. Lett.*, Vol. 38, pp. 1376-1377.

Ito, H.; Nagatsuma, T.; Hirata, A.; Minotani, T.; Sasaki, A.; Hirota, Y. & Ishibashi, T. (2003a). High-power photonic millimetre-wave generation at 100 GHz using matching-circuit-integrated uni-travelling-carrier photodiodes, *IEE Proc. Optoelectron.*, Vol. 150, No. 2, pp. 138-142.

Ito, H.; Ito, T.; Muramoto, Y.; Furuta, T. & Ishibashi, T. (2003b). Rectangular waveguide output uni-traveling-carrier photodiode module for high-power photonic millimeter-wave generation in the F-band, *IEEE J. Lightwave Tech.*, Vol. 21, pp. 3456-3462.

Ito, H.; Kodama, S.; Muramoto, Y.; Furuta, T.; Nagatsuma, T. & Ishibashi, T. (2004a). High-speed and high-output InP/InGaAs uni-traveling-carrier photodiodes, *IEEE J. Selected Topics in Quantum Electronics*, Vol. 10, No. 4, pp. 709-727.

Ito, H.; Furuta, T.; Nakajima, F.; Ito, T.; Muramoto, Y. & Ishibashi, T. (2004b). Photonic generation of millimetre- and submillimetre-waves using rectangular-waveguide-output uni-travelling-carrier photodiode module, *Electron. Lett.*, Vol. 40, pp. 387-388.

Ito, H.; Furuta, T.; Kosugi, T.; Takahashi, H.; Hirata, A.; Muramoto, Y.; Tokumitsu, M.; Sato, Y.; Nagatsuma, T. & Ishibashi, T. (2005a). Over-10-dBm output uni-traveling-carrier photodiode module integrating a power amplifier for wireless transmission in the 125-GHz band," *IEICE Electron. Express*, Vol. 2, No. 16, pp. 446-450.

Ito, H.; Furuta, T.; Nakajima, F.; Yoshino, K. & Ishibashi, T. (2005b). Photonic generation of continuous THz wave using uni-traveling-carrier photodiode, *IEEE J. Lightwave Tech.*, Vol. 23, pp. 4016-4021.

Ito, H.; Furuta, T.; Muramoto, Y.; Ito, T. & Ishibashi, T. (2006). Photonic millimetre- and sub-millimetre-wave generation using the J-band rectangular-waveguide-output uni-travelling-carrier photodiode module, *Electron. Lett.*, Vol. 42, pp. 1424-1425.

Ito H. & Kodama, S. (2008). Uni-traveling-carrier photodiode (UTC-PD) and PD-EAM optical gate integrating a UTC-PD and a traveling wave electro-absorption modulator, In: *Ultrafast All-Optical Signal Processing Devices*, H. Ishikawa Ed., pp. 89-153, Wiley, West Sussex.

Ito, H.; Yoshino, K.; Muramoto, Y.; Yamamoto, H. & Ishibashi, T. (2010). Sub-terahertz-wave transceiver module integrating uni-traveling-carrier photodiode, Schottky barrier diode, and planar circulator circuit, *Proc. Int. Conf. on Infrared, Millimeter, and Terahertz Waves*, Tu-P.55.

Jun, D.-H.; Jang, J.-H.; Adesida, I. & Song, J.-I. (2006). Improved efficiency-bandwidth product of modified uni-traveling carrier photodiode structures using an undoped photo-absorption layer, *Jpn. J. Appl. Phys.*, vol. 45, No. 4B, pp. 3475-3478.

Kato, K.; Hata, S.; Kawano, K. & Kozen, A. (1993). Design of ultrawide-band, high-sensitivity p-i-n photodetectors, *IEICE Trans. Electron.*, Vol. E76-C, No. 2, pp. 214-221.

Kohjiro, S.; Kikuchi, K.; Maezawa, M.; Furuta, T.; Wakatsuki, A.; Ito, H.; Shimizu, N.; Nagatsuma, T. & Kado, Y. (2008). A 0.2-0.5 THz single-band heterodyne receiver based on a photonic local oscillator and a superconductor-insulator-superconductor mixer," *Appl. Phys. Lett.*, Vol. 93, 093508.

Kroemer, H. (1982). Heterostructure bipolar transistors and integrated circuits, *Proc. IEEE*, Vol. 70, No. 1, pp. 13-25.

Li, X.; Li, N.; Zheng, X.; Demiguel, S.; Campbell, J. C.; Tulchinsky, D. A. & Williams, K. J. (2003). High-saturation-current InP-InGaAs photodiode with partially depleted absorber, *IEEE Photonics Technol. Lett.*, Vol. 15, No.9, pp. 1276-1278.

Li, N.; Li, X.; Demiguel, S.; Zheng, X.; Campbell, J. C.; Tulchinsky, D. A.; Williams, K. J.; Isshiki, T. D.; Kinsey, G. S. & Sudharsansan, R. (2004). High-saturation-current charge-compensated InGaAs-InP uni-traveling-carrier photodiode, *IEEE Photonics Technol. Lett.*, Vol. 16, No. 3, pp. 864-866.

Malcoci, A.; Stöhr, A.; Sauerwald, A.; Schulz, S. & Jäger, D. (2004). Waveguide and antenna coupled traveling-wave 1.55-μm photodetectors for optical (sub)millimeter-wave generation, *Proc. SPIE*, No. 5466, pp. 202-209.

Muramoto, Y.; Kato, K.; Mitsuhara, M.; Nakajima, O.; Matsuoka, Y.; Shimizu, N. & Ishibashi, T. (1998). High-output voltage, high speed, high efficiency uni-traveling-carrier waveguide photodiode," *Electron. Lett.*, Vol. 34, pp. 122-123.

Muramoto Y. & Ishibashi, T. (2003). InP/InGaAs pin photodiode structure maximizing bandwidth and efficiency, *Electron. Lett.*, Vol. 39, pp. 1749-1750.

Nagatsuma, T.; Song, H.-J.; Fujimoto, Y.; Miyake, K.; Hirata, A.; Ajito, K.; Wakatsuki, A.; Furuta, T.; Kukutsu, N. & Kado, Y. (2009a). Giga-bit wireless link using 300-400 GHz bands, *Tech. Dig. IEEE International Topical Meeting on Microwave Photonics*, Th.2.3.

Nagatsuma, T. ; Kumashiro, T.; Fujimoto, Y.; Taniguchi, K.; Ajito, K.; Kukutsu, N.; Furuta, T.; Wakatsuki, A & Kado, Y. (2009b). Millimeter-wave imaging using photonics-based noise source, *Proc. Int. Conf. on Infrared, Millimeter, and Terahertz Waves*, M3C05.0088 .

Nagatsuma, T.; Song H.-J. & Kado, Y. (2010a). Challenges for ultrahigh-speed wireless communications using terahertz waves," *Terahertz Science and Technology*, Vol.3, No.2, pp. 55-65.

Nagatsuma, T.; Kaino, A.; Hisatake, S.; Ajito, K.; Song, H.-J.; Wakatsuki, A.; Muramoto, Y.; Kukutsu, N. & Kado, Y. (2010b). Continuous-wave terahertz spectroscopy system based on photodiodes, *PIERS Online*, Vol. 6, No. 4, pp: 390-394.

Nakajima, F.; Furuta, T. & Ito, H. (2004). High-power continuous-terahertz-wave generation using a resonant-antenna-integrated uni-traveling-carrier photodiode," *Electron. Lett.*, Vol. 40, pp. 1297-1299.

Oh, K.-H.; Shimizu, N.; Kohjiro, S.; Kikuchi, K.; Wakatsuki, A.; Kukutsu, N. & Kado, Y. (2009). High-sweeping-speed optically synchronized dual-channel terahertz-signal generator for driving a superconducting tunneling mixer and its application to active gas sensing, *OPTICS EXPRESS*, Vol. 17, No. 21, pp. 18455-18461.

Redo-Sanchez, A.; Karpowicz, N.; Xu, J., & Zhang X.-C. (2006). Damage and defect inspection with terahertz waves", *Proc. 4th Intern. Workshop on Ultrasonic and Advanced Methods for Nondestructive Testing and Material Characterization*, pp. 67-77.

Rouvalis, E.; Renaud, C. C.; Moodie, D. G.; Robertson, M. J. & Seeds, A. J. (2010). Traveling-wave uni-traveling carrier photodiodes for continuous wave THz generation, *Optics Express*, Vol. 18, pp. 11105-11110.

Shi, J.-W.; Wu, Y.-S.; Wu, C.-Y.; Chiu, P.-H. & Hong, C.-C. (2005). High-speed, high-responsivity, and high-power performance of near-ballistic uni-traveling-carrier photodiode at 1.55-μm wavelength, *IEEE Photonics Technol. Lett.*, Vol. 17, No. 9, pp. 1929-1931.

Shimizu, N.; Watanabe, N.; Furuta, T. & Ishibashi, T. (1998). InP-InGaAs uni-traveling-carrier photodiode with improved 3-dB bandwidth of over 150 GHz, *IEEE Photonics Technol. Lett.*, Vol. 10, No. 3, pp. 412-414.

Song, H.-J.; Shimizu, N.; Furuta, T.; Suizu, K.; Ito, H. & Nagatsuma, T. (2008a). Broadband-frequency-tunable sub-terahertz wave generation using an optical comb signal,

AWGs, optical switches, and uni-travelling carrier photodiode for spectroscopic applications, *IEEE J. Lightwave Technol.*, Vol. 26, No. 15, pp. 2521-2530.

Song, H.-J.; Shimizu, N.; Furuta, T.; Wakatsuki, A. & Nagatsuma, T. (2008b). Subterahertz noise signal generation using a photodetector and wavelength-sliced optical noise signals for spectroscopic measurement, *Appl. Phys. Lett.*, Vol. 93, 241113.

Stöhr, A.; Malcoci, A.; Sauerwald, A.; Mayorga, I. C.; Güsten, R. & Jäger, D. S. (2003). Ultra-wide-band traveling-wave photodetectors for photonic local oscillators, *IEEE J. Lightwave Tech.*, Vol. 21, pp. 3062-3070.

Suzuki, H.; Fujiwara, M.; Iwatsuki, K.; Hirata, A. & Nagatsuma, T. (2005). Photonic millimetre-wave generator using intensity and phase modulators for 10 Gbit/s wireless link, *Electron. Lett.*, Vol. 41, No. 6, pp. 355-356.

Tebbenham R. L. & Walsh, D. (1975). Velocity/field characteristics of n-type indium phosphide at 110 and 330 K, *Electron. Lett.*, Vol. 11, pp. 96-97.

Wakatsuki, A.; Furuta, T.; Muramoto, Y.; Yoshimatsu, T. & Ito, H. (2008). High-power and broadband sub-terahertz wave generation using a J-band photomixer module with rectangular-waveguide output port, *Proc. Int. Conf. on Infrared, Millimeter, and Terahertz Waves*, pp. 1999-1-1999-2.

Williams, K. J.; Esman, R. D. & Dagenais, M. (1994). Effects of high space-charge fields on the response of microwave photodetectors, *IEEE Photonics Technol. Lett.*, Vol. 6, No. 5, pp. 639-641.

15

n-Type β-$FeSi_2$/p-type Si Near-infrared Photodiodes Prepared by Facing-targets Direct-current Sputtering

Mahmoud Shaban[1] and Tsuyoshi Yoshitake[2]
[1]Sauth Valley University,
[2]Kyushu University,
[1]Egypt
[2]Japan

1. Introduction

Light detection in the near-infrared (NIR) region is an important issue in long distance optical fiber communication systems, local area networks, and optical interconnections in high speed computers. In recent years, the development of devices that are compatible with silicon electronic-photonic integrated circuits and can be operated at the telecommunication wavelengths of 1.3 and 1.55 μm has been the subject of intense research. The majority of optical fiber transmission systems operate at these wavelengths in order to take advantage of the low loss and low dispersion characteristics of silica-based optical fibers at these wavelength windows. Therefore, extraordinary technological innovations are increasingly adopted within the telecommunication industry as means of developing optoelectronic devices which operate at these wavelengths and are compatible with the current silicon technology (Kang et al, 1984; Luray et al, 2003; Masini et al, 2008). The key requirements for future integrated optoelectronics are light sources and light detectors that operate at the desired wavelength range and are compatible with the current Si technology.
The iron disilicide phase, β-$FeSi_2$, that can be grown on Si with 2%–5% lattice mismatch, is one of the best candidates for NIR photo-detection owing to several merits: (i) a large absorption coefficient, which is at least two orders of magnitude greater than that of crystalline silicon at 1.5 eV; (ii) a direct band gap of 0.85 eV; (iii) compatibility with Si technology; (iv) abundance of its elements (Fe and Si) in earth's crust; and (v) ecologically friendly material due to its non-toxicity. Therefore, β-$FeSi_2$ is a prime candidate material for optoelectronic and photovoltaic applications.
Various deposition methods such as; ion beam synthesis (Leong et al, 1996), reactive deposition epitaxy (Suemasu et al, 2000), and molecular beam epitaxy (Mahan et al, 1990) have been adapted for the growth of β-$FeSi_2$. However, thermal treatments such as annealing at temperatures greater than 800 °C are generally carried out after film deposition. This causes an easy condensation of β-$FeSi_2$ due to its high surface energy as compared to that of Si, which leads to a discontinuous surface structure. In addition, Fe atoms diffuse into Si. Although the solid solubility of Fe in Si is known to be low, for example, 2×10^{15} cm^{-3} at 900 °C, the diffusion coefficient is extremely large. This prevents the construction of β-

$FeSi_2$/Si heterojunctions with sharp interfaces. Fe atoms migrated into a depletion layer within Si, thereby producing a leak current. To overcome this problem, the template method has been suggested in which a thin template layer is grown at a low temperature before the general film fabrication procedure (Liu et al, 2004; Suemasu et al, 1999). We suggest another method in which thin films are epitaxially as-grown at a substrate temperature of 600 °C without annealing. Both the diffusion coefficient and the solid solubility of Fe in Si at 600 °C are a fifth to a tenth of those at 800 °C (Kendall & Devries, 1969).
Sputtering methods have been applied for a variety of film preparations and is easily available for industrial applications. In this study, we have adapted a facing-targets direct-current sputtering (FTDCS) method, in which a couple of targets are positioned in parallel and the substrate is set in the direction perpendicular to the two targets. As compared to the ordinary sputtering method, FTDCS has the following features due to a magnetic field between the targets: (i) high plasma density, (ii) less damage, (iii) fewer rises in substrate temperature, and (iv) lower stoichiometric difference from a target since the substrate is free of the plasma. On the other hand, species arriving at the substrate have a kinetic energy of several electron-volts; this value is larger than those observed in other preparation methods such as molecular beam epitaxy. These are suitable for growing β-$FeSi_2$ directly at low substrate temperatures by providing both Fe and Si by using $FeSi_2$ alloy targets.

2. Experimental procedure

Sputtering has proven to be a successful method of coating a variety of substrates with thin films of electrically conductive or non-conductive materials. One of the most striking characteristics of sputtering is its universality. Since the coating material is passed into the vapor phase by a mechanical rather than a chemical or thermal process, virtually any material can be deposited. Generally, direct current (DC) is used to sputter conductive materials, while radio frequency (RF) is used for non-conductive materials.
FTDCS apparatus is configured such that a pair of targets is mounted facing each other separated by a prespecified distance away from each other. A magnetic field extending in the space between the targets, called a discharge space, from one target to the other is provided by two permanent magnets, as shown in Fig. 1(a). The magnetic field flux, which has a perpendicular direction to the target surfaces, uniformly surrounds the discharge space to confine plasma within this space and to form a film on a substrate disposed at a position beside the discharge space under vacuum condition. The DC plasma can be adjusted over a wide range to control the deposition rate. The DC power supply supplies sputtering power to the apparatus while the shields of the vacuum chamber walls serve as an anode (ground) and the targets serve as a cathode (Kadokura, 2005).
The film preparation chamber is evacuated by means of an evacuation system, consisting of turbo-molecular pump connected to a rotary pump. A sputtering gas, such as argon, is introduced into the vacuum chamber through a gas inlet by a gas flow rate control system. Generally, inert gases are usually employed as the sputtering gas because they tend not to react with the target material or combine with any process gases and because they produce higher sputtering and deposition rates due to their high molecular weight. The sputtering plasma affects sputtering of the targets, thereby forming a thin film with composition corresponds to that of the targets on the substrate, as shown in Fig. 1(b). The sputtering plasma accelerates ionization of the sputtering gas, thereby increasing the sputtering rate and thus forming a film on the substrate at high deposition rates.

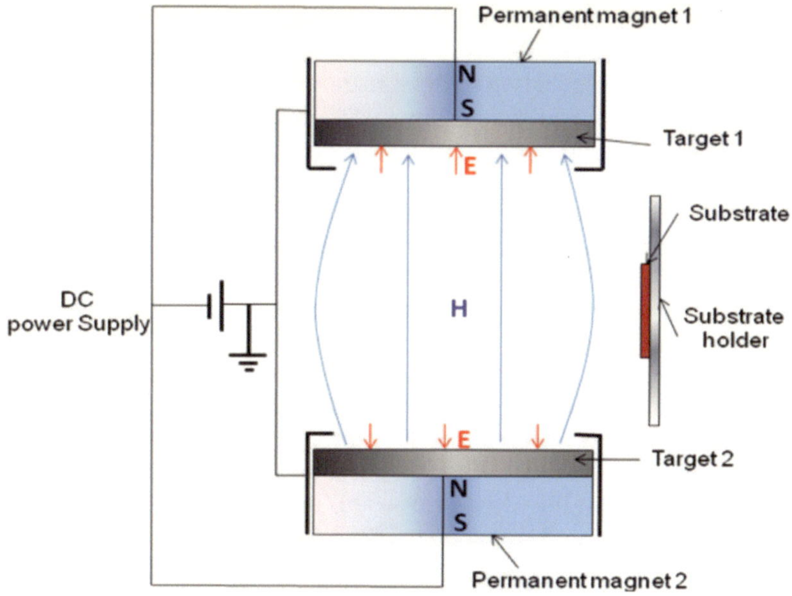

(a)

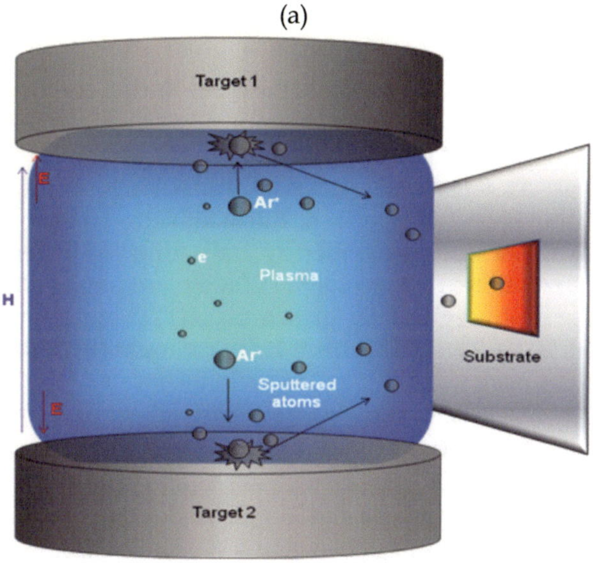

(b)

Fig. 1. Illustrations of; (a) FTDCS configuration, and (b) sputtering mechanism in FTDCS apparatus

The experiment procedure can be summarized as follows: β-$FeSi_2$ thin films of thickness 300 nm were deposited on Czochralski Si(111) substrates (resistivity: 10 Ω cm) of thickness 100 μm using $FeSi_2$ targets (purity: 4 N) with atomic ratio of Fe : Si = 1 : 2. The substrates were placed in the film preparation chamber after their native oxide layers were etched in dilute HF solution (1% HF). The chamber, which was equipped with a turbo-molecular pump, was evacuated to a base pressure of less than 1×10^{-5} Pa. Ar gas (purity: 6 N) was introduced into the chamber at a flow rate of 15 SCCM (SCCM denotes standard cubic centimeter per minute). The β-$FeSi_2$ films were deposited at a constant pressure of 1.33×10^{-1} Pa. The operating DC voltage and current were 1 KV and 1.5 mA, respectively. After the β-$FeSi_2$ films were deposited at the substrate-temperature of 600 °C, they were transferred to a radio-frequency magnetron sputtering apparatus where Pd and Al electrodes were deposited on the top (Si) and bottom (β-$FeSi_2$) surfaces of the films, respectively. The electrodes were deposited at room temperature. The formation of Pd/β-$FeSi_2$ and Al/Si ohmic contacts were experimentally confirmed.

The structural characteristics of the prepared β-$FeSi_2$/Si heterojunctions were investigated by x-ray diffraction (XRD) measurements (Rigaku, RINT-2000/PC) and scanning electron microscopy (SEM) observations (JEOL, JSM-6340F). The optical properties were measured by a spectrometer (JASCO, V-570) with an integrating sphere (JASCO ISN-470) in the photon energy range of 0.6–1.2 eV. The current-voltage (*I-V*) characteristics were measured (Keithley, source-meter 2400) in the dark and under illumination by a 6 mW laser diode (LD) (Neoark, TC20) at a wavelength of 1.31 μm. The photoresponse properties were measured using a Xe lamp (Ushio, UXL-300D) and a monochromator (Oriel 77250) with a focal length of 125 mm and a line density of 600 l/mm. The light intensity was calibrated using a commercial photodiode (Hamamatsu G8372-1).

3. β-$FeSi_2$ film characterization

3.1 Epitaxial growth

The epitaxial growth of β-$FeSi_2$ films on Si(111) was confirmed on the basis of the following: XRD measurements in a $2\theta-\theta$ scan mode, grazing incidence method (2θ scan; at a fixed incidence angle of 4°), φ-scan and pole figure analysis. Figure 2 shows a typical $2\theta-\theta$ XRD pattern of β-$FeSi_2$ films. An intense 220/202 peak and a weak 440/404 peak due to β-$FeSi_2$ were observed near the 111 and 222 peaks of the Si substrate, respectively. No peaks were observed in the 2θ-scan measurement, thereby indicating the absence of polycrystalline elements in the film. Crystallites comprising the film are predominantly 101-oriented, which is similar to the β-$FeSi_2$ film epitaxially grown on Si(111) by chemical vapor deposition (CVD) (Akiyama et al, 2001). The full width at half maximum (FWHM) of the rocking curve corresponding to the 404 peak was 1.54°, as shown in the inset of Fig. 2.

The φ-scan measurement was performed in order to confirm the orientation of the crystalline plane parallel to the substrate plane. The rotation axis was normal to the substrate surface, that is, the [111] direction of the Si(111) substrate. The diffraction peaks due to the 422 planes for the Si substrate and 313 planes for the β-$FeSi_2$ film were detected in this measurement. The φ-scan diffraction patterns are shown in Fig. 3(a). The twin peaks due to β-$FeSi_2$-313 were observed between the Si-422 peaks; this indicates that three types of epitaxial variants were rotated at an angle of 120° with respect to each other (Akiyama et al, 2001). The epitaxial relationship is predominantly as follows:

$$\beta(101)/\mathrm{Si}(111) \text{ with } \beta[101]//\mathrm{Si}[0\bar{1}1],\ [\bar{1}01], \text{ and } [\bar{1}10].$$

The pole figure, shown in Fig. 3(b), shows existence of the three types of epitaxial variants. The azimuth angles of two variants are equal, while the azimuth angle of the third variant is slightly shifted, owing to the orthorhombic crystallographic structure of β-$FeSi_2$ (Akiyama et al, 2001). It was confirmed that the β-$FeSi_2$ films were epitaxially grown not only in a direction perpendicular to but also in-plane with the Si(111) substrate.

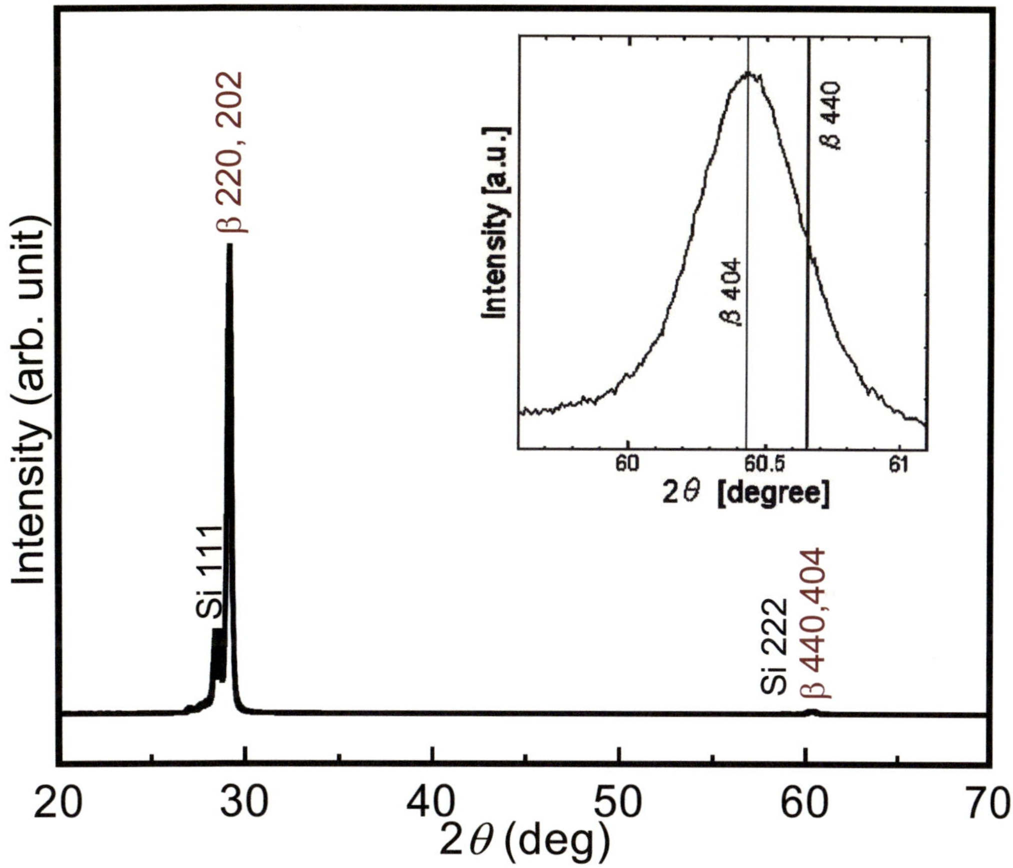

Fig. 2. X-ray diffraction pattern of the β-FeSi$_2$ thin film measured by a *2θ-θ* method. The inset is the extension of β-404, 440 peaks.

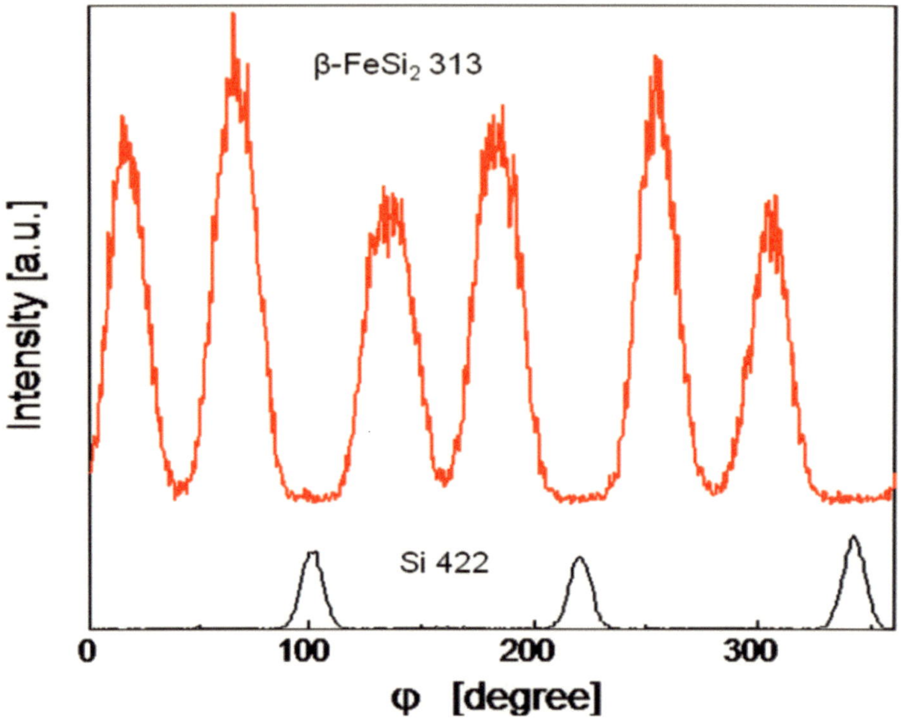

(a)

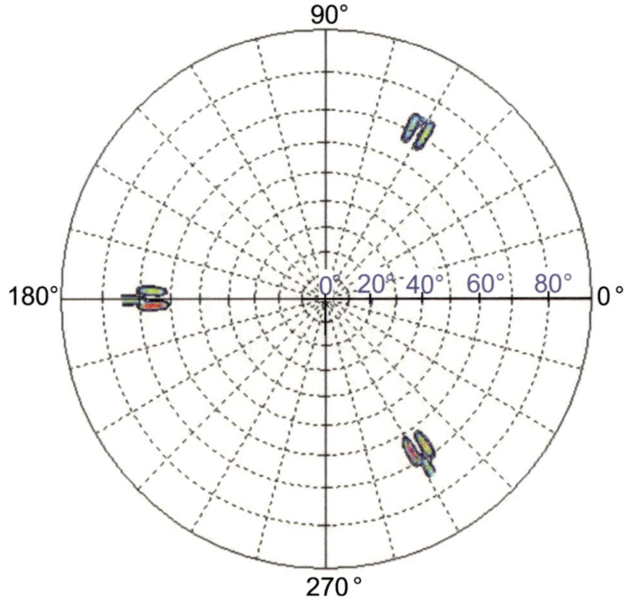

(b)

Fig. 3. (a) X-ray diffraction pattern of the β-$FeSi_2$ thin film measured by a φ-scan method; (b) pole figure plot of the 440/404 diffraction peak of β-$FeSi_2$

3.2 Optical characterization

The absorption coefficient is a strong function of the wavelength or photon energy. Near the absorption edge, the absorption coefficient for direct bandgap transitions can be expressed as;

$$\alpha \propto (hv - E_g)^{1/2}$$

where hv is the photon energy and E_g is the value of the bandgap. The absorption coefficient of β-$FeSi_2$ thin film, with thickness of 200 nm, was measured to be ~ 7×10^4 cm^{-1} at 1 eV. The interpolated indirect and direct optical band gaps were ~ 0.74 and 0.86 eV, respectively (Yoshitake et al, 2006). These values are in good agreement with those of the single crystalline bulk (Udono et al 2004). The absorption coefficient exceeds of 10^4 cm^{-1} for energies above 0.78 eV, which is close to the indirect optical bandgap of β-$FeSi_2$. From this measurement, the corresponding cut-off wavelength (λ_c) of photodiode devices with β-$FeSi_2$ thin film is expected to be ~1.55 μm.

Figure 4(a) shows the photoelectron spectrum of a β-$FeSi_2$ film. The threshold of the incident photon energy, which corresponds to the ionization potential of β-$FeSi_2$ (qV_{ip}), was measured to be 4.71 eV. From this value and the well-known parameters of Si and β-$FeSi_2$, the energy band diagram of the n-type β-$FeSi_2$/p-type Si heterostructure was derived, as shown in Fig. 4(b). Here, the Fermi level of β-$FeSi_2$ was assumed to be close to the conduction band because its high carrier density ranged from 10^{17} to 10^{18} cm^{-3} and its donor level was located only 20 meV below the conduction band (Yoshitake et al, 2006). The built-in potential (qV_{ip}) of the heterojunction was estimated to be ~ 0.9 eV. This value is sufficiently large for efficient photogenerated carrier collection in the heterojunction.

3.3 Electrical characterization

Figure 5 shows the temperature dependence of the electric conductivity of the film. The activation energy above 400 K was estimated to be around 0.41 eV, which corresponds to approximately half the optical band gaps. The Hall coefficient was negative, indicating n-type conduction. The activation energy between 280 and 320 K was estimated to be 0.02 eV. Since this corresponds to the energies between the donor level of Co and the bottom level of the conduction band (Tani et al, 1998), we believe that Co, which is a dominant impurity in the $FeSi_2$ target, are incorporated from the $FeSi_2$ target into the film. Although the carrier concentration was estimated to be approximately 10^{17} cm^{-3} at 300 K from the measured Hall coefficient and electric conductivity, the influence of the substrate on the electric conductivity of the film is not completely negligible. The carrier concentration might approximately lie between 10^{17} and 10^{18} cm^{-3} at 300 K (Yoshitake et al, 2006). It was confirmed that the deposited β-$FeSi_2$ was n-type since the heterojunction with p-type Si exhibited rectification current that is typically observed in a p-n junction. The conduction type of β-$FeSi_2$ is known to be dependent on the stoichiometry between Fe and Si. In this study, since the stoichiometric change from the targets to the film was estimated to be at a maximum of 1%, we believe that the Co incorporation from the targets is the primary reason for n-type conduction.

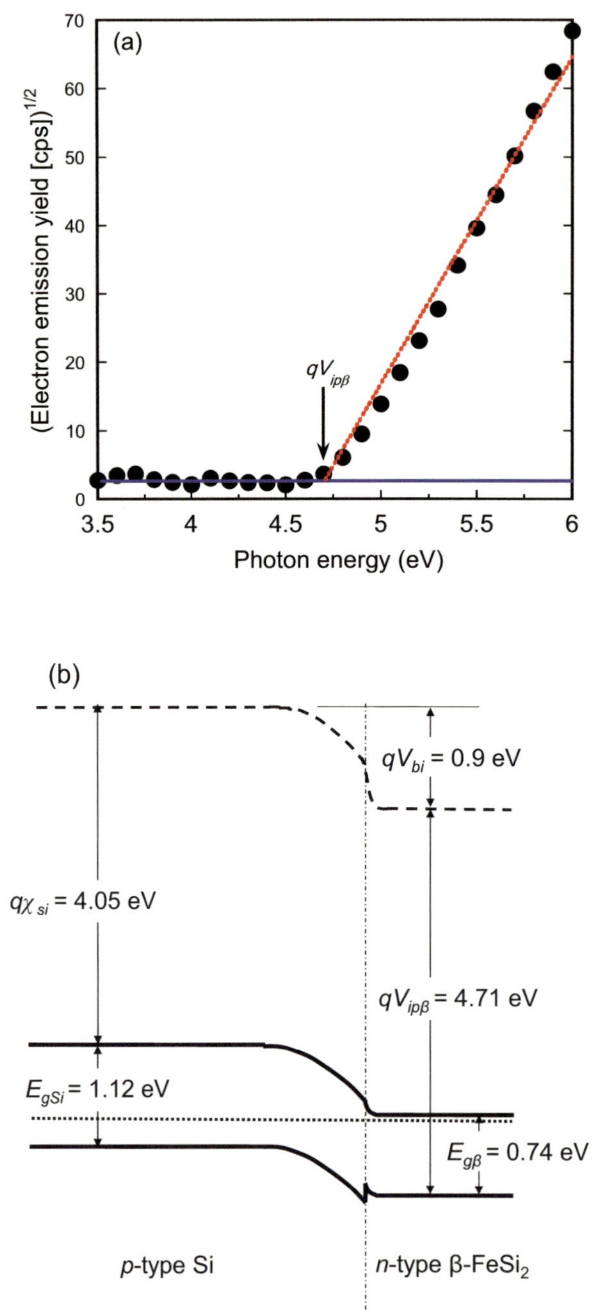

Fig. 4. (a) Photoelectron spectrum of a β-$FeSi_2$ film, wherein the threshold value of the photon energy corresponds to the ionization potential and (b) the energy band diagram of n-type β-$FeSi_2$/p-type Si heterojunction.

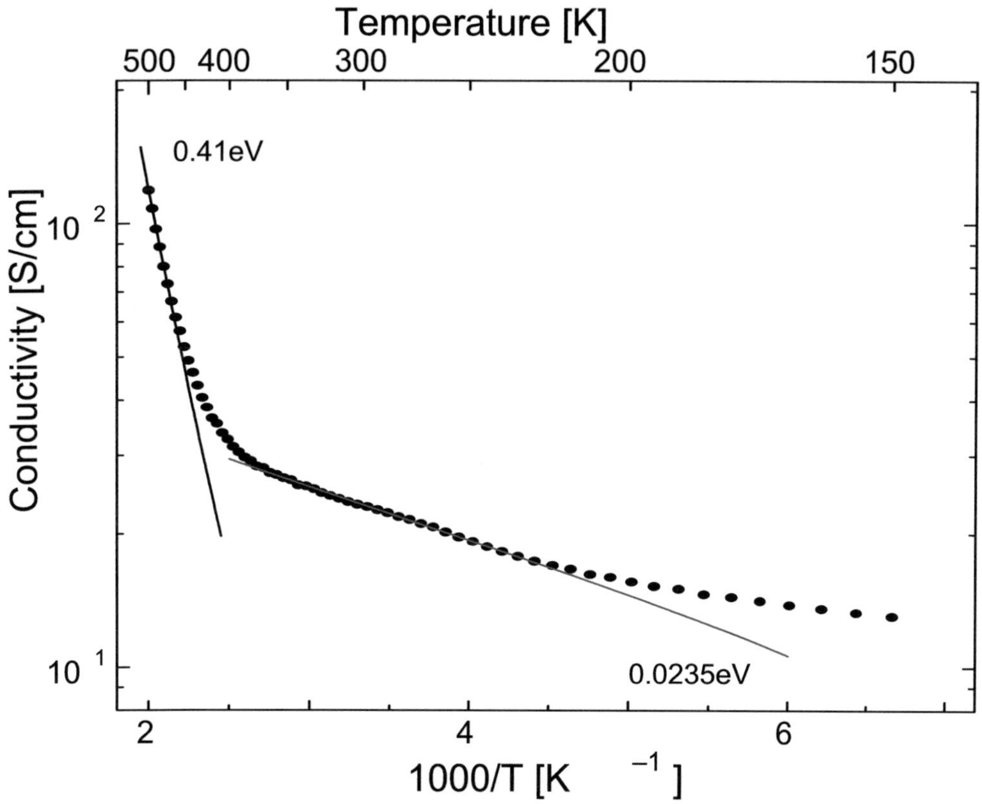

Fig. 5. Temperature dependence of the electric conductivity of the β-$FeSi_2$ film.

4. Device characterization

The fabricated device is a simple p–n heterostructure grown without annealing or passivation. Figure 6(a) illustrates a cross section schematic diagram of n-β-$FeSi_2$/p-Si heterojunction photodiode. The light is irradiated from front side of the Si substrate, which acts as the window material and helps in the epitaxial growth of crystalline β-$FeSi_2$. In this experiment we used Si substrates with thicknesses of 100 µm, however any smaller thickness can be used without affecting the device performance. Figure 6(b) shows a cross section SEM image in which a continuous and uniform β-$FeSi_2$ film without cracks was observed. In addition, the interface between the film and the substrate was extremely sharp which is essential to obtain good junction quality.

4.1 Current-voltage characteristics

Figure 7 shows the device characteristics measured in the dark and under the illumination of the 1.31 µm LD. The device exhibited good rectifying properties similar to those of conventional p-n junctions. The forward current was approximately two orders of magnitude greater than the reverse current. The dark leakage current at a reverse bias of -1 V was 5 µA, and this value increased to 20 µA at -5 V. The main source of the leakage current might be due to the Fe atoms that diffused from the β-$FeSi_2$ film to the Si substrate

during the film deposition process. The diffused atoms result in leakage centers and trap centers for the photogenerated carriers in addition to interface defects.
The measured photocurrent at zero bias was ~ 20 μA, and this value increased to 70 μA at -5 V. The ratio of the photocurrent to the dark leakage current at -1 V was approximately one order of magnitude. The current responsivity at zero bias was estimated to be ~ 3.3 mA/W. The inset of Fig. 3 shows the dynamic resistance deduced from the dark *I-V* characteristics. Because the major noise component at zero bias is mainly the Johnson noise, the device detectivity was estimated from the values of dynamic resistance and responsivity at zero bias. The shunt resistance, which corresponds to the dynamic resistance at zero bias, can be estimated to be approximately 50 kΩ from the inset. The detectivity at room temperature for the 1.31 μm illumination was estimated to be 1.5×10^9 cm √Hz W (Shaban et al, 2009a). This value is the largest in the previous β-$FeSi_2$ reports.

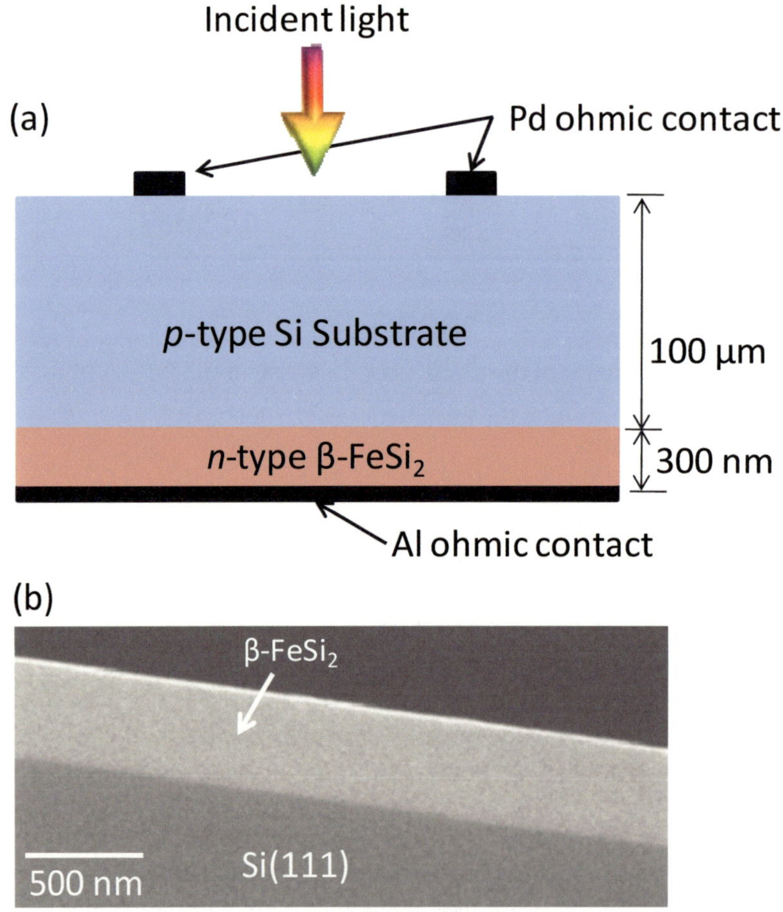

Fig. 6. (a) Schematic illustration of an n-type β-$FeSi_2$ (300 nm)/p-type Si (100 μm) heterojunction photodiode with Pd and Al electrodes on top (Si) and bottom (β-$FeSi_2$) surfaces, respectively. (b) Cross-sectional SEM image of β-$FeSi_2$ film deposited on Si(111) substrate.

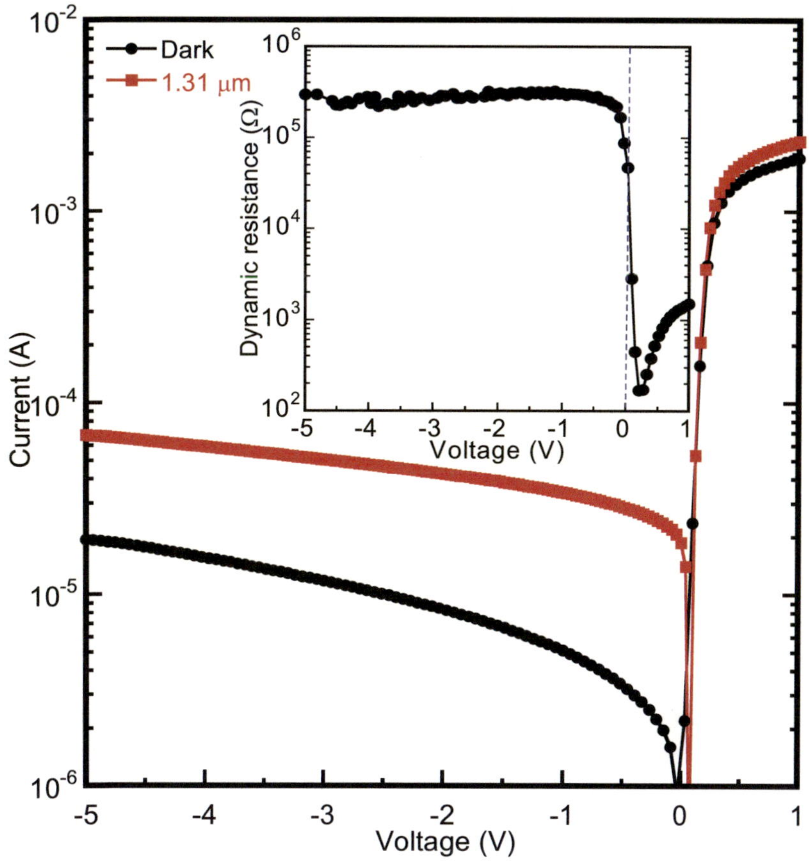

Fig. 7. *I-V* characteristics measured in the dark and under illumination with a 6 mW, 1.31 μm LD at room temperature. The inset shows the dynamic junction resistances deduced from the *I-V* characteristics.

4.2 Photoresponse spectrum

Figure 8 illustrates the photoresponse spectrum measured in the NIR range at zero bias and room temperature. The device evidently exhibited photoresponse beyond the Si cut-off wavelength up to the wavelength (approximately 1.5 μm) corresponding to the bandgap of β-$FeSi_2$. The inset of Fig. 4 shows the external quantum efficiency as a function of the reverse bias at 1.31 μm. The efficiency increased from 0.32% to 1.07% with an increase in the reverse bias from zero to -5 V. The reduction in the efficiency is attributed to light reflection and photocarrier recombination at the β-$FeSi_2$/Si interface. Further improvement in the device performance is expected to be achieved by reducing the surface reflection and passivation of the heterojunction interfaces in addition to suppressing the leakage current.

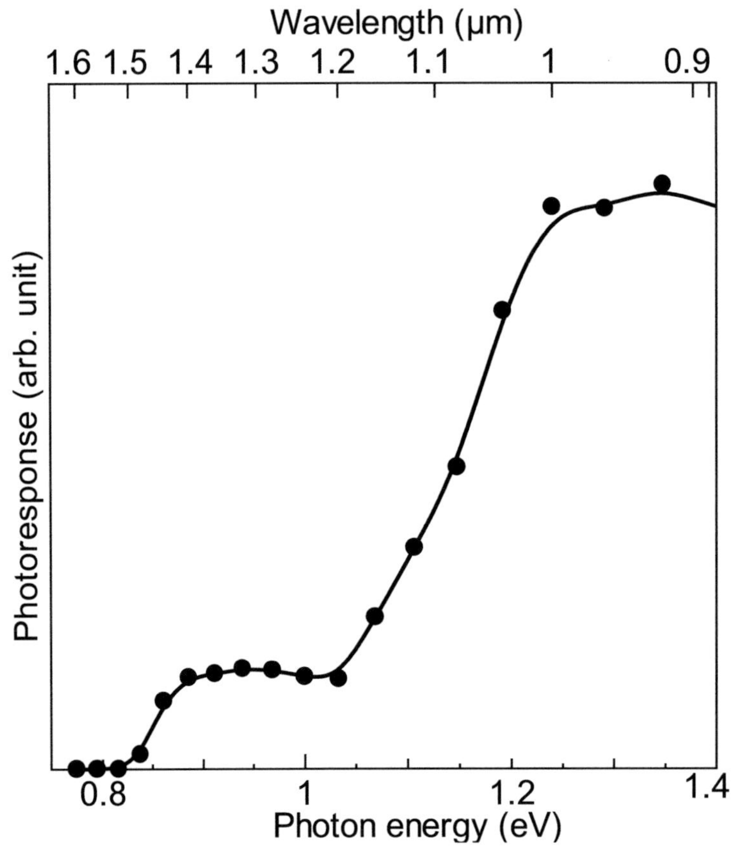

Fig. 8. Photoresponse spectrum of β-$FeSi_2$/Si heterojunction photodiode measured in the NIR range at zero bias and room temperature.

4.3 β-$FeSi_2$/Si photodiode with leakage-blocking layer

In order to improve the device performance, the leakage current should be reduced. This was achieved by inserting intrinsic-Si (*i*-Si) layer between the β-$FeSi_2$ film and Si substrate. Therefore, thin intrinsic-Si layers (thickness: 300 nm) were homoepitaxially grown on Czochralski-Si(111) substrates (resistivity: 10 Ω cm) at a substrate temperature of 700 °C. β-$FeSi_2$ layers (thickness: 300 nm) were successively grown on the Si layers at a substrate temperature of 600 °C by the FTDCS method. After that, the samples were transferred to a radio-frequency magnetron sputtering apparatus for depositing electrodes.

Figure 10(a) shows the *I-V* characteristics of the n-type β-$FeSi_2$/intrinsic-Si/p-type Si photodiodes, with an area of 3 mm^2, measured in the dark and under the illumination of the 6-mW 1.31 μm laser. The device exhibited good rectifying behavior with a rectifying ratio of greater than two orders of magnitude (Shaban et al, 2009b). The leakage current did not saturate and increased monotonously with the reverse voltage until the device experienced a thermal breakdown. Owing to the *i*-Si layer, the leakage current measured in this device was less than that measured in n-type β-$FeSi_2$ / p-type Si photodiodes (Shaban et al, 2009a).

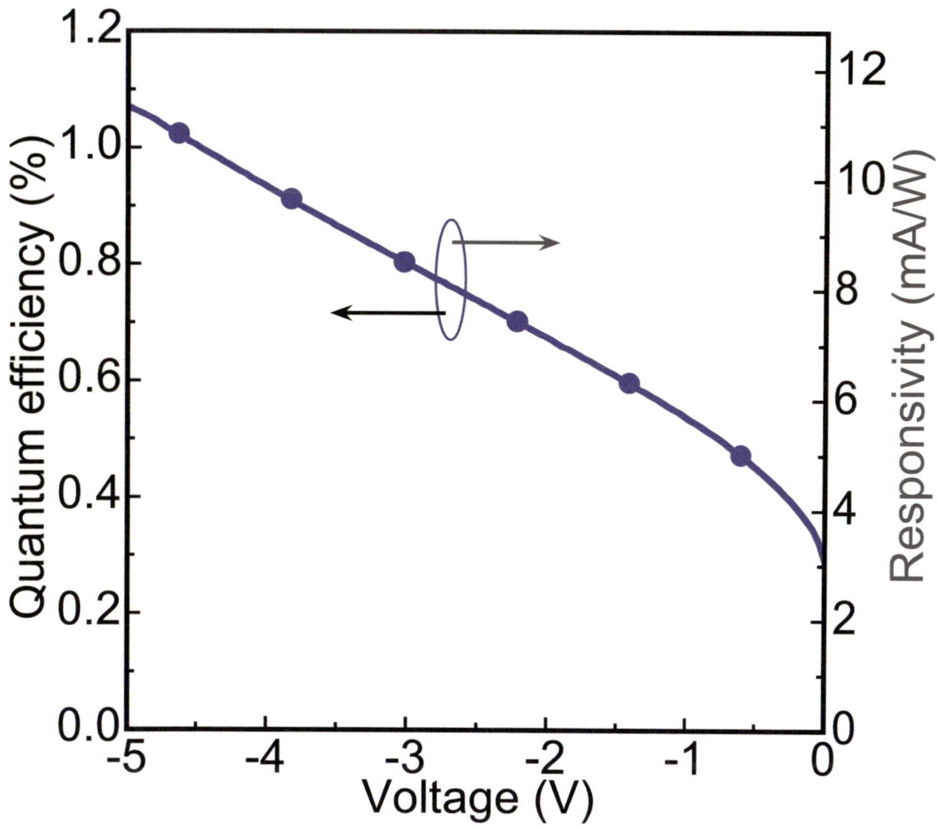

Fig. 9. Quantum efficiency and current responsivity of β-$FeSi_2$/Si heterojunction photodiode plotted against reverse bias under illumination with a 6 mW, 1.31 μm.

The photocurrent, resulted from the 1.31-μm illumination, increased from 40 to 840 μA with an increase in the reverse voltage from 0 to 5 V. The ratio of the photocurrent to the dark leakage current over the entire measured range of reverse voltages was approximately two orders of magnitude (Shaban et al, 2009b). The current responsivity increased from 6.6 to 140 mA/W with the increase in the reverse voltage from 0 to 5 V, as shown in Fig. 10(b). The Johnson-noise limited detectivity, which was measured at 1.31 μm and 300 K, was deduced to be 2.8×10^9 cm√Hz/W. This indicated the ability of the diode to detect light in the NIR region at room temperature.

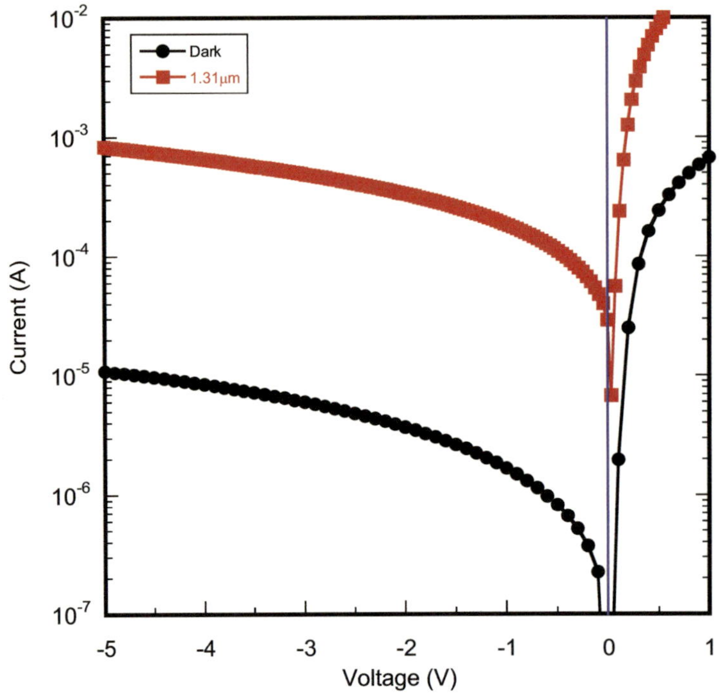

(a)

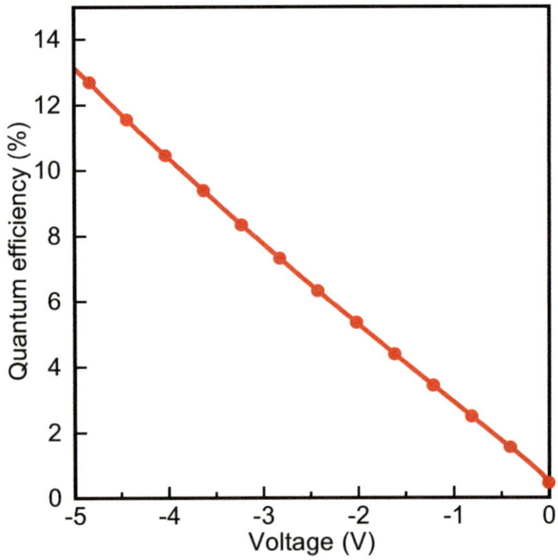

(b)

Fig. 10. (a) *I-V* characteristics of a heterojunction measured in the dark and under the illumination of a 6-mW 1.31 μm laser at 300 K. (b) External quantum efficiency vs reverse voltage measured under illumination at a wavelength of 1.31 μm.

5. Conclusions

A FTDCS method using $FeSi_2$ alloy targets, in which Fe and Si atoms are provided on the substrate at a low deposition rate comparable with that of molecular beam epitaxy, is a simple and successful method for the direct epitaxial growth of β-$FeSi_2$ thin films. The deposited β-$FeSi_2$ thin films show nearly the same optical and electric properties as the single crystalline bulk.

n-type β-$FeSi_2$/p-type Si heterojunctions in which the β-$FeSi_2$ layers were heteroepitaxially grown on Si(111) substrates with sharp interfaces were fabricated by FTDCS without carrying out postannealing. The band diagram of the heterostructure was derived, and the built-in potential was measured to be~ 0.9 V.

The *I-V* characteristics showed good rectifying properties with a shunt resistance of approximately 50 kΩ. The photoresponsivity under illumination at 1.31 μm was approximately 3.3 mA/W at room temperature. The specific detectivity was estimated to be 1.5×10^9 cm√HzW and the quantum efficiency, estimated at –5 V, was ~ 1.07%.

In order to improve the device performance, a leakage-blocking (*i*-Si layer) layer was suggested. The n-Type β-$FeSi_2$/*i*-Si/p-type Si heterojunction photodiodes showed an improved performance. They exhibited a current responsivity of 140 mA/W and quantum efficiency of 13% at –5 V.

The results suggest that these devices can operate efficiently at room temperature in the photoconductive mode and therefore they are promising candidates for NIR detectors that are compatible with Si.

6. References

Akiyama, K.; Ohya, S.; Takeno, H.; Kieda, N. & Funakubo, H. (2001). Growth of β-$FeSi_2$ Thin Film on Si (111) by Metal-Organic Chemical Vapor Deposition, In: *Jpn. J. Appl. Phys.*, Vol. 40, No. , pp.460-462.

Kadokura, S. (2005). Facing-targets-type sputtering apparatus and method, In : *U.S. Patent*, No. 6911123.

Kang, Y. ; Zadka, M. ; Litski, S. ; Sarid, G. ;Morse, M. ; Paniccia, M. J. ; Kuo, Y. H.; Bowers, J. ; Beling, A. ; Liu, H. D. ; McIntosh, D. C. ; Campbell, J. & Pauchard, A. (2008). Epitaxially-grown Ge/Si avalanche photodiodes for 1.3 microm light detection, In : *Optics Express*, Vol. 16, No. , pp. 9365-9368.

Kendall, D. L& Devries, D. B. (1969). Diffusion in silicon, In: *Semiconductor Silicon*, edited by R. R. Haberecht and E. L. Kern, Electrochemical Society, pp. 358-421, New York.

Leong, D. N.; Harry, M. A.; Reeson, K. J. & Homewood, K. P. (1996). On the origin of the 1.5 μm luminescence in ion beam synthesized β-$FeSi_2$. In: *Appl. Phys. Lett.*, Vol. 68, No. , pp. 1649-1651.

Luryi, S. ; Kastalsky, A. & Bean, J. C. (1984). New infrared detector on silicon chip, In : *IEEE Trans. Electron Devices*, Vol. 31, No. , pp.1135–1139.

Mahan, J. E.; Geib, K. M. & Robinson, G. Y. (1990). Epitaxial films of semiconducting $FeSi_2$ on (001) silicon, In : *Appl. Phys. Lett.*, Vol. 56, pp. 2126.

Masini, G. ; Colace, L. & Assanto, G. (2003). 2.5 Gbit/s polycrystalline germanium-on-silicon photodetector operating from 1.3 to 1.55 μm, In : *Appl. Phys. Lett.*, Vol. 82, No. , pp. 2524-2526.

Ootsuka, T. ; Fudamoto, Y. ; Osamura, M. ; Suemasu, T. ; Makita, Y. ; Fukuzawa, Y. & Nakayama, Y. (2007). Photoresponse properties of Al/n-β-$FeSi_2$ Schottky diodes using β-FeSi2 single crystals, In : *Appl. Phys. Lett.*, Vol. 91, No. 142114.

Shaban, M.; Nakashima, K.; Yokoyama, W. & Yoshitake, T. (2007). Photovoltaic Properties of n-type β-FeSi2/ p-type Si Heterojunctions, In : *Jpn. J. Appl. Phys.*, Vol. 46, pp. 667-669.

Shaban, M.; Nomoto, K.; Izumi, S. & Yoshitake, T. (2009a). Characterization of near-infrared n-type β-$FeSi_2$ /p-type Si heterojunction photodiodes at room temperature, *Appl. Phys. Lett.*, Vol. 94, No. 222113.

Shaban, M.; Izumi, S.; Nomoto, K. & Yoshitake, T. (2009b). n-Type β-$FeSi_2$/intrinsic-Si/p-type Si heterojunction photodiodes for near-infrared light detection at room temperature, *Appl. Phys. Lett.*, Vol. 95, No. 162102.

Suemasu, T.; Negishi, Y. ; Takakura, K. & Hasegawa, F. (2000). Room temperature 1.6 μm electroluminescence from a Si-Based light emitting diode with β-$FeSi_2$ active region, In: *Jpn. J. Appl. Phys.*, Vol. 39, pp. 1013-1015.

Tani J. & Kido, H. (1998). Electrical properties of Co-doped and Ni-doped β-$FeSi_2$, In: J. Appl. Phys., Vol. 84, pp. 1408-1410.

Udono, H.; Kikuma, I.; Okuno, T.; Matsumoto, Y.; Tajima, H. & Komuro, S. (2004). Optical properties of β-$FeSi_2$ single crystals grown from solutions, In: Thin Solid Films, Vol. 461, pp. 182-187.

Yoshitake, T.; Inokuchi, Y.; Yuri, A. & Nagayama, K. (2006). Direct epitaxial growth of semiconducting β-$FeSi_2$ thin films on Si(111) by facing targets direct-current sputtering, In: *Appl. Phys. Lett.*, Vol. 88, No. 182104.

16

GaN-based Photodiodes on Silicon Substrates

L.S. Chuah[1] and Z. Hassan[2]

[1]*Physics Section, School of Distance Education, Universiti Sains Malaysia, 11800 Penang,*
[2]*Nano-Optoelectronics Research and Technology Laboratory, School of Physics, Universiti Sains Malaysia, 11800 Penang, Malaysia*

1. Introduction

GaN-based or the III-V nitrides materials are wide band gap semiconductor materials with dormant utilizations in optoelectronic as well as in electronic devices operating at high power and high temperature conditions. Silicon (Si) is one of the most common elements of the earth crust and the substrates are of very low price and are available in very large size due to its mature development and large-scale production. The thermal conductivity is higher than that of sapphire and is close to that of GaN. The crystal perfection of Si is better than that of any other substrate material and it has good thermal stability under GaN epitaxial growth condition. The growth of GaN on Si enables the possibility of integrating GaN optoelectronics devices with Si-based electronics. However, despite much effort, there is no significant breakthrough that has been obtained because of the high density of dislocations in these materials leading to a rapid degradation of all devices fabricated so far. In contrast, the performances of GaN-based devices are known to be quite acceptable despite the large density of dislocations present in the films. There are only few reports on GaN photodetectors on Si(111), including PN heterojunction photodiode, metal-semiconductor-metal (MSM) photodetectors, and Schottky photodiodes. This chapter will present the fabrication and characterization aspects of GaN-based photodiodes on Si substrates.

Photodetectors operating in the UV and with a visible blind behavior have drawn great attention in recent years, with a number of applications in both civil and military industries which include detection of missile plumes, flame sensors, engine control, solar UV monitoring, source calibration, UV astronomy, and secure space-to-space communications (Keem et. al., 2004; Hassan et. al., 2004).

Contrast to Si, gallium nitride (GaN) has wide band gap, outstanding thermal stability, small dielectric constant, chemical inertness and radiation hardness. These features make it an typical choice for use in fabricating high frequency, high power electronic devices. Furthermore, the essential preference of III-V nitrides detectors above competing devices based on semiconductors with smaller bandgaps is the long wavelength response cut-off, which is straightly associated to the bandgap of the material in the dynamic area and consequently, does not involve external filters (BenMoussa et. al., 2008). Gallium nitride (GaN) is of distinguished interest because of its great UV photoresponse, well-founded mixture techniques, and the capability of operating at high temperature and in harsh circumstances. Dissimilar to wide band gap photodetector, the forward surface of a vacuum ultraviolet silicon photodiode is coated with filter materials to limit the sensitivity of the silicon to a much narrower band in the vacuum ultraviolet (VUV) because one significant shortcoming of Si photodiodes for certain VUV applications is the inherent broadband response extending from x-rays to the near infrared, which is undesirable.

Mainly, photodetectors can be assorted in accordance with the type of optical to electrical conversion effect. A photoelectric detector is based on the process of photon absorption by the material with the release of an electron-hole pair. In the case where the photo-generated electron is further emitted out of the material in which they become available for collection or multiplication, the device is called a photo emissive detector which is based on the external photoelectric effect. Some examples of photo emissive detectors are like photomultipliers, vacuum photodiodes, and image intensifier tubes. However, if there is no emission taking place but the photo generated electron-hole pair is available for the current circulation in the external circuit; then this is called an internal photoelectric effect detector which is the characteristic of all semiconductor photodetectors.

A p-n junction photodiode operates under reverse bias by the absorption of incident photons with energy larger than the energy gap (E_g), therefore producing electron-hole pairs in the material on both sides of the junction. The holes and electrons produced in a diffusion length from the junction get to the depletion region by diffusion. Over there, the electron-hole pairs are separated by the firm electric field and minority carriers are readily accelerated to become majority carriers on the other side, generating a photocurrent. The carrier diffusion step in this process is time consuming and should be eliminated if possible. Consequently, it is useful that the distance across of the depletion region be wide enough so that nearly all of the photons are absorbed within the depletion region rather than in the neutral p and n regions. Several types of GaN ultraviolet detectors have been reported, for example, photoconductive, p-n junction, p-п-n, Schottky barrier, MSM, and heterojunction in recent years. For metal-semiconductor contacts, the ohmic contact is needed for connections to devices because of its linear current-voltage (I-V) characteristics in both biasing directions.

Aluminium nitride, AlN (wurtzite structure type) has specific physical properties like high thermal conductivity (91-190 W/m-K), large breakdown electric field, high electrical resistivity (10^{11}-10^{14} Ωcm), high melting point and large energy gap (Qioa et. a., 2000). Its application as a component of refractory ceramics or buffer layers for GaN epilayers grown on sapphire are widely known. In the last decade considerable attention arose in the use of thin films of AlN for various applications, from coatings for magneto-optic media, to thin films transducers and GHz-band surface acoustic wave devices. AlN is a broad band gap semiconductor; its characteristics are similar to those of an insulator with a high dielectric constant, large breakdown electric, and field good conductivity. Consequently, AlN could be the best insulator to block the leakage current. Before that, it has been demonstrated that one can decrease gate leakage current and interface state density in GaN-based effect transistor by using the AlN thin layer (Yu. et. al., 2006).

In the present work, we fabricated the p-GaN/n-Si heterojunction photodiode to observe the photoelectric effects. Besides that, the impacts of thermal annealing on GaN Schottky barrier photodiode with a thin AlN cap layer are also examined, principally because of the high thermal stability of GaN that has prompted us to bring out the best from thermal treatment to the electrical characteristics of the UV photodetectors. Thermal annealing has been effective in reducing the dark current in Schottky photodiodes which is the discussion that we are addressing in this text. We also have investigated growth of $Al_{0.09}Ga_{0.91}N$ epilayers grown on silicon (111). Also reported is our attempt to fabricate and characterize metal-semiconductor-metal (MSM) photodiode based on these films. The effect of post annealing in oxygen ambient on the electrical properties of Ni/$Al_{0.09}Ga_{0.91}N$ is studied by *I-V* measurement.

2. Operation principle

Photodiodes are illuminated devices utilized to identify optical signals straight throughout electronics processes, ordinarily performance under photoconductive method. The performance contains a couple of steps: (i) charge carrier transport; (ii) carrier formation by light absorption, of which the photon need only have greater energy than that of the band gap of device materials; and (iii) the interplay of current with the external circuit to provide the output voltage signal. The supreme wavelength of the light that can generate electrons and holes is found as (Sze,1981):

$$\lambda = hc/E_g \quad (1)$$

where h is the Planck's constant; c, the velocity of light; and E_g is the bandgap of the semiconductor material. The devices function under reverse bias for the photoconductive mode for the purpose of decreasing the heterojunction capacitance and the giving up of carriers because of the larger depletion width (Luo et. al., 2006).

The I-V characteristic of a photodiode without incident light intensity is similar to a curve of a conventional diode rectification. When a reverse bias is applied, a small reverse current appears which is related to dark current. When the photodiode is irradiated with optical distribution, the *I-V* curve was changed by the number of photocurrent. A schematic view of the typical energy band diagram of the p-GaN/n-Si structure and the theory of the photodiode is shown in Fig. 1. When the photon energy of the incident light is higher than the bandgap of GaN (3.4 eV), electron-holes pairs are generated inside GaN by light absorption. Simultaneously, the electron-hole pairs are separated by the electric field inside the depletion region of GaN to create photocurrent. The illumination also produces electrons and holes in the inert *n*-type and *p*-type zones, a few of which can diffuse to the depletion layer, contributing to the photocurrent (Philippe et. al., 1999). In real devices, a barrier layer at the interface of GaN might be existing due to the very thin of layer native silicon dioxide that is not drawn in Fig.1.

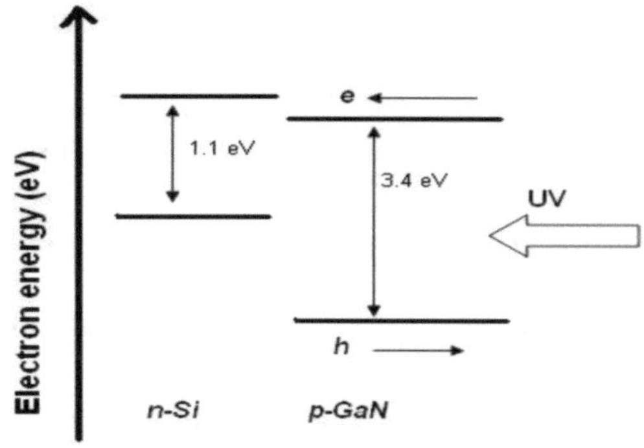

Fig. 1. Schematic energy band diagram of the p-GaN/n-Si structure.

3. Experiment

For the first sample, a heterojunction of Mg-doped p-type crystalline gallium nitride (GaN) were grown on n-type Si(111) by radio frequency nitrogen plasma-assisted MBE (PAMBE). Sample nominally consisted of 0.10 µm AlN as a buffer layer followed by 0.23 µm Mg-doped GaN with carrier concentration of ~ 2×10^{18} cm^{-3} as determined by Hall Effect measurement and the electrical conduction of all the samples was p-type.

For the wafer cleaning process prior to metallization of the contact metal, the GaN samples were dipped in a 1:20 NH_4OH:H_2O solution for 15 s followed by a 10 s dip in a 1:50 HF:H_2O solution. Then, it was rinsed with distilled water and blown dry with a nitrogen gas blower. Metal contact was sputtered by a sputtering system onto the backside and front-side, respectively. Here, to provide a low contact resistance of electrode, Ni/Ag (250 nm/600 nm) dots (250 µm in diameter) was deposited on the Mg-doped GaN layer at a corner, as shown in Fig. 2. In the same manner, a large area ohmic contact of Al/Ti (250 nm/600 nm) was also deposited on the backside of the Si substrate to form the second electrode of the p-n diode. A couple of contacts were patterned on the back part of an isolated piece of the sample to verify ohmic contact formation.

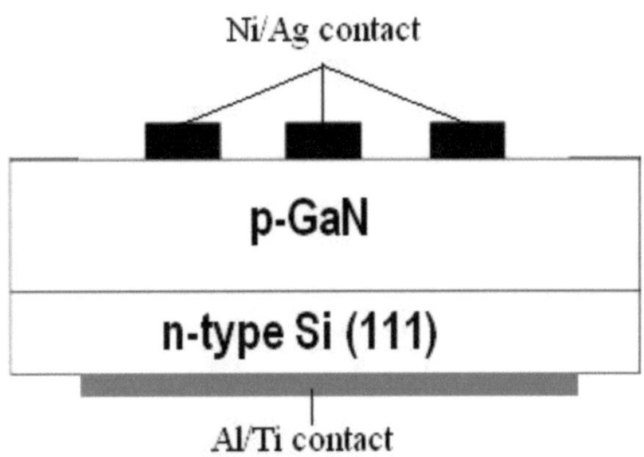

Fig. 2. Cross section view of the p-n heterojunction diode sample.

After metallization, the samples were annealed in a tube furnace at 400°C under flowing nitrogen gas environment for duration of 10 minutes. The setup of the furnace is shown in Fig. 3. Nitrogen gas was purged into the tube during annealing to displace the room ambient inside the tube. Room ambient contains nitrogen molecules that may interfere with the ohmic contact formation. The gas was purged at a mass flow rate of approximately 4 $L.min^{-1}$. The electrical behaviors of the contacts were analyzed by current-voltage (I-V) measurements. For the reverse bias configuration, negative bias was applied to the p-GaN film. The measurements were performed with or without an Hg-lamp onto GaN surface. The Hg-lamp illumination was placed vertically and 3 cm away from the sample. The light beam size was large enough to barely cover the GaN surface.

For the second set sample, unintentionally doped n-type GaN (800 nm) layers were grown, followed by subsequent thin AlN cap layer (50 nm) grown on the GaN surface. GaN-based Schottky photodiodes with a thin AlN cap layer were then fabricated.

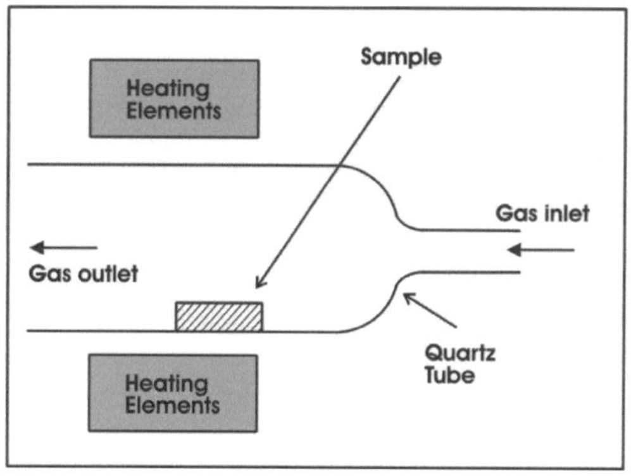

Fig. 3. Setup of tube furnace for thermal annealing.

Following surface treatment, aluminium (100 nm) metal stripes were coated on the sides of the samples as ohmic contacts by thermal evaporation technique. Contacts were then annealed in the furnace at 400 °C for 10 min under flowing nitrogen gas environment. *I–V* measurement was performed to confirm the ohmic behavior of the contacts. Consequent to the ohmic contacts deposition, Ni (150 nm) was deposited as the Schottky contact metal for all of the fabricated devices. The metal mask applied for Schottky contacts fabrication encompasses a pattern of dots with diameter of 200 μm. Both contacts were formed on the uppermost surface of the sample.

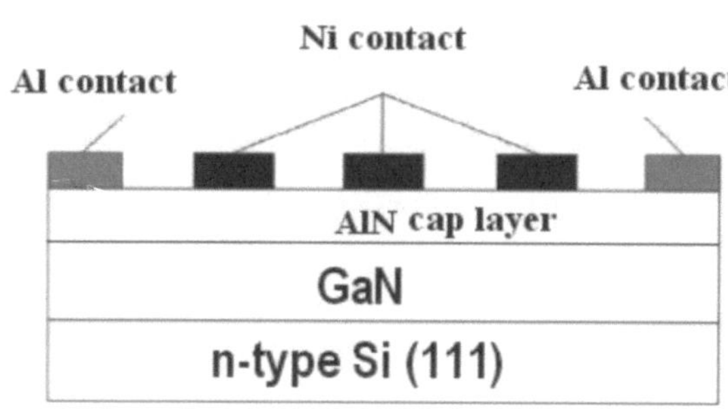

Fig. 4. Cross section view of ohmic and Schottky contacts of a typical GaN-based Schottky photodiodes using a thin AlN cap layer.

The fabricated photodiodes were then annealed in nitrogen ambient at temperatures from 500 - 700 °C in a typical tube furnace. For sample (photodiode) annealed at 500 °C, the annealing duration was 15 min, while the 600 °C sample was annealed for 5 min and 2 min for the 700 °C sample (Lee et. al., 2005). Photocurrent and dark current of the fabricated photodiodes were then measured by Keithley high-voltage-source-measure-unit model 237-semiconductor parameter analyzer. The ideality factors (n) and effective Schottky barrier heights (SBHs) are deduced from the I-V measurement.

For the third set sample, sample nominally consisted of 0.20 μm low temperature nucleation layer AlN followed by 0.23 μm of unintentionally doped n-type AlGaN (Fig. 5). The Al content of the sample was measured to be 9 % from high resolution XRD (PANalytical X'pert MRD) simulation. Following surface treatment, Ni Schottky contacts were deposited by thermal evaporation using a metal mask in patterning of the contact structure.

MSM photodiode is a planar device consisting of two fork-shaped interdigitated contacts on the semiconductor surface. These contacts performs as back to back Schottky contacts with finger spacing of 400 μm, fingers width of 230 μm, and the length of each electrode was about 3.3 mm (Lee et. al., 2005). It consists of 4 fingers at each electrode as shown in Fig. 6. The fabricated photodiodes were then annealed at temperatures from 400 - 700 °C in a conventional tube furnace in flowing oxygen environment. The annealing duration for the samples annealed at temperatures from 400 °C to 500 °C was 15 minutes, while the 600 °C samples were annealed for 5 minutes, and 2 minutes for the 700 °C samples. The effective Schottky barrier heights (SBHs) are deduced from *I-V* measurement assuming the results can be described by the standard thermionic emission equation.

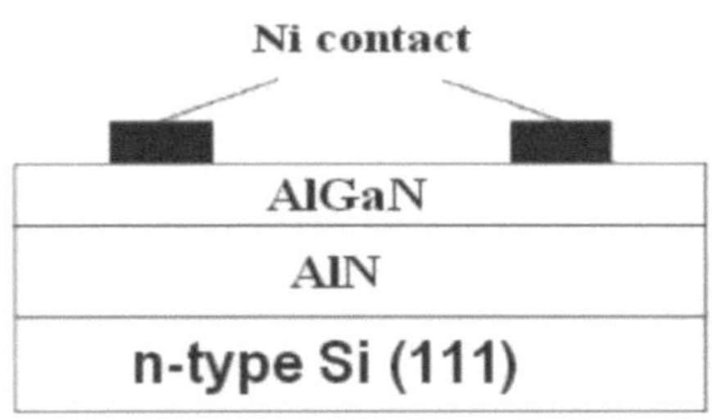

Fig. 5. Cross section view of the Ni Schottky contacts on AlGaN sample.

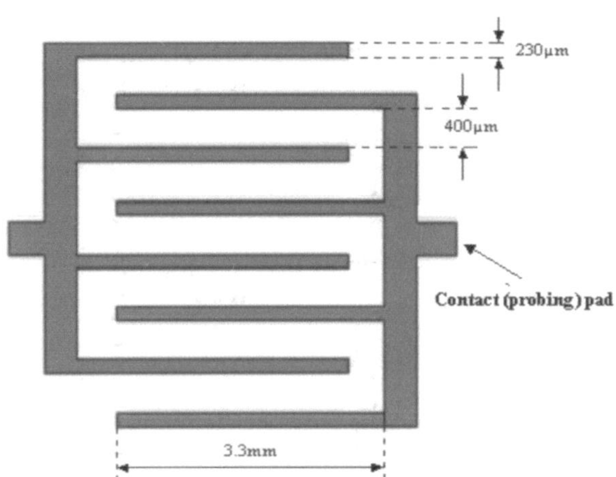

Fig. 6. Schematic diagram of the metal-semiconductor-metal (MSM) structure (Lee et. al., 2005)

4. Results and discussions

Studies of the I-V characteristics of the fabricated p-GaN/n-Si structure revealed a good p-n heterojunction as shown in Fig. 7. The I-V relationship for a heterojunction is given by (BenMoussa et. al., 2008):

$$I=I_s\left[\exp\left(\frac{qV}{K_BT}\right)-1\right] \tag{2}$$

where I is the current; V is the applied voltage across the heterojunction from p-side to n-side; K_B is the Boltzmann constant; I_s is the saturation current; and T is the absolute temperature. The typical *I-V* rectifying characteristic in the dark (without light illumination) at room temperature is observed in Fig. 7 to evaluate the turn on voltage. A pronounced rectifying diode-like behavior with turn on voltage of 0.8 V can be seen. The forward current was as high as 0.012 A at 8 V forward bias.

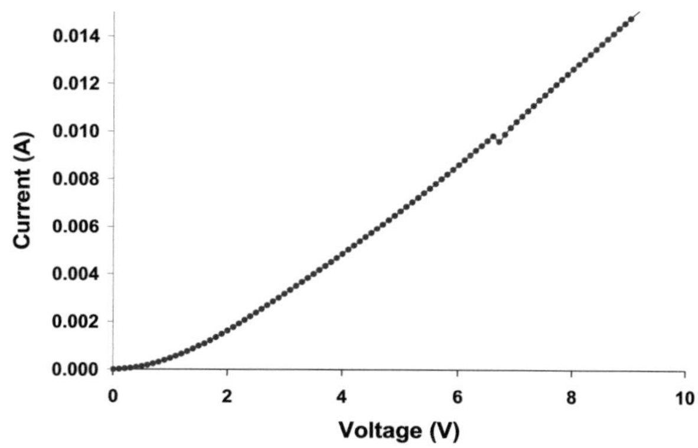

Fig. 7. I-V characteristics of the heterojunction in dark at room temperature (Chuah et. al., 2008a)

At the wavelength of 365 nm, the UV photo measurement was performed using an Hg-lamp. The mechanism of the crystalline-based p-GaN/n-Si diode is described as follows. Since the photon energy is larger than the bandgap of GaN, UV light source was absorbed by the GaN thin film producing electron-hole pairs, which were isolated by the electric field inside the GaN thin film contributing to the raise of the external current. Before p-GaN an n-Si are in contact, p-GaN has high concentration of holes and few electrons compared to that of n-Si. On contacting, diffusion of carriers will take place at concentration gradient near to the junction. The holes will diffuse from p-GaN to n-Si and negative space charge will remain behind in p-GaN near the junction.

From the p-GaN side, holes diffuse into the n-Si side, and electrons diffuse from n-Si into p-GaN. Although the electrons and holes can move to the opposite side of the junction, the donors and acceptors are fixed in space. When electrons diffuse from the n-Si to the p-GaN side, they drop behind uncompensated donor ions in the n-Si material, and holes leaving the p-GaN region produce uncompensated acceptors. The diffusion of electrons and holes from the vicinity of the junction establishes a region of positive space charge near the n side of the junction and negative charge near the *p* side. An electric field is constructed at the interface. Photocurrents are directly gained (An Doan et. al., 2002).

At zero voltage bias, the energy gaps (E_g) for GaN and Si are 3.4 eV (Miskys et. al., 2003) and 1.12 eV, respectively. The electron affinity for GaN is approximated as 4.10 eV (Yeh et. al., 2007). However the electron affinity for Si is 3.95 eV (Edgar et. al., 1999). The energy barrier ΔE_c for electrons is $\Delta E_c = \chi_{GaN} - \chi_{Si} = (4.10 - 3.95)$ eV = 0.15 eV, while the energy barrier ΔE_v for holes is $\Delta E_v = E_{g,GaN} + \Delta E_c - E_{g,Si} = (3.4 + 0.15 - 1.12)$ eV = 2.43 eV. Consequently, the energy barrier for holes (ΔE_v) is 16 times more than the barrier for electrons (ΔE_c) (He et. al., 2007).

When a reverse bias is used, holes confront ΔE_v, producing a low current. In adverse, when a forward bias is applied, the electron only require to surmount a much smaller potential barrier (ΔE_c), hence presenting rise to the rectifying effect. These arguments are for the ideal case, and direct measurements are required to ascertain the exact band structure of the heterojunction (He et. al., 2007).

In addition, crystalline GaN photodiodes were designed to be responsive to optical source. The UV photocurrent measurement was performed using an Hg-lamp. Because the photon energy is higher than the bandgap of GaN (~ 364 nm wavelength), UV light was absorbed by the crystalline GaN creating electron-hole pairs, which were further separated by the electric field inside GaN contributing to the increase of the external current. Since the bandgap of p-GaN layer is larger than that of the substrate, band-to-band photo excitation cannot take place in this layer (Luo et. al., 2006; Soci et. al., 2010).

In this work, reverse bias was used for the photodiode operation to maximize the depletion width, reduce the transit time, and the carrier loss because of recombination process in diffusion area. Photodiodes typically have high resistance under reverse bias. This resistance is reduced when light of an appropriate frequency illuminates on the junction to create additional free electron-hole pairs. Consequently, a reverse biased diode may be applied as a detector by observing the current running through it (He et. al., 2007).

The I-V characteristics are shown in Fig. 8, where clear rectifying behavior can be seen both in the dark and under UV illumination conditions. Clear response to UV illumination can be seen from Fig. 7 in the reverse-biased condition because of the photogeneration of additional electron-hole pairs. The magnitude of photocurrent improves with the increase of applied reverse bias because of enhanced carrier collection. The photocurrent is 2.0 mA with -3 V reverse bias. On the other hand, the dark leakage current for the GaN photodiodes is weak (0.8 mA in the -3 V reverse bias). This behavior shows that GaN photodiodes can sensitively distinguish UV light to produce the measurable photocurrent response. The capability of the device is influenced by a few factors, including the presence of a thin native silicon dioxide barrier layer at the interface of the heterojunction that will decrease the photocurrent; and light loss by reflection and absorption of the top metal contact (Fahrettin et. al., 2008).

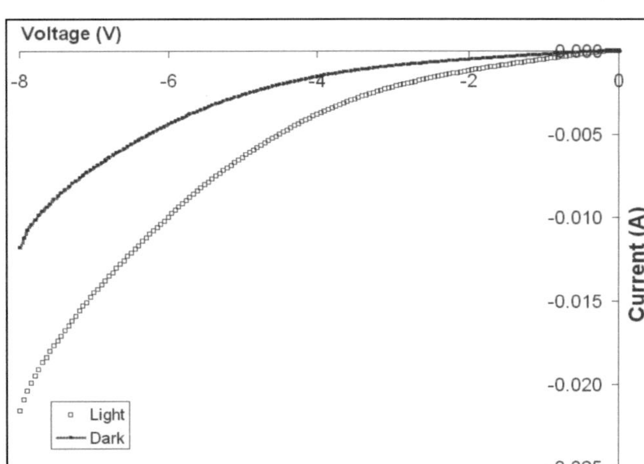

Fig. 8. The I-V characteristics of p-GaN/n-Si photodiodes. Photo-currents under illumination (light) are shown from the diodes (Chuah et. al., 2008a)

For the second set sample, Fig. 9 show the *I-V* characteristics of the Ni/AlN/GaN/AlN Schottky barrier photodiodes annealed at different temperatures. The higher turn-on voltage observed for all measurements under dark currents could be attributed to the highly resistive nature of the AlN cap layer. The dark current was 2.37 x 10^{-3} A under 10 V applied bias for as deposited Ni/AlN/GaN/AlN Schottky diode. On the other hand, for Schottky diodes annealed at 500 °C, 600 °C, and 700 °C, the dark currents were 3.25 x 10^{-4}, 4.97 x 10^{-5}, and 5.05 x 10^{-5} A, respectively, under 10 V applied bias.
By referring to Table 1, annealed samples exhibited more significant changes to the dark current characteristics compared to the as deposited Ni/AlN/GaN/AlN Schottky diode. This finding was further confirmed by measuring the contrast ratio of photo-current and dark current at 10 V. The contrast ratio for annealed sample (at 700 °C) and as deposited Schottky diode were found to be 25 and 2, respectively. High temperature (600 °C and 700 °C) annealing treatment increased the barrier height as well as reduced the dark current as compared to the low temperature annealing (500 °C).
The *I-V* characteristics of the Schottky diode, Φ_B and n, were determined assuming thermionic emission (Monroy et. al., 1998):

$$I = I_o [\exp \{qV/nkT\}-1] \quad (3)$$

$$I_o = A^*AT^2 \exp\{-q\,\Phi_B/(kT)\} \quad (4)$$

where I_o is the saturation current, n is the ideality factor, k is the Boltzmann's constant, T is the absolute temperature, Φ_B is the barrier height, A is area of the Schottky contact and A^* is the effective Richardson coefficient. The theoretical value of A^* can be calculated using

$$A^* = 4\pi m^* q k^2/h^3 \quad (5)$$

where h is Planck's constant and $m^* = 0.27 m_0$ is the effective electron mass for AlN (Chen at. Al., 2004). The value of A^* is determined to be 32.4 $Acm^{-2}K^{-2}$.
Using equation (3) and the theoretical value of A*, under dark condition, the Schottky barrier height derived by the I-V method is 0.48 eV for as deposited Ni/AlN/GaN/AlN Schottky diode. On the other hand, the effective barrier heights of 0.52 eV, 0.55 eV, and 0.57 eV were obtained for Schottky diodes annealed at 500 °C, 600 °C, and 700 °C, respectively.
In our earlier study, a thin AlN cap layer was incorporated in GaN Schottky diode to improve the effective Schottky barrier height and decreases the dark current. A barrier height of 0.52 eV for typical Ni/GaN Schottky diode was increased to the effective barrier height of 0.63 eV for Ni/GaN Schottky diode with thin AlN cap layer. The low dark currents in GaN Schottky barriers with the thin AlN cap layer might be attributed to the AlN cap layer, resulting in a higher potential barrier, compared to standard sample. The deleterious effect of the interface states near the metal/semiconductor interface may be reduced for the sample owing to the insertion of the AlN cap layer. It was established that AlN cap layers would completely suppress the dark current of the GaN Schottky diodes and resulted in improved device characteristics (Chuah et. al., 2008b).
In the last few years, various kinds of nitride-based photodetectors have been reported (Vivian et. al., 2010). Nevertheless, an opposing condition usually occurs, when large differences in the lattice constant and thermal expansion coefficient of GaN and silicon inevitably lead to high dislocation density in the GaN epitaxial layer. Such a result contributes to a large dark current and smaller photocurrent/dark current ratio for nitride-

based Schottky barrier photodetctors. Despite, we attained low dark current from the GaN metal-semiconductor-metal photodiode with a 50-nm thick low temperature GaN barrier enhancement layer (Chuah et. al., 2008c).

With a panorama to decreasing the dark current of Schottky barrier photodetectors, metal-insulator-semiconductor (MIS) structures are for the present deeply investigated. In past reports, a number of gate dielectrics for example SiO_2 have been used in MIS structures (Casey et. al., 1996; Gaffey et. al., 2001). Nevertheless, these insulators were all ex-situ deposited and the contamination might occur at the insulator/semiconductor interface. Consequently, we deposited an in-situ AlN cap layer by MBE in our study. Nevertheless, not much report on GaN-based Schottky barrier photodiodes capped with an AlN layer can be detected in the literature, to our knowledge.

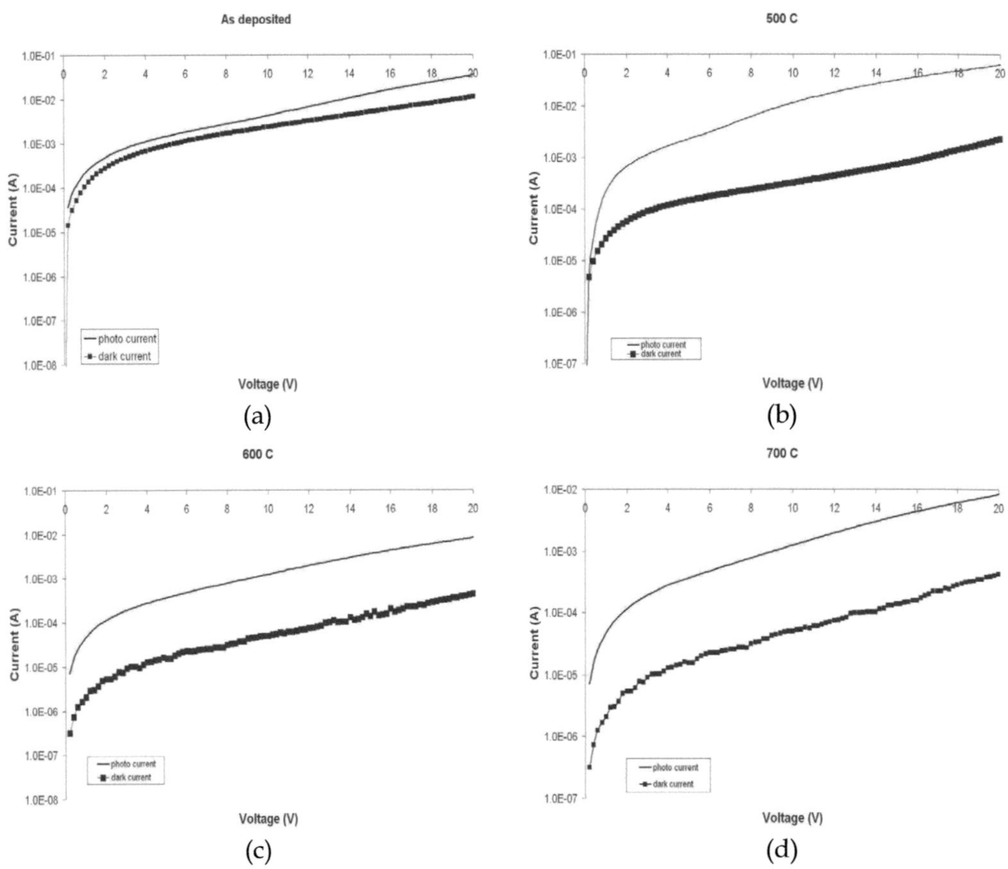

Fig. 9. I-V characteristics of the fabricated photodiodes annealed at different temperatures (a) as deposited, (b) 500°C, (c) 600°C, and (d) 700°C (Chuah et. al., 2009a)

According to Reddy (Reddy et. al., 2007), they hypothesized that the temperature dependence of Schottky barrier height maybe due to the changes of surface morphology of Pt films on the n-GaN and variation of nonstochiometric defects at the interface. Khanna et al. (Khanna et. al., 2006) investigated temperature dependence of W_2B_5-based rectifying contacts to n-GaN and showed that the Schottky barrier height (0.65 eV) increased with annealing temperature up to

200 °C. Wang et al. (Wang et. al., 2003) investigated that the variation of barrier height upon annealing could be ascribed to variation of surface morphology and inequality of nonstochiometric defects at the interface vicinity. In the current work, the increase of the Schottky barrier height could be because of the decrease of nonstochiometric defects in the metallurgical interface (Wang et. al., 2003). The region involving the defects could be decreased because of the interdiffusion of Ni and Al. Hence, the consumption of the defect region is followed by a raise in the value of the Schottky barrier height extracted from the I-V characteristics for the samples annealed from 500 ° C to 700 °C.

Temperature (°C)	Samples (photodiodes)	Ideality factor, n	Barrier height, Φ_b (eV)	Current at 10V (A)
As deposited	dark	1.08	0.48	2.37×10^{-3}
	illumination	1.10	0.46	4.29×10^{-3}
500°C	dark	1.07	0.52	3.25×10^{-4}
	illumination	1.08	0.50	1.15×10^{-2}
600°C	dark	1.03	0.55	4.97×10^{-5}
	illumination	1.04	0.53	1.20×10^{-3}
700°C	dark	1.05	0.57	5.05×10^{-5}
	illumination	1.04	0.53	2.00×10^{-3}

Table 1. Summary of the dark and photo-current (I-V) characteristics of the samples annealed at different temperatures.

As generally known, chemical reaction between the metal and the semiconductor interfaces can play an important role in the electrical properties of metal/semiconductor contact. The change in the barrier height of Ni/AlN/GaN/AlN Schottky contact with annealing temperature may also be ascribed to the interfacial reaction occurring between metals and AlN and their alloys which extend to AlN films. These interfacial layers might have dissimilar work functions than the Ni/AlN contact layers, which is responsible for the observed increase of barrier height (Lee et. al., 2006).

From the value of the ideality factors that we have obtained, the deviation of the diodes' ideality factor from unity may be due to tunneling effects, image force, and edge leakage. Furthermore, a unique value of ideality factor could be associated with a given set of diode conditions like temperature, bias, and doping. From the literature, the Schottky barrier heights varied widely from 0.50 to 1.15 eV for Ni, depending on the measuring methods, doping concentration and quality of the GaN (Schmitz et. al., 1996; Kalinina et. al., 1997). This also suggests that the metal work function should not be the only factor affecting the Schottky characteristics of the diodes (Kalinina et. al., 1997).

For the third set sample, a Schottky contact behaviour could be more closely described by the equation which takes into account the barrier height lowering because of electric field, tunneling effects, the presence of an interfacial layer, and carrier recombination in the space charge region of the metal-semiconductor contact as given by (Rideout et. al., 1975; Abdulmecit et. al., 1992)

$$I = I_0 \exp\left(\frac{eV}{nkT}\right)\left[1 - \exp\left(\frac{-eV}{kT}\right)\right] \tag{6}$$

where I is the current, I_0 is the saturation current, V is the bias voltage, and n is the ideality factor. The expression for the saturation current, I_0 is

$$I_0 = AA^*T^2\exp(\frac{-q\Phi_b}{kT}) \quad (7)$$

where A is the Schottky contact area, and the theoretical value (Hacke et. al., 1993) of A* for GaN is ~ 26.4 $Acm^{-2}K^{-2}$. Here, the theoretical value of A* for $Al_{0.09}Ga_{0.91}N$ can be estimated by using the relation $A^* = 4\pi qm^*k^2/h^3$ where h is Planck's constant and m* is the effective electron mass for AlGaN. As there has not been any references or reported experimental results of the effective mass of electron in AlGaN, therefore m* for $Al_xGa_{1-x}N$ with different x is estimated by a linear interpolation from the theoretical value of $m^* = 0.35\ m_0$ for AlN (Kim et. al., 1997) and $m^* = 0.22\ m_0$ for GaN (Piotr et. al., 1997). Here, m_0 is the free electron mass. From the equation as being mentioned above, A* for $Al_{0.09}Ga_{0.91}N$ is estimated to be ~ 27.9 $Acm^{-2}K^{-2}$. However, it should be noted that a large variation in A* does not have a significant influence on the Φ_B value that is to be determined (Liu et. al., 1998). Equation (6) can be written in the form of

$$\frac{I\exp(eV/kT)}{\exp(eV/kT)-1} = I_0\exp(eV/nkT) \quad (8)$$

At T $\leq$ 370 K and when V $\leq$ -0.5 V, equation (8) can be simplified to (Averine et. al., 2000)

$$I\exp(\frac{eV}{kT}) = I_0\exp(\frac{eV}{nkT}) \quad (9)$$

$$\ln[\,I\exp(\frac{eV}{kT})\,] = \ln I_0 + \frac{eV}{nkT} \quad (10)$$

The plot of ln [Iexp(eV/kT)] vs V should give a straight line with the slope = e/nkT and y-intercept at $\ln I_0$. Using equation (7), the ideality factor (n) and Schottky barrier height (SBH) were determined by the *I-V* method.

Figures 10 show the I-V characteristics of Ni/AlGaN photodetectors under dark and illumination conditions. Table 2 summarizes dark and illumination current measured at 10V, as well as the ideality factor and SBH of the samples determined from the I-V measurements. By referring to Table 2, it is found that high temperature annealing in oxygen ambient (600 °C and 700 °C) resulted in more significant changes to the dark current characteristics compared to the lower temperature annealing treatment. High temperature annealing treatment increased the barrier height as well as reduced the dark current level. For lower annealing temperature (400 °C and 500 °C), the barrier height for annealed samples (photodiodes) increased. As shown in Fig. 10, it can be seen that dark currents became significantly smaller after annealing. Such a reduction can again be attributed to the formation of NiO layers since NiO is p-type in nature (Averine et. al., 2000). As a result, we achieved larger Schottky barrier heights from the thermally annealed samples.

In the literatures, it can been seen that the bi-layer of Ni/Au film annealed in oxygen by using photo-CVD would transform the metallic Ni into NiO along with Au grains and amorphous Ni-Ga-O phases (Chen et. al., 1999). Hence, the Ni/Au semitransparent contact

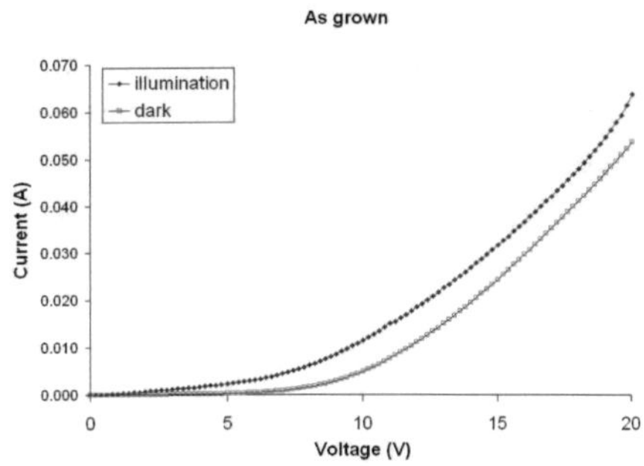

(a) As grown samples.

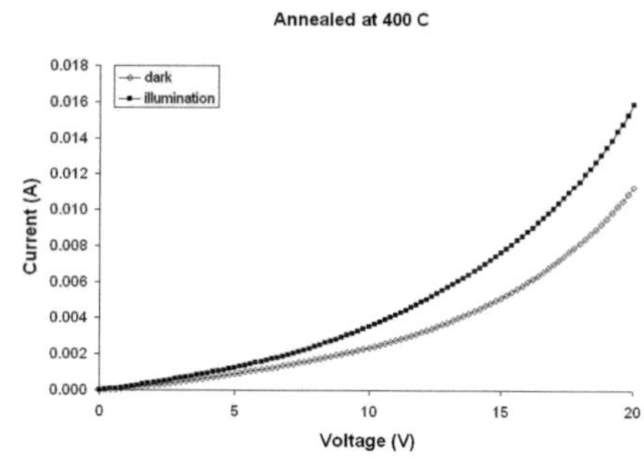

(b) Annealed at 400 °C.

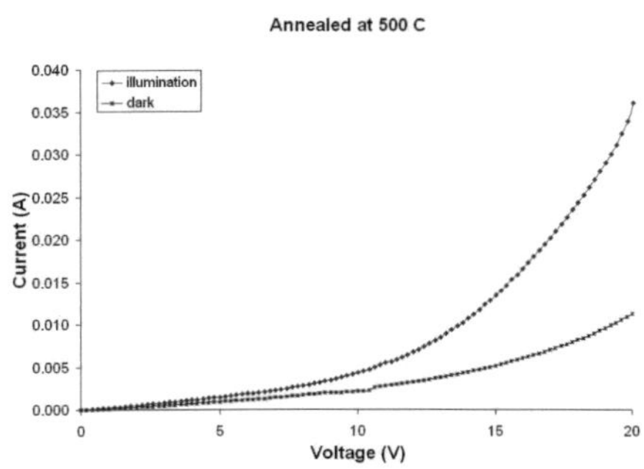

(c) Annealed at 500 °C.

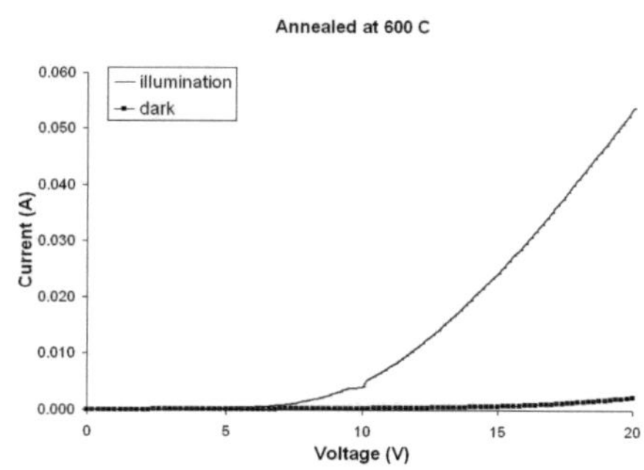

(d) Annealed at 600 °C

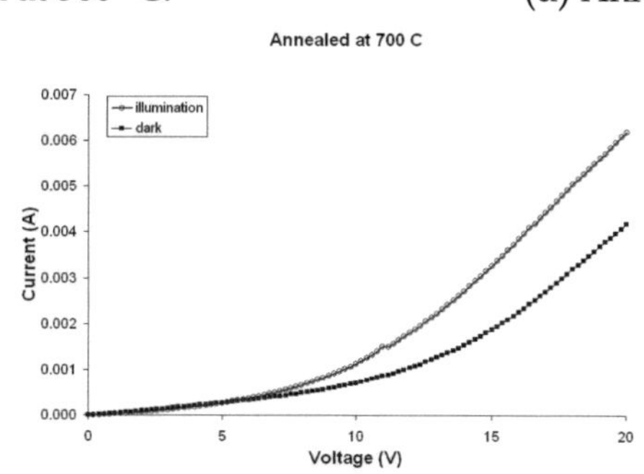

(e) Annealed at 700 °C.

Fig. 10. Current-voltage (I-V) characteristics of Ni/AlGaN photodetectors under dark and illumination conditions (Chuah et. al., 2009b)

can be used to form good Schottky contacts with superior Schottky barrier height. In this study, the Ni/$Al_{0.09}Ga_{0.91}N$ samples were annealed at a range of temperatures in flowing oxygen by using a tube furnace. NiO, which is reported to behave as a p-type semiconductor with nickel vacancies and/or oxygen interstitials, has also been reported as a kind of passivation layer on the AlGaN interface (Kim et. al., 2006). As the annealed temperature increases, the transmittance increases. The increase of transmittance also indicates the formation of NiO. We are able to achieve excellent photo-current to dark-current contrast ratio from the AlGaN MSM photodetectors, which can be attributed to the formation of the transparent NiO in the Ni contacts. It was found that dark current of the detector became significantly smaller after annealing.

With a 10V applied bias, it was found that we can achieve a photocurrent to dark current contrast ratio of 12 from the photodetectors with 600 °C annealed Ni contacts. This could be attributed to the more transparent nature of NiO formed after annealing so that more photons can be absorbed by the underneath AlGaN epitaxial layer. Such an effect should compensate the effect of larger Schottky barrier height. As the annealing temperature was increased to 700 °C, it was found that we can achieve a photocurrent to dark current contrast ratio of 2, which may be due to the formation of a rough Ni layer surface. Such a rough surface could increase light scattering losses and thus degraded the film transparency.
On the other hand, for Ni/AlGaN Schottky contact, the effect of "surface patches" which originates from surface defects such as dislocations and micropipes with residual oxides may not be neglectable (Sawada et. al., 2000). Postannealing is an effective method to repair the surface patches. However, considering that the annealing temperature is relatively low and the time is short, the surface patches effect could not be regarded as the main factor to the enhancement of the barrier height. When the sample was under illumination condition, the change of current was significant for the annealed sample as compared to the as grown sample. Both as grown and annealed samples originated from the same AlGaN wafer, and both of the devices were fabricated using same processing tools and under identical parameters, however, significant difference in dark current was observed. High dark current in the as grown sample could be attributed to the low barrier height of the metal contact with the AlGaN thin film.
Since the Ni film became more transparent after annealing in O_2, more photons should be absorbed by the underneath semiconductor. The oxidation layer near the interface provides an increasing energy barrier for carrier injection into AlGaN. As the oxidation temperature increases (annealed at 800 °C), the rectifying characteristics is deteriorated by excess oxidation. There may be some interfacial reaction between oxygen and AlGaN. The real interaction is still under investigation.
The effect of the reduction of leakage current or dark current, barrier height or ideality factor enhancement in the MSM photodiodes after heat treatment can mainly be explained by the chemical reaction of the metal with the interfacial oxide layer of the semiconductor

Temperature (°C)	Samples (MSM photodiodes)	Ideality factor, n	Barrier height, Φ_b (eV)	Current at 10V (A)
As grown	illumination	1.23	0.51	1.15×10^{-2}
	dark	1.20	0.53	4.94×10^{-3}
400°C	illumination	1.15	0.55	3.54×10^{-3}
	dark	1.10	0.57	1.37×10^{-3}
500°C	illumination	1.09	0.60	6.25×10^{-3}
	dark	1.08	0.61	2.15×10^{-3}
600°C	illumination	1.05	066	4.04×10^{-3}
	dark	1.04	0.69	2.20×10^{-4}
700°C	illumination	1.02	0.67	1.01×10^{-3}
	dark	1.01	0.70	7.09×10^{-4}

Table 2. Summary of the dark current characteristics of the samples annealed in oxygen ambient at different temperatures.

(Abdulmecit et. al., 1992) or macroscopic interaction between the metal and the semiconductor (Liu et. al, 1998). Experimental results also revealed that reactive contact metals decrease the interfacial layer and react with the semiconductor during annealing and hence it can cause a low or high barrier height for the MSM photodiodes.
It can be viewed that dark current increased gradually with the applied reverse bias and does not exhibit any impact of saturation. The absence of saturation for a Schottky contact under reverse bias could be clarified in terms of barrier height which is dependent on the electric field strength in the barrier as a result of the existence of an interfacial layer between the metal and the semiconductor. Besides that, the lack of saturation might be caused by image force lowering of the barrier height and because of the generation of electron-hole pairs in the depletion region as generation current is more pronounced at low temperatures than high temperatures because it has lower activation energy than the thermionic emission component (Rhoderick et. al., 1988).
$Al_{0.09}Ga_{0.91}N$ is known to experience from a high volume of defect densities due to the reasons like the difference between the thermal expansion coefficient and large lattice mismatch between the silicon substrates and the $Al_{0.09}Ga_{0.91}N$ material. During the reverse bias of the Schottky contact, the effect of the applied bias can be much greater when compared to the forward bias. Thus, tunneling current can also be observed in a semiconductor with a lower doping concentration during reverse biases which mean that the tunnel current cannot be left out when investigating Schottky contacts under reverse bias voltage.
Consequently, apart from the usual thermionic emission current that survive within a metal-semiconductor contact, the high amount of dark current observed can be attributed to the tunneling of carriers across the barrier. This effect can be assisted by traps generated by defects (threading dislocations that reach the surface) and the interfacial layer to produce trap assisted tunnel currents. Deep level bulk states that are within a tunneling distance of the interface can be another kind of traps. Therefore, thermionic emission is the primary transport mechanism in these MSM photodiodes.
Another contribution to the high dark current in $Al_{0.09}Ga_{0.91}N$ is dominated by other current mechanism like leakage current generated by a high defect density present in $Al_{0.09}Ga_{0.91}N$ as well as the traps assisted tunnel current. The high defect density in $Al_{0.09}Ga_{0.91}N$ justifies the presence of inhomogeneity at Schottky contacts, which causes a local enhancement of the tunnel current. The rotation of the layer during the growth and the different positions of the III-element sources could produce alloy inhomogeneities, as already reported.
From the results, the value of the ideality factor is quite near to unity, thus indicating the high quality Schottky contact under investigation and the absence of a thick interfacial layer. However, the existence of a thin interfacial layer cannot be ruled out unless the semiconductor is cleaved in an ultra high vacuum (UHV) condition (Guo et. al., 1996). The variation in barrier height values may be due to formation of alloy, different thickness of interfacial layer present on the film, variation in surface roughness of samples used, defects present in films, presence of several transport mechanism, and the variations in the local stoichiometry (Pearton et. al., 1999).

5. Conclusions

In the previous years, GaN growth on Si has expanded fastly and a few solutions are emerging to adjust the stress because of the thermal mismatch. In recent years, many groups have obtained thick, crack-free GaN on Si (111) by molecular beam epitaxy (MBE) or metalorganic chemical vapor deposition (MOCVD). There are also some reports about devices fabricated on GaN/Si (111) such as light emitting diodes (LEDs) and high electron mobility transistors (HEMTs). PN heterojunction photodiode, metal-semiconductor-metal (MSM) photodetectors, and Schottky photodiodes are now well developed and expected to be commercialized soon. However, reports of photodetectors based on crack-free GaN on Si (111) are limited. In the near future, Si may be the predominant substrate material for GaN growth because of its low price and availability in large diameters. The further interest is that the Si substrate can be etched away for laser or LEDs and for development of GaN substrates.

6. Acknowledgement

Support from RU grant and Universiti Sains Malaysia is gratefully acknowledged.

7. References

Abdulmecit Turut, Sebahattin Tuzemen, Muhammet Yildirim, Bahattin Abay, and Mustafa Saglam, (1992) Barrier height enhancement by anealing Cr/Ni/Co alloy Schottky contacts on LEC GaAs, *Solid-State Electron.*, 35(10), pp. 1423 - 1426.

An Doan, (2002) Studies of Minority Carrier Recombination Mechanisms in Be Doped GaAs for Optimal High Speed LED Performance.

Averine, S., Chan, Y. C., Lam, Y. L.; (2000) Evaluation of Schottky contact parameters in metal-semiconductor-metal photodiode structures, *Appl. Phys. Lett.*, 77(2), pp. 274 - 276.

BenMoussa, A.; Hochedez, J.F.; Dahal, R.; Li, J.; Lin, J. Y.; Jiang, H.X.; Soltani, A.; Jaeger, J-C. De; Kroth, U.; Richter, M.; (2008) Characterization of AlN metal-semiconductor-metal diodes in the spectral range of 44-360 nm: Photoelission assessments, *Appl. Phys. Lett.*, 92, pp. 022108-0221011.

Casey Jr., H.C.; Fountain, G.G.; Alley, R.G.; Keller, B.P.; DenBaars, S.P.; (1996) Low interface trap density for remote plasma deposited SiO_2 on n-type GaN, *Appl. Phys. Lett.*, Vol. 68, Issue 13, pp. 1850-1852.

Chang, P.C.; Chen, C.H.; Chang, S.J.; Su, Y.K.; Yu, C.L.; Chen, P.C.; Wang, C. H.; (2004) AlGaN/GaN MSM photodetectors with photo-CVD aneealed Ni/Au semitransparent contacts, *Semicond. Sci. Technol.* Vol. 19, No. 12, pp. 1354-1357.

Chen, L.C. ; Chen, F.R. ; Kai, J.J. ; Chang, Li; Ho, J.K,; Jong, C. S.; Chiu, C.C.; Huang, C. N.; Chen, C. Y.; Shih, K. K. (1999) Microstructural investigation of oxidized Ni/Au ohmic contact to p-type GaN, *J. Appl. Phys.* 86, pp.3826- 3832.

Chen, L.C.; Fu, M.S.; Huang, I.L. ; (2004) Metal-semiconductor-Metal AlN Mid-Ultraviolet Photodetectors grown by Magnetron reactive sputtering deposition, *Japanese Journal of Applied Physics*, vol. 43, No. 6A, pp. 3353-3355.

Chuah, L.S.; Hassan, Z.; Abu Hassan, H.; Hussein Mourad, M.; (2008a) p-GaN/n-Si heterojunction photodiodes, *Surface Review and Letters*, Vol. 15, pp. 699-703.

Chuah, L.S.; Hassan, Z.; Abu Hassan, H.; Yam, F.K.; Chin, C.W.; Thahab, S. M.; (2008b) Barrier height enhanced GaN Schottky diodes using a thin AlN surface layer, *International Journal of Modern Physics B*, Vol. 22, pp. 5167-5173.

Chuah, L.S.; Hassan, Z.; Abu Hassan, H.; Chin, C.W.; Thahab, S.M.; (2008c) Large area GaN metal semiconductor metal (MSM) photodiode using a thin low temperature GaN cap layer, *Journal of Nonlinear Optical Physics and Materials*, Vol. 17, pp. 59 - 69.

Chuah, L.S.; Hassan, Z.; Abu Hassan, H.; Ahmed, N. M.; (2009a) GaN Schottky barrier photodiode on Si(111) with low-temperature-grown cap layer, *Journal of Alloys and Compounds*, Vol. 481, pp. L15-L19.

Chuah, L.S.; Hassan, Z.; Abu Hassan, H.; (2009b) The effects of annealing treatment in oxygen ambient on Ni/$Al_{0.09}Ga_{0.91}N$ UV photodetectors, *Journal of Optoelectronics and Advanced Materials*, Vol. 11, No. 5, pp. 76-82.

Edgar, J.H.; Strike, S.; Akasaki, I.; Amano H., Wetzel C. (1999) : Properties: Processing and application of Gallium Nitride and related semiconductor, INSPEC, London.

Fahrettin Yakuphanoglu, (2008) analysis of interface states of metal-insulator-semiconductor photodiode with n-type silicon by conductance technique, Sensors and Actuators A: Physical , vol. 147 pp. 104-109.

Gaffey, B.; Guido, L.J. ; Wang, X.W. ; Ma, T.P. ; (2001) High-quality oxide/nitride/oxide gate insulator for GaN MIS structures, *IEEE Trans. Electron Dev.*, Vol. 48, pp. 458 - 464.

Guo, J. D.; Pan, F. M.; Feng, M. S.; Guo, R. J.; Chou, P. F.; Chang, C. Y.; (1996) Schottky contact and the thermal stability of Ni on n-type GaN, *J. Appl. Phys.*, 80(3), pp. 1623-1628.

Hacke, P.; Detchprohm, T.; Hiramatsu, K.; Sawaki, N.; (1993) Schottky barrier on n-type GaN grown by hydride vapor phase epitaxy, *Appl. Phys. Lett.*, 63, pp. 2676 -2678.

Hassan, Z.; Lee, Y.C.; Yam, F.K.; Abdullah, M.J.; Ibrahim, K.; Kordesch, M.E. (2004) Microcrystalline GaN film grown on Si(100) and its application to MSM photodiode, *Mater. Chem. Phys.*, Vol. 84, pp. 369-374.

He, J. H.; Ho, S. T.; Wu, T. B.; Chen, L.J.; Wang, Z. L.; (2007) Electrical and photoelectrical performances of nano-photodiode based on ZnO nanowires, *Chemical Physics Letters*, vol. 435, pp. 119-122.

Kalinina, E.V., Kuznetsov, N.I., Babanin, A.I., Dmitriev, V.A.; Shchukarev, A.V.; (1997) Structural and electrical properties of Schottky barriers on n-GaN, *Diamond Relat. Mater.* Vol. 6, Issue 10, pp. 1528- 1531.

Keem, K.; Kim. H.; Kim, G.T.; Lee, J.S.; Min, B.; Cho. K.; Sung, M.Y.; Kim, S.; (2004) Photocurrent in znO nanowires grown from Au electrodes, *Appl. Phys. Lett.*, 84, pp. 4376-4378.

Khanna, R.; Pearton, S.J. ; Ren, F.; Kravchenko, I. (2006) Annealing and measurement temperature dependence of W2B5-based rectifying contacts to n-GaN, *Appl. Surf. Sci.*, 252, pp. 5814- 5819.

Kim, K.; Lambrecht, W. R. L.; Segall, B.; Schilfgaarde, M. V.; (1997) Effective masses and valence-band splittings in GaN and AlN, *Phys. Rev. B*, 56, pp. 7363-7375.

Kim, H.; Schuette, M.; Jung, H.; Song, J.H.; Lee, J.; Lu, W. (2006) Passivation effects in Ni/AlGaN/GaN Schottky diodes by annealing, *Appl. Phys. Lett.*, 89, pp. 053516-053519.

Lee, Y.C.; Hassan, Z.; Abdullah, M.J.; Hashim, M.R.; Ibrahim, K. (2005) Dark current characteristics of thermally treated contacts on GaN-based ultraviolet photodetectors, Microelectronics Engineering, vol 81, pp. 262-267.

Lee, M.L.;. Sheu, J.K; Lin, S.W.; (2006) Schottky barrier heights of metal contacts to n-type gallium nitride with low-temperature-grown cap layer, *Applied Physics letters*, 88, pp. 032103-032106.

Liu, Q. Z.; Yu, L. S. ; Deng, F.; Lau, S. S.; Redwing, J. M.; (1998) Ni and Ni silicide Schottky contacts on *n*-GaN, *J. Appl. Phys.*, 84, pp. 881-887.

Luo, L.; Zhang, Y.F.; Mao, S. S.; Lin, L.W.; (2006) Fabrication and characterization of ZnO nanowires based UV photodiodes, *Sensors and Actuators A*, 127, pp. 201-206.

Miskys, C. R.; Kelly, M. M; Ambacher, O.; Stutmann, M., (2003) Freestading GaN-substrates and devices, *Phys. Stat. Sol. (C)*, 6, pp. 1627 - 1650.

Monroy, E.; Munoz, E.; Sanchez, F.J.; Calle, F.; Calleja, E.; Beaumont, B.; Gibart, P.; Munoz, J.A.; Cusso, F.; (1998) High-performance GaN p-n junction photodetectors for solar ultraviolet applications, *Semicond. Sci. Tech.*, 13, pp. 1042 - 1046.

Pearton, S. J.; Zolper, J. C.; Shul, R. J. ; Ren, F. (1999) GaN: Processing, defects, and devices, *J. Appl. Phys*, vol. 86, Issue 1, pp. 1-78.

Piotr Perlin, Tadeusz Suski, Michal Leszczynski, and Henryk Teisseyre (1997) *GaN and Related Materials*, edited by S. J. Pearton, (Gordon and Breach Science Publishers, Amsterdam.

Philippe, A.; Bru-chevallier, C.; Gamez-cuatzin, H.; Guillot, G.; Martinez-Guerrero, E.; Feuillet, G. ; Daudin, B.; Aboughe-Nze, P.; Monteil, Y.; (1999) Optical study of cubic gallium nitride band-edge and relation with residual strain, *Phys. Stat. Sol. (B)*, vol. 216, Issue 1, pp. 247- 252.

Reddy, V.R.; Rao, P.K.; Ramesh, C.K.; (2007) Annealing effects on structural and electrical properties of Ru/Au on n-GaN Schottky contacts, *Material Science and Engineering B*, 137, pp. 200 - 204.

Rhoderick, E. H.; Williams, R. H.; (1988) Metal-Semiconductor Contacts, second ed., Oxford University Press, New York.

Rideout, V. L.; (1975) A review of the theory and technology for ohmic contacts to group III-V compound semiconductors, *Solid-State Electron.*, Vol. 18, Issue 6, pp. 541- 550.

Qioa, D.; Yu, L.S.; Lau, S. S.; Redwing, J. M.; Lin, J. Y. Lin; Jiang, H.X.; (2000) Dependance of Ni/AlGaN Schottky barrier height on Al mole fraction, *Journal of Applied Physics*, 87 (2000) pp. 801-804.

Vivian K.X. Lin; Esther Alarcon-Llado; Kim, H.H.; Arulkumaran, S.; Ng, G.I.; Raman, A.; Tripathy, S.; (2010) Structural and electrical characterization of AlxGa1-xN/GaN Interfaces for UV Photodetectors, *Electrochem. Solid-State Lett.*, Vol. 13, Issue 9, pp. H301-304.

Wang, J.; Zhao, D.G.; Sun, Y.P.; Duan, L.H.; Wang, Y.T.; Zhang, S.M.; Yang, H.; Zhou, S.; Wu, M.; (2003) Thermal annealing behavior of Pt on n-GaN Schottky contacts, *J. Phys D: Appl. Phys.*, vol. 36, pp. 1018-1022.

Sawada, T.; Ito, Y.; Imai, K.; Suzuki, K.; Tomozawa, H.; Sakai, S.; (2000) Electrical properties of metal/GaN and SiO_2 /GaN interfaces and effects of thermal annealing, *Appl. Surf. Sci*, 159/160, pp. 449-455.

Schmitz, A.C.; Ping, A.T.; Asif Khan, M.; Chen, Q.; Yang, J.W.; Adesida, I.; (1996) Schottky barrier properties of various metals on n-type GaN, *Semicond. Sci. Technol.*, 11, pp. 1464 - 1467.

Soci, C., Zhang, A., Bao, X.Y.; Kim, H.K.; Lo, Y.; Wang, D.; (2010) Nanowire photodetectors, *Journal of Nanoscience and Nanotechnology*, Vol. 10, pp. 1-20.

Sze, S.M. (1981) Physics of Semiconductor Devices, 2nd ed., Wiley, New York.

Yeh, D.H.; Hsieh, L.Z.; Chang, L.B.; Jeng, M. J.; (2007) The characteristics of platinum diffusion in n-type GaN, *Applied Surface Science*, 253, pp. 6910-6914.

Yu, C.L.; Chang, S.J.; Chang, P.C.; Lin, Y.C.; Lee, C. T.; (2006) Nitride-based ultraviolet Schottky barrier photodetectors with LT-AlN cap layers, *Superlattices and Microstructures*, vol. 40, pp. 470-475.

17

Gas Source MBE Grown Wavelength Extending InGaAs Photodetectors

Yong-gang Zhang and Yi Gu
State Key Laboratory of Functional Materials for Informatics, Shanghai Institute of Microsystem and Information Technology, Chinese Academy of Sciences
China

1. Introduction

Photodetectors (PDs) responding to the light in short wave infrared (SWIR) of 1-3 μm have attracted much attention because of the unique spectral features in this band. For example, the water molecules, which are the most important component of this world, have 4 important absorption bands around 1.1, 1.4, 1.9 and 2.7 μm respectively as shown in Fig.1. (Rothman et al., 2005), between them clear transmission windows exist. The high contrast characteristics make it fascinating for the observation of our earth from the satellite to get diversified information. On the earth the spectroscopy character of numerous water containing substances as fruitage, beverage and medicament in SWIR band also arouses much interests. Besides, CO_2 also has significant spectral fingerprints in this band especially those around 1.6 and 2.1 μm. Those absorption features just located in the water windows with distinct intensities, which makes them quite suitable for CO_2 distribution mapping or monitoring from space or at ground. Furthermore, many areas including spectroscopy of characteristic absorption of gases or liquids, night and fog penetrate vision, thermophotovoltaic energy conversion, thermal imaging of high temperature objects, wind detection lidar, etc. also presume upon photodetectors in this band.

To detect the light in this band for the applications mentioned above arouses mainly two trait of concern, the sensitivity and response speed. For those applications the response frequencies in the range of tens of KHz to hundreds of MHz are common despite for special communication purpose. Among different categories of photodetectors, the thermal detectors are suffering from their inherent lower response speed and sensitivity. Therefore, the photon or quantum types of detectors should play an important role in this field. Quantum type PDs in SWIR band, which are the devices concerned in this chapter, may relate to different conformations and could be constructed using different type of materials. In this chapter a review on the material issues will appear first, then concentrated on the PDs using ternary III-V material of $In_xGa_{1-x}As$ on InP substrate grown by gas source molecular beam epitaxy (GSMBE) with different type of buffer structures and configurations. The buffer layer optimization will be emphasized and wafer growth, chip processing and device characterization will be introduced in detail.

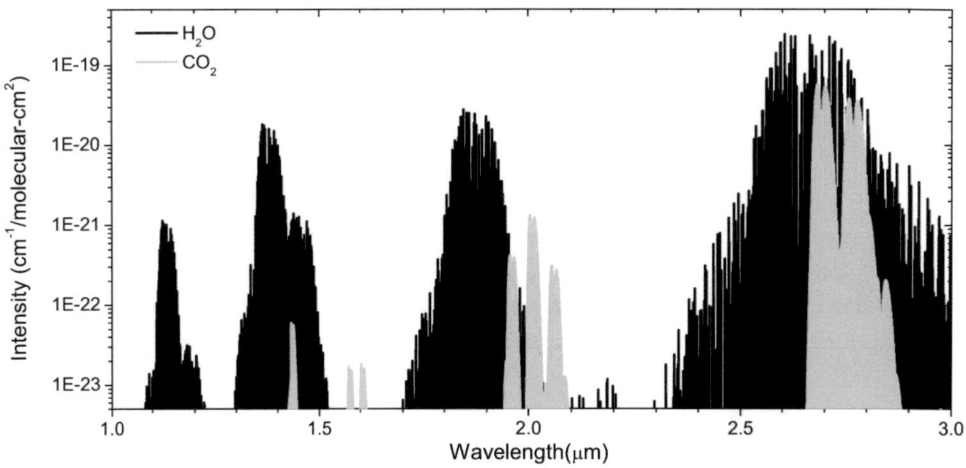

Fig. 1. Absorption characteristics of H_2O and CO_2 in short wave infrared band of 1-3 μm.

2. Material issues

2.1 Material system considerations

Quantum or photon type PDs may have different types such as photo-conductor or photovoltaic; from practical point of view photovoltaic are more preferable because the detectors can operate at zero bias with lower dark current. In SWIR band this type of detectors mainly rely on two kinds of semiconductor materials II-VI or III-V. The HgCdTe (MCT), a representative variable gap semiconductor of II-VIs, has worked successfully in SWIR band. It may be epitaxial grown on expensive but lattice matched CdZnTe substrate and excellent performances of the detectors and focal plane arrays (FPAs) have been reached. However, motivations to replace MCT especially in SWIR band always exist because of some technological problems of this material. The most important reason is the weak Hg-Te bond in MCT, which results in instabilities of the surface, interface and even the bulk material. Some other issues related to this are the difficulties in material growth and chip processing at lower temperatures with lower yield, as well as the lower operating temperature and weak radiation hardness. In III-Vs, the antimonides form some lattice matched systems covering SWIR band. The binary InAs and GaSb show nearly the same lattice constant around 6.1 Å, on those substrates different systems as quaternary InAsPSb or InGaAsSb could be grown. The quaternary InAsPSb on InAs substrate, which is the only quaternary with three group-V elements studied so far, covers the bandgap from InAs of 0.36 eV to about 0.7 eV (considering the miscibility gap) at 300K matches the SWIR perfectly. On this system liquid phase epitaxy (LPE) grown SWIR PDs have been demonstrated (e.g.: Zhang et al., 1990, 1992; Afrailov, 2010), whereas the difficulty in the control of three group-V elements As, P and Sb simultaneously to grow device quality wafers using metal-organic vapor phase epitaxy (MOVPE) or MBE remains a challenge. The other lattice matched antimonide quaternary system of InGaAsSb on GaSb, which covers the bandgap from InAs rich corner of 0.283 eV to GaSb rich corner of 0.727eV at 300K also matches the SWIR quite well. On this system LPE (e.g. Piotrowski et al., 2001), MOVPE (e.g.: Bhagwat et al., 2006) and solid source MBE (e.g.: Li et al., 1995; Liang et al., 2010) grown SWIR PDs also have been demonstrated. In this system it seems no severe limitations exist, but in practice the difficulties in the treatment of Sb containing materials in both epitaxy and chip processing aspects still exist. For example, the compatibility of Sb source with other III-V sources in the MOVPE or MBE growth remains a problem; the wet or dry etching of certain Sb containing materials is difficult. Besides, comparing to GaAs or InP, the availability of high quality GaSb substrates at moderate price is still an issue.

In addition to MCT or antimonide materials, the III-V ternary $In_xGa_{1-x}As$ has been a good candidate to cover the SWIR band. The $In_xGa_{1-x}As$, namely the alloy of two direct bandgap binaries with direct bandgap in full region, covers the whole bandgap range of 0.36-1.43 eV of InAs and GaAs. At Indium composition x of 0.53, it is lattice matched to InP substrate with bandgap about 0.75 eV at 300K. The photodiodes using this composition with cutoff wavelength of about 1.7 μm have been sufficiently developed for decades and widely used in the optical communication wavelength of 1.31 or 1.55 μm wavelength, and their excellent performances have been definitely proved. For the ternary $In_xGa_{1-x}As$ at x from 0.53 to 1, the bandgap could be varied from 0.75 to 0.36 eV at 300K, corresponding to cutoff wavelength of 1.7 to 3.4 μm fitting the SWIR well. To shift the response to longer wavelength, the Indium content in the InGaAs alloy should be increased. For instance, to move the cutoff wavelength of the photodiodes grown on InP substrate from 1.7 μm to about 2.4 μm, the Indium contents of the InGaAs alloy have to be increased from 53% to about 80%, which introduces a quite large lattice mismatch of about +1.85% between InGaAs layer and InP substrate, in this case an extraordinary buffer layer should be inserted to prevent the degradation of the material quality. On the other hand, the full maturity of the growth and processing technology of this material system, which have been validated from mass production, could compensate the residual degradation caused by mismatch adequately, makes this ternary very attractive from the application point of view. In the lattice mismatched system of $In_xGa_{1-x}As$ ($x>0.53$) on InP, some options of different buffer materials exist with different essential points of concern.

Fig.2 shows the lattice constant and bandgap energy contours of the InGaAsP quaternary, where InGaAs is at the top. As a natural choice, the InGaAs itself is a buffer layer option. The Indium composition x of $In_xGa_{1-x}As$ could be increased gradually from 0.53 to a desired value of y (corresponding to desired cutoff wavelength) in a continuous or step manner, on this buffer the PD structure could be grown. Generally, this scheme should be easier to realize without adopting other sources. Wavelength extending InGaAs photodetectors using $In_xGa_{1-x}As$ buffer have been grown by using MOVPE (e.g.: Moseley et al., 1986) or GSMBE (e.g.: Zhang et al., 2005, 2006a, 2006b) methods with different types of buffer and cap layer structures, and their performances were well evaluated. For the PDs normally a cap layer lattice matched to the light absorption layer is needed, in this scheme using also the composition y for the cap layer material with the same bandgap as that of absorption layer to form a homo pn junction is a rational choice. However, normally the cap layer using wider bandgap material is preferable to reduce the carrier loss due to the surface recombination, also to enhance the response at short wavelength side. Furthermore, in array applications the chip is often flip chip bonded to readout integrated circuit (ROIC), therefore back-side illumination is often adopted thus wider bandgap materials are also needed for the buffer layer to eliminate the absorption. In the InGaAsP system though using quaternary material as buffer is possible, taking care of four elements in the growth certainly is unnecessary, therefore the ternary InAsP is a good choice for the wider bandgap graded buffer. InAsP is also an alloy of two direct bandgap binaries InAs and InP with direct bandgap in full region. The lattice constant of InAsP can be changed from that of InP to InAs thus it can be lattice matched to $In_xGa_{1-x}As$ for x>0.53 and have a wider bandgap than that of InGaAs lattice matched to it, so it is suitable for both the buffer and cap layers. The wavelength extending devices are used mainly for sensing purpose where response speed is not an important trait of concern, so the carrier accumulation at the heterointerface is not important.

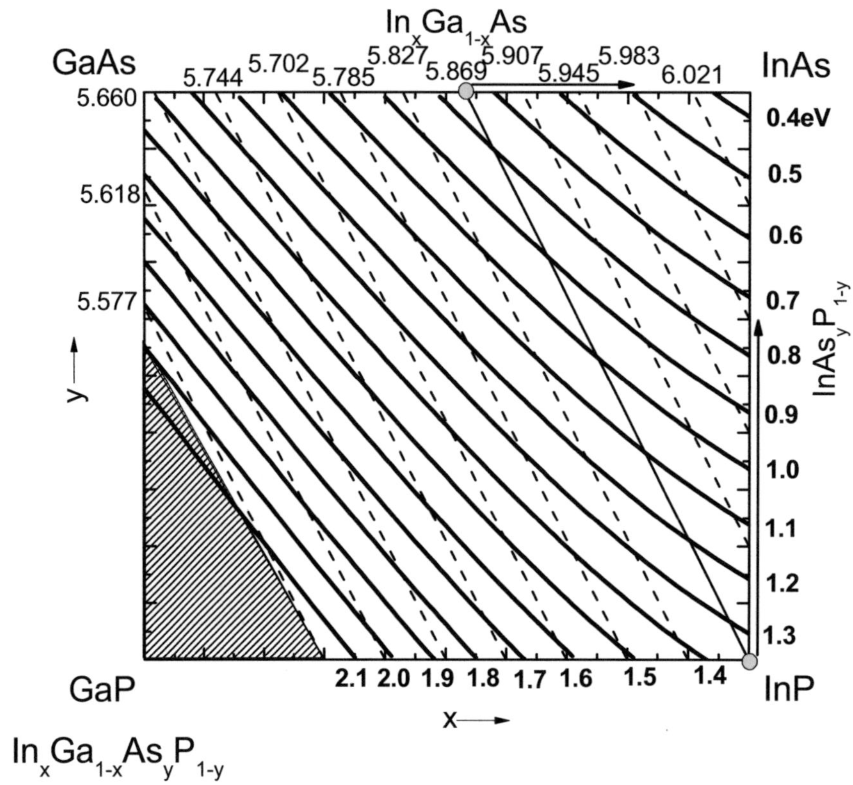

Fig. 2. Lattice constant and bandgap energy of InGaAsP quaternary system, the left bottom corner indicates the indirect bandgap zone.

From Fig.2 an example could be given. As buffer layer of $In_{0.8}Ga_{0.2}As$ active absorption layer with bandgap of about 0.5 eV cutoff at about 2.5 μm, the As composition in InAsP should be increased gradually from 0 to about 0.6; at this composition the bandgap of InAsP is about 0.75 eV, much wider than that of $In_{0.8}Ga_{0.2}As$. The InAsP is composed of two group-V elements As and P, indubitably in the growth these two sources should be needed and their ratio needs to be controlled effectively. Wavelength extending InGaAs photodiodes grown by using hydride vapor phase epitaxy (HVPE) (e.g.: Makita et al., 1988; Hoogeveen et al., 2001) or MOVPE (e.g.: di Forte-Poisson et al., 1992; Hondt et al., 1998) methods with different type of buffer and cap layer structures have been reported, and their performances were well evaluated. Note that in the lattice mismatched PD structure normally the buffer layer needs to be thick enough to reach a full relax. The growth of such materials with two group V components using vapor phase epitaxy is quite feasible, however, using molecular beam epitaxy (MBE) to grow such materials should be a challenge especially for thick layers. In MBE, setting the parameters to control two group V sources continuously and simultaneously will be very difficult in spite of using gas source or solid source, regardless various other demands of the photodiode structures.

Other options can be explored from the $Al_zGa_xIn_yAs$ (x+y+z=1) quaternary system, Fig.3 shows the lattice constant and bandgap energy contours of this quaternary system. This quaternary system is composed of three group-III elements of Al, Ga and In, therefore As is the only group-V element. In this system the ternaries $In_xGa_{1-x}As$ and $In_xAl_{1-x}As$ are at the right and left ridge, their direct bandgap at 300K could be written as $E_g=0.36x+1.43(1-x)-0.477x(1-x)$ and $E_g=0.36x+3.03(1-x)-0.7x(1-x)$ respectively. From Fig.3 it is remarkable that, the lattice constant grid is almost horizontal in this system, which means for AlGaInAs quaternary with the same Indium composition the lattice constant is almost the same. In the AlGaInAs system though using quaternary material as buffer is also possible, but work with three group-III elements simultaneously will increase the complexity in the growth without benefit, therefore the ternary InAlAs becomes a superior buffer choice. The ternary $In_xAl_{1-x}As$ is the alloy of two binaries of direct bandgap InAs and indirect bandgap AlAs; therefore it becomes indirect bandgap material when the Al composition exceeds about 60%. At Indium composition x of 0.52% this ternary lattice matched to InP substrate with a quite wide bandgap of 1.47 eV at 300K, even wider than that of GaAs. From Fig.3 it could be seen, using $In_xAl_{1-x}As$ as buffer layer of $In_{0.8}Ga_{0.2}As$ active absorption layer with bandgap about 0.5 eV cutoff at about 2.5 μm, the In composition in InAlAs should be increased gradually from 0.52 to 0.8 likewise; at this composition the bandgap of InAlAs is about 0.78 eV, even wider than that of InAsP at the same lattice constant. The InAlAs is composed of two group-III elements In and Al, indubitably in the growth these two sources should be

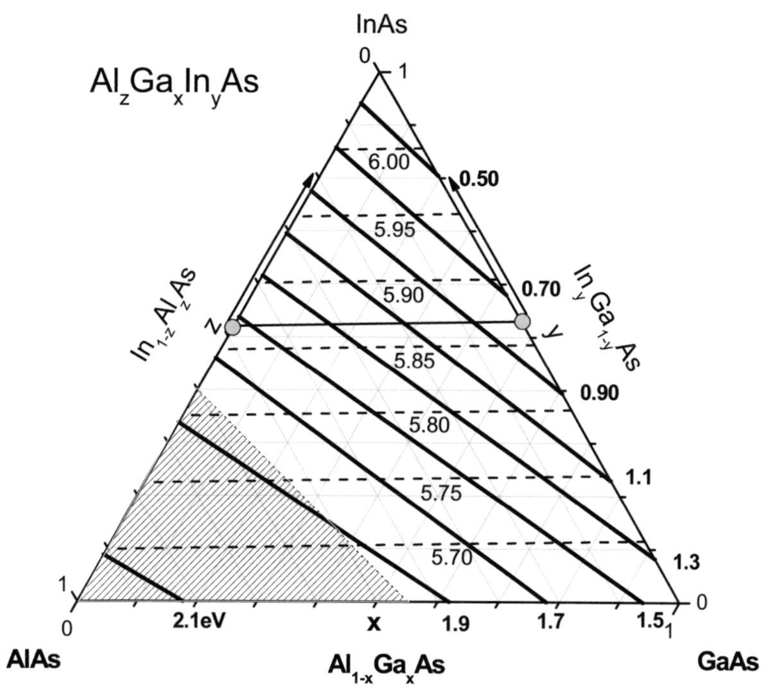

Fig. 3. Lattice constant and bandgap energy of AlGaInAs quaternary system, left bottom corner is the indirect bandgap zone.

needed and their ratio needs to be controlled effectively. Note that this is quite feasible in MBE. Besides, in this PD structure only InGaAs and InAlAs are adopted and in the epitaxial growth the phosphorus source is only used at pre-heat and desorption stage to protect InP substrate, therefore it is also feasible for the solid source MBE system without phosphorus source where arsenic flux may also be used for protection purpose. At the moment the solid source MBE system adopting solid phosphorus source is becoming popular, where this PD structure is also beneficial to save the phosphorus source. The lower Al composition <52% used in the wavelength extending PD structure will cause no reliability problem, and therefore it should be suitable for both the buffer and the cap layers. Wavelength extending InGaAs PDs grown on InP by using GSMBE (e.g.: Zhang et al., 2008a, 2008b, 2009a, 2009b, Tian et al., 2008b; Wang et al., 2009; Li et al., 2010; Gu et al., 2010) methods with different type of InAlAs buffer and cap layer structures have been reported, and their performances were well evaluated. These types of PDs even have been grown on GaAs substrate with even higher mismatch (e.g.: Zimmermann et al., 2003).

2.2 Gas source MBE growth of PD structures

In our works, two kinds of PD structures denoted as homostructure (where all the buffer, absorption and cap layers are InGaAs alloy) or heterostructure (where buffer and cap layers are InAlAs alloy, absorption layer is InGaAs alloy) were grown. All the epi-wafers in this work were grown by using a VG Semicon V80H GSMBE system. Photodetector wafers of various structures with different doping level in the $In_yGa_{1-y}As$ active absorption layer have been grown, where the In composition y is at desired value in the range from 0.6 to 0.9. In the growth, the elemental In, Ga and Al were used as group III sources, and their fluxes were controlled by changing the cell temperatures. Arsine (AsH_3) and phosphine (PH_3) high pressure cracking cells were used as group V sources (phosphorus source is only used for substrate protection purpose at pre-heat treatment stage), their fluxes were pressure controlled, and the cracking temperature was around 1000°C. Standard Be and Si effusion cells were used as p and n type doping sources, and doping level was also controlled by changing the cell temperatures. The fluxes of Al, Ga and In were calibrated with an in situ ion gauge to reach expected composition as well as moderate growth rate around 1 μm/hour. Calibration growths and X-ray measurements were carried out for lattice matched conditions of $In_{0.53}Ga_{0.47}As$ and $In_{0.52}Al_{0.48}As$ and then the parameters of Ga/In and Al/In fluxes were determined for further growth. Certainly, for the homostructure Al source is not used.

Before growth, the pre-heat treatment of the substrate was carried out in phosphor pressure at desorption. The process of pre-heat treatment was in situ monitored by reflection high energy electron diffraction patterns to see desorption of the surface oxygen, and then the substrate temperature was 10-30 °C decreased to begin buffer layer growth. In the growth of continuously graded $In_xGa_{1-x}As$ or $In_xAl_{1-x}As$ buffer layer, the In/Ga or In/Al flux ratio were increased continually until the In content increases from 53% to the desired value y (80% here as an example), then the growth of the $In_{0.8}Ga_{0.2}As$ absorption layer started. The $In_{0.8}Ga_{0.2}As$ absorption layer was doped with Si at designed level in the range between 1E16 cm^{-3} to 1E17 cm^{-3} according to the calibration data. After that, the Be doped $In_{0.8}Ga_{0.2}As$ or $In_{0.8}Al_{0.2}As$ cap layer was grown. The grown samples all have quite similar morphologies as shown in Fig.4 regardless of the final compositions. Under eye it has shiny surface, but with optical microscope the cross ripple could be seen along the $<011>$ and $<01\bar{1}>$ directions, which are typical morphology related to the existing of mismatch dislocations, especially at large positive mismatched conditions.

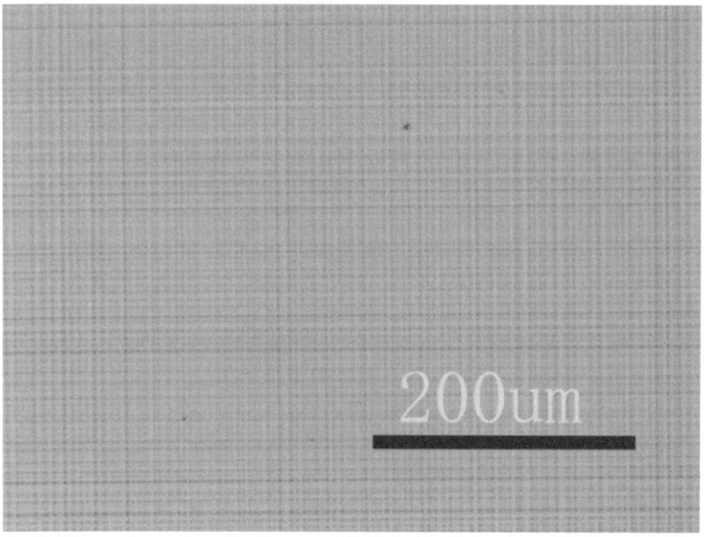

Fig. 4. Typical normasky micrograph of a GSMBE grown wavelength extending InGaAs detector epi-wafer with cross ripple pattern morphology, this wafer using 1 μm InGaAs continuously graded buffer with Indium composition of 0.6 in the absorption layer (Zhang et al., 2005).

In the growth of homostructure or heterostructure photodetectors with quite large lattice mismatch to the substrate, normally thicker layer with large lattice mismatch is desired, in this case the calibration growth of bulk layer will be quite difficult, and even possibly the calibrated data is not precise. We noticed that in the MBE growth the Ga and Al have almost the same flux characteristics and stick efficiency, but quite different from those of In. Based on this, a convenient and reliable growth procedure with excellent feasibility is developed, in which only the growth parameters of lattice matched $In_{0.53}Ga_{0.47}As$ and $In_{0.52}Al_{0.48}As$ are needed, that is: In the growth of all layers the As flux was kept constant, the growth of continuously graded $In_xGa_{1-x}As$ or $In_xAl_{1-x}As$ buffer begins from lattice matched condition of In/Ga or In/Al cell temperatures, T_{In}/T_{Ga} or T_{In}/T_{Al} (here T_{In} is the same for both InGaAs and InAlAs), then increase and decrease the cell temperature of In/Ga or In/Al simultaneously for ΔT_1 and ΔT_2 respectively at a very slow but fixed ramp rate. When the cell temperatures of In/Ga and In/Al reach $T_{In}+\Delta T_1/T_{Ga}-\Delta T_2$ or $T_{In}+\Delta T_1/T_{Al}-\Delta T_2$ at the same time, the growth of InGaAs absorption layer begins with $T_{In}+\Delta T_1/ T_{Ga}-\Delta T_2$ to a desired thickness, then growth the InGaAs or InAlAs cap with $T_{In}+\Delta T_1/ T_{Ga}-\Delta T_2$ or $T_{In}+\Delta T_1/T_{Al}-\Delta T_2$ to fulfill the whole structure. In this procedure the desired cutoff wavelength of the PD is only determined by ΔT, in our experience high quality wafers could be grown just by simply setting $\Delta T_2=2\Delta T_1$, no other careful calibration was needed.

After growth, the wafers were processed into mesa type photodetectors. The mesas with different diameters were defined by using photolithography and wet etching, and then passivated using plasma enhanced chemical vapor deposition of Si_3N_4. The contacts were formed using photolithography, evaporation of metals and lift-off. After an alloy step, the wafer was lapped and diced into chips, then packaged for further measurements.

3. Buffer layer optimization for mismatched InGaAs on InP

Unlike some other devices where only layers with sub-critical thickness are needed and dislocation free layers can be obtained, normally the growth of wavelength extending PD structures with thicker layers is accompanied by the generation of misfit dislocations and therefore the degradation of crystalline quality. Accordingly, a suitable buffer layer between InGaAs active absorption layer and InP substrate plays a prominent role on the growth of wavelength extending PD. The material system, modality and thickness of the buffer layer are the most important traits of concern. An appropriate buffer layer should relax the strain sufficiently, prevent the propagation of threading dislocations formed during the relaxation process into the active layers, form a moderate smooth surface morphology suitable for further device processing, and retain the intrinsic structural, optical and electrical qualities of the active layers.

As discussed in section 2.1, InGaAs homojunction, InAsP and InAlAs heterojunction buffer layers are some possible options for wavelength extending InGaAs PD structures. Efforts have continued to optimize the modality and growth schemes of buffer layers. Some people prefer a continuously graded buffer as it inhibits the dislocations propagating toward the active layer effectively (e.g.: Zimmermann et al., 2003; Zhang et al., 2008a, 2008b, 2009a, 2009b); others choose a step graded buffer since the interfaces between the consequent layers could bend the dislocation into the plane of growth (e.g.: Moseley et al., 1986; Makita et al., 1988; di Forte-Poisson et al., 1992; Hondt et al., 1998); strained-layer superlattice between steps is also adopted to filter threading dislocations (e. g.: Wada et al., 1993).

3.1 Continuously graded buffer of $In_xGa_{1-x}As$ versus $In_xAl_{1-x}As$

By MBE technology, the temperature ramping of the solid source beam flux to control the composition at high precision is more effective than the mass flow or pressure control in vapor phase epitaxy. Thus the $In_xGa_{1-x}As$ or $In_xAl_{1-x}As$ continuously graded buffer layers are feasible schemes in MBE. Comparing between $In_xGa_{1-x}As$ and $In_xAl_{1-x}As$ graded buffer layers, InAlAs heterojunction buffer with wider bandgap should be more appropriate for the reduction of carrier loss and FPA applications, whereas InGaAs homojunction buffer is generally considered to get better material quality if only from the material point of view. To evaluate the material quality of wavelength extending InGaAs PD structures using $In_xGa_{1-x}As$ or $In_xAl_{1-x}As$ continuously graded buffer layers in experiments, two PD structures with In content of about 0.78 in the InGaAs active absorption layer for the cutoff wavelength around 2.4 μm were grown on InP substrate by GSMBE with a grading rate of 0.6%μm^{-1} in the buffer layers, their characteristics were investigated and compared by the measurements of AFM, x-ray diffraction, TEM and PL.

The AFM images with 40 × 40 μm^2 scan area are shown in Fig. 5. The cross-hatch pattern could be attributed to the two types of misfit dislocations A and B oriented along the [1-10] and [110] directions, corresponding to group V and III atom-based cores, respectively (Wel et al., 1992). Along the [1-10] direction, the primary ridges are in parallel, and the period of primary ridges is around 5 μm. On the ridges small undulations exist periodically and the oval-like defects pop out on the top of the ridges. The period of the small undulations is around 1 μm. The surface morphology of sample (b) with InAlAs buffer appears to be more irregular than that of sample (a) with InGaAs buffer, which is due to the lower surface mobility of the Al atoms (Chyi et al. 1996). In spite of the more irregular morphology of sample (b), the root mean square (RMS) roughnesses of the two samples are nearly the same with 8.2 nm and 8.7 nm for sample (a) and (b), respectively. It is inadequate to judge the qualities of the samples from morphology analysis alone, and further evaluations are needed.

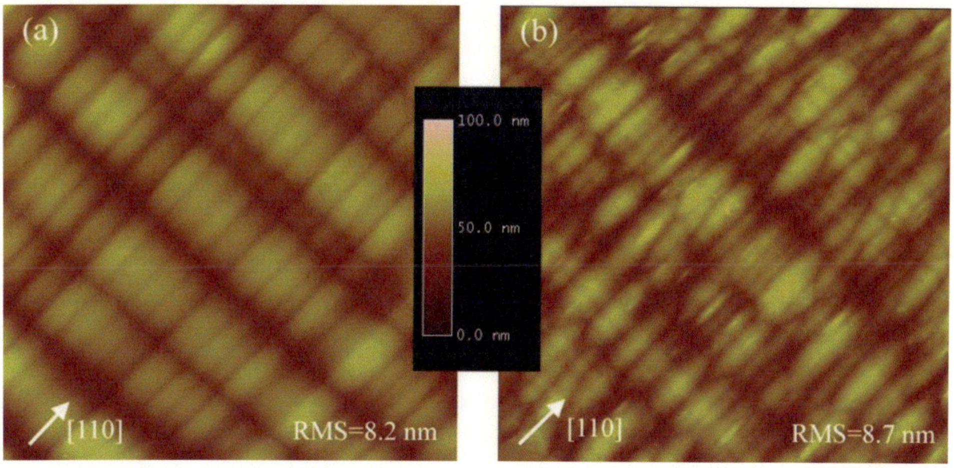

Fig. 5. AFM images of wavelength extending InGaAs PD structures with continuously graded (a) $In_xGa_{1-x}As$ and (b) $In_xAl_{1-x}As$ buffer layers.

To characterize the structural properties of the samples, x-ray diffraction reciprocal space mapping (RSM) and cross-sectional TEM measurements were used. The (004) and (224) RSM reflections were shown on the left side of Fig. 6. The intensities are in the logarithmic scale. In all RSMs, the relatively narrow and circular peaks correspond to the InP substrate (denoted as S). In Fig. 6(a), the relatively elliptical peak corresponds to the InGaAs absorption and cap layer of sample (a) (denoted as L), with a larger diffuse scattering perpendicular to the normal line due to the existence of dislocations. In Fig. 6(b), the diffraction features from InGaAs and InAlAs individual epitaxy layers adjacent with almost the same In contents are distinguished in the RSMs of the sample with InAlAs buffer. The two elliptical layer peaks correspond to the InGaAs absorption layer (denoted as L1) and InAlAs cap layer (denoted as L2) respectively. Those layers show diffusely scattered intensity patterns around reciprocal lattice point maxima, which results from the structural imperfections that can potentially degrade the material quality. L1 and L2 were assigned by measuring the HRXRD rocking curves in the same directions before and after the etching of the InAlAs cap layer. The layer peak was shifted toward the substrate side after the etching. On the (004) reflections, the divergencies of the centers of layer peaks and substrate peaks along the horizontal direction correspond to the macroscopic tilts of the layers to substrates (Fewster, 1993). The tilt angle of InGaAs absorption layer to the substrate is -26.4° and 1.1° for sample (a) and (b), respectively. For the (224) reflections, the intensities of substrate peaks are weaker than layer peaks due to the thick epi-layers and the weaker diffraction intensity from asymmetric (224) diffraction. The parallel mismatch and perpendicular mismatch of the InGaAs absorption layer were extracted from the RSMs, and then the cubic lattice mismatch, In composition, the degrees of relaxation and residual strain of the InGaAs absorption layer have been calculated and listed in Table 1. The degrees of relaxation are both larger than 90% for sample (a) and (b). It is noted that the residual strain in sample (a) is significantly larger than the calculated value from Turnoff's model (Tersoff, 1993). This suggests that a high tilt angle may be concurrent with a high residual strain. A possible explanation is that if the material has a significant residual strain, the lattice needs to tilt itself in order to reduce the elastic strain energy. This is consistent with the proposition of some people (e. g. Lee et al., 2007) that elastic strain can be released by lattice tilt. On the other hand, the tilt angle and residual strain of sample (b) is really slight. The different behavior between sample (a) and (b) may be due to the large composition variation of sample (a) caused by the phase separation (Quitoriano et al., 2007).

Sample	Tilting Angle (°)	Indium content (%)	Perpendicular mismatch (%)	Parallel mismatch (%)	Cubic mismatch (%)	Degree of relaxation (%)	Residual strain (10^{-3})	
							XRD	Tersoff
(a)	-26.4	77.5	2.31	0.58	1.69	93. 7	-11	3.1
(b)	1.1	77.0	1.62	1.69	1.65	96.1	0.34	3.0

Table 1. Results from XRD RSM measurements for InGaAs absorption layer of the two samples.

The occurrence of phase separation for sample (a) was confirmed by TEM measurements as shown in right side of Fig. 6. The defects have been both reduced significantly in InGaAs absorption layer on the $In_xGa_{1-x}As$ and $In_xAl_{1-x}As$ graded buffer layers. In the graded buffer layers, most misfit dislocations are localized at the internal interfaces. Less misfit dislocations appear to propagate vertically through the structure in the $In_xAl_{1-x}As$ buffer

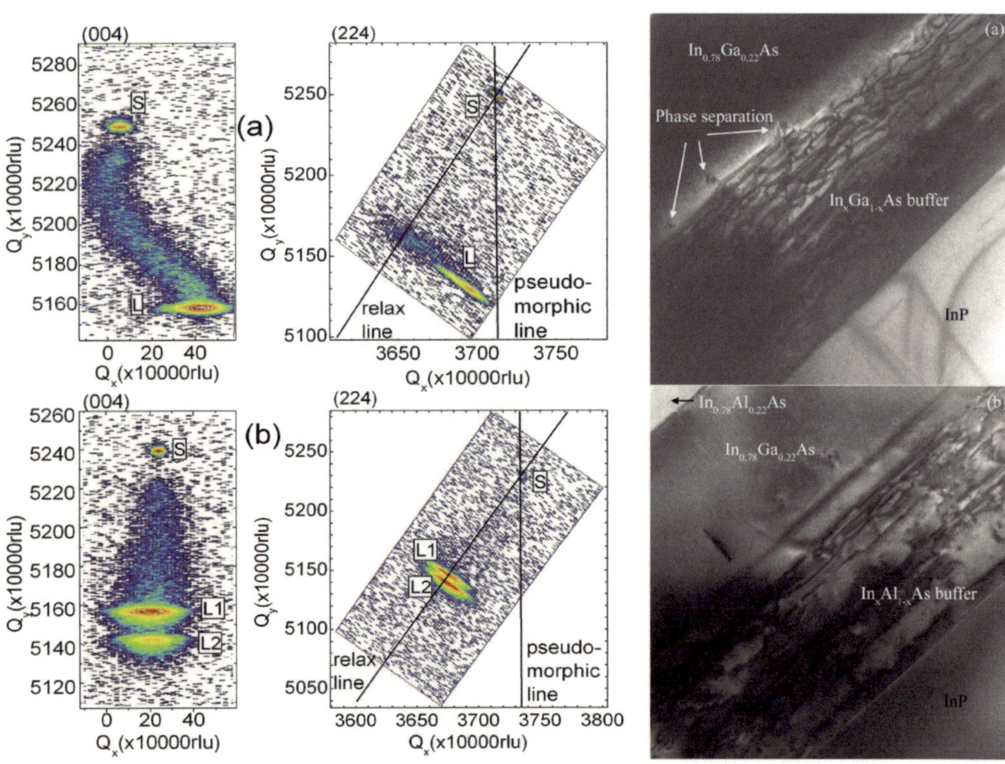

Fig. 6. RSM and cross-sectional TEM images of wavelength extending InGaAs PD structures with continuously graded (a) $In_xGa_{1-x}As$ and (b) $In_xAl_{1-x}As$ buffer layers.

layer than in $In_xGa_{1-x}As$ buffer layer. At the interface of $In_xGa_{1-x}As$ graded buffer and InGaAs absorption layer in sample (a), some diffraction contrast regions can be observed, which corresponds to the composition fluctuation induced by phase separation, whereas no obvious phase separation is observed in sample (b). It is probably due to the lower phase separation critical temperature of InGaAs than that of InAlAs (Quitoriano et al., 2007). The growth temperature is possibly located in the phase separation temperature region of InGaAs, whereas below the region of InAlAs.

The optical characteristics of the samples, which are correlated to the performance of the optoelectronic device in a more straightforward way, were evaluated using PL measurements at room temperature after etching away the cap layers. As shown in Fig. 7, both of the two samples show a PL peak from InGaAs absorption layer at about 2.41 μm. The PL intensity of sample (b) is more than twofold of sample (a) at room temperature. This indicates the improved optical characteristics and less presence of non-radiation recombination centers for the PD structure with $In_xAl_{1-x}As$ graded buffer layer.

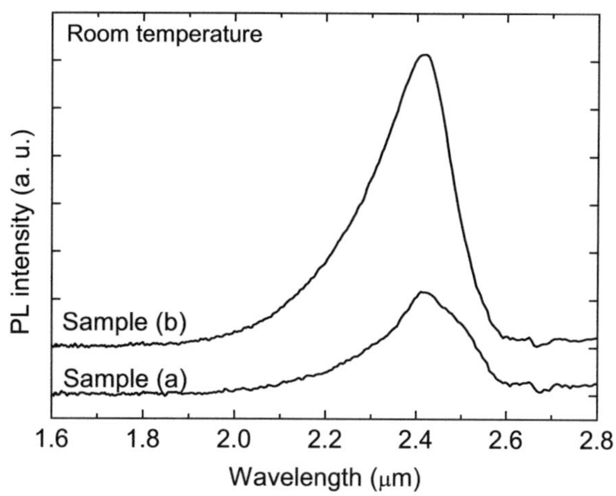

Fig. 7. PL spectra at room temperature of wavelength extending InGaAs PD structures with continuously graded (a) $In_xGa_{1-x}As$ and (b) $In_xAl_{1-x}As$ buffer layers, after etching away the cap layers.

As depicted in section 4.2, lower dark current and higher zero bias resistance area products R_0A could be obtained by using $In_xAl_{1-x}As$ heterojuction instead of $In_xGa_{1-x}As$ homojunction graded buffer and cap layer, which was attributed to the reduced carrier loss as the wider bandgap of InAlAs. It is believed that the reduced residual strain and decreased non-radiation recombination centers of the PD structure with $In_xAl_{1-x}As$ graded buffer layer should be also beneficial for the improvement of device performances.

3.2 $In_xAl_{1-x}As$ buffers at different grading rate

Generally, the buffer layers should be thick enough so that they become favourable for misfit dislocation to form and relax the buffer toward its freestanding lattice parameter. In vapour phase epitaxy, the achievable high growth rate makes the use of a thick buffer appropriate; buffer thickness exceeding 10 μm is ordinary. However, the growth of an optically and electrically inactive buffer layer with excessive thickness is not only cost ineffective but also impractical sometimes, especially for the MBE where the actual growth rate is limited to around 1 μmh^{-1}. Therefore from practical point of view a thin buffer should be more preferable.

To evaluate how thin the buffer layer can be shrunk, $In_{0.78}Ga_{0.22}As$ PD structures for the cutoff wavelength around 2.4 μm were grown on InP substrate by GSMBE with continuously graded $In_xAl_{1-x}As$ (x=0.52 to 0.78) buffer and $In_{0.78}Al_{0.22}As$ cap layer. The thicknesses of the buffer layers are set to be 0.7 μm, 1.4 μm and 2.8 μm for sample (a), (b) and (c), corresponding to the lattice mismatch grading rate of 2.4, 1.2 and 0.6 $\%\mu m^{-1}$ respectively. As a reference, a lattice matched PD structure annotated as sample (d) was also grown, with a 0.5 μm N^+ InP buffer, a 2.5 μm n^- $In_{0.53}Ga_{0.47}As$ absorption layer and a 0.6 μm P^+ InP cap layer.

Fig. 8 up right shows the AFM images with 40×40 μm^2 scan area measured in contact mode. In AFM images the cross-hatch pattern along [1-10] and [110] directions can be seen clearly. The RMS roughness for sample (a) with 0.7 μm buffer is 6.6 nm. With the double of buffer thickness, the RMS roughness increases to 9.5 nm for sample (b), and keeps almost unchanged of 9.7 nm for samples (c) with fourfold buffer thickness. Compare to the RMS roughness below 1 nm for the lattice matched structure sample (d), the mismatch induced morphology degradation is obvious. Results show that for the mismatch grading rate range from 2.4 to 0.6$\%\mu m^{-1}$ moderate roughness below 10nm could be reached, for further device processing this roughness is tolerable. Nevertheless it is inadequate to judge the quality of the wafer from the morphology analysis alone.

To determine the structural properties of the grown wafers, X-ray diffraction RSM analysis was applied. Fig. 8 left shows the RSMs from the symmetric (004) and asymmetric (224) reflections. In all RSMs, the relatively narrow and symmetric circular peaks correspond to the InP substrate (denoted as S). Other two predominant elliptical peaks, which are belong to the InGaAs absorption layer (denoted as L1) and InAlAs cap layer (denoted as L2) can be clearly distinguished. From the left side of Fig. 8, the (004) diffraction intensity maxima for each layer is almost centered on the substrate reciprocal lattice points along the vertical line, indicating minimum lattice tilt with respect to the substrate. The tilt angles of L1 to the substrate are -1.8°, -5.6° and -1.1° for sample (a), (b) and (c) respectively. From the asymmetric (224) reflection the contours of layers are all far from the pseudomorphic line, indicating that the strain has been relaxed through the continuously graded buffer. For sample (b) and sample (c) the intensity contours corresponding to InGaAs absorption layer L1 are centered on the relaxation line symmetrically, indicating the full relaxation of the layer. However, for sample (a) with the highest mismatch grading rate of 2.4$\%\mu m^{-1}$ the intensity contours corresponding to the InGaAs absorption layer L1 is apart from the relaxation line, makes an angle of 30.7° with respect to the vertical substrate reciprocal lattice intensity contours, which indicating that this layer is not relaxed sufficiently, since the angle between (004) and (224) should be 35.2° for a full relaxed layer. From the (224) reflection the In content, perpendicular mismatch and parallel mismatch of the layer L1 are extracted as listed in Table 2. The cubic lattice mismatch and degree of relaxation of the layer L1 are then calculated as also listed in Table 2. For both sample (b) and (c) the degree of relaxation reaches 102% whereas for sample (a) the value is 95%, regardless of the experimental error. In Table 2 the calculated residual strain at the interface of graded In_xAl_{1-}

$_x$As buffer to $In_yGa_{1-y}As$ (L1) absorption layer by using Turnoff's model were also listed (Tersoff 1993). Comparing the residual strain and In composition data in Table 2, an alloy concentration setback effect could be clearly seen. For sample (a) with higher grading rate the residual strain at the interface pullback the In composition distinctly, the trend consists with Turnoff's model well.

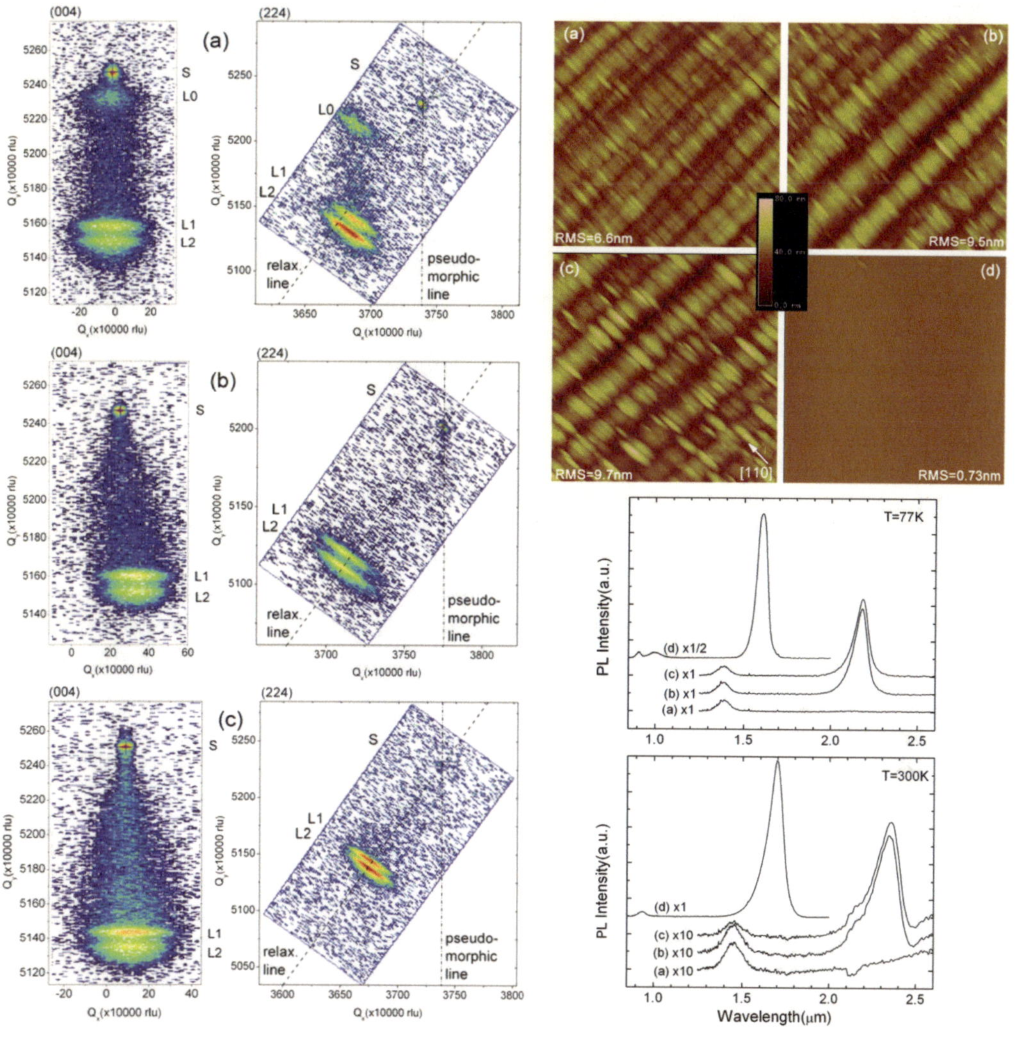

Fig. 8. RSM, AFM and PL spectra of wavelength extending InGaAs PD structures with continuously graded $In_xAl_{1-x}As$ buffer at different grading rate.

It is noticed from the RSMs that two types of relax mode may exist. From (004) reflections, the presence of continuously graded buffer can be noticed in the region between S and L1. For sample (b) and sample (c) in this region a pattern along Q_y direction appears diffusely, the mosaic structure is observable through a gradual extension along the Q_x directions, indicating a gradual relaxation process of the lattice. However, for sample (a) in this region the pattern is more uniform, a thin layer (denoted as L0) appears, which shows a weaker reflecting signal compare to L1 and L2. Note that this layer is unintentionally grown. From the (004) reflection L0 is quite close to the substrate. According to (224) reflection, a very weak pattern between L0 and L1 exists, which is almost parallel to the pseudomorphic line respect to L0 and L1. Detailed analysis from (224) reflection shows that L0 is almost fully relaxed in the parallel direction with respect to the substrate, while remaining strained along the perpendicular direction. This phenomenon prompts that when the grading rate during the growth becomes too high, an intermediate layer may form to release the strain in two steps. In this case, plenary relaxation of the subsequent layer will be difficult. The results reveal that in this type of structures containing thicker mismatched active layers, the grading rate is extremely important to the totally relaxation.

The optical qualities of the epi-layers are evaluated using PL measurements. Fig. 8 down right shows the 300K and 77K results. The PL of the reference sample (d) is also shown, which show 5~10 times enhancement in the intensity. At room temperature, sample (b) and (c) show fair PL signals from the InGaAs absorption layer peaked at about 2.35 μm with the FWHM of about 50 meV, whereas for sample (a) with the mismatch grading rate exceeding 2%μm^{-1} no PL signal could be observed. For these three samples, the PL signals from the InAlAs cap peaked at about 1.45 μm are almost identical. At 77 K, the PL peaks blue shifted to around 2.18 μm with the FWHM of about 35 meV and the intensities increase more than 6 folds, but for sample (a) still no PL signal from the InGaAs layer could be found. Therefore at this exorbitant grading rate the relaxation process are definitely insufficient; the residual strain and dislocation give rise to nonradiative centers and defect-assisted minority carrier recombination, degrading the optical quality of the layer notably. Moreover, for sample (c) with the lowest grading rate of 0.6%μm^{-1}, no further improvement of the PL could be noticed compared to sample (b) with a doubled grading rate, indicating that for this wavelength extending PD structure with a continuously graded buffer layer, a mismatch grading rate around 1%μm^{-1} should be slow enough to reach favorable structural and optical qualities.

Sample	Buffer thickness /grading rate (μm)/(% μm^{-1})	Indium content (%)	Perpendicular mismatch (%)	Parallel mismatch (%)	Cubic mismatch (%)	Degree of relaxation (%)	Residual strain (10^{-3})
(a)	0.7/2.4	73.75	1.662	1.512	1.586	95.33	6.388
(b)	1.4/1.2	74.42	1.607	1.682	1.645	102.2	4.517
(c)	2.8/0.6	74.53	1.616	1.691	1.654	102.2	3.194

Table 2. Results from XRD RSM measurements for InGaAs absorption layer of samples (a), (b) and (c).

4. Performance of GSMBE grown wavelength extending InGaAs PDs

4.1 Homostructure PDs

The response spectra of the detector were measured by using a Nicolet Magna 760 Fourier transform infrared (FTIR) spectrometer; in the measurement CaF2 beam splitter and Ever-Glo IR source were used. The DUT detector chip was mounted on a Cu block and packing into a metal Dewar and its output signal was fed into the pre-amplifier of the spectrometer. Fig.9 shows the measured relative response spectra of detectors cutoff at about 1.9, 2.2 and 2.5 μm at 300K, denoted as (a), (b) and (c) respectively (1.9μm, 2.2μm and 2.5μm detector hereafter), the detectors were under zero bias. For those detectors, homostructures with relatively thin graded layer were used. The epitaxy structure of the photodetectors consists of a 0.5μm n- InP buffer layer, a 1.0 μm (for a) or 1.5 μm (for b and c) n^- $In_xGa_{1-x}As$ continuously graded buffer layer with x changes from 0.53 to y of 0.6, 0.7 or 0.8, a 2.5μm n^- $In_yGa_{1-y}As$ absorbing layer and a 0.5μm p^+ $In_yGa_{1-y}As$ top layer. The samples were grown on exactly (100) oriented n^+ InP epi-ready substrates doped with S, their carrier concentration were about 2E18 cm^{-3}. From Fig.9 it could be seen that, at 300K the detectors show response peaks at 1.86, 2.10 and 2.36 μm, with 50% cutoff wavelength of 1.90, 2.17 and 2.45 μm respectively. At 77K, the peaks blue shift to 1.68, 1.92 and 2.12 μm and cutoff at 1.72, 1.95 and 2.17 μm respectively, with temperature coefficients of 0.78, 0.99 and 1.25 nm/K in this temperature range. In the spectra, fluctuations around 1.87 μm as well as 1.38 μm, which are

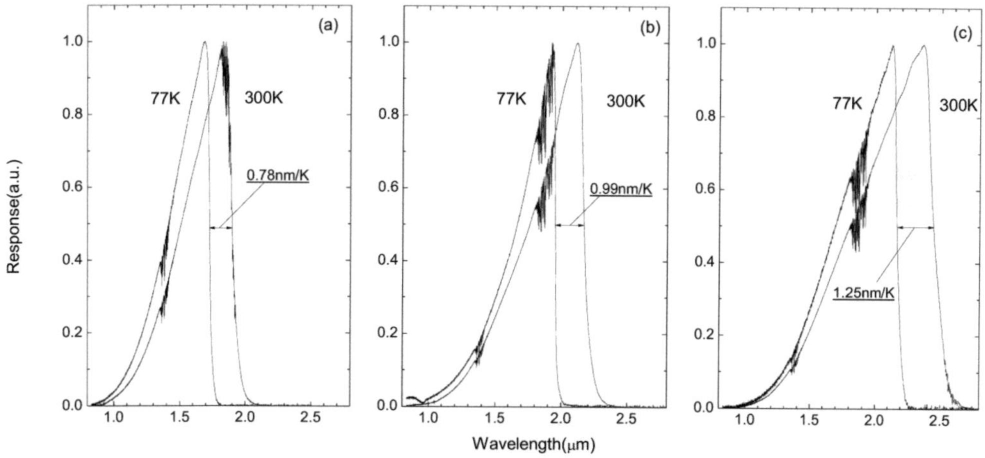

Fig. 9. Measured response spectra of the homostructure InGaAs PDs with cutoff wavelengths about (a)1.9μm , (b)2.2μm and (c)2.5 μm at room temperature.

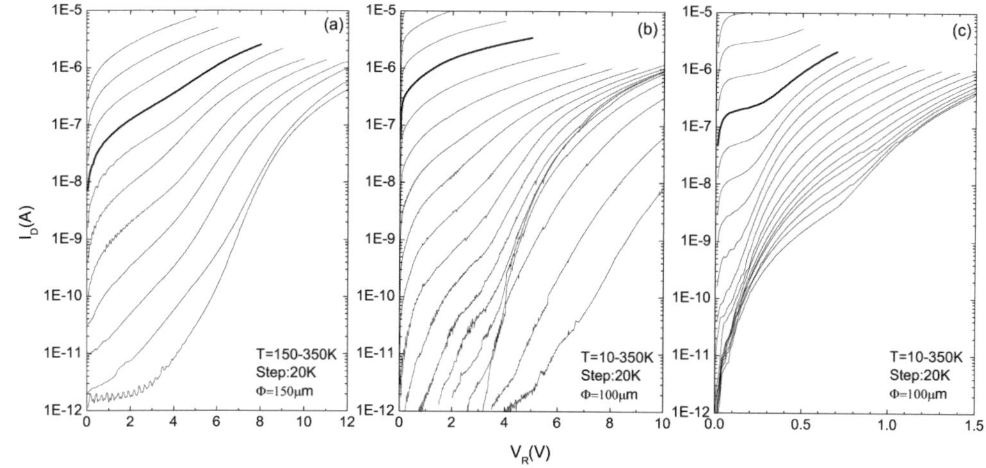

Fig. 10. Measured temperature dependent I-V characteristics of the homostructure InGaAs PDs with cutoff wavelengths about (a)1.9, (b)2.2μm and (c)2.5 μm at room temperature.

corresponding to the water vapor absorption band caused by trace vapor in the air along the optical path, could be clearly seen.

The temperature behaviors of the dark current, which are directly correlated to the performance of the detector, have been measured using a HP4156A precise semiconductor analyzer, in the measurements the detector chips were mounted into a DIP package and installed on the cold head of a closed cycle He cryopump to control their temperature. Fig.10 shows the typical I-V characteristics of the detectors measured in a wider temperature range up to 350 K with step of 20K and over 7 orders of magnitude in current range. It could be seen from Fig.10 that, the dark current of 1.9, 2.2 and 2.5 μm detectors are 2.9 nA, 57 nA and 67 nA at 290K respectively with reverse bias V_R=10mV, and decrease to 28 pA, 84 pA and 161 pA at 210K. To see the temperature dependence of the dark current more clearly, an Arrhenius plot was done with reverse bias voltages of 10 mV. At temperature range from about 210 K to 350 K, a fixed slope could be seen for those detectors, so in this temperature range the thermally activated dark currents, which could be expressed as $I_d \propto \exp(-E_a/kT)$, are dominant, which means in this temperature range the Johnson noise is the main noise source to determine the detector performance. At V_R=10mV, activation energies E_a of 0.488eV, 0.447eV and 0.404eV could be deduced for 1.9 μm, 2.2 μm and 2.5 μm detectors respectively. Based on this data, an order of decrease in the dark current could be expected for those detectors by using one stage TE cooling. We also noticed that this activation energy is comparable but lower than the band gap of the material, so in this ternary alloys the detectors may have similar temperature features. At lower temperature, the dark current drops further but with a lower slope, and becomes constant below about 150K.

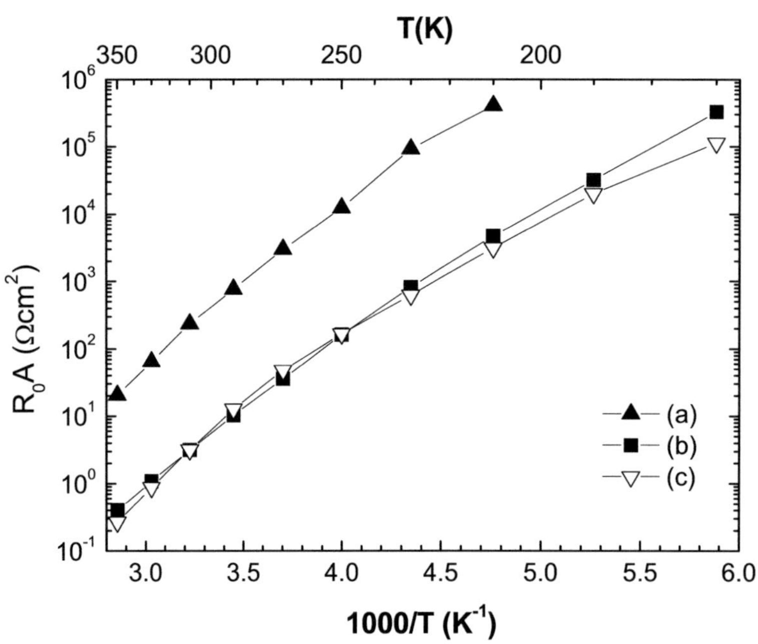

Fig. 11. Measured resistance area product R_0A versus reciprocal temperature of the homostructure InGaAs PDs with cutoff wavelengths about (a)1.9, (b)2.2μm and (c)2.5 μm at room temperature.

The shunt resistances of the detector at 0V bias R_0 are also measured, Fig.11 plots the resistance area product R_0A versus temperature of the detectors. At 290 K, R_0A of 765, 10.3 and 12.7 Ωcm^2 were measured for 1.9 μm 2.2 μm and 2.5 μm detector respectively. When the detectors were cooled down to 210K, the R_0A increased more than 2 orders to 404, 4.70 and 3.12 $K\Omega cm^2$ respectively. The calculated peak detectivity from the R_0A reached 2.3E11, 3.0E10 and 3.7E10 $cmHz^{1/2}/W$ at 290K, and increased to 5.6E12, 7.2E11 and 6.6E10 $cmHz^{1/2}/W$ at 210K respectively, in the calculation the quantum efficiency η=0.7 was supposed. From the measured results we also noticed that, at lower bias conditions and higher temperature region (e.g. T>200K), the performances of the this 2.5 μm detector are comparable to (or even slightly better than) that of 2.2 μm detector, this could be attributed to the slightly higher Si doping (higher electron concentration) in the InGaAs absorption layer of the 2.5 μm detector. From our analysis (Hao et al. 2006), at lower bias and higher temperature the dark current of the detector was dominated by the generation-recombination mechanism, so slightly higher carrier concentration in the InGaAs absorption layer could result in lower dark current and therefore higher R_0A. Whereas at higher bias and lower temperature, the tunneling mechanism becomes dominant and higher carrier concentration results in higher dark current. It means that the performance of the detectors can benefit from slightly higher doping in the absorption layer, especially for detectors mostly working at lower bias (or zero bias) conditions and higher temperature region. However, higher carrier concentration in the InGaAs absorption layer will decrease the depletion layer thickness, and therefore the quantum efficiency of the PDs (Tian et al. 2008a, 2008c). Notice that signal/noise ratio related detectivity, which is a figure of merit of the PDs, is decided by both R_0A and the quantum efficiency; therefore a tradeoff of the doping level in the absorption layer should be taken to optimize the performance. In most case, doping level of 2E16 - 5E16 cm^{-3} in the absorption layer is appropriate. In an optoelectronic system normally the PDs are connected to the ROIC or pre-amplifier directly, hence the noise of the electronics should be considered with those of PDs together. In the case of wavelength extending InGaAs PDs especially with longer cutoff wavelength and operating at higher temperature, the noise from the PDs will be dominant. Consequently, for the optimization of the system performance the R_0A or dark current is the first factor of concern. Those types of epi-wafers cutoff at 2.4 μm have been used for array applications where moderate performances have been reached (Zhang et al. 2007).

4.2 Heterostructure PDs

Fig. 12 shows the response spectra of two heterostructure PDs at zero bias, the PDs are with the cutoff wavelengths about 2.9, and 2.4 μm respectively at room temperature (denoted as sample A and sample B or 2.9μm and 2.4μm detector hereafter).

For those PDs, InAlAs graded layer were used. The epitaxy structure of the PDs consists of a 0.15 μm InP buffer, a 2.4μm N^+ $In_xAl_{1-x}As$ continuously graded buffer layer, the composition x is graded from 0.52 to y of about 0.9 or 0.78 for sample A and sample B respectively and carrier concentrations were about 2E18 cm^{-3}; a 2.0μm n^- $In_yGa_{1-y}As$ absorbing layer and a 0.6μm P^+ $In_yAl_{1-y}As$ cap layer with carrier concentration of about 2E18 cm^{-3}. The samples were grown on exactly (100) oriented S. I. InP epi-ready substrates. The response spectra of the PDs are similar to the homostructure PDs without noticeable difference. The response peaks are at 2.70 μm and 2.24 μm with 50% cutoff wavelength of 2.88 μm and 2.38 μm for sample A and sample B respectively at 300K.

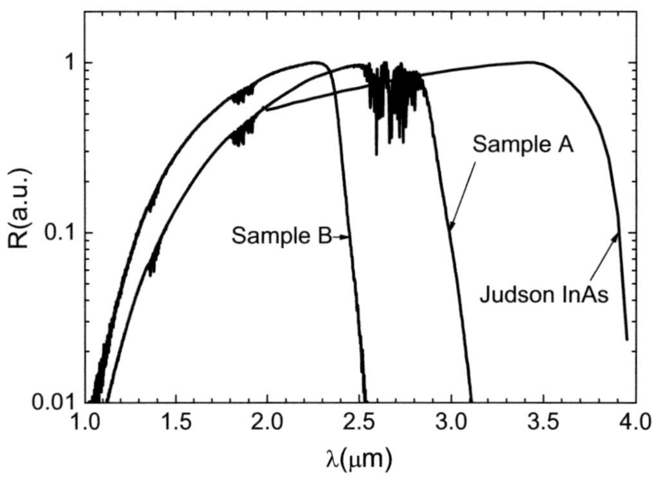

Fig. 12. Measured response spectra of the heterostructure InGaAs PDs with cutoff wavelengths about 2.9 μm (sample A) and 2.4 μm (sample B) at room temperature. (Judson InAs PD is for comparison).

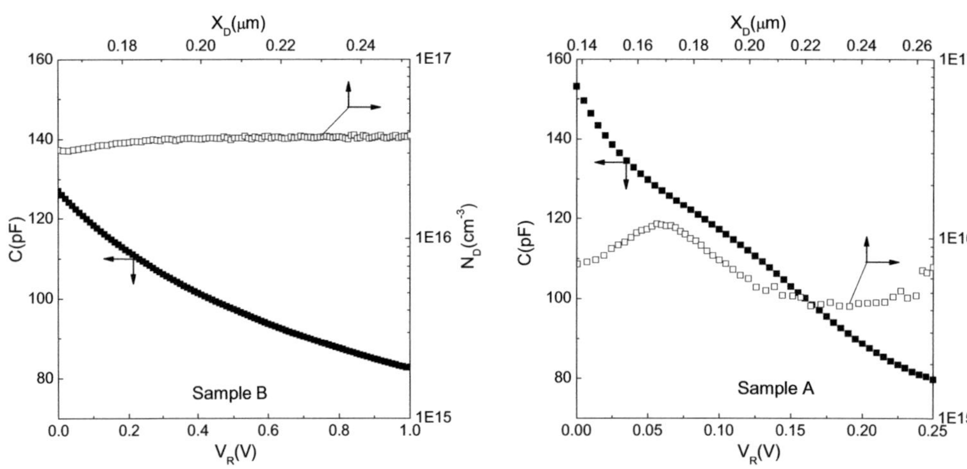

Fig. 13. Measured C-V characteristics of the PDs and calculated carrier concentration in the absorption layer of sample A and sample B.

Fig. 13 shows the C-V characteristics of the detectors measured at 300 K and 1 MHz, the diameters of the detector mesas are 500 μm. From Fig. 13 the capacitances at 0 V bias are 153 pF (78.1nF/cm^2) and 127 pF (64.5nF/cm^2) for sample A and sample B respectively. The electron concentrations educed from the C-V date are below 1E16 cm^{-3} for sample A and around 3E16 cm^{-3} for sample B by using $N = (1/q\varepsilon_0\varepsilon_r A^2)\ [C^3/(dC/dV)]$ and $X_d = \varepsilon_0\varepsilon_r A/C$, where A and C are the area and capacitance of the detectors. It can be seen that at the same Si doping temperature sample A shows lower carrier concentration in the InGaAs absorption layer. One possible reason is due to the carrier depletion caused by the electrical inactivation of dopant and electron trap, for example, formed by misfit dislocations (Uchida et al. 1993).

Typical reverse I-V characteristics of the two heterostructure PDs measured at different temperature are shown in Fig. 14. The detector diameters of sample A and B are 300 μm and 500 μm respectively. The purpose to use a larger diameter for sample B is that the minimum measurable current on our cryostat system is on the order of 10 pA limited by EM interference, therefore a lower temperature could be reached if a larger diameter is used. The current features of those two samples with different cutoff wavelengths are quite different. For sample A at the reverse bias V_R=10 mV, the dark current decreases only about an order of magnitude from 2.0 μA (2.8E-3 A/cm^2) at 290 K to 0.14 μA (2.0E-4 A/cm^2) at 190 K, whereas for sample B the dark current decreases three orders of magnitude from 0.67 μA (3.4E-4 A/cm^2) at 290 K to 456 pA (2.3E-7 A/cm^2) at 190 K, so the current transport mechanisms for those two samples should also be different. To see the dark current characteristics more clearly, Arrhenius plots of the dark current at -10 mV are made for both samples as shown in Fig. 15. For sample A, a fixed slope can be only seen at higher temperature range from about 250 K to 350 K, whereas current drop saturation can be clearly seen at lower temperatures. For sample B, fixed slope extends in whole measurement temperature range, no current drop saturation can be seen until 190 K. This means that, for sample B with lower Indium composition of x=0.78 and therefore shorter cutoff wavelength, at reverse bias of V_R=10 mV the thermally activated dark current, which is expressed as $I_d \propto \exp(-E_a/kT)$, is dominant in a wider temperature range. Activation energy of 0.46 eV can be deduced. However, for sample A with higher Indium composition of x=0.9 and cutoff wavelength at 2.9 μm, at reverse bias of V_R=10 mV the thermally activated dark currents only dominant at higher temperature range, where activation energy of 0.33 eV can be deduced.

The band gaps E_g of the InGaAs layers at room temperature are about 0.43 eV and 0.52 eV for sample A and B respectively. From the results it can be inferred that, at lower bias of 10 mV, the diffusion current may be the main current from above room temperature to lower

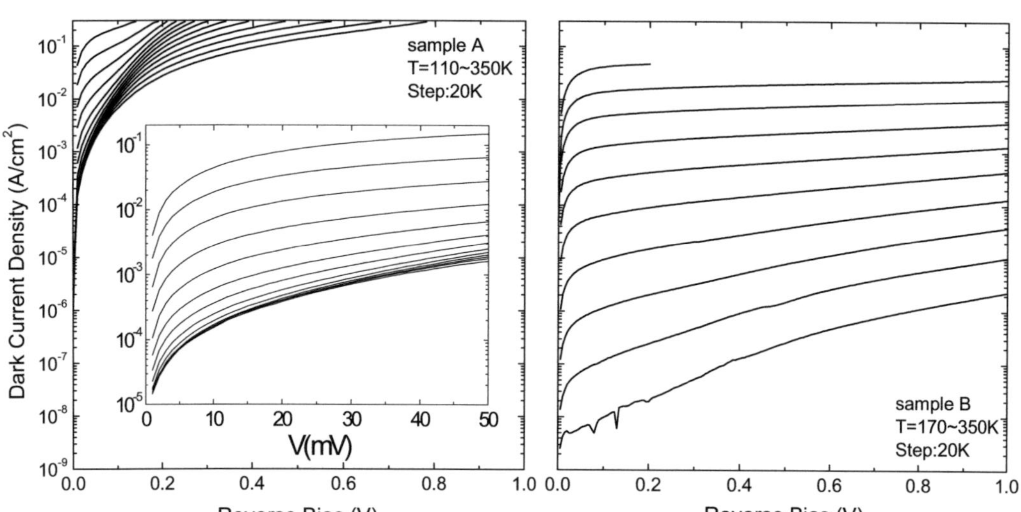

Fig. 14. Measured temperature dependent I-V characteristics of the heterostructure InGaAs PDs with room temperature cutoff wavelengths about 2.9 μm (sample A) and 2.4 μm (sample B).

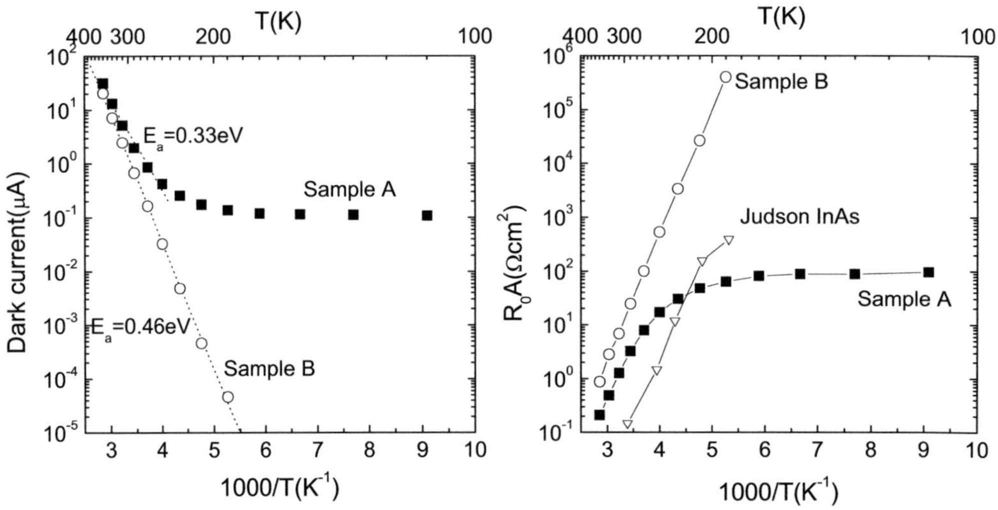

Fig. 15. Arrhenius plots of the dark current versus reciprocal temperature at reverse bias of 10 mV (left) and measured resistance area product R_0A versus reciprocal temperature (right) of the heterostructure InGaAs PDs with cutoff wavelengths about 2.9μm (sample A) and 2.4μm (sample B), for comparison the R_0A data of Judson InAs PD was also shown in the right, it cutoff at 3.6 μm at room temperature.

than 190 K for sample B because its E_a is more close to E_g, while for sample A from 250 K to 350 K, the G-R current and diffusion current both have important contributions. At temperature lower than about 250 K, the dark current of sample A decreases slowly. The thermal activated current at lower temperature decreases rapidly, also for our processing the ohmic leakage is not on this high level, meanwhile, at this lower bias of 10mV the direct tunnelling could also be eliminated. This means the trap assisted tunnelling current, which represents electrons tunnel from the occupied trap states to the empty band states, begins to play an important role. From above results, the performances of 2.9 μm InGaAs detectors seem still quite competitive at higher operation temperatures until about 250 K, but the effects of further decreasing the operation temperature to improve the performances of the detectors will be limited, mainly because of the trap states caused by the large lattice mismatch related defects. Despite the lower energy gap of the material, larger lattice mismatch accelerates the degradation of the detector performances further.

The shunt resistances R_0 of the detectors at 0 V bias were also measured. Fig. 15 right shows the typical resistance area product R_0A versus reciprocal temperature. At 290 K, R_0A of 3.2 Ωcm^2 (R_0=4.5 kΩ) and 24 Ωcm^2 (R_0=12 kΩ) are measured for sample A and sample B, respectively. When the detectors are cooled down to 250 K, the R_0A increases exponentially to 17 Ωcm^2 (R_0=24 kΩ) and 534 Ωcm^2 (R_0=272 kΩ) for sample A and sample B. The R_0A of commercial Judson InAs detector is also shown in Fig. 4, with 0.15 Ωcm^2 (R_0=0.3 kΩ) at 295 K and 1.5 Ωcm^2 (R_0=3 kΩ) at 253 K respectively (detector diameter of 250 μm). It can be seen that the R_0A of 2.9 μm InGaAs detector is higher than lattice matched Judson InAs detector until about 220 K.

The black-body response measurements of the detectors were also performed at room temperature with a T_B=900 K black body. The measured peak detectivity $D^*_{\lambda p}$ of 2.9 μm detector (sample A) is 6.6E9 cmHz$^{1/2}$/W, while for 2.4 μm detector (sample B) is 3.1E10

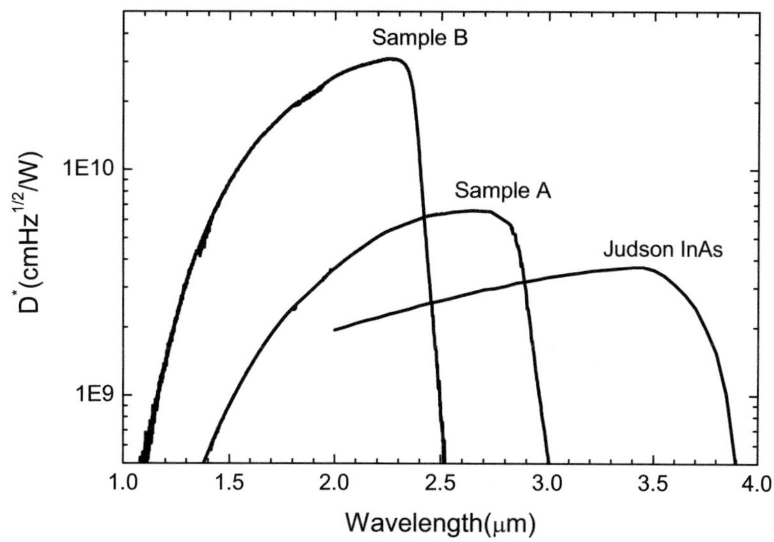

Fig. 16. Measured black body and peak detectivity versus wavelength of sample A and sample B (the spectra are smoothed) at room temperature. For comparison the data of Judson InAs PD was also shown.

cmHz$^{1/2}$/W, as listed in Table 3. Moreover, the wavelength dependent detectivity of the detectors for sample A and sample B at room temperature are shown in Fig.16. In Fig.16 the response spectra are smoothed to eliminate the water absorption bands. The detectivity spectrum of Judson InAs detector at room temperature is also shown in Fig.16 for comparison, which has a value of 3.7E9 cmHz$^{1/2}$/W at response peak about 3.45 μm. At wavelength of 2.65 μm, the detectivity of sample A reaches 6.6E9 cmHz$^{1/2}$/W, twice that of Judson InAs detector of 2.9E9 cmHz$^{1/2}$/W at the same wavelength. From Fig.16 it could be inferred that, extending the cutoff wavelength of InGaAs detectors on InP substrate up to about 3 μm is still quite valuable. Therefore for certain applications where the cutoff wavelength less than 3 μm is enough, this detector is still a good choice especially at high operation temperatures. Furthermore, the wavelength extending InGaAs detectors grown on InP substrate are very suitable for the array integration illuminated from the substrate (back) side, where the detector arrays are often connected to the read-out circuits using flip-chip bounding. The S. I. InP substrate is transparent to the detected light, whereas in this case the InAs substrate is opaque. However, for applications using discrete device at even longer wavelength, lattice matched InAs detector may be more suitable especially at lower operation temperatures, further extending the cutoff wavelength of InGaAs detectors remains a challenge. The PDs have been used for gas sensing purpose where good performances have been reached (Zhang et al., 2007b, 2008c).

Sample	V_s/V_n	g-factor	D^*_{bb} (cmHz$^{1/2}$/W)	$D^*_{\lambda p}$ (cmHz$^{1/2}$/W)
A (2.9μm)	1.4E3	7.0	9.4E8	6.6E9
B (2.4μm)	3.3E3	14	2.2E9	3.1E10

Table 3. Measured black body and peak detectivity of sample A and B at room temperature using a 900K Black body

4.3 n on p configuration.

In most array applications, illuminating from back side with Indium bump flip chip bounding to CMOS readout integrated circuit (ROIC) is much preferable, in this case the p-on-n configuration may not be a good choice. In p-on-n configuration, when the light is illuminated from substrate side (backside), the stronger absorption is also occurs at the absorption layer near the backside, where are away from the pn junction depletion region if the absorption layer is too thick, thus restrict the collection of the generated carriers. Furthermore, the response speed may also be affected because of the slower diffusion process. However, if the absorption layer is too thin, the full absorption of the illuminating light becomes impossible. Through the tradeoff of the absorption layer thickness in conjunction with the doping level the situation can be improved (Tian et al. 2008a, 2008c), but the optimization still be quite difficult. In certain array applications including earth observation the quantum efficiency or sensitivity are the most important trait of concern. Notice that normally the ROIC is suitable for both p-on-n and n-on-p polarities especially at small signal conditions, so instead of p-on-n the n-on-p configuration should be more preferable. Based on those considerations, heterostructure PDs with n-on-p configuration were constructed. Two photodiode wafers (Sample A of 2.0 μm device and Sample B of 2.4 μm device hereafter) with different In content y in the $In_yGa_{1-y}As$ absorption layer were grown. Epitaxial structure of the photodiode wafers began from a heavy Be doped P^+ $In_xAl_{1-x}As$ continuously graded buffer layer (thickness of 3 μm), its composition x was graded from 0.52 to a setting value of y and carrier concentration was about 2E18 cm^{-3}; afterwards, a n- $In_yGa_{1-y}As$ absorbing layer (thickness of 2.2 μm) with light doping of Si optimized to abort 3E16 cm^{-3} and a N^+ $In_yGa_{1-y}As$ contact layer (thickness of 0.2 μm) with heavy doping of Si to about 2E18 cm^{-3} were grown. In this experiment the samples were grown on exactly (100) oriented Zn doped P^+ InP epi-ready substrates, in fact for back illumination application the S. I. InP substrate is better for its much lower free carrier absorption.

Fig.17 left shows the I-V characteristics of the photodiodes measured in the temperature range from 150 K to 350 K with step of 20K and over 7 orders of magnitude in current range, the photodiode mesa diameters were 300 μm. For sample A at reverse bias V_R=10mV, the dark current was 11.3 nA(1.6E-5 A/cm^2) at 290K and decreased to 30.7 pA(4.33E-8 A/cm^2) at 210K; for sample B at reverse bias V_R=10mV, the dark current was 75.1 nA(1.06E-4 A/cm^2) at 290K and decreased to 102 pA(1.44E-7 A/cm^2) at 210K.

Meantime, an Arrhenius plot was done as shown in Fig.18 at reverse bias voltages of 10 mV. At temperature range from 210 K to 350 K, a fixed slope could be seen for both samples. In this temperature range the thermally activated dark currents, which could be expressed as $I_d \propto exp(-E_a/kT)$, are dominant. This implies that in this temperature range the Johnson noise is the main noise source to determine the performance. At reverse bias of 10mV, activation energies E_a of 0.4510eV and 0.4070eV were deduced for sample A and sample B respectively. The shunt resistances of the detector at 0V bias R_0 are also measured; Fig.18 plots the typical resistance area product R_0A versus temperature. At 290 K, R_0A of 760 Ωcm^2 (R_0= 1.07 MΩ) and 104 Ωcm^2 (R_0=147 KΩ) were measured for sample A and sample B respectively. When the photodiodes were cooled down to 210K, the R_0A increased more than two orders to 250 KΩcm^2 (R_0=354 MΩ) and 76.2 KΩcm^2 (R_0=107 MΩ) for sample A and sample B. Some measured data were listed in Table 5. Results showed that those detectors with n-on-p configuration exhibited better performance than those of p-on-n configuration. The dark current decreased more than one order of magnitude at room temperature for the detectors with similar composition and doping level, which were mainly owing to the better quality of the absorption layer because of the heavy Be doped InAlAs graded buffer layer.

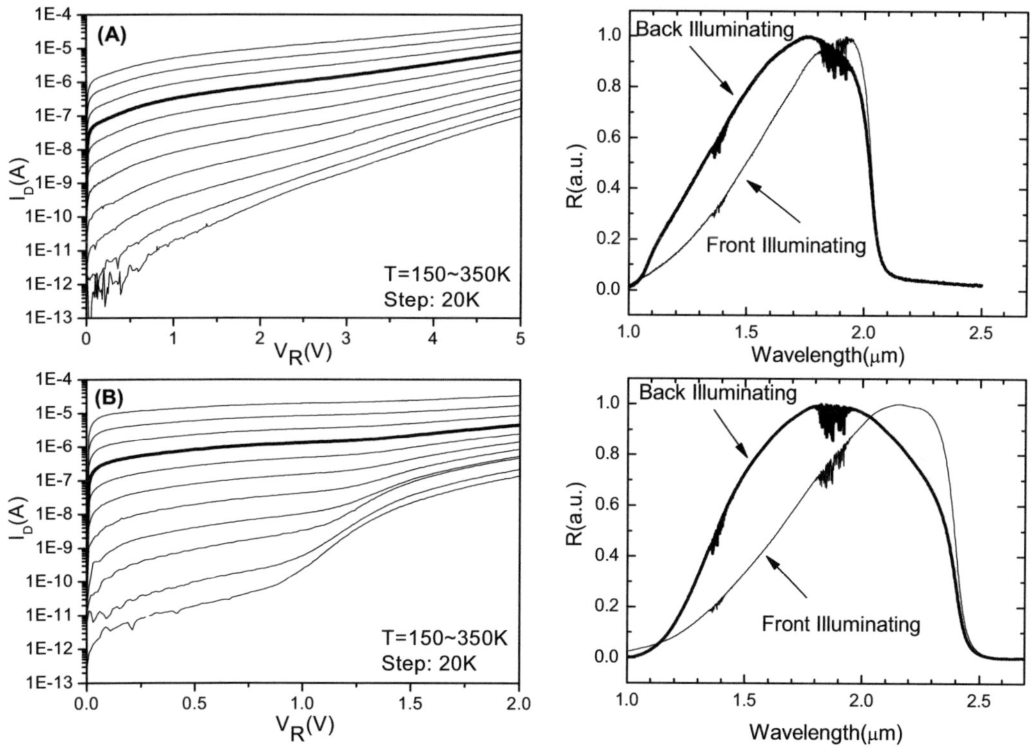

Fig. 17. Measured I-V (left) and room temperature response spectra (right) of the sample A (cutoff at 2.0 μm) and Sample B (cutoff at 2.4 μm) photodiodes.

Φ=300μm N_D~3E16cm^{-3}	Temp. (K)	I_D (nA) V_R=10mV	R_0A (Ωcm^2)	$D^*_{\lambda p}$(R_0A) (cmHz$^{1/2}$/W) Back Illumn.	$D^*_{\lambda p}$(R_0A) (cmHz$^{1/2}$/W) Front Illumn.
Sample A	290	11.3	760	1.534E11 (λp=1.75μm)	1.700E11 (λp=1.94μm)
	210	0.0307	250K	3.17E12 (λp=1.7μm)	3.53E12 (λp=1.89μm)
Sample B	290	75.1	104	5.933E10 (λp=1.83μm)	7.132E10 (λp=2.2μm)
	210	0.102	76.2K	1.80E12 (λp=1.75μm)	2.19E12 (λp=2.12μm)

Table 4. Measured dark current I_D and zero bias resistance area products R_0A data of the photodiodes at 290K and 210K. The detectivity at peak wavelength $D^*_{\lambda p}$ were calculated from the R_0A data (suppose external quantum efficiency η=0.5)

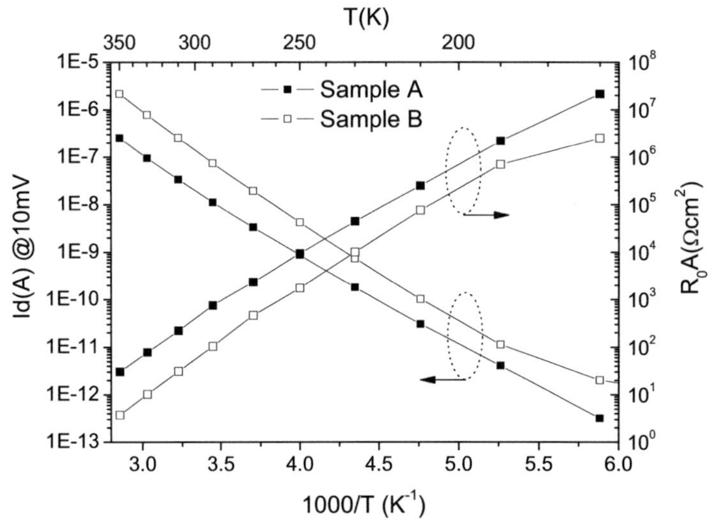

Fig. 18. Arrhenius plots of the dark current versus reciprocal temperature of the sample A (cutoff at 2.0 μm) and Sample B (cutoff at 2.4 μm) photodiodes at reverse bias voltage of 10mV. The zero bias resistance area products R_0A of the photodiodes were also plotted.

	Sample A		Sample B	
	Front Illumination	Back Illumination	Front Illumination	Back Illumination
Signal(mV) @10^{-6}A/V	10.8	4.32	27.5	10.3
Noise(μV) @10^{-8}A/V	220	220	540	540
S/N	4909	1963	5092	1907
G-Factor	27.27	25.59	13.07	13.19
$R_{\lambda p}$ (A/W)	0.65721	0.24618	0.79204	0.30205
$D^*_{\lambda p}$ ($cmHz^{1/2}/W$)	1.31E11	4.91E10	6.50E10	2.46E10

Table 5. Measured black body response of the Φ=500μm detectors at room temperature (~295K) using 900K black body source under front or back illumination. The detectivity at peak wavelength $D^*_{\lambda p}$ and current responsibility $R_{\lambda p}$ from the signal/noise ratio S/N and spectra G-Factor were also listed.

To evaluate the performance of the detector more directly, the black-body response measurements were performed at room temperature (~295 K) with a 900K black-body source. From the measured signal/noise ratio S/N and calculated G-factor from the response spectra, the peak detectivity $D^*_{\lambda p}$(BB) were calculated as listed in Table 5. Notice that for front illumination the $D^*_{\lambda p}$(BB) in consists with the calculated $D^*_{\lambda p}(R_0A)$ quite well, whereas for back illumination the $D^*_{\lambda p}$(BB) is only about half of the $D^*_{\lambda p}(R_0A)$, the signal was weaker but noise keeps unchanged. This was mainly because of the free carrier absorption of the P^+ InP substrate and diffuse reflection caused by imperfectness of the back side polishing. Therefore for back illumination the S. I. InP substrate should be used and backside polishing becomes important. The peak responsibilities R_λ of the detectors from the black body measurements were 0.657 A/W and 0.792 A/W for sample A and sample B respectively at front illumination condition, corresponding to an external quantum efficiency of about 42% and 45%. For those detectors without AR coating this quantum efficiency is moderate.

5. Conclusion

In conclusion, benefited from the developments of InP based optoelectronic devices, wavelength extending $In_xGa_{1-x}As$ (x>0.53) photodetectors have been a remarkable candidate in SWIR band especially for the applications at higher operation temperatures and robust circumstances. For growing this lattice mismatched ternary thick layer on InP substrate the buffer layer is ineluctable. Though $In_xGa_{1-x}As$ itself could be used as graded buffer material (Zhang et al., 2006a), whereas the ternary $In_xAl_{1-x}As$(x>0.52), which is a compatible mate of $In_xGa_{1-x}As$ in MBE, is an excellent option especially for its wider bandgap needful to both performances and utilization of concern (Zhang et al., 2008b). Based on this advisement a convenient and reliable growth procedure with excellent feasibility has been developed for GSMBE (Zhang et al., 2008a), using this procedure a series of devices with anticipant performances have been demonstrated. Our efforts show that in front illumination conditions the $In_yAl_{1-y}As/In_yGa_{1-y}As/In_xAl_{1-x}As$ heterostructure device with p-on-n configuration is an outstanding choice, whereas for back illumination using n-on-p configuration becomes advisable (Zhang et al., 2009a). Our attempt also show that for this PDs the cutoff wavelength at room temperature could be extended to about 3 μm, at this wavelength the performances of this PDs are still competitive for certain applications at higher operation temperatures (Li et al., 2010).

6. References

Afrailov M. A. (2010). Photoelectrical characteristics of the InAsSbP based uncooled photodiodes for the spectral range 1.6-3.5 μm. *Infrared Physics & Technology,* 53(1), 29-32, ISSN: 1350-4495

Bhagwat V; Xiao Ye-gao; Bhat I et al. (2006). Analysis of leakage currents in MOCVD grown GaInAsSb based photodetectors operating at 2 μm. *J. Electronic Materials,* 35(8), 1613-1617, ISSN: 0361-5235

Chyi J I; Shieh J L; Pan J W et al. (1996). Material properties of compositional graded $In_xGa_{1-x}As$ and $In_xAl_{1-x}As$ epilayers grown on GaAs substrates. *J. Appl. Phys.* 79(11), 8367-8370, ISSN: 0021-8979

di Forte-Poisson M.A; Brylinski C; di Persio J et al. (1992). GaInAs/InAsP/InP photodiode for the 1.6 to 2.4 μm spectral region grown by low pressure MOCVD, *J. Crystal Growth,* 124(1-4), 782-791, ISSN: 0022-0248

Fewster P F, (1993). X-ray diffraction from low-dimensional structures. *Semicon. Sci. Technol.* 8(11), 1915-1934, ISSN: 0268-1242

Gu Yi; Li Cheng; Wang Kei et al. (2010). Wavelength extended InGaAs/InP photodetector structures with lattice mismatch up to 2.6%. *J. Infrared Millim. Waves,* 29(4), 81-86, ISSN: 1001-9014

Hao Guo-qiang; Zhang Yong-gang; Gu Yi et al. (2006). Performance analysis of extended wavelength InGaAs photovoltaic detectors grown with gas source MBE. *J. Infrared Millim. Waves,* 25(4), 241-245, ISSN: 1001-9014

Hondt M.D; Moerman I; Daele P.V et al. (1998) Dark current optimization for MOVPE grown 2.5 μm wavelength InGaAs photodetectors, *Electron. Lett.*, 34(9), 910-912, ISSN: 0013-5194

Hoogeveen R.W.M; van der A R. J; Goede A.P.H et al. (2001). Extended wavelength InGaAs infrared (1.0-2.4 μm) detector arrays on SCIAMACHY for space-based spectrometer of the Earth atmosphere. *Infrared Phys. & Tech.* 42(1),1-16, ISSN: 1350-4495

Lee D; Park M S; Tang Z et al. (2007). Characterization of metamorphic InAlAs/GaAs buffer layers using reciprocal space mapping. *J. Appl. Phys.* 101(6), 063523(8pp), ISSN: 0021-8979

Li Ai-zhen, Zhong Jin-quan, Zheng Yan-lan et al. (1995). Molecular beam epitaxial growth, characterization and performance of high-detectivity InGaAsSb/GaSb PIN detectors operating at 2.0 to 2.6 μm. *J. Crystal Growth,* 150(2), 1375-1378 , ISSN: 0022-0248

Li Cheng; Zhang Yong-gang; Wang Kai et al. (2010). Distinction investigation of InGaAs photodetectors cutoff at 2.9 μm. *Infrared Physics & Technology*, 53(3), 173-176, ISSN: 1350-4495

Liang Bang-li; Chen Dian-yong; Wang Bo et al. (2010). Analysis, optimization and design of 2-2.8μm stacked multiple-junction PIN GaInAsSb/GaSb photodetectors for future O/E interconnections. *IEEE Trans. on Electron Devices,* 57(2), 361-367, ISSN: 0018-9383

Makita K; Torikai T; Ishihara H et al. (1988). GaInAs/InAsP pin photodiode for long wavelength region (λ>2μm) grown by hydride vapor phase epitaxy. *Electron. Lett.*, 24(7), 379-380, ISSN: 0013-5194

Moseley A. J; Scott M. D; Moore A. H et al. (1986). High-efficiency, low-leakage MOCVD grown GaInAs/AlInAs heterojunction photodiodes for detection to 2.4μm. *Electron. Lett.*, 22(22), 1206-1207, ISSN: 0013-5194

Piotrowski T. T; Piotrowska A; Kaminska E et al. (2001). Design and fabrication of GaSb/InGaAsSb/ AlGaAsSb mid-IR photodetectors. *Proc. SPIE,* 4413, 339, ISBN: 0-8194-4116-3

Quitoriano N J & Fitzgerald E A (2007). Relaxed, high-quality InP on GaAs by using InGaAs and InGaP graded buffers to avoid phase separation. *J. Appl. Phys.* 102(3), 033511(17pp), ISSN: 0021-8979

Rothman L. S; Jacquemart D; Barbe A et al. (2005) The HITRAN 2004 molecular spectroscopic database. *J. Quantitative Spectroscopy & Radiative Transfer,* 96(2), 139-204, ISSN: 0022-4073

Tersoff J. (1993). Dislocations and strain relief in compositionally graded layers. *Appl. Phys. Lett.* 62(7), 693-695, ISSN: 0003-6951

Tian Zhao-bing; Gu Yi & Zhang Yong-gang. (2008a). Quantum efficiency optimization of InP based $In_{0.53}Ga_{0.47}As$ photodetectors. *J. Infrared Millim. Waves,* 27(2), 81-85, ISSN: 1001-9014

Tian Zhao-bing; Gu Yi; Wang Kai et al. (2008b). Gas source MBE grown metamorphic InGaAs photodetectors using InAlAs buffer and cap layers with cou-off wavelength up to 2.7 μm. *Chin. Phys. Lett.*, 25(6), 2292-2295, ISSN: 0256-307X

Tian Zhao-bing; Gu Yi; Zhang Xiao-Jun et al. (2008c). Quantum efficiency optimization of wavelength extended InGaAs photodetectors" *Chin. J. Semicon. Optoelectronics,* 29(6), 851-854, ISSN: 1001-5868 (in Chinese)

Uchida Y; Kakibayashi H; Goto S. (1993) Electrical and structural properties of dislocations confined in a InGaAs/GaAs heterostructure. *J. Appl. Phys.* 74(11), 6720-6725, ISSN: 0021-8979

Wada M & Hosomatsu H (1993). Wide wavelength and low dark current lattice-mismatched InGaAs/InAsP photodiodes grown by metalorganic vapour-phase epitaxy. *Appl. Phys. Lett.* 64(10), 1265-1267, ISSN: 0003-6951

Wang Kai; Zhang Yong-gang; Gu Yi et al. (2009). Improving the performance of extended wavelength InGaAs photodetectors by using digital graded heterointerface superlattice. *J. Infrared Millim. Waves,* 28(6), 405-409, ISSN: 1001-9014 (in Chinese)

Wel P J; Nijenhuis J; Eck E R H et al. (1992). High-spatial-resolution photoluminescence studies on misfit dislocations in lattice-mismatched III-V heterostructures. *Semicon. Sci. Technol.* 7(1A), A63-A70, ISSN: 0268-1242

Zhang Ke-feng; Tang Heng-jing; Wu Xiao-li et al. (2007a). Performance analysis of 256 element linear 2.4 μm InGaAs photovoltaic detector arrays. *Proc. SPIE,* 6835, 683506, ISBN: 9780819470102

Zhang Yong-gang; Zhou Ping; Chen Hui-ying et al. (1990). LPE growth of InAsPSb on InAs: melt composition, lattice mismatch and surface morphology. *Chin. J. Rare Metals,* 9(1), 48-51, ISSN: 1001-0521

Zhang Yong-gang; Zhou Ping; Shan Hong-kun et al. (1992). InAsPSb/InAs mid-infrared photodetectors. *Chin. J. Semiconductors,* 13(10), 623-628, ISSN: 0253-4177 (in Chinese)

Zhang Yong-gang; Hao Guo-qiang; Gu Yi et al. (2005). 1.9 μm InGaAs PIN Photodetectors Grown by Gas Source MBE. *Chin. Phys. Lett.,* 22(1), 250-253, ISSN: 0256-307X

Zhang Yong-gang; Gu Yi, Zhu Cheng et al. (2006a). Gas source MBE grown wavelength extended 2.2 and 2.5 μm InGaAs PIN photodetectors. *Infrared Physics & Technology,* 47(3), 257-262, ISSN: 1350-4495

Zhang Yong-gang; Gu Yi, ZHU Cheng et al. (2006b). Fabrication of short wavelength infrared InGaAs/InP photovoltaic detector series. *J. Infrared Millim. Waves,* 25(1), 6-9, ISSN: 1001-9014 (in Chinese)

Zhang Yong-gang; Zhang Xiao-jun; Zhu Xiang-rong et al. (2007b). Tunable diode laser absorption spectroscopy detection of N_2O at 2.1 μm using antimonide laser and InGaAs photodiode. *Chin. Phys. Lett.,* 24(8), 2301-2303, ISSN: 0256-307X

Zhang Yong-gang; Gu Yi; Tian Zhao-bing et al. (2008a). Wavelength extended 2.4 μm heterojunction InGaAs photodiodes with InAlAs cap and linearly graded buffer layers suitable for both front and back illuminations. *Infrared Physics & Technology,* 51(4), 316-321, ISSN: 1350-4495

Zhang Yong-gang; Gu Yi; Wang Kai et al. (2008b). Properties of gas source molecular beam epitaxy grown wavelength extended InGaAs photodetector structures on linear graded InAlAs buffer. *Semicon. Sci. Technol.* 23(12), 125029(7pp), ISSN: 0268-1242

Zhang Yong-gang; Gu Yi; Zhang Xiao-jun et al. (2008c). Gas sensor using a robust approach under time multiplexing scheme with a twin laser chip for absorption and reference" *Chin. Phys. Lett.,* 25(9), 3246-3249, ISSN: 0256-307X

Zhang Yong-gang; Gu Yi; Tian Zhao-bing et al. (2009a). Wavelength extended InGaAs/InAlAs/InP photodetectors using n-on-p configuration optimized for back illumination. *Infrared Physics & Technology,* 52(1), 52-56, ISSN: 1350-4495

Zhang Yong-gang; Gu Yi; Tian Zhao-bing et al. (2009b). Performance of gas source MBE grown wavelength extended InGaAs photodetectors with different buffer structures. *J. Crystal Growth,* 311(7), 1881-1884, ISSN: 0022-0248

Zimmermann L; John J; Degroote S et al. (2003). Extended wavelength InGaAs on GaAs using InAlAs buffer for back-side-illuminated short-wave infrared detectors. *Appl. Phys. Lett.* 82(17), 2838-2840, ISSN: 0003-6951

18

Use of a-SiC:H Photodiodes in Optical Communications Applications

P. Louro[1,2], M. Vieira[1,2,3], M. A. Vieira[1,2],
M. Fernandes[1,2] and J. Costa[1,2]
[1]*Electronics Telecommunications and Computer Dept, ISEL, Lisbon*
[2]*CTS-UNINOVA, Lisbon*
[3]*DEE-FCT-UNL, Quinta da Torre, Monte da Caparica, 2829-516, Caparica*
Portugal

1. Introduction

1.1 Short range optical communications

Silica single mode optical fiber is widely used in long distance communication systems for high speed data transmission (Gbit/s) because of its high bandwidth and low attenuation coefficient. The use of this fiber for short-distance interconnection is not preferred due to the small core diameter of the single-mode fiber, and consequently high requirements for coupling and adjustments. Therefore, in local systems it is preferable to use multi-mode fiber which has greater diameter. The plastic optical fiber (POF) is a suitable, promising solution as transmission medium for short range communications [M. Kagami 2007, M. Kuzyk 2008, O. Zieeman et al 2007]. Its core diameter (250-1000 μm) offers the possibility to use inexpensive polymer connectors and its flexibility enables bending radii, which are by far more critical for glass fibers. Besides, it is more resilient to damage than glass due to its intrinsic material characteristics. It is easier to terminate, polish, and connect as well, which can reduce the cost of installation and maintenance.

There are many uses for POF. A few possibilities include short range networking, e.g. indoor and LAN applications, which are directly related to the simplicity of connection and use and therefore to a significant reduction in installation costs. Another sector where POF displaces the traditional communication medium is in-house communication networks (light switches, door bell, temperature measurement, smoke detection, moisture measurement, counters) [W. Stallings 2007], although the possibilities of application are not confined inside of the house itself. Today, Internet technologies are used to transmit more and more images and develop new services that require more data (high definition TV, movies on demand communications videos that may lead to applications such as tele-assistance of persons alone, sick and elderly,...). The use of the data transmission technology DSL pair Telephone wire is no longer sufficient to meet these new demands. This fiber added at the user provides a data transmission speed Internet 10 times greater than current supply. Note that the approach of wiring the building using fiber Adds value to the appartmenmts and allows residents to enjoy the services supplied by the fiber. In the future POF will most likely displace copper cables for the so-called last mile between the last distribution box of the telecommunication company and the end-consumer. Today, copper cables are the most significant bottleneck for high-speed internet "Triple Play", the combination of VoIP, IPTV and the classical internet, is introduced in the market forcefully and therefore high-speed connections are essential. It is highly expensive to realize any VDSL system using copper components, thus the future will be FTTH. POF can be applied in the house itself for different scenarios, such as "A/V Server Network" (communication between e.g. television, hi-fi-receiver and DVD-player), "Control Server Network" (messaging between e.g. refrigerator and stove) or "Data Server Network" (data exchange between e.g. notebook and printer).

Another field of recent application is the auto industry, where the main benefits from POF arise from the tight bending radius that makes POF well-suited to the automobile environment. The installation of multiple networks in modern cars, and the expansion of these systems have quickly accelerated. Sensors, in a car network (electric windows, electric mirrors, locks and even power-seat controls on the door) can be planned to be coupled over wireless interfaces with other active devices (such as lights, shutters, local and remote displays, alarms).

Other possible industrial sectors include consumer electronics, the aerospace industry, due to the lighter weight, or the medical sector, namely medical imaging for image-transfer applications. But all these applications have one thing in common – they all need high-speed data transmission capabilities. To increase bandwidth for this technology the only possibility is to increase the data rate, which lowers the signal-to-noise ratio and therefore can only be improved in small limitations.

1.2 Wavelength division multiplexing technique

The explosion in demand for network bandwidth is largely due to the growth in data traffic. Leading service providers report bandwidths doubling on their backbones about every six to nine months. This is largely in response to the 300 percent growth per year in Internet traffic, while traditional voice traffic grows at a compound annual rate of only about 13 percent [D. Nolan et al 2002]. At the same time that network traffic volume is increasing, the nature of the traffic itself is becoming more complex.

Faced with the challenge of dramatically increasing capacity while constraining costs, carriers have two options: Install new fiber or increase the effective bandwidth of existing fiber. Laying new fiber is the traditional means used by carriers to expand their networks. Deploying new fiber, however, is a costly proposition. Laying new fiber makes sense only when it is desirable to expand the embedded base.

Increasing the effective capacity of existing fiber can be accomplished in two ways, which include the increase of the bit rate in the existing systems or the increase on the number of wavelengths transmitted by the fiber. This latter option to enlarge the transmission capacity of the waveguide is designated the Wavelength-Division-Multiplexing (WDM) techniques, which in glass fiber technology within the infrared range has long been an established practice.

The standard communication over POF uses only one single channel [W. Daum et al 2002]. To increase the bandwidth, and consequently the data transmission speed, the only possibility is to increase the data rate, which lowers the signal-to-noise ratio and therefore offers limited improvements. One approach for increasing the capacity of the waveguide is to use WDM techniques. In glass fiber technology, the use of WDM in the infrared range has

long been an established practice. This multiplexing technology uses multiple wavelengths to carry information over a single fiber.
For WDM two key-elements are indispensable, a multiplexer (MUX) and a demultiplexer (DEMUX). The multiplexer takes optical wavelengths from multiple fibers and makes them converge into one beam. At the receiving end the system must be able to separate out the components of the light so that they can be detected. Demultiplexers perform this function by separating the received beam into its wavelength components and coupling them to individual fibers (Fig. 1).

Fig. 1. Multiplexing and Demultiplexing operations.

The standard MUX/DEMUX devices are well known for infrared telecom systems [M. Bas 2002]. These devices can be either passive or active in design. Passive designs are based on interference filters and plane diffraction gratings, while active designs combine passive devices with tunable filters. Most of these devices require additional collimating and focusing optics that need alignment and lead to complicated designs. Figure 2 shows the sketch of the basic concept of a DEMUX device for optical communication where the different wavelengths are separated using a prism.

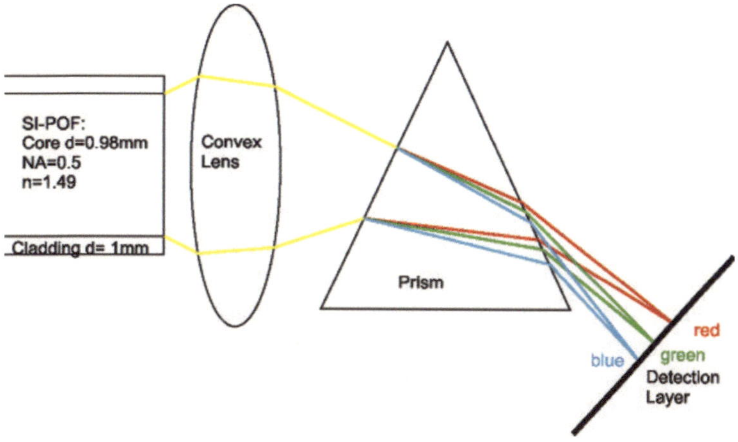

Fig. 2. Demultiplexer device based on an optical component.

The primary challenges of these devices are to minimize cross-talk and maximize channel separation. Cross-talk is a measure of how well the channels are separated, while channel separation refers to the ability to distinguish each wavelength. Between multiplexing and demultiplexing points in a DWDM system, there is an area in which multiple wavelengths exist. It is often desirable to be able to remove or insert one or more wavelengths at some point along this span. An optical add/drop multiplexer (OADM) performs this function.

Rather than combining or separating all wavelengths, the OADM can remove some while passing others on. OADMs are a key part of moving toward the goal of all-optical networks.
Usually, demultiplexing must be done before the light is detected, because standard photodetectors (based on crystalline materials) are inherently broadband devices that cannot selectively detect a single wavelength.
This basic concept of the use of WDM can also be assigned to POF. However POF shows a different attenuation behavior, with better performance in the visible window of the spectrum. For this reason, only the visible spectrum can be applied when using POF for communication. This limitation demands the design of new devices for the implementation of this technique using POF technology [S. Randel et al 2007, M. Haupt 2006]. Several technical solutions for this problem are available, but none of them can be efficiently utilized in the POF application scenario described here, mostly because these solutions are all afflicted with high costs and therefore not applicable for mass production. A solution to overcome the bandwidth bottleneck of standard POF communication is to adapt WDM for the visible wavelength range. Therefore newly designed multiplexers and demultiplexers are essential.

2. WDM based on amorphous technology

2.1 WDM over POF

The use of multilayered structures based on a-SiC:H alloys as photo-sensing or wavelength sensitive devices in the visible range, has been widely studied in the past for different applications, namely, solar cells, color sensors, optical sensors [G. Cesare et al 1995, A. Zhu at al 1998, M. Topic et al 2000, M. Mulato et al 2001]. In these multilayered devices the light filtering is achieved through the use of different band gap materials, namely a-$Si_{1-x}C_x$:H. In these devices the spectral sensitivity in the visible range is controlled by the external applied voltage. Thus, proper tuning of the device sensitivity along the visible spectrum allows the recognition of the absorbed light wavelength [H. Tsai 1987, H. Stiebig et al 1995]. Using this operation principle, other applications were developed, such as, the optical image sensor LSP (Laser Scanned Photodiode) reported by M. Vieira et al 2002, the CLSP (Color Laser Scanned Photodiode) by P. Louro et al 2007 and M. Vieira et al 2008 or the currently under development optical WDM device [P. Louro et al 2009, M. Vieira et al 2010].
In this chapter the wavelength sensitive transducers are optimized to take advantage of the whole visible spectrum. The challenges in this development are mainly related to the ability of the transducer to discriminate wavelengths, which is directly related to the bandwidth of the communication system. Other technical issues related to the signal attenuation and to the light-to-dark sensitivity of the optical device will also be analysed.
Those transducers present several advantages as they:

- are active optical transmission devices and receive-side opto-electronic devices that convert light pulses into electrical signals.
- amplify all the wavelengths at once and can be also used to boost signal power.
- can be active designs combining passive devices with tunable filters and are able to remove or insert one or more wavelengths at some point. An optical add/drop multiplexer (OADM) performs this function.
- combine several lower-bandwidth streams of data into a single beam of light and they are therefore required to be wavelength selective, allowing for the transmission, recovery, or routing of specific wavelengths, where the device behaves like a wavelength converter.

2.2 Device configuration and operation

2.2.1 Configuration

The WDM device is a double heterostructure (Fig. 3b) produced by PECVD (Plasma Enhanced Chemical Vapor Deposition). Deposition conditions are described elsewhere [M. Vieira et al 2007]. The thickness and the absorption coefficient of the front photodiode are optimized for blue collection and red transmittance, and the thickness of the back one adjusted to achieve full absorption in the greenish region and high collection in the red spectral one. As a result, both front and back diodes act as optical filters confining, respectively, the blue and the red optical carriers, while the green ones are absorbed across both [Louro et al 2007]. In Fig. 3a it is displayed the recombination profiles (straight lines) under red (λ_R =650 nm) green (λ_G =550 nm) and blue (λ_B =450 nm) optical bias and different electrical bias (-6V<V<0V). The generation profiles are also shown (symbols). We used a device simulation program ASCA-2D [A. Fantoni et al 1999] to analyze the profiles in the investigated structures.

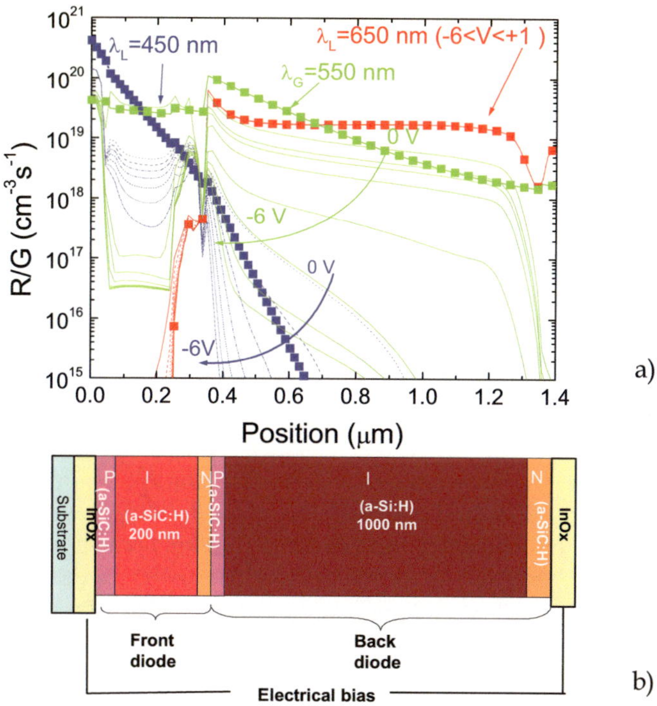

Fig. 3. a-SiC:H WDM device. a) Recombination profiles (straight lines) under red (λ_R =650 nm) green (λ_G =550 nm) and blue (λ_B =450 nm) optical bias and different applied voltages (-6V<V<0V). The generation profiles are also shown (symbols). b) Device configuration.

The WDM device consists of a glass/ITO/a-SiC:H (p-i-n) photodiode which faces the incoming modulated light followed by an a-SiC:H(-p) /Si:H(-i')/SiC:H (-n')/ITO heterostructure that allows the optical readout. By reading out, under different applied bias, the total photocurrent generated by all the incoming optical carriers the information (wavelength, modulation frequency) is multiplexed or demultiplexed and can be transmitted or recovered again [M. Vieira et al 2008].

2.2.2 WDM working principle

In Figures 4 the device configuration is depicted in both multiplexing and demultiplexing modes. Here, multiple monochromatic (Figure 20) or a single polychromatic (Figure 21) beams are directed to the device where they are absorbed, accordingly to each wavelength, giving rise to a time and wavelength dependent electrical field modulation across it [M. Vieira et al 2001].

In the multiplexing mode the device faces the modulated light incoming together from the fibers, each with a wavelength in a specific range (R, G, B channels). The combined effect of each input channel is then converted to an electrical signal via the WDM device (Figure 19a) keeping the memory of the input channels (wavelength and bite rate).

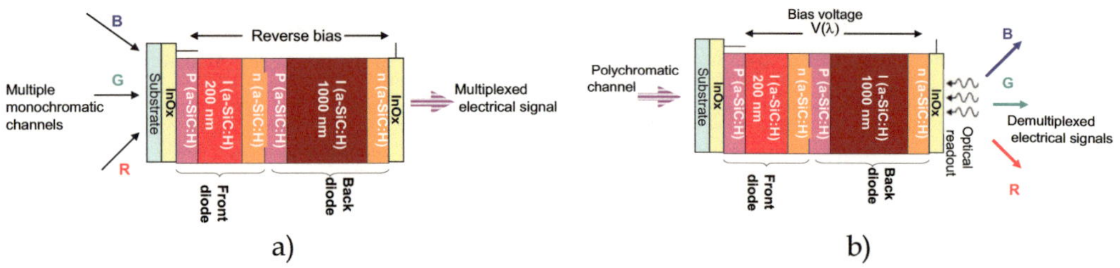

Fig. 4. WDM device configuration : a) multiplexing mode, b) demultiplexing mode.

In the demultiplexing mode a polychromatic light beam (mixture of different wavelength) is projected onto the device and the signal measured at appropriate applied voltages. Here, the spectral sensibility of the device is voltage controlled allowing the recognition of the RGB channels (Figure 4b).

2.3 Optical characterization

2.3.1 Spectral response

The devices were characterized through spectral response measurements (400-800 nm), under different modulated light frequencies (15 Hz to 2 KHz) and electrical bias (-10V to +3V). In Figure 5a it is displayed, under reverse bias, the spectral photocurrent at different frequencies and, in Figure 5b, the trend with the applied voltages is shown at 2 KHz.

Different trends with the frequency are observed. Under reverse bias and low frequencies (f<400Hz), the spectral response increases with the frequency in the reddish region while in the blue/green spectral regions the photocurrent remains constant (Figure 5a). For higher frequencies (f>400Hz) the spectral response does not depend on the modulated light frequency and, as the applied voltage changes from forward to reverse (Figure 5b), the blue/green spectral collection is enlarged while the red one remains constant.

2.3.2 Voltage controlled sensitivity

In Figure 6a it is displayed the measured spectral photocurrent and in Figure 6b the ac current-voltage characteristics under illumination are shown. In this last measurement three modulated monochromatic lights: R (λ_R=626 nm); G (λ_G=520 nm) and B (λ_B=470 nm), and their polychromatic combinations; R&G (Yellow); R&B (Magenta); G&B (Cyan) and R&G&B (White) illuminated separately the device and the photocurrent was measured as a function of the applied voltage. As light sources ultra-bright LEDs were used with a 20 nm spectral bandwidth. The output optical power was adjusted for each independent wavelength at 19 $\mu W/cm^2$.

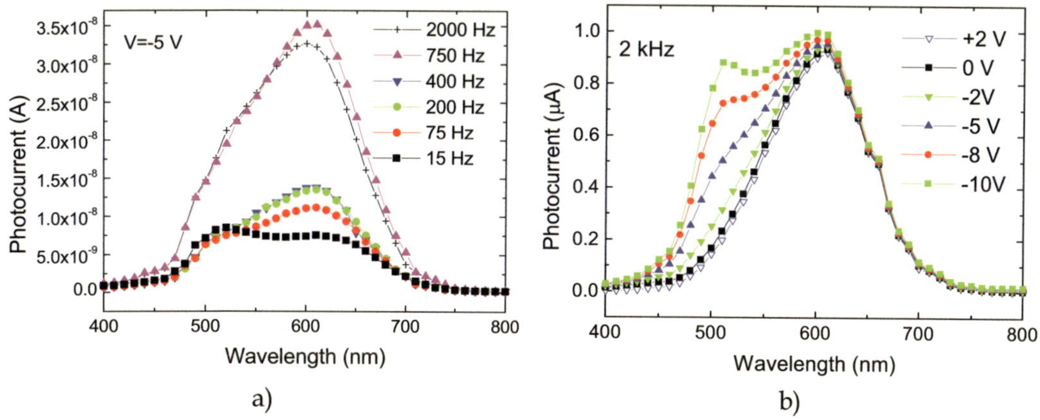

Fig. 5. Spectral photocurrent under: a) reverse bias (-5V) and different frequencies; b) different applied voltages and at a modulated frequency of 2000 Hz.

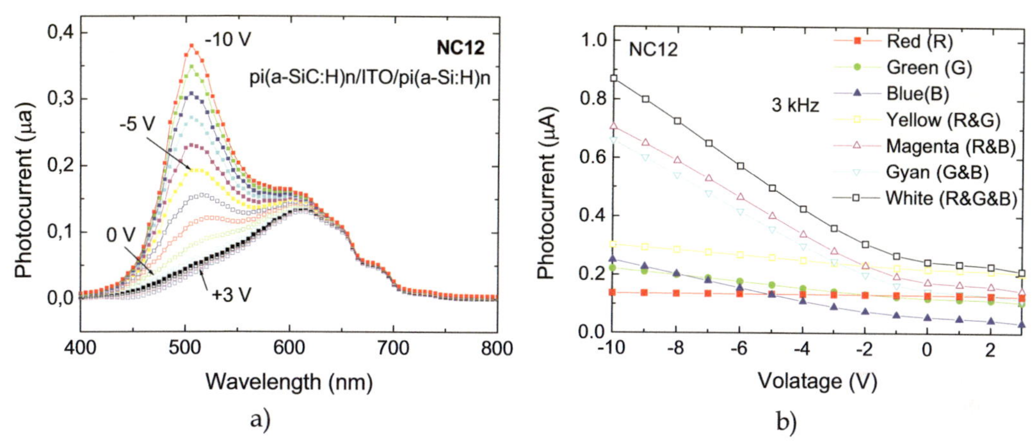

Fig. 6. a) Spectral response at different applied bias. b) Photocurrent voltage characteristics under different light wavelengths.

Data from Figure 6 confirm that, as the applied voltage changes from forward to reverse the blue/green spectral collection is enlarged while the red one remains constant (Figure 6a). The photocurrent under red modulated light (Figure 6b) is independent on the applied voltage while under blue, green or combined irradiations; it increases with the reverse bias. If the blue spectral component is present (B&R, B&G), a sharp increase with the reverse bias is observed. Under positive bias the blue signal becomes negligible and the R&B, the G&B and the R&G&B multiplexed signals overlap, respectively with the R, the G and the R&G signals. This behavior illustrates, under forward bias, the low sensitivity to the blue component of the multiplexed signal. It is interesting to notice that under reverse bias the green signal has a blue-like behavior, while under forward bias its behavior is red-like confirming the green photons absorption across both front and back diodes.
In Figure 7 (a, c) the spectral photocurrent under different electrical bias and its trend (b, d) with the applied voltage, under specific wavelengths, are displayed separately for the front, p-i' (a-SiC:H)-n, and back p-i (a-Si:H)-n, photodiodes. Here, the internal transparent ITO contact was used to apply the voltage (Figure 4).

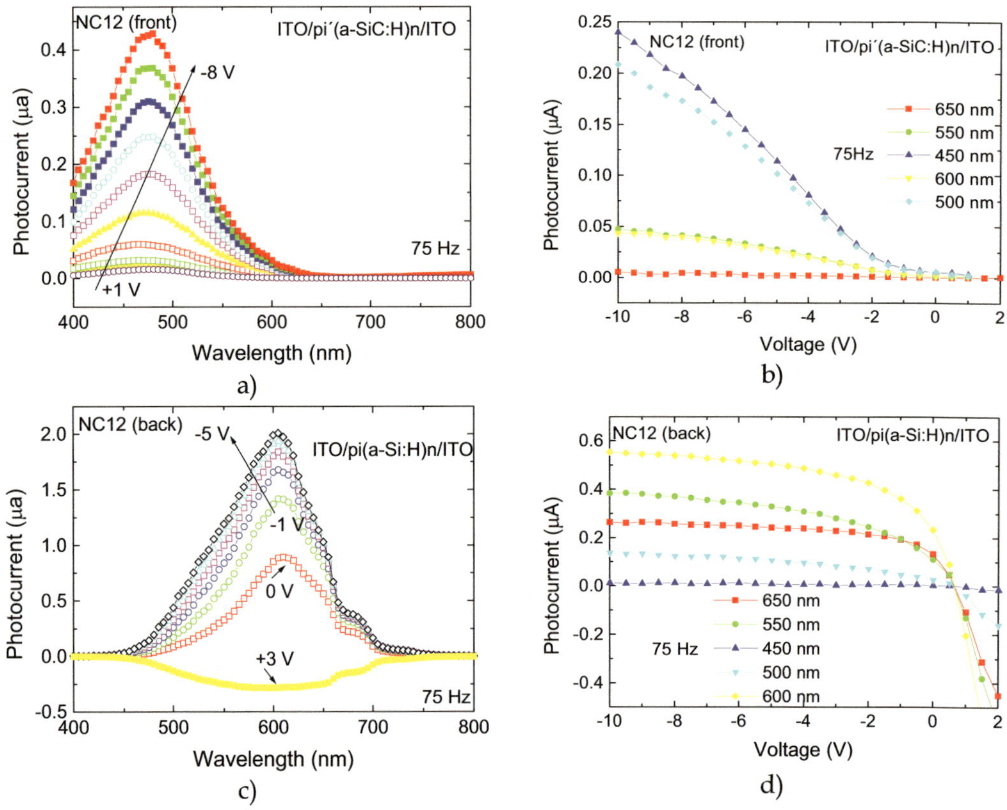

Fig. 7. Spectral photocurrent under different applied bias (a, c) and its trend with the applied voltage, at different wavelengths (b, d), for the front, p-i' (a-SiC:H)-n, and back, p-i (a-Si:H)-n.

Results confirm that the front and back photodiodes act as optical filters, respectively in the blue and red spectral regions. The front diode, based on a-SiC:H, cuts the red component of the spectrum while the back one, based on a-Si:H, cuts the blue component. Each diode separately presents the typical responses of single p-i-n structures while the stacked configuration (Figure 6a) shows the influence of both front and back diodes modulated by its interconnection through the internal n-p junction.

2.3.3 Selective wavelength discrimination

In Figure 8 it is displayed, the spectral photocurrent under negative (-10V) and positive (+3V) external bias and its differences (symbol), the dot lines show the multi peak curve fit at -10V.

Under negative bias the contribution of both front and back diodes (dotted lines) is clear. Two peaks centered, respectively at 500 nm and 600 nm are observed. Under positive bias, the response around 500 nm disappears while the one around 600 nm remains constant as expected from Figure 6a. So, under forward bias the device becomes sensitive to the red region and under reverse bias to the blue one working as a selective optical device in the visible range.

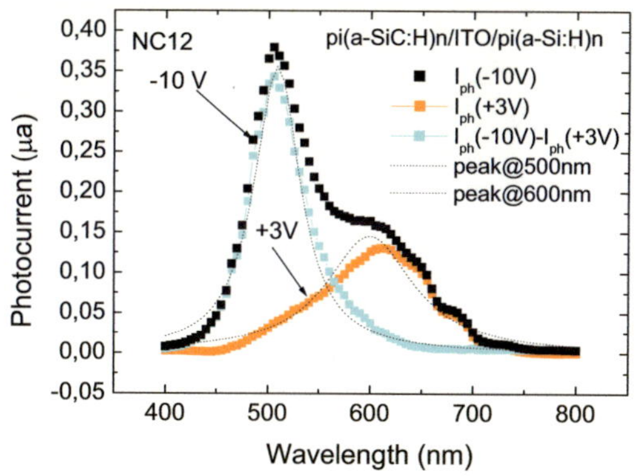

Fig. 8. Spectral response at under negative and positive electrical bias.

2.3.4 Frequency dependence

In Figure 9 the ac current-voltage characteristics under different wavelengths: 650 nm (R); 450 nm (B); 650 nm & 450 nm (R&B), and at the low and high frequency regimes is displayed.

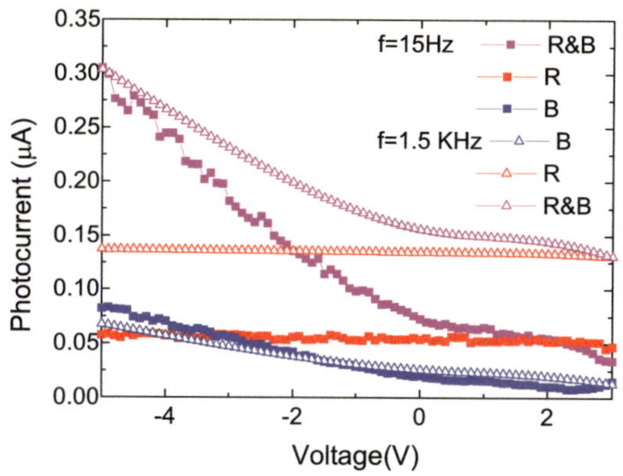

Fig. 9. ac IV characteristics under R (λL =650 nm); B(L =450 nm), R(L =650 nm) & B L =450 nm) modulated light and different light frequencies (15Hz;1.5KHz).

Results show that in both regimes, under red modulated light, the collection efficiency remains always independent on the applied voltage being higher at high frequencies, as expected from Figure 6a. Under blue irradiation the collection does not depend on the frequency regime. Those effects suggest different capacitive effects in both front and back diodes.

3. Wavelength division multiplexing device

3.1 Voltage controlled device

The effect of the applied voltage on the output transient multiplexed signal is analyzed. To readout the combined spectra, the generated transient photocurrent due to the simultaneous effect of two (λ_R =650 nm, λ_B =450 nm) and three (λ_R =650 nm, λ_G =550 nm, λ_B =450 nm) pulsed monochromatic channels was measured, under different applied voltages (Fig. 10).

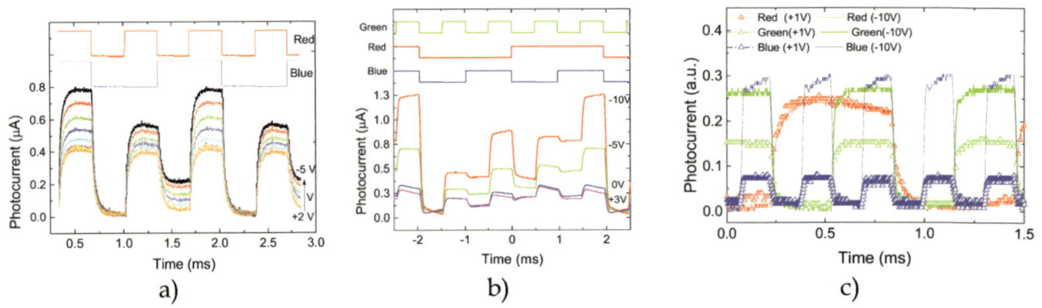

Fig. 10. Transient multiplexed signals at different applied voltages and input wavelengths: a) R&B (λR,B=650nm, 450 nm). The highest frequency of the input signal is 1.5 kHz. b) R&G&B (λR,G, B=650 nm, 550 nm, 450 nm). The highest frequency of the input signal is 1 kHz. c) dependence of the input colour channel with the applied voltage

The input wavelength channels are superimposed in the top of the figures to guide the eyes. The reference level was assumed to be the signal when all the input channels were OFF (dark level). In Figure 10a the red frequency was 1.5 KHz and the blue one half of this value while in Figure 10b the ratios between the three frequencies were always one half.

In Figure 10c the dependence of each pulsed single channel with the applied voltage is also displayed. As expected from Figures 5 and 6 the red signal remains constant while the blue and the green decrease as the voltage changes from negative to positive. The lower decrease in the green channel when compared with the blue one is related with to the lower bias dependence of the back diode where a part of the green photons is absorbed.

Data show that the multiplexed signal depends on the applied voltage and on the wavelength and transmission rate of the each input channel. Under reverse bias, there are always four (Figure 28a) or eight (Figure 28b) separate levels depending on the number of input channels. The highest level appears when all the channels are ON and the lowest if they are OFF. Furthermore, the levels ascribed to the mixture of two input channels (R&B, R&G, G&B) are higher than the ones due to the presence of only one (R, G, B). The step among them depends on the applied voltage and channel wavelength. As expected from Figure 22 and Figure 23, as the reverse bias increases the signal exhibits a sharp increase if the blue component is present. Under forward bias the blue signal goes down to zero, so the separated levels are reduced to one half.

In Figure 11a it is displayed the multiplexed signals (solid lines) due to two input channels in the low and high frequency regimes. The transient signals were acquired at different applied voltages (-5V <V<+2V). The blue (λ_L=450 nm; dotted line) and the red (λ_L =650 nm; dash line) input channels are superimposed to guide the eyes across the monochromatic R and B input channels. The red signal frequency was 1.5 KHz and the blue one was half of this value. In Figure 11b the same output signals of Figure 11a are shown, but using for the input frequencies two orders of magnitude lower.

Both figures show that the multiplexed signal depends on the applied voltage and on the frequency regime of the input channels. Results show also that in the high frequency regime (Figure 11a) the multiplexer acts as a charge integrator device while in low frequency regime it works as a differentiator. In the high regime, the output signals (multiplexed signals) show the potentiality of using the device for WDM applications since it integrates every wavelength to a single one retaining the input information.

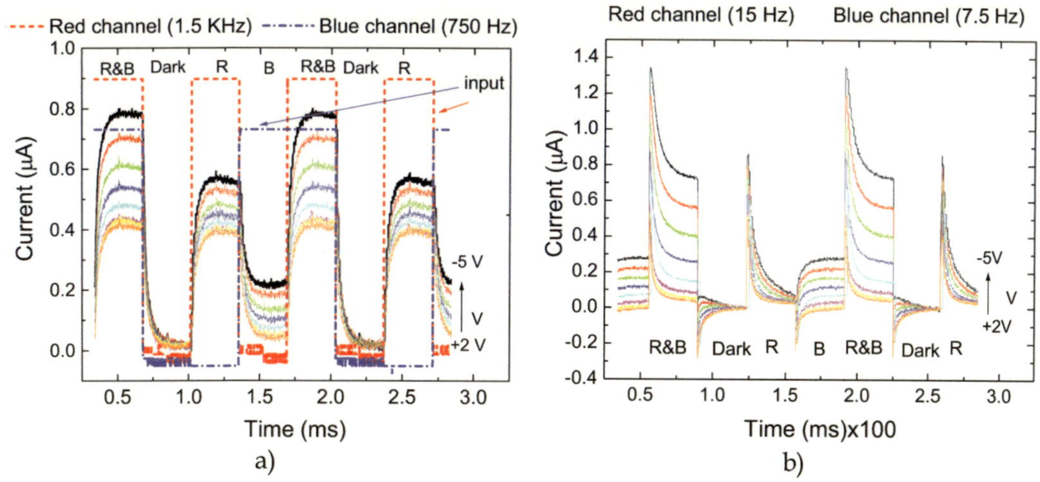

Fig. 11. Wavelength division multiplexing (solid lines) at different applied voltages, obtained using the WDM device: a) High frequency regime. The blue (dotted blue line) and the red (dash-dot red line) guide the eyes into the input channels; b) Low frequency regime.

3.2 Bias sensitive multiplexing technique

Figure 12 displays the photocurrent signal obtained with the WDM device under single and combined modulated light bias: red (R: 626 nm), green (G: 524 nm) and blue (B: 470nm) from the glass side. The generated photocurrent is measured under negative (-8V; solid arrow) and positive (+1V, dotted arrow) bias to readout the combined spectra. The light modulation frequency of each channel was chosen to be multiple of the others to ensure a synchronous relation of ON-OFF states along each cycle. For each independent wavelength, the output optical powers were adjusted to give different signal magnitudes at -8V (solid arrows). The correspondent photocurrent signals at +1V are also displayed (dotted arrows). The reference level was assumed when all the input channels were OFF.

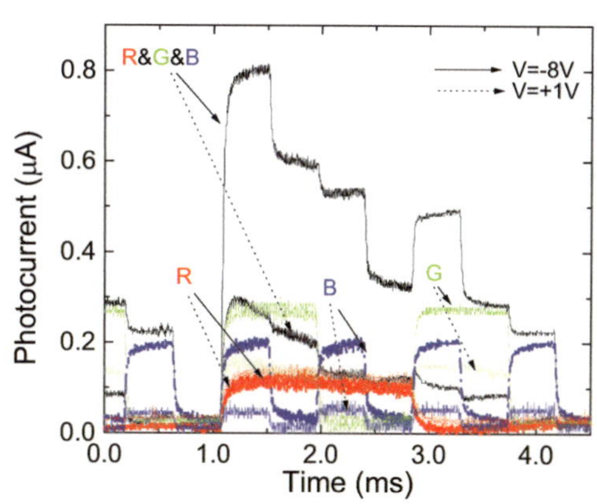

Fig. 12. Multiplexed signals obtained under reverse (solid arrow) and forward (dotted arrow) bias using single (R, G and B) and combined (R&G&B) optical bias of different wavelengths.

As it was expected from Figure 10b, under reverse bias, there are eight separate levels while under positive bias they were reduced to one half. Also, the highest level appears when all the channels are ON and the lowest if they are OFF. Under forward bias the device becomes blind to the front photodiode and the blue component of the combined spectra falls into the dark level, allowing the tuning of the red and green input channels.

In Figure 13 it is shown the multiplexed signals, under reverse and forward bias, obtained with two RGB bit sequences and the same bit rate (2000 bps).

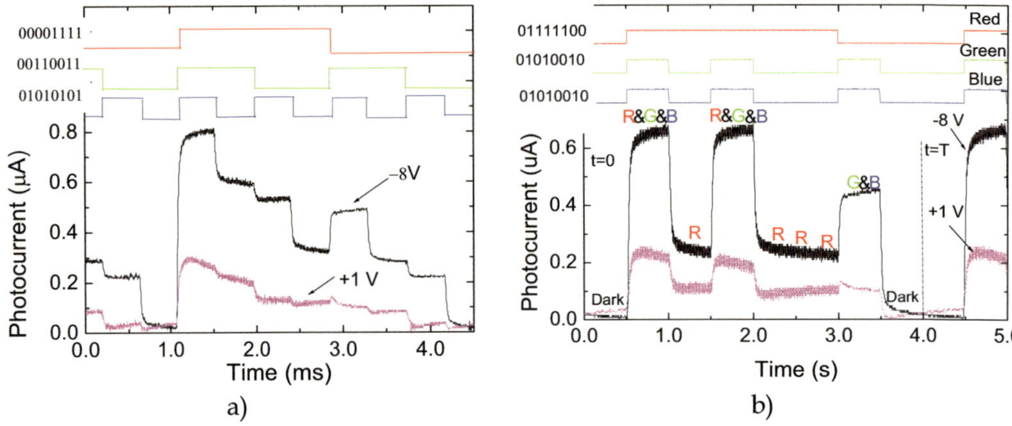

Fig. 13. Multiplexed signals under negative and positive bias using two different bit sequences: a) R [00111100], G [01010010], B[00110011]; b): R [01111100], G [01010010], B[01010010]. On the top, the optical signal used to transmit the information guide the eyes on the different ON-OFF states.

In the bit sequence of Figure 13a (the same of Figure 12) there is a synchronous relation of ON-OFF states along each cycle. The photocurrent under reverse bias, exhibits the expected eight different levels that correspond each to different optical bias states. As the electrical bias goes from reverse to forward the signal amplitude decreases and the levels of the threshold photocurrent associated to each optical state become closer and less defined showing the extinction of the photocurrent caused by the short wavelength optical signals. This mechanism can be used for the identification of the input channels using the photocurrent signal obtained under forward and reverse signals and comparing the magnitude of the variation in each optical state. In the input sequence of Figure 30b the blue and the green channels transmit the same information, and thus the thresholds assigned to the single green or blue channel ON (G or B) and their combination with the red (R&G, R&B) do not appear in the multiplexed signal that contains only four photocurrent levels: R&G&B, G&B, R and Dark.

3.3 Influence of the bit rate

In Figure 14 the acquired signals (solid lines) due to the simultaneous presence of two input channels, respectively λ_L=650 and λ_L=550 nm, under different values of the modulation light frequencies (1k Hz, 10 kHz and 100 kHz), at -5 V are shown. The superimposed red and green dashed lines correspond to the different input channels, and are displayed just to illustrate the different ON-OFF states of the light bias.

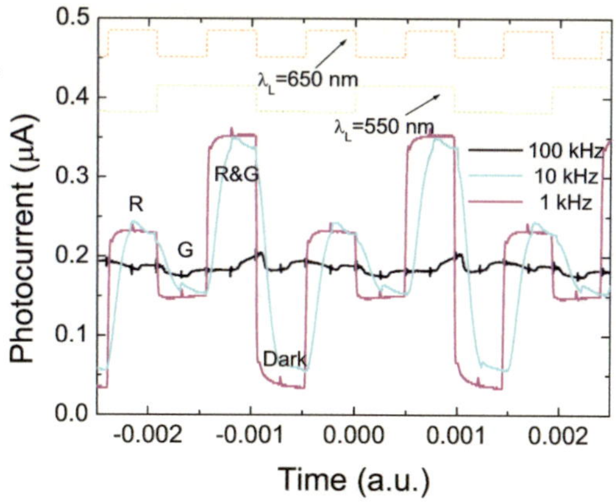

Fig. 14. Wavelength division multiplexing (solid lines) at – 5 V under different values of the modulation light frequency. The red and the green dashed lines guide the eyes into the input channels.

The multiplexed signal shows that in each cycle it is observed for the different frequencies, the presence of the same four levels. The highest occurs when both red and green channels are ON (R&G) and the lower when both are OFF (dark). The green level (G) appears if the red channel is OFF and is lower then the red level (R) that occurs when the green channel is OFF. This behavior is observed even for high frequencies although the measured current magnitude is reduced. So the sensitivity of the device decreases with the increase of the bit rate.
In Figure 15 the multiplexed and the single input channels are shown at different bit rates but with the same bit sequence: a) 2000bps, b) 4000bps. In the top of both figures the bit sequencies are displayed to guide the eyes.

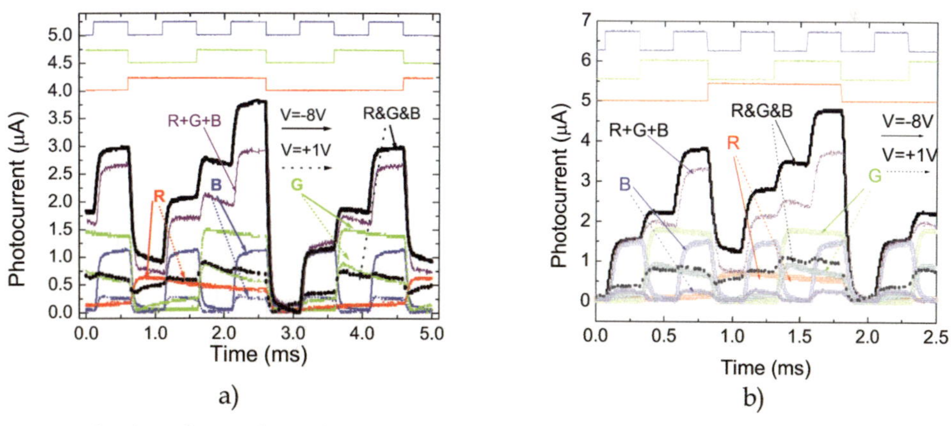

Fig. 15. Multiplexed signals under negative and positive bias using two different bit rates: a) 2000 bps, b) 4000bps. On the top figure, the optical signal used to transmit the information guide the eyes on the different ON-OFF states.

Results show that in the analyzed range, the multiplex signal is independent on the bit rate. It is interesting to notice that in both and under reverse bias, the sum of the input channels is lower than the correspondent multiplexed signals. This optical amplification, mainly on the

ON-ON states, suggests capacitive charging currents due to the time-varying nature of the incident lights.

3.4 Influence of optical signal intensity

The identification of the different input channels requires a previous calibration of the transmission signal in order to know the response of the WDM device to each individual channel as the signal attenuation along the transmission medium causes a reduction of the optical intensity at the reception end (WDM device). In order to analyze the influence of this effect the multiplexed signal was acquired with input signals of different optical intensities at -8V and +1V. Measurements were made with different levels of increasing optical power, up to 140 μWcm^{-2}.

In Figure 16 it is displayed the output photocurrent density variation with the optical bias measured for each optical channel (R: 626 nm, G: 524 nm and B: 470 nm) at -8 V and + 1V.

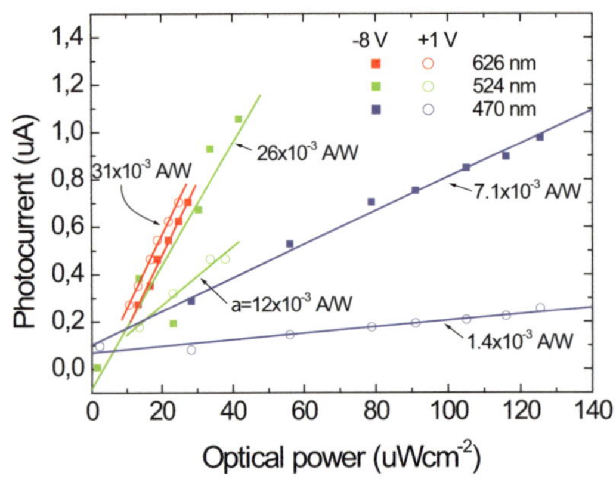

Fig. 16. Photocurrent density variation with the optical bias measured for each optical channel (R: 626 nm, G: 524 nm and B: 470 nm) at -8 V (solid symbols) and + 1V (open symbols). The solid lines correspond to linear fits of the experimental data. Slopes (□) of each plot are also displayed.

Results show that the multiplexed signal magnitude under red illumination and the same intensity conditions is almost independent on the applied bias. Its magnitude increases with the optical power intensity, exhibiting a linear behavior with a growth rate around 31×10^{-3} A/W. Under green and blue light the dependence of the photocurrent magnitude is strongly dependent on the polarity of the applied bias as already demonstrated before. It increases with the optical intensity of each channel either for reverse and forward bias and the growth rate depends on the applied voltage, being higher at reverse bias. Under blue light (470 nm) the growth rate is 5 times higher at reverse than under forward bias, while under green illumination (524 nm) this ratio is only of a factor of 2. This is due to the strong reduction of the device sensitivity for the shorter wavelengths under forward bias.

4. Wavelength division demultiplexing device

4.1 Bias sensitive wavelength division demultiplexing technique

The major propose of the WDM device is to detect the signal that is being communicated. So, different wavelengths which are jointly transmitted must be separated to regain all the information. These separators are called demultiplexers. A chromatic time dependent wavelength combination of red and blue or red, green and blue of different transmission rates, were shining on the device. The generated photocurrent was measured under negative and positive bias to readout the combined spectra (Fig. 17).

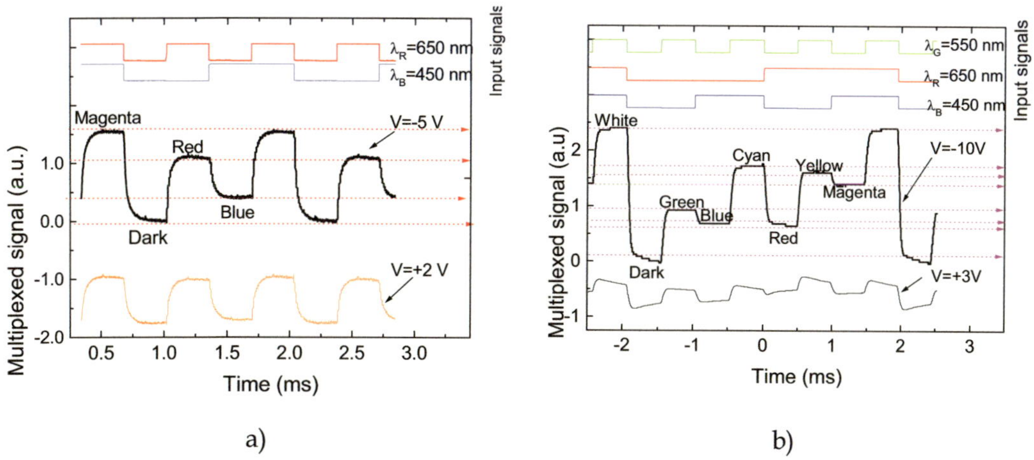

Fig. 17. Transient multiplexed signals under negative and positive bias. a) Polychromatic red and blue time dependent mixture. b) Polychromatic red, green and blue time dependent mixture. The digital wavelength demultiplexed signal is displayed on the top of both figures.

If only two R and B channels are involved (four levels; Figure 17a), under forward bias, the blue component of the combined spectra falls into the dark level, tuning the red input channel. Thus, by switching between reverse and forward bias the red and the blue channels were recovered and the transmitted information regained. If three R, G and B input channels with different transmission rates are being used (eight levels, Figure 17b), under reverse bias, the levels can be grouped into four main thresholds, ascribed respectively to the simultaneous overlap of three (R&G&B), two (R&B, R&G, B&G), one (R, G, B) and none (dark) input channels. Since under forward bias, the blue component of the multiplexed signal approaches the dark level the R, the G and the R&G components are tuned. By comparing the multiplexed signals under forward and reverse bias and using a simple algorithm that takes into account the different sub-level behaviors under reverse and forward bias (Figure 17b) it is possible to split the red from the green component and to decode their RGB transmitted information. The digital wavelength division demultiplex signals are displayed on the top of both figures.

In Figure 18, it is displayed the photocurrent generated by the combination of two red and blue input channels in short circuit (open symbol) and under reverse bias (line). Results show that under short circuit the blue component of the combined spectra falls into the dark level, tuning the red input channel. Thus, by switching between short circuit and reverse bias the red and the blue channels were recovered.

The device acts as a charge integrator, keeping the memory of the input channel. So, it can be used as a non selective division wavelength multiplexing device, WDM.

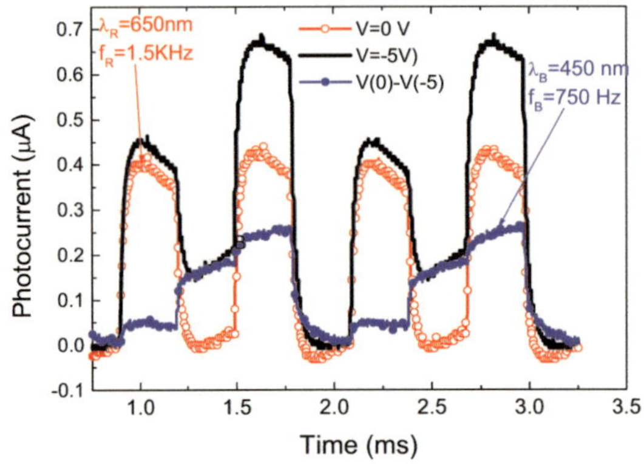

Fig. 18. Blue and red wavelength division demultiplexing output channels (dot lines) for the input signal (solid line).

4.2 Signal recovery

In Figure 19 it is displayed, under reverse and forward bias, the multiplexed signals due to the simultaneous transmission of three independent bit sequences, each one assigned to one of the RGB color channels. On the top, the optical signal used to transmit the information is displayed to guide the eyes on the different ON-OFF states. The bit sequence obtained for the demultiplexed signal is also shown for comparison.

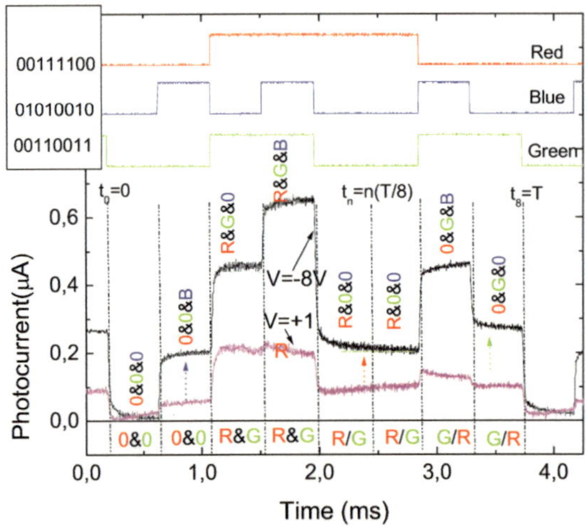

Fig. 19. Multiplexed signals under reverse and forward bias. On the top, the optical signal used to transmit the information guide the eyes on the different ON-OFF states. The bit sequence for the demultiplexed signal is shown for comparison.

To recover the transmitted information (8 bit per wavelength channel) the multiplexed signal, during a complete cycle ($0<t<T$), was divided into eight time slots, each corresponding to one bit where the independent optical signals can be ON (1) or OFF (0). As, under forward bias, the WDM device has no sensitivity to the blue channel, the red and green transmitted information can be identified from the multiplexed signal at +1V. The

highest level corresponds to both channels ON (R&G: R=1, G=1), and the lowest to the OFF-OFF stage (R=0; G=0). The two levels in-between are related with the presence of only one channel ON, the red (R=1, G=0) or the green (R=0, G=1) (see horizontal labels in Figure 19). To distinguish between these two situations and to decode the blue channel, the correspondent sub-levels, under reverse bias, have to be analyzed. From Figure 10, 12 and 15 it is observed that the green channel is more sensitive to changes on the applied voltage than the red, and that the blue only appears under reverse bias. So, the highest increase at -8V corresponds to the blue channel ON (B=1), the lowest to the ON stage of the red channel (R=1) and the intermediate one to the ON stage of the green (G=1) (see colour arrows and vertical labels in Figure 36). Using this simple algorithm the independent red, green and blue bit sequences can be decoded as: R[00111100], G[00110011] and B[01010010].

To validate the ability of this device to multiplex and demultiplex optical signals, the multiplexed signals, in Figure 12, were analysed, in a time window between 0.7ms and 4.2ms, resulting in the bit sequences: R[11110000], G[11001100] and B[10101010] which are in agreement with the signals acquired for the independent channels.

A demux algorithm was implemented in Matlab that receives as input the measured photocurrent and derives the sequence of bits that originated it. The algorithm makes use of the variation of the photocurrent instead of its absolute intensity to minimise errors caused by signal attenuation. A single linkage clustering method is applied to find automatically eight different clusters based on the measured current levels in both forward and reverse bias. This calibration procedure is performed for a short calibration sequence. Each cluster is naturally bound to correspond to one of the known eight possible combinations of red, green and blue bits. Following this procedure the sequence of transmitted bits can be recovered in real time by sampling the photocurrent at the selected bit rate and finding for each sample the cluster with closest current levels. In Figure 20 we present an example of the output obtained for two different sequences at 4000 bps.

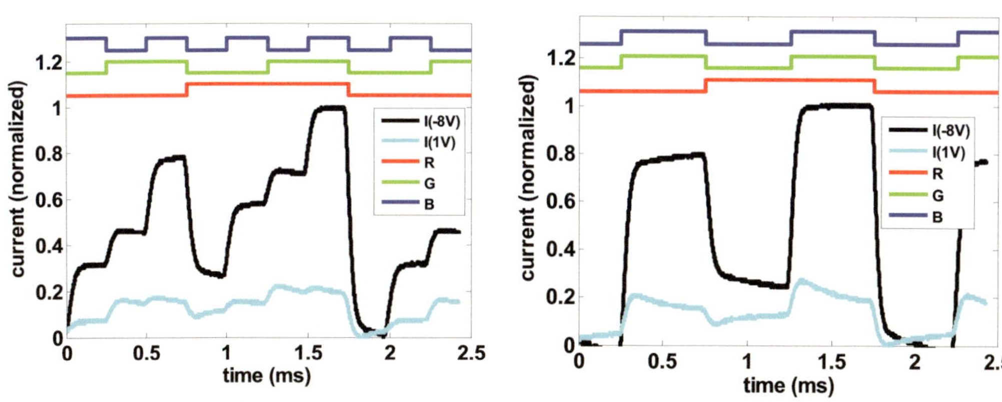

Fig. 20. A snapshot of the output from the MatLab routine used to demux the transmitted sequence of bits. The sequence of red, green and blue bits (shown at the top) was derived from the measured currents.

Output data of the demux algorithm show that the derived sequences are R [00011100], G [011001100] and B [10101010] for the multiplexed signal of Figure 20a and R [00011100], G

[011001100] and B [011001100] for Figure 20b. Both were found to be in exact agreement with the original sequences of bits that were transmitted. These observations indicate that it is possible to improve the performance of the WDM device without adding any optical pre-amplifier or optical filter, which is an advantage when compared with the standard p-i-n cells. The impulsive response of the nano-optical filter is bias controlled and the recovery algorithm is very simple, without the limiting factors associated with the operation of an optical complex receiver structure when using a standard p-i-n cell.

5. Signal attenuation

The identification of the different input channels requires a previous calibration of the transmission signal in order to know the response of the WDM device to each individual channel as the signal attenuation along the transmission medium causes a reduction of the optical intensity at the reception end (WDM device). In order to analyze the influence of this effect the multiplexed signal was acquired with input signals of different optical intensities.

In Figure 21 the output multiplexed signal is displayed at -8V and +1V for different optical power intensities of the different input channels. Measurements were made with three levels of increasing optical power, starting with R= 11 μWcm^{-2}; G= 14 μWcm^{-2}; B= 28 μWcm^{-2} (I=1×) and then for twice (I=2×: R = 13 μWcm^{-2}; G = 28 μWcm^{-2};B = 56 μWcm^{-2}) and five times (I=5×: R= 22 μWcm^{-2}; G= 38 μWcm^{-2};B = 105 μWcm^{-2}) more current to drive the LEDs.

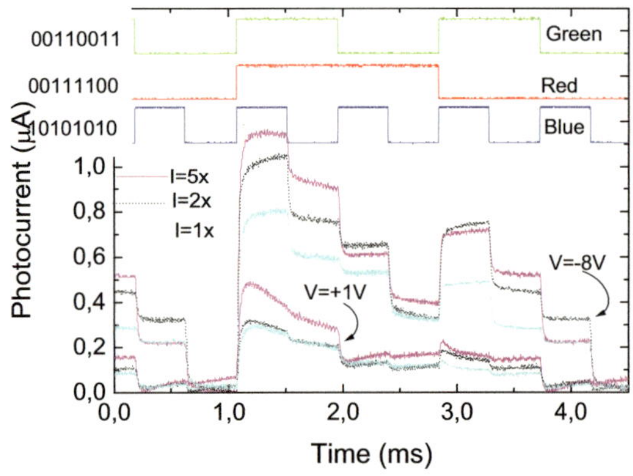

Fig. 21. Multiplexed signal at -8V and +1V under different optical power intensities of the RGB input channels.

Results show that under reverse bias, as the channel intensity decreases, the multiplexed output signal also decreases. Under forward bias the signal is almost unaltered for the lowest intensities.

6. Linearity

In Figure 22 it is displayed the multiplexed signal obtained at reverse bias, under monochromatic light bias of different wavelengths (λ_R=650 nm, λ_G=550 nm, λ_B=450 nm) and dual wavelengths combinations (λ_R=650 nm & λ_G=550 nm and λ_R=650 nm & λ_B=450 nm). The light modulation frequency of one the light bias was chosen to be half of the other in order to ensure a synchronous relation of ON-OFF states along each cycle.

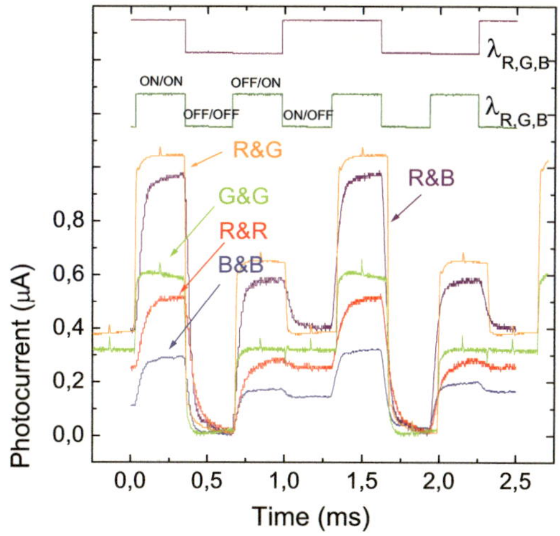

Fig. 22. Multiplexed signals at – 5V and under different input light bias: R&R ($\lambda_{R,R}$=650,650 nm); G&G($\lambda_{G,G}$=550,550 nm), B&B ($\lambda_{B,B}$= 450,450 nm), R&B($\lambda_{R,B}$=650,450 nm) and R&G($\lambda_{R,G}$=650,550 nm).

Under monochromatic illumination the ON-ON state corresponds to the maximum intensity of light bias, while the ON-OFF and OFF-ON to a lower intensity and the OFF-OFF to the dark conditions. Results suggest that for every wavelength the photocurrent signal exhibits linearity, as its magnitude is maximum in the ON-ON state, lower and constant in both OFF-ON and ON-OFF states (half of the observed in the ON-ON state) and minimum in the OFF-OFF state. The highest photocurrent signal under monochromatic illumination is obtained with green light (λ_G=550 nm), followed by the red (λ_R=650 nm) and then by the blue (λ_B=450 nm). The multiplexed signal obtained under red and green bias (λ_R=650 nm & λ_G=550 nm) or red and blue bias (λ_R=650 nm & λ_B=450 nm) shows four different magnitude levels of photocurrent corresponding each to the conditions of states of the light bias.

7. Self optical amplification

7.1 Self bias amplification under transient conditions and uniform irradiation

When an external electrical bias (positive or negative) is applied to a double pin structure, its main influence is in the field distribution within the less photo excited sub-cell [M. Vieira et al 2008]. The front cell under red irradiation, the back cell under blue light and both under green steady state illumination. In comparison with thermodynamic equilibrium conditions (dark), the electric field under illumination is lowered in the most absorbing cell (self forward bias effect) while the less absorbing reacts by assuming a reverse bias configuration (self reverse bias effect).

In Figure 23, the spectral photocurrent at different applied voltages is displayed under red (a), green (b) and blue (c) background irradiations and without it (d).

Results confirm that a self biasing effect occurs under unbalanced photogeneration. As the applied voltages changes from positive to negative the blue background enhances the spectral sensitivity in the long wavelength range. The red bias has an opposite behavior since the spectral sensitivity is only increased in the short wavelength range. Under green background the spectral photocurrent increases with the applied voltage everywhere.

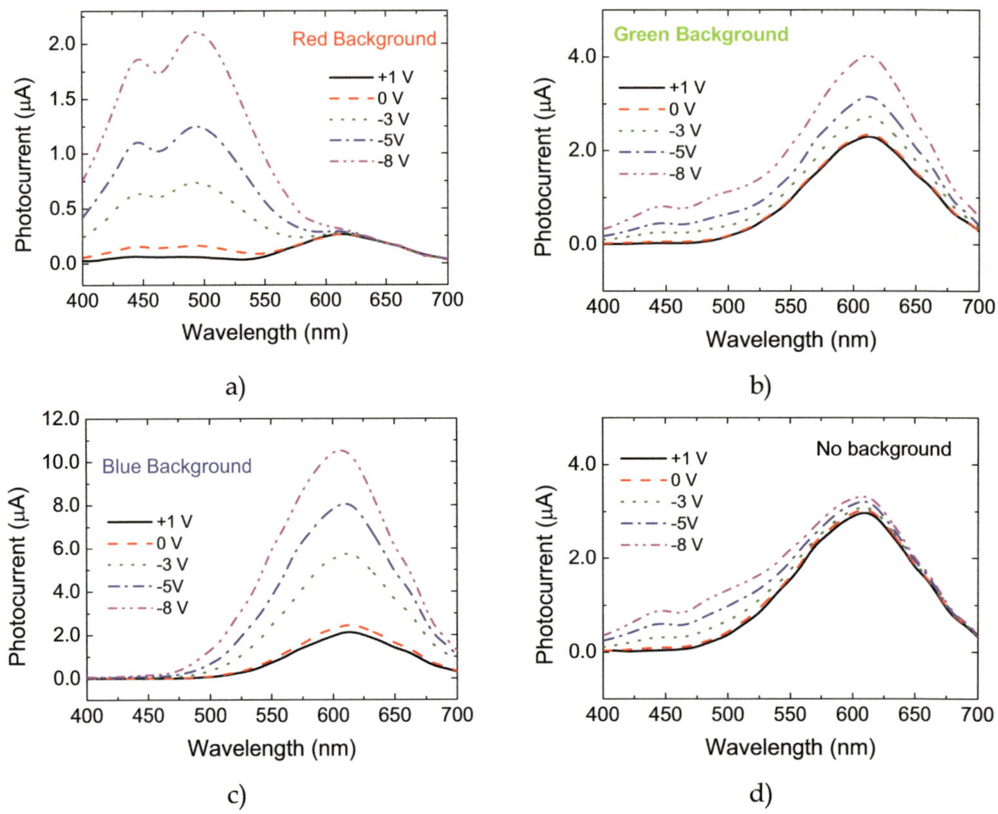

Fig. 23. Spectral photocurrent under reverse and forward bias measured without and with (λ_L) background illumination.

As expected, results show that the blue background enhances the light-to-dark sensitivity in the long wavelength range and quenches it in the short wavelength range. The red bias has an opposite behavior; it reduces the ratio in the red/green wavelength range and amplifies it in the blue one.

In Figure 24 the spectral is displayed at negative and positive electrical bias; without and under red (626 nm), green (524 nm) and blue (470 nm;) background illumination from the p side.

We have demonstrated that when an external electrical bias is applied to a double pin structure, its major influence is in the field distribution within the less photo excited sub-cell. The front cell when under red irradiation, the back cell when under blue light and both when under green steady state illumination. In comparison with thermodynamic equilibrium (no background), the electric field under illumination is lowered in the most absorbing cell (self forward bias effect) while in the less absorbing cell reacts by assuming a reverse bias configuration (self reverse bias effect).

Results confirm that, under negative bias, a self biasing effect occurs under an unbalanced photogeneration. Blue optical bias enhances the spectral sensitivity in the long wavelength ranges and quenches it in the short wavelength range. The red bias has an opposite behavior: it reduces the collection in red/green wavelength ranges and amplifies in the blue

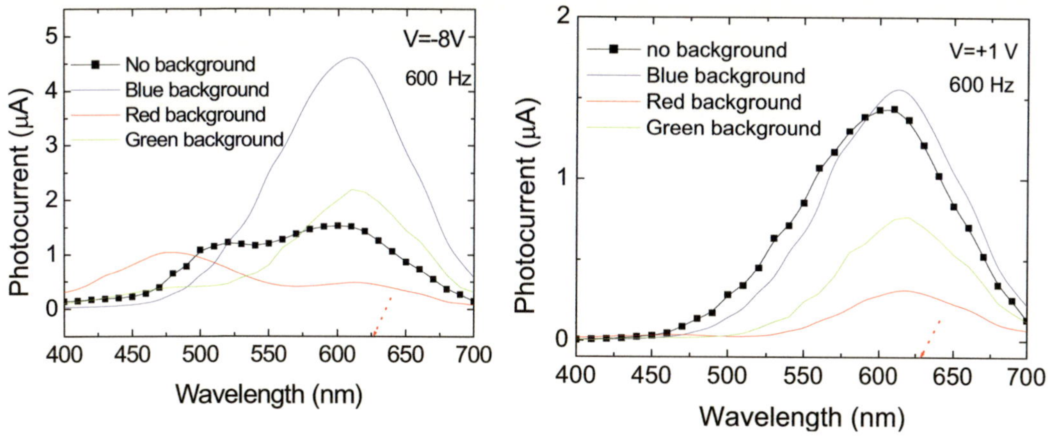

Fig. 24. Spectral photocurrent @ +1 V, -8 V without (dark) and under red, green and blue optical bias.

range. The green optical bias only reduces the spectral photocurrent in the medium wavelength range keeping the other two almost unchangeable. Under positive bias no significant self bias effect was detected. This voltage controlled light bias dependence gives the sensor its light-to-dark sensitivity allowing the recognition of a color image projected on it. In Figure 25 the ratio between the spectral photocurrents under red, green and blue steady state illumination and without it (dark) is plotted.

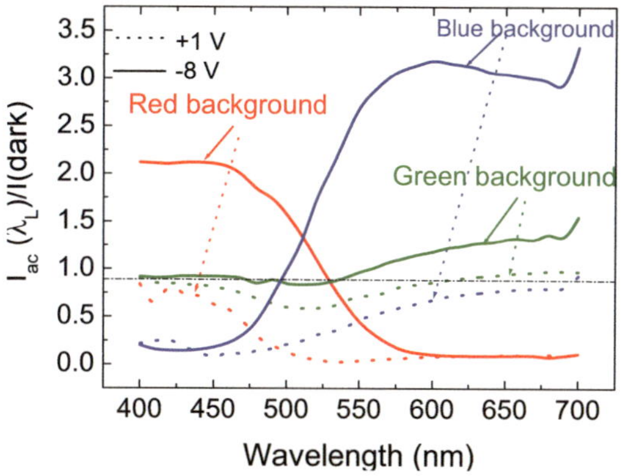

Fig. 25. Ratio between the photocurrents under red, green and blue steady state illumination and without it (dark).

The sensor is a wavelength current-controlled device that makes use of changes in the wavelength of the optical bias to control the power delivered to a load, acting as an optical amplifier. Its gain, defined as the ratio between the photocurrent with and without a specific background depends on the background wavelength that controls the electrical field profile across the device. If the electrical field increases locally (self optical amplification) the collection is enhanced and the gain is higher than one. If the field is reduced (self optical quench) the collection is reduced and the gain is lower than one. This optical nonlinearity makes the transducer attractive for optical communications.

7.2 Optical bias controlled wavelength discrimination

Figure 26 shows the time dependent photocurrent signal measured under reverse (-8V, symbols) and forward (+1V, dotted lines) bias using different input optical signals without (no bias) and with (λ_L) red, green and blue steady state additional optical bias. Both optical signals and steady state bias were directed onto the device by the side of the a-SiC:H thin structure. The optical signals were obtained by wave square modulation of the LED driving

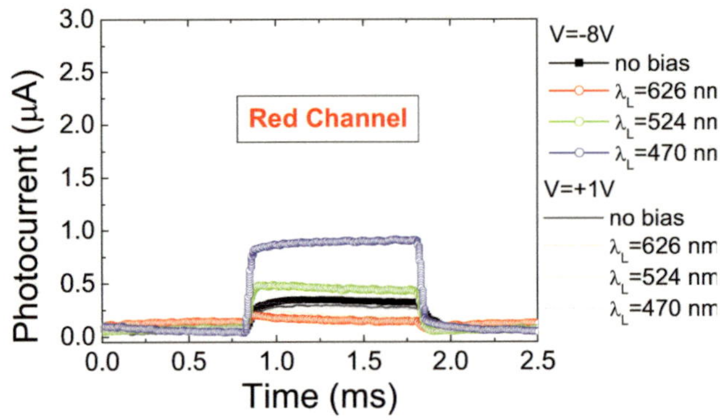

a)

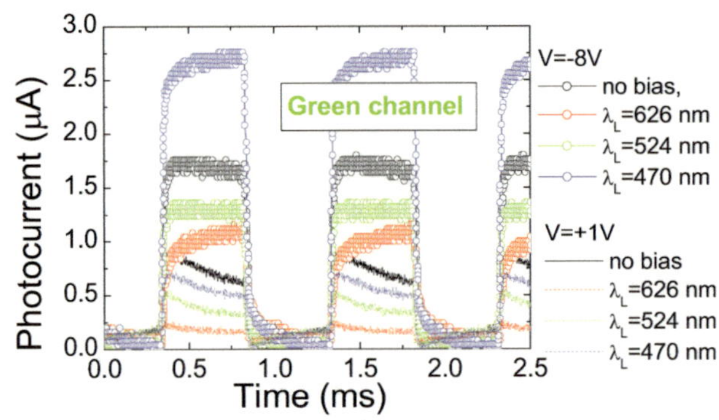

b)

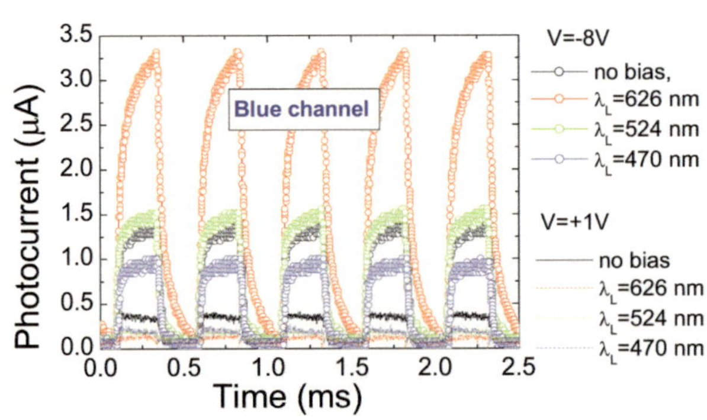

c)

Fig. 26. Red (a), green (b) and blue (b) channels under reverse and forward voltages without and with (λ_L) red, green and blue steady state bias.

current and the optical power intensity of the red, green and blue channels adjusted to 51, 90, 150 μW/cm², respectively. The steady state light was brought in LEDS driven at a constant current value (R: 290 μW/cm², G: 150 μW/cm², B: 390 μW/cm²).
Results show that the blue steady state optical bias amplifies the signals carried out by the red (Figure 26a) and the green channels (Figure 26b) and reduces the signal of the blue channel (Figure 26c). Red steady state optical bias has an opposite behavior, reinforcing the blue channel and decreasing the blue and the green channels. The green optical bias mainly affects the green channel, as the output signal is reduced while the signals of the red and blue channels show negligible changes.
When an optical bias is applied, it mainly enhances the field distribution within the less photo excited sub-cell: the back under blue irradiation and the front under red steady bias. Therefore, the reinforcement of the electric field under blue irradiation and negative bias increases the collection of carriers generated by the red channel and decreases the blue one. Under red optical bias, the opposite behavior is observed. The green bias absorption is balanced in both front and back cells and the collection of carriers generated by the green channel is strongly reduced. This optical nonlinearity makes the transducer attractive in optical communications. This nonlinear effect, under transient conditions, is used to distinguish a wavelength, to read a color image, to amplify or to suppress a color channel add and drop an optical signal.
In Figure 27, for the same bit sequence of Figure 19, the multiplexed signal is displayed.

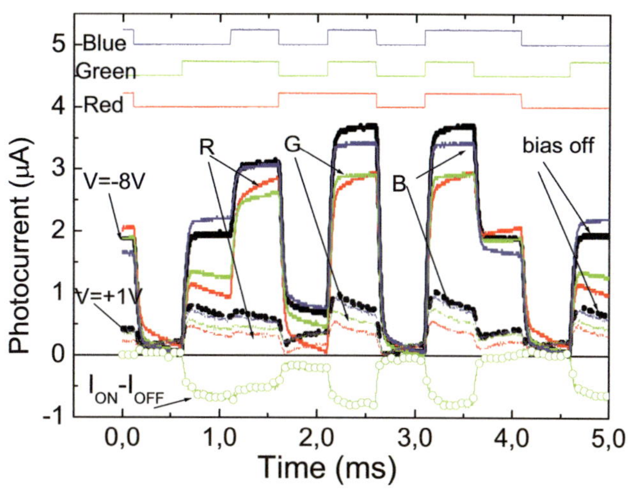

Fig. 27. Multiplexed signals at -8V/+1V (solid /dot lines); without (bias off) and with (R, G, B) optical bias.

Results confirm that under appropriated homogeneous wavelength irradiation it is possible to select or to suppress a color channel without having to demultiplex the stream.

7.2 Self bias amplification under transient conditions and uniform irradiation

A chromatic time dependent wavelength combination (4000 bps) of R (λ_R=624 nm), G (λ_G=526 nm) and B (λ_B=470 nm) pulsed input channels with different bit sequences, was used to generate a multiplexed signal in the device. The output photocurrents, under positive (dot arrows) and negative (solid arrows) voltages with (colour lines) and without (dark lines) background are displayed in Figure 28. The bit sequences are shown at the top of the figure.

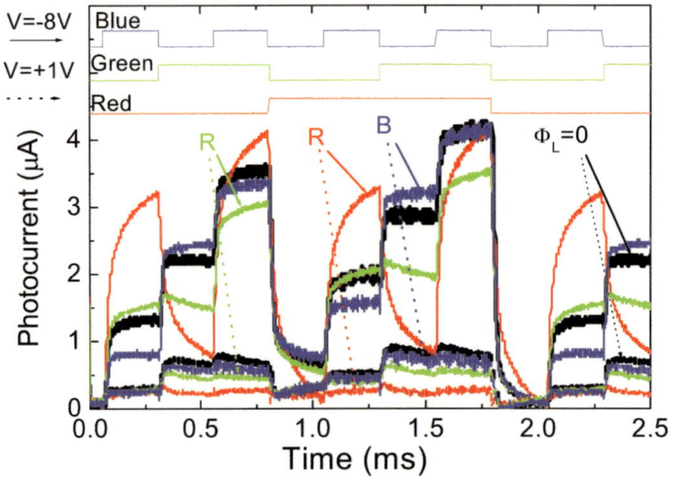

Fig. 28. Single and combined signals at -8V; without (solid arrows) and with (dotted arrows) red, green and blue optical bias.

Results show that, even under transient input signals (the input channels), the background wavelength controls the output signal. This nonlinearity, due to the transient asymmetrical light penetration of the input channels across the device together with the modification on the electrical field profile due to the optical bias, allows tuning an input channel without demultiplexing the stream.

8. Conclusions

In this chapter we present results on the optimization of multilayered a-SiC:H heterostructures for wavelength-division (de) multiplexing voltage control.
The non selective WDM device is a double heterostructure in a glass/ITO/a-SiC:H (p-i-n) /a-SiC:H(-p) /a-Si:H(-i?)/a-SiC:H (-n?)/ITO configuration. The single or the multiple modulated wavelength channels are passed through the device, and absorbed accordingly to its wavelength, giving rise to a time dependent wavelength electrical field modulation across it. The effect of single or multiple input signals is converted to an electrical signal to regain the information (wavelength, intensity and frequency) of the incoming carriers. Here, the (de) multiplexing channels is accomplished electronically, not optically. This approach has advantages in terms of cost since several channels share the same optical components; and the electrical components are typically less expensive than optical ones. An electrical model gives insight into the device operation.
Stacked structures that can be used as wavelength selective devices, in the visible range, are analysed. Two terminal heterojunctions ranging from p-ií-n to p-i-n-p-i′-n configurations are studied. Three terminals double staked junctions with transparent contacts in-between are also considered to increase wavelength discrimination. The color discrimination was achieved by ac photocurrent measurement under different externally applied bias.

Experimental data on spectral response analysis and current –voltage characteristics are reported. A theoretical analysis and an electrical simulation procedure are performed to support the wavelength selective behavior. Good agreement between experimental and simulated data was achieved. Results show that in the single p-i-n configuration the device acts mainly as an optical switch while in the double ones, due to the self bias effect, the input channels are selectively tuned by shifting between positive and negative bias. If the internal terminal is used the inter-wavelength cross talk is reduced and the S/N increased.
Results show that by switching between positive and negative voltages the input channels can be recovered or removed. So, this optical device allows to add and drop one or several channels in a WDM optical network (OADMS) and can be used in optical communications.
Three modulated input channels were transmitted together, each one located at different wavelength and frequencies. The combined optical signal was analyzed by reading out the photocurrent generated across the device.
A physical model supported by an electrical and a numerical simulation gives insight into the device operation. Voltage controlled multiplexing devices, in multilayered a-SiC:H pin architectures, were compared.
Experimental and simulated results show that the device acts as a charge transfer system. It filters, stores and transports the minority carriers generated by current pulses, keeping the memory of the input channels (color and transmission speed). In the stacked configuration, both front and back transistors act separately as wavelength selective devices, and are turned on and off sequentially by applying current pulses with speed transmissions dependent off the on-off state of all the input channels.
To enhance the bandwidth of the optical transmission system more work has to be done in order to enlarge the number of input channels and to improve the frequency response.

9. References

A. Fantoni, M. Vieira, R. Martins, Mathematics and Computers in Simulation, Vol. 49. (1999) 381-401.

A. Zhu, S. Coors, B. Schneider, P. Rieve, M. Bohm, IEEE Trans. on Electron Devices, Vol. 45, No. 7, July 1998, pp. 1393-1398.

D. Nolan K. O. Hill et all, Fiber Optics HandbookFiber, Devices, and Systems for Optical Communications McGraw Hill 2002

G. de Cesare, F. Irrera, F. Lemmi, F. Palma, IEEE Trans. on Electron Devices, Vol. 42, No. 5, May 1995, pp. 835-840.

H. Stiebig, J. Gield, D. Knipp, P. Rieve, M. Bohm, Mat. Res. Soc. Symp. Proc 337 (1995) 815-821.

H.K. Tsai, S.C. Lee, IEEE electron device letters, EDL-8, (1987) pp.365-367. .

M. Bas, Fiber Optics Handbook, Fiber, Devices and Systems for Optical Communication, Chap, 13, Mc Graw-Hill, Inc. 2002.

M. G. Kuzyk, Polimer Fiber Optics, Materials Physics and Applications, Taylor and Francis Group, LLC; 2007.

M. Haupt, C. Reinboth and U. H. P. Fischer. "Realization of an Economical Polymer Optical Fiber Demultiplexer", Photonics and Microsystems, 2006 International Students and Young Scientists Workshop, Wroclaw, 2006.

M. Kagami. "Optical Technologies for Car Applications Innovation of the optical waveguide device fabrication". Optical communications - perspectives on next generation technologies, October 23-25, 2007 in Tokyo, Japan.

M. Mulato, F. Lemmi, J. Ho, R. Lau, J. P. Lu, R. A. Street, J. of Appl. Phys., Vol. 90, No. 3 (2001), pp. 1589-1599.

M. Topic, H. Stiebig, D. Knipp, F. Smole, J. Furlan, H. Wagner, J. Non Cryst. Solids 266-269 (2000) 1178-1182.

M. Vieira, A. Fantoni, M. Fernandes, P. Louro, G. Lavareda and C.N. Carvalho, Thin Solid Films, 515, Issue 19, 2007, 7566-7570.

M. Vieira, A. Fantoni, P. Louro, M. Fernandes, R. Schwarz, G. Lavareda, and C. N. Carvalho, Vacuum, Vol. 82, Issue 12, 8 August 2008, pp: 1512-1516.

M. Vieira, A. Fantoni, P. Louro, M. Fernandes, R. Schwarz, G. Lavareda, C.N. Carvalho "Self-biasing effect in colour sensitive photodiodes based on double p-i-n a-SiC:H heterojunctions" Vacuum, Volume 82, Issue 12, 8 August 2008, pp 1512-1516.

M. Vieira, M. Fernandes, A. Fantoni, P. Louro, M. A. Vieira, "Large area a-SiC:H WDM devices for signal multiplexing and demultiplexing in the visible spectrum", Thin Solid Films, (In Press).

M. Vieira, M. Fernandes, J. Martins, P. Louro, R. Schwarz, and M. Schubert, IEEE Sensor Journal, 1, 2001, 158-167.

M. Vieira, M. Fernandes, P. Louro, A. Fantoni, M. Barata, M A Vieira, "Multilayered a-SiC:H device for Wavelength-Division (de)Multiplexing applications in the visible spectrum" Symposium A: S. Francisco, USA, 24 -29 March, 2008, in Amorphous and Polycrystalline Thin-Film Silicon Science and Technology, edited by A. Flewitt, J. Hou, S. Miyazaki, A. Nathan, and J. Yang (Mater. Res. Soc. Symp. Proc. Volume 1066, Warrendale, PA, 2008), pp.225-230 A08-01.

O. Ziemann, J. Krauser, P.E. Zamzow, W. Daum, POF Handbook, Optical Short Range Transmission Systems, Springer, 2nd Ed., 2008.

P. Louro, M. Vieira, M A Vieira, M. Fernandes, A. Fantoni, C. Francisco, M. Barata "Optical multiplexer for short range application" Physica E: Low-dimentional Systems and Nanostructures , In Press, Corrected Prof, Available online20, August 2008

P. Louro, M. Vieira, Yu. Vygranenko, A. Fantoni, M. Fernandes, G. Lavareda, N. Carvalho Mat. Res. Soc. Symp. Proc., 989 (2007) A12.04.

P. Louro, Y. Vygranenko, J. Martins, M. Fernandes and M. Vieira "Colour sensitive devices based on double p-i-n-i-p stacked photodiodes" Thin Solid Films Vol. 515, Issue 19, 16 July 2007, pp 7526-7529.

S. Randel, A.M.J. Koonen, S.C.J. Lee, F. Breyer, M. Garcia Larrode, J. Yang, A. Ng'Oma, G.J Rijckenberg, H.P.A. Boom. "Advanced modulation techniques for polymer optical fiber transmission". proc. ECOC 07 (Th 4.1.4). Berlin, Germany (2007) 1-4.

W. Daum, J. Krauser, P.E. Zamzow, O.Ziemann."Polimer Optical Fibers for Data Communication" Springer-Verlag, 2002.

W. Stallings, "Data and Computer Communications", Pearson Education, Inc., 2007. p.247-248

19

Three Transducers Embedded into One Single SiC Photodetector: LSP Direct Image Sensor, Optical Amplifier and Demux Device

M. Vieira[1,2,3], P. Louro[1,2], M. Fernandes[1,2], M. A. Vieira[1,2], A. Fantoni [1,2] and J. Costa[1,2]
[1]*Electronics Telecommunications and Computer Dept, ISEL, Lisbon,*
[2]*CTS-UNINOVA, Lisbon,*
[3]*DEE-FCT-UNL, Quinta da Torre, Monte da Caparica, 2829-516, Caparica, Portugal*

1. Introduction

Amorphous Si/SiC photodiodes working as photo-sensing or wavelength sensitive devices in the visible range, have been widely studied in the past for different applications (solar cells, color sensors, image sensors).

The term "amorphous" is commonly applied to non-crystalline materials prepared by deposition from gases. Research into amorphous silicon began, nearly fifty years ago (1960). At that time amorphous silicon was grown by evaporation or sputtering and exhibited a large defect density. In 1969 occurred the growth of amorphous silicon from plasma of silane by Chittiwick et al. (1969). Another research development occurred in 1975, with the demonstration of substitutional doping by Spear e LeComber (1975).

In early studies of amorphous silicon, it was determined that plasma-deposited amorphous silicon contained a significant percentage of hydrogen atoms bonded into the amorphous silicon structure. When amorphous silicon is deposited under hydrogenation conditions the hydrogen atoms binds to dangling bond defects and remove the corresponding electronic states in the band gap, which eliminates most of the trapping and recombinations centers. This process of introducing hydrogen into silicon is usually designated as passivation of the dangling bonds. These atoms were discovered to be essential to the improvement of the electronic properties of the material. Amorphous silicon is generally known as "hydrogenated amorphous silicon", or a-Si:H. Hydrogenated amorphous silicon (a-Si:H) has a sufficiently low amount of defects to be used within devices. However, the hydrogen is unfortunately associated with light induced degradation of the material, termed the Staebler-Wronski Effect (D. E. Carlson &C. R. Wronski ,1976).

Amorphous alloys of silicon and carbon (amorphous silicon carbide, also hydrogenated, a-Si1-xCx:H) are an interesting variant to this material. Introduction of carbon adds extra freedom for controlling the properties of the material. Increasing concentrations of carbon in the alloy widen the electronic gap between conduction and valence bands (also called "optical gap" and bandgap), in order to potentially increase the light efficiency of solar cells

made with amorphous silicon carbide layers. On the other hand, the electronic properties as a semiconductor (mainly electron mobility), are badly affected by the increasing content of carbon in the alloy, due to the increased disorder in the atomic network.
While a-Si suffers from lower electronic performance compared to c-Si, it is much more flexible in its applications. It may also produce savings on silicon material cost, as a-Si layers can be made thinner than c-Si.
One advantage is that a-Si can be deposited at very low temperatures on glass. Once deposited, a-Si can be doped in a fashion similar to c-Si, to form p-type or n-type layers and ultimately to form electronic devices.
Another advantage is that a-Si can be deposited over large areas by Plasma Enhanced Chemical Vapor Deposition (PECVD). The design of the PECVD system has great impact on the production cost of such panel, therefore most equipment suppliers put their focus on the design of PECVD for higher throughput that leads to lower manufacturing cost.
Amorphous silicon has become the material of choice for the active layer in thin-film transistors (TFTs), which are widely used in large-area electronics applications, mainly for liquid-crystal displays (LCDs).
Nowadays, modern optical networks use Arrayed Waveguide Grating (AWG) as optical wavelength (de)multiplexers (M. Bas, 2002). There has been much research on semiconductor optical amplifiers as elements for optical signal processing, wavelength conversion, clock recovery, signal demultiplexing and pattern recognition (M. J. Connelly, 2002). Here, a specific band or frequency needs to be filtered from a wide range of mixed signals. Active filter circuits can be designed to accomplish this task by combining the properties of high-pass and low-pass into a band-pass filter. Amorphous silicon carbon tandem structures, through an adequate engineering design of the multiple layers' thickness, absorption coefficient and dark conductivities (P. Louro et al., 2007) can accomplish this function.

2. Amorphous Si/SiC photosensitive devices

2.1 Device configuration and sample preparation

Voltage controlled devices, were produced by PECVD in different architectures, as displayed in Fig. 1, and tested for a proper fine tuning of the visible spectrum.
The simplest configuration is a p-i-n photodiode (#1) where the active intrinsic layer is based on an a-Si:H thin film. In the other two (#2a and #2b), the active device consists of a p-i'(a-SiC:H)-n / p-i(a-Si:H)-n heterostructures (NC5 sample code).
To test the efficiency of the internal n-p junction, a third transparent contact was deposited in-between in sample #2b. The thickness (200 nm) and the optical gap (2.1 eV) of all a-SiC:H intrinsic layers (i'-) are optimized for blue collection and red transmittance. The thickness (1000 nm) of the a-Si:H i-layers was adjusted to achieve full absorption in the green and high collection in the red spectral ranges. As a result, both front and back diodes act as optical filters confining, respectively, the blue and the red optical carriers, while the green optical carriers are absorbed across both (P. Louro et al., 2007).
The deposition conditions of the i- and i'- intrinsic layers were kept constant in all the devices. They present good optoelectronic properties with conductivities between 10^{-11} and $10^{-9}\Omega^{-1}$ cm^{-1} and photosensitivity higher than 10^4 under AM1.5 illumination (100 mW/ cm^2).
To decrease the lateral currents which are crucial for device operation (P. Louro et al., 2001; M. Vieira et al., 2005), low doping levels were used and methane was added during the

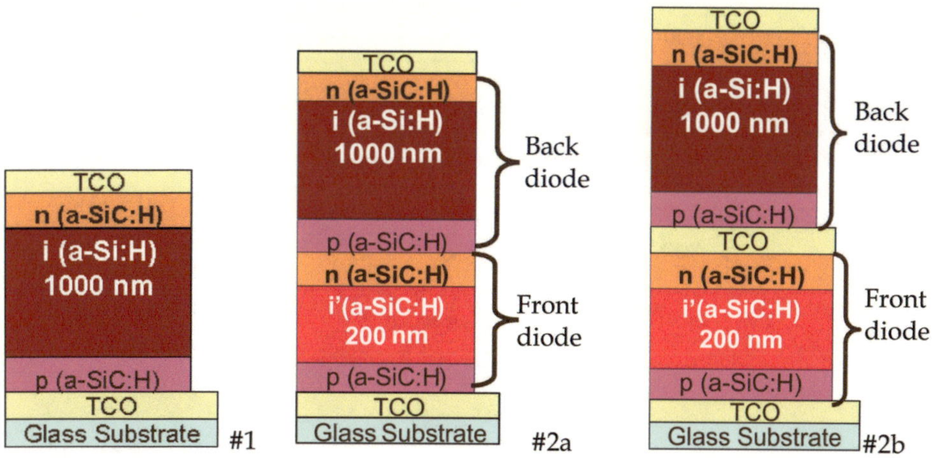

Fig. 1. Sensor element configuration.

deposition process. The doped layers (20 nm thick) have high resistivity (>$10^7 \Omega$cm) and optical gaps around 2.1 eV. Transparent contacts have been deposited on front and back surfaces to allow the light to enter and leave from both sides. The back contact defines the active area of the sensor (2×2 cm^2). The front and back contacts are based on ZnO:Al or ITO and have an average transmission around 80% from 425 nm to 700 nm and a resistivity around 9×10^{-4} Ωcm.

Besides, the stacked devices, the simplified test a-SiC:H p-i-n and a-Si:H p-i-n structures have also been deposited during the same deposition process. The film layers were deposited using a parallel-plate PECVD reactor. Deposition conditions such as the RF power, partial pressure and gas flow rates are shown in Table I. The substrate temperature was held at 260 °C.

Type	RF Power (W)	Pressure (mTorr)	Gas flow(sccm)			
			SiH_4	1% TMB+99% H_2	2%PH_3+98% H_2	CH_4
p (a-SiC:H)	4	600	10	25	–	15
i (a-SiC:H)	4	500	10	–	–	15
n (a-SiC:H)	4	500	10	–	5	15
i' (a-SiC:H)	2	400	20	–	–	–
p' (a-SiC:H)	2	400	20	10	–	–

Table I. Deposition conditions of the a-Si:H and a-SiC:H films.

2.2 Ligth-to-dark sensitivity

The typical I-V characteristics in dark and under red, blue and red&blue irradiation are displayed in Fig. 2 for the sample #1. Results show that the shape of the I-V characteristics are controlled by the background light and by the applied voltage.

To improve the light-to-dark sensitivity in sensor #1, the doped layer resistivity and optical gap were optimized. In Fig. 3 the sensitivity as a function of the wavelength is displayed under no optical bias (Φ_L=0 Wcm^{-2}) and under uniform illumination (530 nm, 2 mWcm^{-2}), respectively. Three samples were deposited keeping constant the deposition conditions for

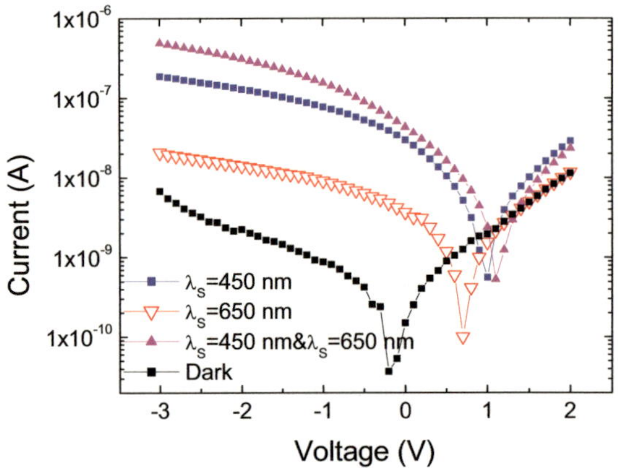

Fig. 2. I-V characteristics in dark and under red (650 nm), blue (450nm) and red&blue irradiation for the sample #1.

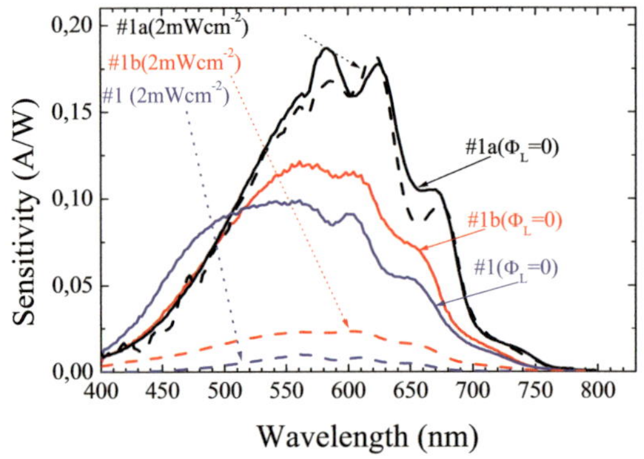

Fig. 3. Spectral sensitivity with (dash) and without (solid) applied optical bias.

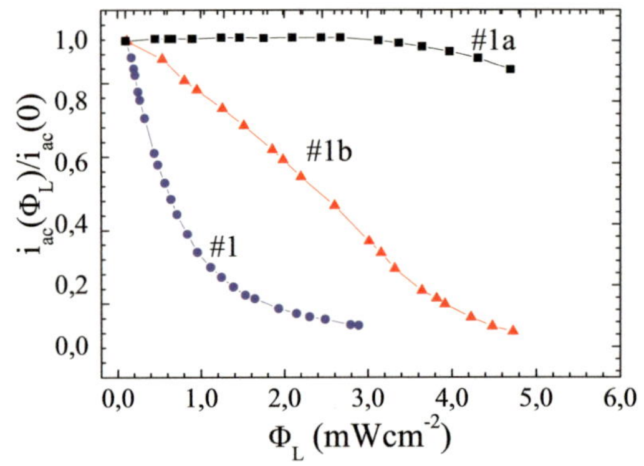

Fig. 4. $iac(\Phi_L)/iac(0)$ ratio dependence with Φ_L.

all i-layers, while they varied in the doped layers by adding methane during the deposition process. All the layers on sample #1a are based a-Si:H; (homostructure), while the p-layer in #1b and the p- and n-layers in #1 are based on a-SiC:H alloy (heterostructures).
Data reveals that, when wide band gap doped layers are used, the sensitivity is lower and decreases significantly with the optical bias. Under steady state irradiation the band misalignment reduces the electrical field in the bulk and increases the recombination at the interfaces decreasing the carrier collection.
In Fig. 4 we plot the light to dark sensitivity as a function of the applied optical bias, Φ_L. Results show also that the light-to-dark ratio depends strongly on the material of the doped layers. In the heterostructures the responsivity for low fluxes is high, while in the homostructure only a small signal could be detected in the flux range analyzed. When the sensor has both doped layers based on a-SiC:H layers (#1) the signal ratio steeply decreases. If only one layer is based on a-SiC:H (#1b) the signal ratio also decreases but at slower rate. Finally, in the homostructure (#1a) the sensor remains "blind" to the optical bias and only at higher light fluxes the signal ratio gently decreases. This light bias dependence enables different applications of the device depending on the readout technique. If a light scan is used to readout the generated carriers it can recognize a color pattern projected on it. If the photocurrent generated by different monochromatic pulsed channels or their combination is readout the information is multiplexed or demultiplexed and can be transmitted, tuned or recovered again.

2.3 Light filtering

The characterization of the devices was performed through the analysis of the photocurrent dependence on the applied voltage and spectral response under different optical and electrical bias conditions. The responsivity was obtained by normalizing the photocurrent to the incident flux. To suppress the *dc* components all the measurements were performed using the lock-in technique.

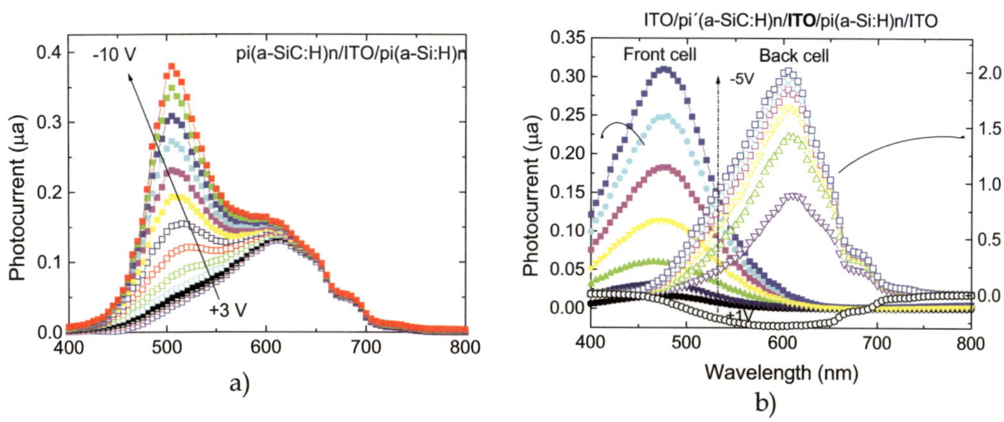

Fig. 5. a) p-i'-n-p-i-n spectral photocurrent under different applied voltages b) Front, p-i' (a-SiC:H)-n, and back, p-i (a-Si:H)-n spectral photocurrents under different applied bias.

Fig. 5a displays the spectral photocurrent of the sensor #2b under different applied bias (+3V<V<-10V), the internal transparent contact was kept floating in all measurements. In Fig. 5b the spectral photocurrent, under different electrical bias is displayed for the front, p-i' (a-SiC:H)-n, and the back p-i (a-Si:H)-n, photodiodes.

Results confirm that the front and back photodiodes act, separately, as optical filters. The front diode, based on a-SiC:H heterostructure, cuts the wavelengths higher than 550nm while the back one, based on a-Si:H, cuts the ones lower than 500nm. Each diode, separately, presents the typical responses of single p-i-n cells with intrinsic layers based on a-SiC:H or a-Si:H materials, respectively. Since the current across the device has to remain the same, in the stacked configuration, it is clearly observed the influence of both front and back diodes modulated by its serial connection through the internal n-p junction.

2.4 Optical amplification

Three monochromatic pulsed lights (input channels): red (R: 626 nm), green (G: 524 nm) and blue (B: 470nm) illuminated separately device #2a. Steady state red, green and blue optical bias was superimposed separately and the photocurrent generated measured at -8V and +1 V. In Fig. 6 the signal is displayed for each monochromatic input channel.

Results show that blue steady state optical bias amplifies the red channel and that the red light amplifies the blue channel. When an optical bias is applied it mainly enhances the field distribution within the less photo excited sub-cell: the back under blue irradiation and the front under red steady bias (see Fig. 8). So, the reinforcement of the electric field under blue irradiation and negative bias increases the collection of the carriers generated by the red channel and decrease the blue one. Under red optical bias an opposite behaviour is observed. The green bias absorption is balanced in both front and back cells. So, the green channel collection is reduced while the red and blue collections are almost insensitive to the green irradiation. This effect can be used either to amplify the red or blue channels or to tune the green one since the others remain almost constant.

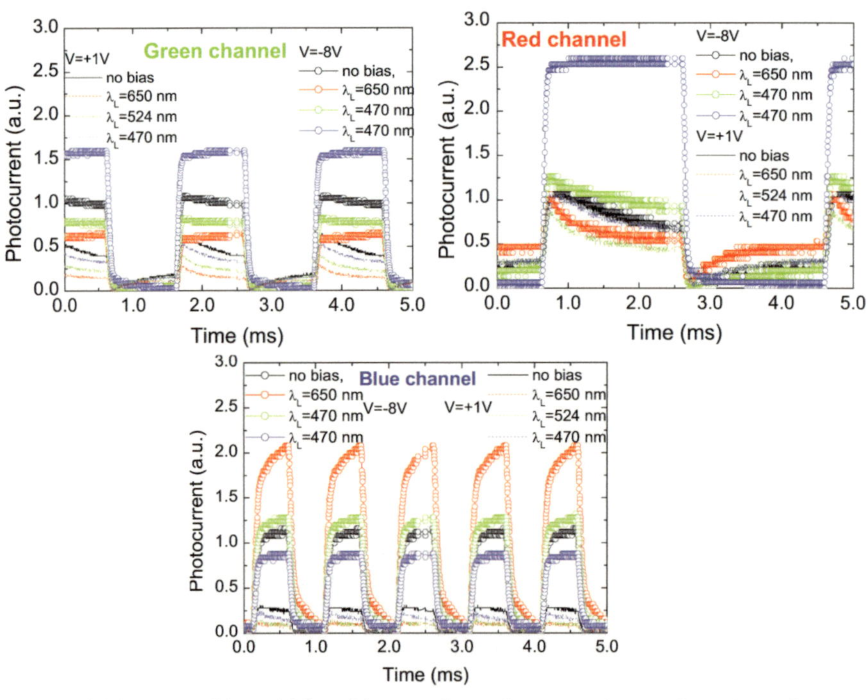

Fig. 6. Input red (a) green (b) and blue (c) signals under negative and positive bias without and with red, green and blue steady state optical bias for the #2a device.

3. Self bias effect in pinpin photodiodes

3.1 Numerical simulation

A device simulation program ASCA-2D (A. Fantoni et al., 1999) was used to analyze the potential profiles in the investigated structure. Typical values of band tail and gap state parameters for amorphous materials were used. In the films the optical band gaps were chosen in compliance with the obtained experimental values (Table I). Band discontinuities were equally distributed over the valence and conduction band offsets ($\Delta E_V = \Delta E_c = 0.15$ eV).

3.2 Generation/recombination profiles

The photogeneration/recombination profiles used in this simulation are depicted in Fig. 7 for a tandem p-i(a-SiC:H)-n/p-i(a-Si:H)-n -cell having, respectively, 200 nm and 1000 nm thick absorbers.

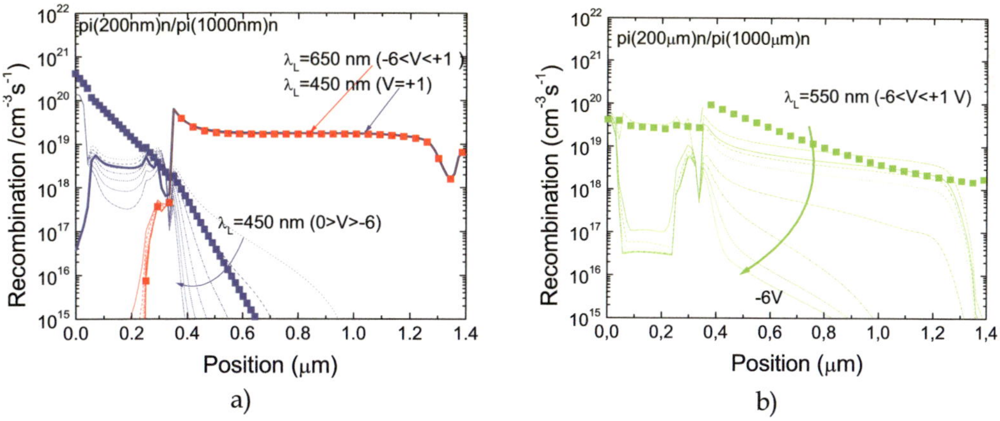

Fig. 7. Numerical simulation under different background light: (a) generation (solid lines)/recombination (dash lines) rates.

The thickness of the front diode and its optical gap was optimized for high conversion efficiency in the blue/green light and transparency of the red photons coming either from the image or from the scanner. In the a-SiC:H and a-Si:H absorbers an optical band gap of 2.1 eV and 1.8 eV and a thickness of 200 nm and 1000 nm were, respectively, chosen in compliance with the obtained experimental values for the single films. The doping level was adjusted in order to obtain approximately the same conductivity of the layers as in the tested samples.

3.3 Electrical field profiles

Fig. 8 a), b) and c) reports the simulated electric field profile within a #2-like structure under different optical bias wavelengths and for different values of the external electrical bias. In Fig. 8d) is displayed the electric field profile under thermodynamic equilibrium.

Simulated results show that the shallow penetration of the blue photons into the front diode, the deep penetration of the red photons into the back absorber or the decay of the green absorption across both, control the internal electrical field. The balance between the electrical field adjustments due to the non uniform absorption throughout the structures depends on the generation/recombination ratio profiles at each applied voltage (Fig. 7). When an external electrical bias (forward or reverse) is applied, it mainly influences the field distribution within the less photo excited sub-cell. When compared with the electric field profile under thermo-dynamical equilibrium conditions, the field under illumination is lowered in the most absorbing cell, while the less absorbing one reacts by assuming a reverse bias configuration. Consequently, opposite behavior is observed under red and blue background light while under green light condition the redistribution of the field profile is balanced between the two sub-cells.

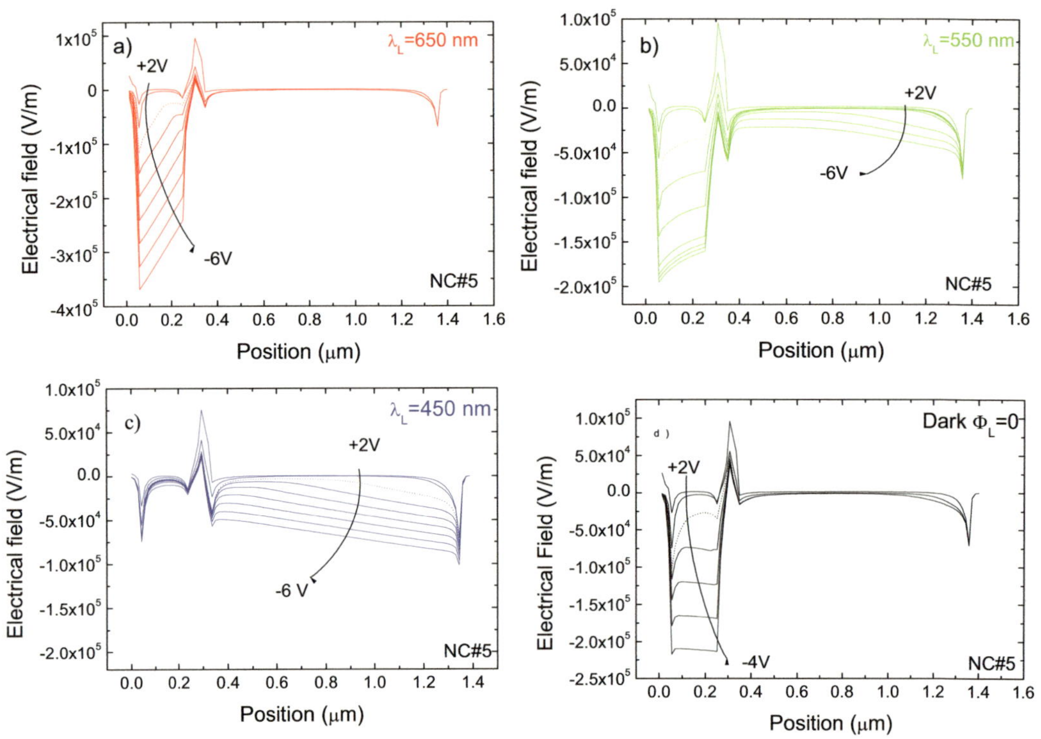

Fig. 8. Electric field profile within the p-i-n/p-i-n tandem structure for different values of the external electrical bias and for different wavelengths of impinging light: a) 650 nm; b) 550 nm; c) 450 nm d) thermodynamic equilibrium (dark) for device #2.

3.4 Potential profiles

Fig. 9 shows the simulated potential profiles at different applied voltages in dark (dash line) and under red (straight line), green (cross +) and blue (cross x) irradiation.

Results confirm that the application of an external electrical bias interferes mainly with the less absorbing cell. In the blue range as the reverse bias increases the potential drop across the non irradiated diode increases while in the front diode it remains almost negligible. In the red range the potential drop occurs across de front diode where no carriers are generated and remains negligible at the absorbing region. Due to the non uniform absorption across the back diode (Fig. 7), under green irradiation, the potential drop is distributed across both diodes and balanced between the blue (in the back diode) and the red (in the front diode) behaviors.

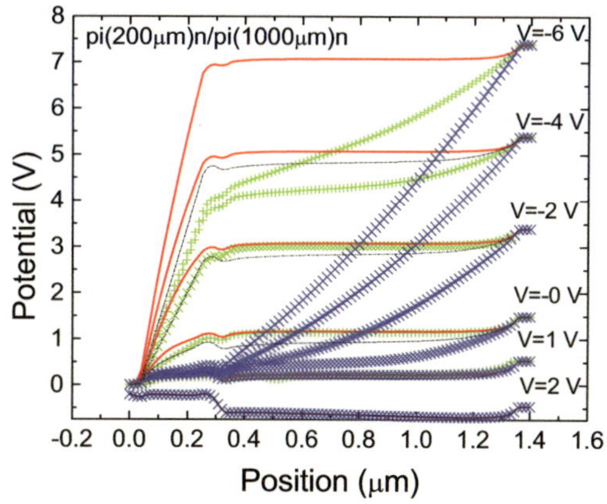

Fig. 9. Simulated potential profile under different applied voltages in dark and under red (λ_L=650 nm) green (λ_L=550 nm) and blue (λ_L=450 nm) irradiation.

3.5 Self-bias effect in p-i-n/p-i-n structures

Taking into account the geometry of the device and since light traverses through the sequence and is absorbed according to its wavelength (Fig. 7), the successive diodes should ensure that they each give the same current:

$$I_1^{pin} = I_2^{pin} = I \quad V = V_1 + V_2 \tag{1}$$

$$I_1^{pin} = I_{01}\left(\exp\left(\frac{V_1}{\eta V_T}\right)-1\right) - I_{ph1} \quad I_2^{pin} = I_{02}\left(\exp\left(\frac{V_2}{\eta V_T}\right)-1\right) - I_{ph2} \tag{2}$$

Where V_1 and V_2 are the voltage drop across each diode and V is the external voltage. $I_{0,1,2}$ and $I_{ph,1,2}$ are, respectively, the leakage and the photo currents, V_T the thermal voltage and η the ideality factor.

Neglecting the series resistance and assuming: $I_1 = I_{01} + I_{ph1}$ and $I_2 = I_{02} + I_{ph2}$ the current across the structure will be given by:

$$I = 0.5\times\left(-(I_1+I_2) + \sqrt{(I_1-I_2)^2 + 4I_{o1}I_{02}\exp\left(\frac{V}{\eta V_T}\right)}\right) \tag{3}$$

$$V = \eta V_T \ln\frac{(I+I_1)(I+I_2)}{I_{01}I_{02}} \quad V_1 = \eta V_T \ln\left(\frac{(I+I_1)}{I_{01}}\right) \quad V_2 = V - V_1 \tag{4}$$

Any diode whose current is bellow the other would have to reduce its bucking current and consequently voltage to try to catch up. This diode may even have to reverse bias itself in its

efforts to get in line with the other. Consequently voltage is decreased, if the diode is reverse biased, it becomes a power sink. This effect is what we call the self bias effect.
In Fig. 10 is displayed the potential across the front (V_1) and back (V_2) diodes as a function of the applied voltage (V) and for different photocurrent values. The η, I_{01} and I_{02} parameters were obtained in compliance with the experimental I-V characteristics ($I_{01} < I_{02}$).

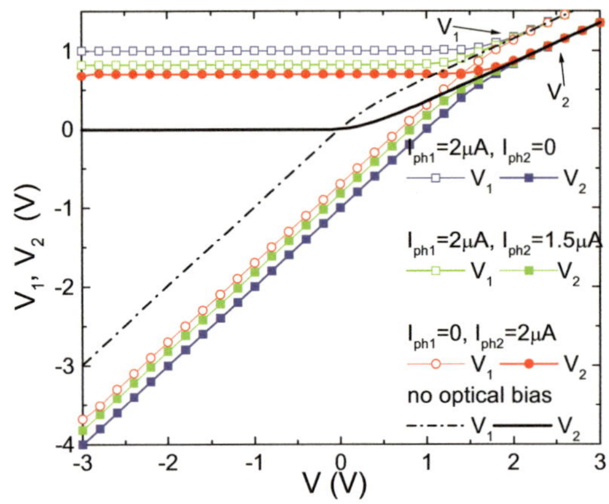

Fig. 10. Potential drop across front (V_1) and back (V_2) diodes as a function of the applied voltage (V) and for different photocurrent values.

Data show that the potential across each diode depends on the level of irradiation of both front and back diodes. Opposite behaviors are observed under red ($I_{ph1}=0$, $I_{ph2}\neq 0$, Fig. 8a) and blue ($I_{ph1}\neq 0$, $I_{ph2}=0$, Fig. 8c) background light. Under green ($I_{ph1}\neq 0$, $I_{ph2}\neq 0$, Fig. 8b) the trend depends on the I_{ph1}/I_{ph2} ratio, it approaches the blue trend if $I_{ph1}>I_{ph2}$, the red if $I_{ph1}<I_{ph2}$ and is the same as in dark if both photocurrent are balanced. Any diode whose current ($I_{1,2}^{pin}$) is below the other would have to reduce its net current (Equation 2) and consequently voltage (Equation 4) in order to try to hold up.

4. Direct LSP image and colour sensor

4.1 Colour sensitive devices

Color sensitivity in crystalline silicon video cameras is obtained by using different detection channels with three CCD arrays, or by using a mosaic of filters deposited directly onto one single solid state sensor, (R.F. Wolffenbuttel, 1987). This method can also be used for large area amorphous silicon sensor arrays and requires three sensors for each pixel. Several attempts to achieve structures capable of modifying their sensitivity spectrum by simply changing the applied bias have been reported in the literature (Y.K. Fang et al., 1987; F. Palma, 2000; M. Vieira et al., 2002). Those approaches simplify the interconnections as only two terminals are necessary.
The conventional capturing technology in an image sensor is called mosaic capture. The way that mosaic capture works is by using just one single layer of photo detectors in a mosaic pattern to capture the red, green, and blue light, or three separate layers which can absorb different wavelengths of light at different depths with each layer capturing a different color. Using three separate layers does not need to interpolate. In contrast to color

filter arrays that use light-absorbing filters, this technology converts light of all colors into useful signal information at every pixel location.

The image sensors that use the mosaic capture system have to rely on interpolation for the missing colors. Making a guess at the colors is complex which means it takes processing power. The more accurate the image sensor is, the more processing power it has to be used. Since guessing that many colors will probably never be 100% accurate, it obviously leads to color artifacts and loss of image detail.

Amorphous silicon-carbon (a-SiC:H) is a material that exhibits excellent photosensitive properties. This feature together with the strong dependence of the maximum spectral response with the applied bias has been intensively used for the development of color devices. Various structures and sequences have been suggested (H.K. Tsai & S.C. Lee, 1987; G. de Cesare, 1995; A. Zhu et al., 1998; M. Topic et al., 2000; M. Mulato et al., 2001). In our group efforts have been devoted towards the development of a new kind of color sensor. Large area hydrogenated amorphous silicon single and stacked p-i-n structures with low conductivity doped layers were proposed as color Laser Scanned Photodiode (LSP) image sensors [M. Vieira et al., 2001 ; M. Vieira et al., 2002; M. Vieira et al., 2003). These sensors are different from the other electrically scanned image sensors as they are based on only one sensing element with an opto-mechanical readout system. No pixel architecture is needed. The advantages of this approach are quite obvious like the feasibility of large area deposition and on different substrate materials (e.g. glass, polymer foil, etc.), the simplicity of the device and associated electronics, high resolution, uniformity of measurement along the sensor and the cost/simplicity of the detector. The design allows a continuous sensor without the need for pixel-level patterning, and so can take advantage of the amorphous silicon technology. It can also be integrated vertically, *i. e.* on top of a read-out electronic, which facilitates low cost large area detection systems where the signal processing can be performed by an ASIC chip underneath.

4.2 Single p-i-n laser scanned Imager

In order to analyze the light-to-dark sensitivity of sensor #1, a pattern composed by two dark regions separated by an illuminated one was projected onto the device. Then, the photocurrent (i_{ac}) generated by a low power moving spot, that scans the device in the raster mode, was measured. In Fig. 11 a line scan of the image under zero bias voltage is

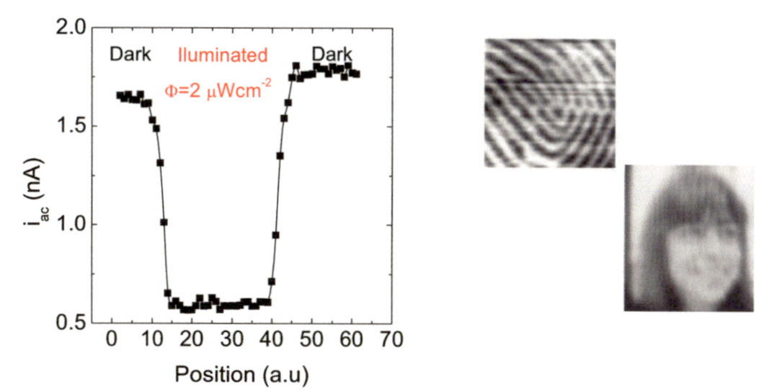

a) b)

Fig. 11. a) The ac photocurrent for one dimension scan at zero bias. b) Fingerprint and grayscale photo representations.

represented. Results show that when low conductive doped layers are used the carriers are confined into the generation regions. At the illuminated regions the band misalignment reduces the electrical field in the bulk decreasing the carrier collection (low signal). In the dark regions the electrical field is high and most of the generated carriers are collected (strong signal). As a possible application, Fig. 11b displays a grayscale photo and a fingerprint representation acquired under short circuit and 10 μWcm^{-2}. No image processing algorithms were used to enhance the image. The images have good contrast and a resolution around 30 μm showing the potential of these devices for biometric applications.

4.3 Optically addressed readout

The image to acquire is optically mapped onto the photosensitive surface and a low-power light spot scans the device. The photocurrent generated by the moving spot is recorded as the image signal, and its magnitude depends on the light pattern localization and intensity.

The image and the scanner are incident on opposite sides (Fig. 12). This approach simplifies the optical system as image and scanner have different optical paths. A low power solid state red laser ($\lambda_S = 650$ nm; Φ_S=10μW/cm^2) is used as scanner. The scanning beam position is controlled by a two axis deflection system. The line scan speed is close to 1 kHz. Two additional photodiodes provide the synchronization signals for the scanner position information needed for real time image reconstruction.

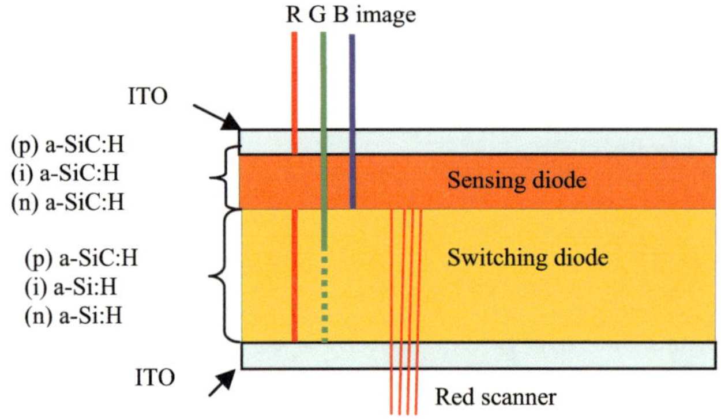

Fig. 12. Schematic of the optically addressed device optimized for colour recognition.

The current from the device is amplified and converted to digital format by a signal acquisition card installed in a computer. The data is stored as a matrix of photocurrent values (electronic image) which provide in real-time, the spatial information on the illumination conditions at the active area of the sensor (optical image). No image processing algorithms are used during the image reconstruction process.

The imaging is performed in a write-read simultaneous process: the write exposure, which converts the optical image into a localized packet of charges and the optical readout which performs the charge to current conversion by detecting the photocurrent generated by a light beam scanner. During the image acquisition process no charge transfer to move the packets of charge within the sensor is needed. This allows a real-time optically addressed readout.

4.4 Double p-i-n laser scanned colour imager

In order to optimize the readout parameters (electrical bias) and to evaluate the sensors responsivity to different light pattern wavelengths, the photocurrent generated by the scanner was measured with a lock-in amplifier under different steady-state illumination conditions bias (Φ_L=200 μWcm^{-2}, λ_L=650 nm; 550 nm; 450 nm) and displayed in Fig. 13a.

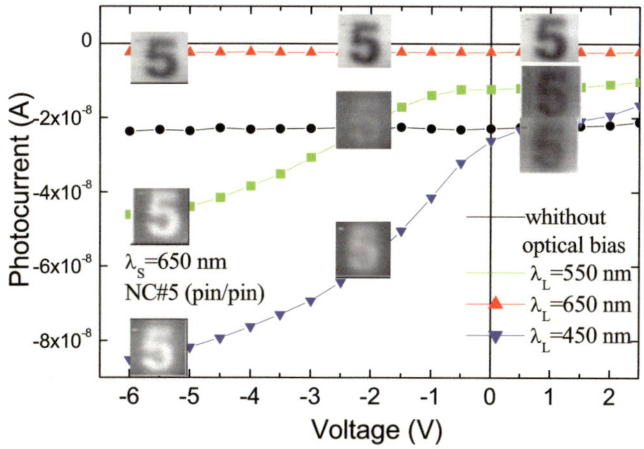

a)

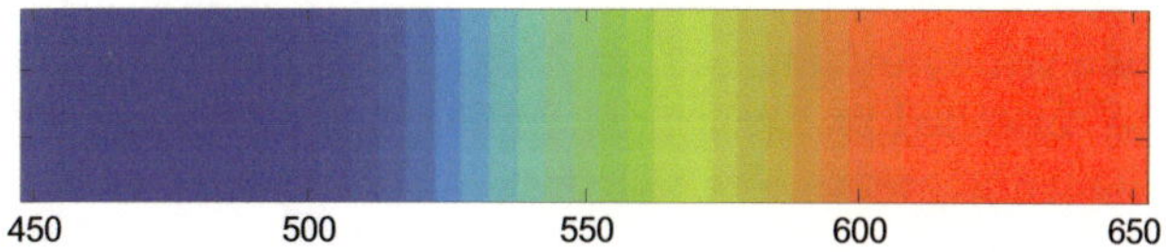

b)

Fig. 13. a) Photocurrent as a function of the applied bias for sensors #2 in dark and under blue, green and red irradiation. (Φ_L=50μW/cm2). The inserts show, at the acquired applied voltages, the images from the same RGB picture (5). b) Digital image representation of the rainbow picture acquired with device #2a.

The images, defined as the difference between the photocurrents with and without optical bias, are shown as inserts, at the acquired applied voltages. Here the same green, red and blue pictures (5) were projected, one by one, onto the front diode and acquired through the back one with a moving red scanner. The line scan frequency was close to 1 kHz and no algorithms where used during the image restoration process. For a readout time of 1 ms the frame time, for a 50 lines image, is around 50 ms.

Results show that under red irradiation or in dark (without optical bias) the photocurrent generated by a red scanner is independent on the applied voltage. Under blue/green irradiation it decreases as the applied voltage changes from reverse to forward bias being higher under blue than under green irradiation. The main difference occurs in the green spectral range. It is interesting to notice that around -2 V the collection with or without green optical image is the same, leading to the rejection of the green image signal. Taking the signal without bias as a reference, and tuning the voltages to -2 V, the red and blue signal are high and opposite and the green signal suppressed allowing blue and red color recognitions. The green information is obtained under slight forward bias (+1 V), where the blue image signal goes down to zero and the red remains constant. Readout of 1000 lines per second was achieved allowing continuous and fast image sensing.

The combined integration of this information allows recording full range of colors at each location instead of just one color at each point of the captured image as occurs with the CCD image sensors. Readout of 1000 lines per second was achieved. Fig. 13b shows the digital image using as optical image a graded wavelength mask (rainbow) to simulate the visible spectrum in the range between 400 and 700 nm. For image acquisition two applied voltages were used to sample the image signal: + 1V and -6 V. The line scan frequency was close to 500 Hz. For a readout time of 2 ms the frame time for a 40 lines image takes around 80 ms. The algorithm used for image color reconstruction took into account that at -6 V the positive signals correspond to the blue/green contribution and the negative ones to the red inputs. The green information was extracted from the image signal sampled at +2 V, where the blue signal is almost suppressed and the green and red signals are negative (Fig. 13a).

4.5 Optical bias intensity and colour rejection

In order to tune correctly the applied voltage that leads to color rejection and to be sure that this value is independent on the image intensity, the photocurrent generated by the scanner (λ_S=650 nm, 10 μWcm^{-2}) was measured with a lock-in amplifier under different steady-state illumination conditions (400 nm$<\lambda_L<$750 nm, $0<\Phi_L<160$ μWcm^{-2}). In these measurements the element sensor was uniformly illuminated through the switching diode with red pulsed light and the different optical bias were applied through the sensing one.
In Fig. 14, for sensor #2, the scanner photocurrent dependence on the applied voltage is shown under blue (a), green (b) and red (c) optical bias and different flux irradiations (Φ_L). Results confirm that for a wide flux range of the blue and green irradiation the amplitude of the image signal can be cancelled by tuning the applied voltage to an appropriated voltage (see arrows) allowing blue and green color rejection.

4.6 Conclusions

A wavelength-optimized optical signal and imaging device for color and image recognition is presented. A trade-off between sensor configuration (thickness and absorption coefficient of the a-SiC:H/ a-Si:H sensing switching absorbers) and readout parameters (light pattern and scanner wavelength) was established in order to improve the resolution and the contrast of the image.
When a thin a-SiC:H sensing absorber, optimized for red transmittance and blue collection, is used the detector behaves itself as a filter giving information about the radiation wavelength and the position where it is absorbed. By sampling the absorption region with three different bias voltages it was possible to extract separately the RGB integrated information with a good rejection ratio. For both sensors readout of 1000 lines per second was achieved allowing continuous and fast image sensing, and color recognition.

5. Wavelength-Division (de)Multiplexing device

5.1 DEMUX devices

Wavelength division multiplexing (WDM) devices are used when different optical signals are encoded in the same optical transmission path, in order to enhance the transmission capacity and the application flexibility of optical communication and sensor systems. Various types of available wavelength-division multiplexers and demultiplexers include prisms, interference filters, and diffraction gratings. Currently modern optical networks use

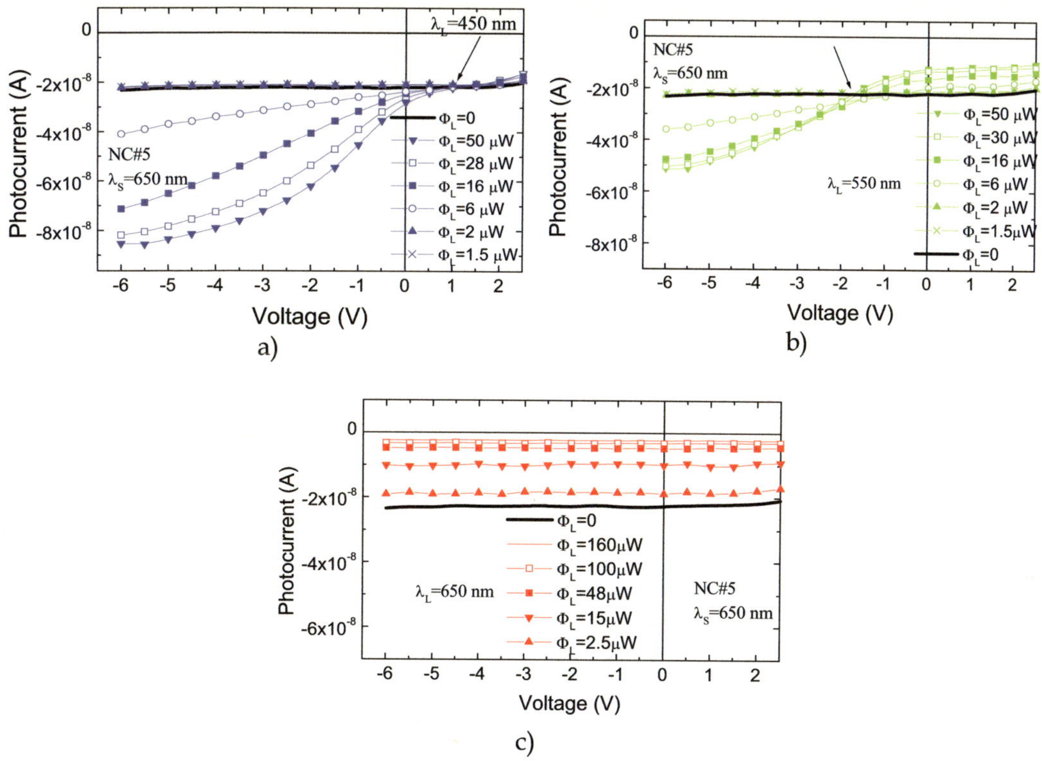

Fig. 14. Photocurrent as a function of the applied bias under blue (a), green (b) and red (c) irradiation, ($0<\Phi_L<160\mu W/cm^2$).

Arrayed Waveguide Grating (AWG) as optical wavelength (de)multiplexers (Michael Bas, 2002) based on multiple waveguides to carry the optical signals. In this paper we report the use of a monolithic WDM device based on an a-Si:H/a-SiC:H multilayered semiconductor heterostructure that combines the demultiplexing operation with the simultaneous photodetection of the signal. The device makes use of the fact that the optical absorption of the different wavelengths can be tuned by means of electrical bias changes or optical bias variations. This capability was obtained using adequate engineering design of the multiple layers thickness, absorption coefficient and dark conductivities (P. Louro et al, 2007, M. Vieira et al., 2005). The device described herein operates from 400 to 700 nm which makes it suitable for operation at visible wavelengths in optical communication applications.

5.2 Device operation

Monochromatic pulsed beams together or one single polychromatic beam (mixture of different wavelength) impinge in the device and are absorbed, according to their wavelength (Fig.s 5 and 7). By reading out, under appropriate electrical bias conditions, the photocurrent generated by the incoming photons, the input information is electrically multiplexed or demultiplexed.

In the multiplexing mode the device faces the modulated light incoming together (monochromatic input channels). The combined effect of the input signals is converted to an electrical signal, via the device, keeping the input information (wavelength, intensity and

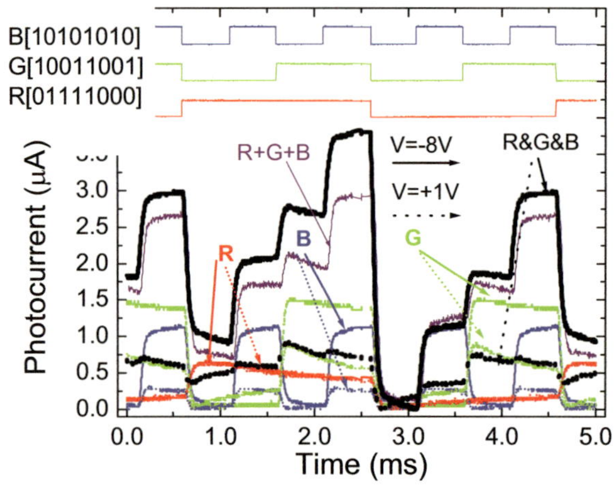

Fig. 15. Single (R, G and B) and combined (R&G&B) signals under -8V (solid arrows) and +1V (dotted arrows).

modulation frequency). The output multiplexed signal, obtained from the combination of the three optical sources, depends on both the applied voltage and on the ON-OFF state of each input optical channel. Under negative bias, the multiplexed signal presents eight separate levels. The highest level appears when all the channels are ON and the lowest if they are OFF. Furthermore, the levels ascribed to the mixture of three or two input channels are higher than the ones due to the presence of only one (R, G, B). Optical nonlinearity was detected; the sum of the input channels (R+B+G) is lower than the corresponding multiplexed signals (R&G&B). This optical amplification, mainly on the ON-ON states, suggests capacitive charging currents due to the time-varying nature of the incident lights. Under positive bias the levels were reduced to one half since the blue component of the combined spectra falls into the dark level, the red remains constant and the green component decreases.

In the demultiplexing mode a polychromatic modulated light beam is projected onto the device and the redout is performed by shifting between forward and reverse bias. Fig. 15 displays the input and multiplexed signals under negative (-8V) and positive (+1V) electrical bias. As expected, the input red signal remains constant while the blue and the green ones decrease as the voltage changes from negative to positive.

To recover the transmitted information (8 bit per wavelength channel) the multiplexed signal, during a complete cycle, was divided into eight time slots, each corresponding to one bit where the independent optical signals can be ON (1) or OFF (0). Under positive bias, the device has no sensitivity to the blue channel, so the red and green transmitted information are tuned and identified. The highest level corresponds to both channels ON (R=1, G=1), and the lowest to the OFF-OFF stage (R=0; G=0). The two levels in-between are related with the presence of only one channel ON, the red (R=1, G=0) or the green (R=0, G=1). To distinguish between these two situations and to decode the blue channel, the correspondent sub-levels, under reverse bias, have to be analyzed. The highest increase at -8V corresponds to the blue channel ON (B=1), the lowest to the ON stage of the red channel (R=1) and the intermediate one to the ON stage of the green (G=1). Using this simple key algorithm the independent red, green and blue bit sequences were decoded as: R[01111000], G[10011001] and B[10101010], as shown on the top of Fig. 15, which are in agreement with the signals acquired for the independent channels.

5.3 Conclusion

A mulilayered device based on a-SiC:H/a-Si:H was used for demultiplexing optical signals operating in the visible range until 4000 bps. The effect of the electrical applied voltage was analyzed. A recovery algorithm to demultiplex the optical signals was proposed and tested.

6. Optical amplifier

6.1 Semiconductor amplifiers

An optical parametric amplifier comprises a material that has nonlinear, i.e., amplitude-dependent, response to each incident light wave. High optical nonlinearity makes semiconductor amplifiers attractive for all optical signals. There has been much research on semiconductor optical amplifiers as elements for optical signal processing, wavelength conversion, clock recovery, signal demultiplexing and pattern recognition (M. J. Connelly, 2002), where a particular band, or frequencies need to be filtered from a wider range of mixed signals. The basic principle of an optical amplifier is very simple. All amplifiers are current- control devices. An optical amplifier uses a changing capacitance to control the power delivered to a load. This section reports results on the use of a double pi'n/pin a-SiC:H WDM heterostructure as an active band-pass filter transfer function depending on the wavelength of the trigger light and device bias.

6.2 Steady state conditions

When an external electrical bias (positive or negative) is applied to a double pin structure, its main influence is in the field distribution within the less photo excited sub-cell (M. Vieira et al., 2008, Fig. 8). In comparison with thermodynamic equilibrium conditions (dark), the electric field under illumination is lowered in the most absorbing cell (self forward bias effect) while the less absorbing reacts by assuming a reverse bias configuration (self reverse bias effect). In Fig. 16, the spectral photocurrent at different applied voltages is displayed under red (a), green (b) and blue (c) background irradiations and without it (d).

Results confirm that a self biasing effect occurs under unbalanced photogeneration. As the applied voltages changes from positive to negative the blue background enhances the spectral sensitivity in the long wavelength range. The red bias has an opposite behavior since the spectral sensitivity is only increased in the short wavelength range. Under green background the spectral photocurrent increases with the applied voltage everywhere.

In Fig. 17 the ratio between the spectral photocurrents at +1V and -8 V, under red, green and blue steady state illumination and without it (dark), are plotted.

As expected, results show that the blue background enhances the light-to-dark sensitivity in the long wavelength range and quenches it in the short wavelength range. The red bias has an opposite behavior; it reduces the ratio in the red/green wavelength range and amplifies it in the blue one.

The sensor is a wavelength current-controlled device that make use of changes in the wavelength of the optical bias to control the power delivered to a load, acting as an optical amplifier. Its gain, defined as the ratio between the photocurrent with and without a specific background (Fig. 17) depends on the background wavelength that controls the electrical field profile across the device. If the electrical field increases locally (self optical amplification) the collection is enhanced and the gain is higher than one. If the field is reduced (self optical quench) the collection is reduced and the gain is lower than one. This optical nonlinearity makes the transducer attractive for optical communications.

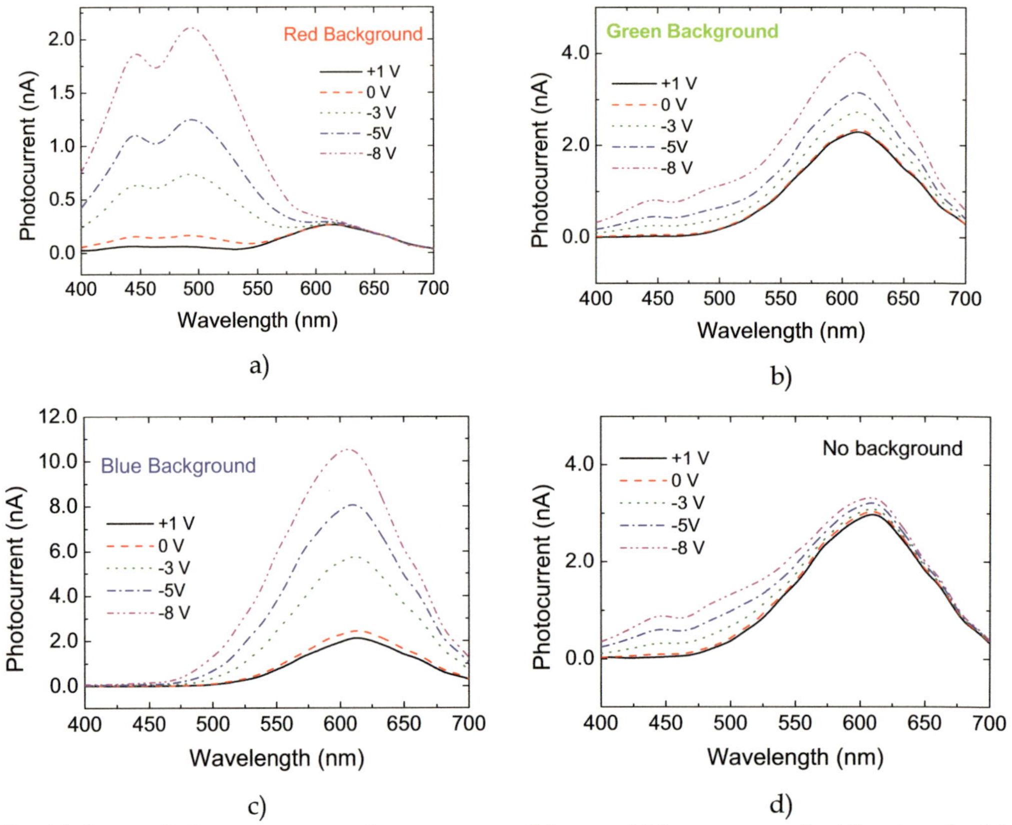

Fig. 16. Spectral photocurrent under reverse and forward bias measured without and with (λ_L) background illumination.

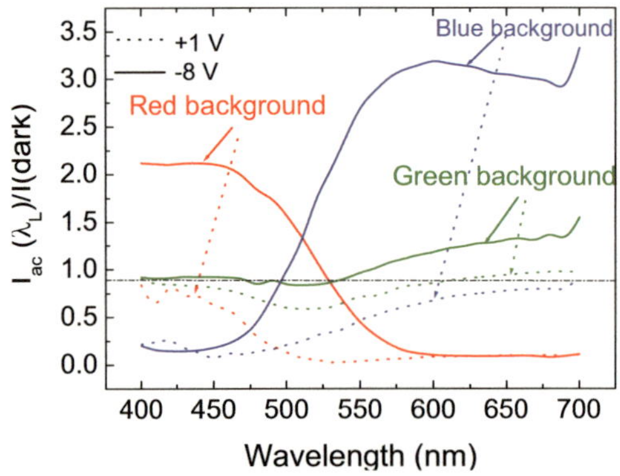

Fig. 17. Ratio between the photocurrents under red, green and blue steady state illumination and without it (no background).

6.3 Transient conditions and uniform irradiation

A chromatic time dependent wavelength combination (4000 bps) of R (λ_R=624 nm), G (λ_G=526 nm) and B (λ_B=470 nm) pulsed input channels with different bit sequences, was used to generate a multiplexed signal in the device. The output photocurrents, under positive (dot arrows) and negative (solid arrows) voltages with (colour lines) and without (dark lines) background are displayed in Fig. 18. The bit sequences are shown at the top of the figure.

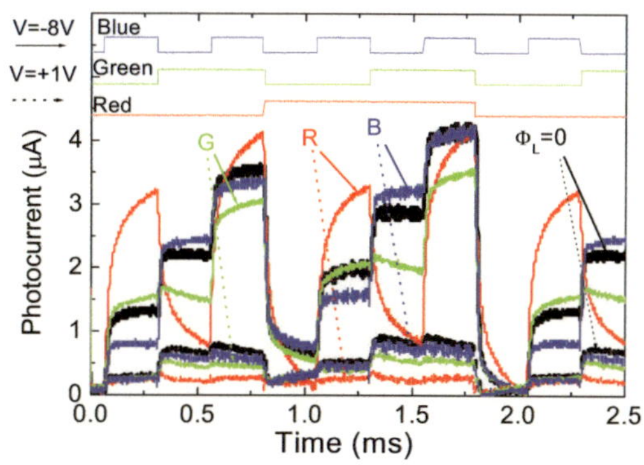

Fig. 18. Single and combined signals @-8V; without (solid arrows) and with (dotted arrows) green optical bias.

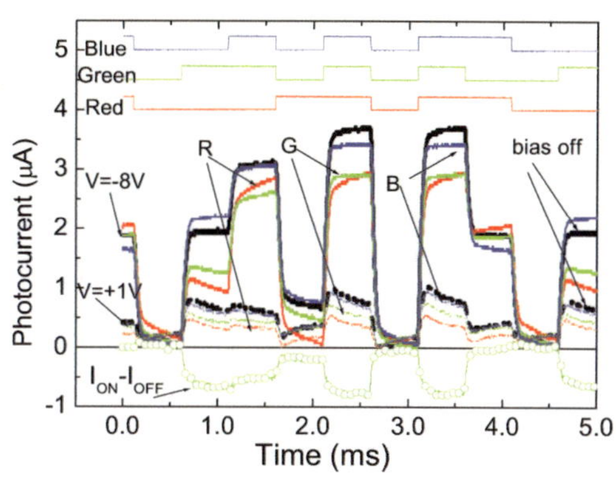

Fig. 19. Multiplexed signals @-8V/+1V (solid/dot lines); without (bias off) and with (R, G, B) green optical bias.

Results show that, even under transient input signals (from the input channels), the background wavelength controls the output signal. This nonlinearity, due to the transient asymmetrical light penetration of the input channels across the device together with the modification on the electrical field profile due to the optical bias, allows tuning an input channel without demultiplexing the stream. In Fig. 19 the green channel is tuned through the difference between the multiplexed signal with and without green bias.

7. The optoelectronic model.

7.1 Two connected transistor model

Based on the experimental results and device configuration an optoelectronic model was developed (M. A. Vieira, 2009) and displayed in Fig. 20. The optoelectronic block diagram

(a) consists of four essential elements: a double pin device for detection, a voltage supply to *dc* voltage bias; optical connections for light triggering and optical bias to *dc* light bias control. These four elements when connected together form the essentials of the optoelectronic circuit.

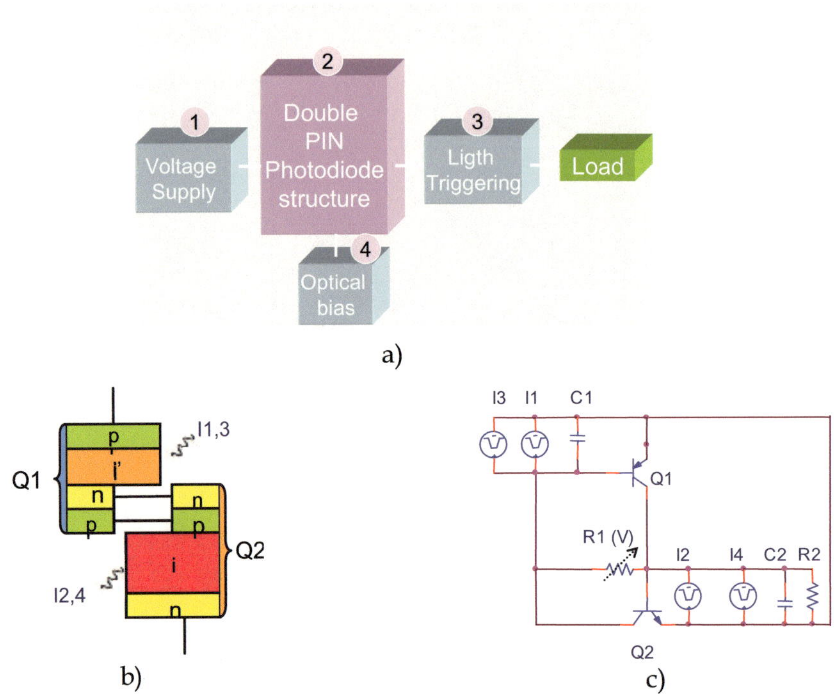

Fig. 20. a) Optoelectronic block diagram. b) Compound connected phototransistor equivalent model. c) *ac* equivalent circuit.

The monolithic device is modelled by the two-transistor model (Q_1-Q_2) (Fig. 20b) that can be seen as a *pnp* (Q_1) and *npn* (Q_2) phototransistors separately. The *ac* circuit representation is displayed in Fig. 20c. Its operation is based upon the following principle: the flow of current through the resistor connecting the two transistor bases is proportional to the difference in the voltages across both capacitors (charge storage buckets). To trigger the device four square-wave current sources with different intensities are used; two of them, I_1 and I_2, with different frequencies to simulate the input blue and red channels and other two, I_3 and I_4, with the same frequency but different intensities to simulate the green channel due to its asymmetrical absorption across both front and back phototransistors. The charge stored in the space-charge layers is modeled by the capacitor C_1 and C_2. R_1 and R_2 model the dynamical resistances of the internal and back junctions under different *dc* bias conditions. The multiplexed signal was simulated by applying the Kirchhoff's laws for the simplified ac equivalent circuit and the four order Runge-Kutta method to solve the corresponding state equations.

Once the *ac* sources are connected in the load loop an *ac* current flows through the circuit, establishing voltage modifications across the two capacitors. During the simultaneous transmission of the three independent bit sequences, the set-up in this capacitive circuit loop is constantly changing in magnitude and direction. This means that the voltage across one capacitor builds up until a maximum and the voltage across the other builds up to a minimum. The system then stops and builds up in the opposite direction. It tends to saturate and then leave the saturation because of the cyclic operation. This results in changes on the reactance of both capacitors. The use of separate capacitances on a single resistance R_1 results in a charging current gain proportional to the ratio between collector currents. The *dc* voltage, according to its strength, aids or opposes the *ac* currents. In Fig. 21 the simulated currents (symbols) under negative bias, with (b) and without (a) red bias, are compared. The current sources are also displayed (dash lines). To simulate the red background, current sources intensities were multiplied by the on/off ratio between the input channels with and without optical bias. The same bit sequence of Fig. 15 was used but at a lower bit rate (2000 bps). To validate the model the experimental multiplexed signals are also shown (solid lines).

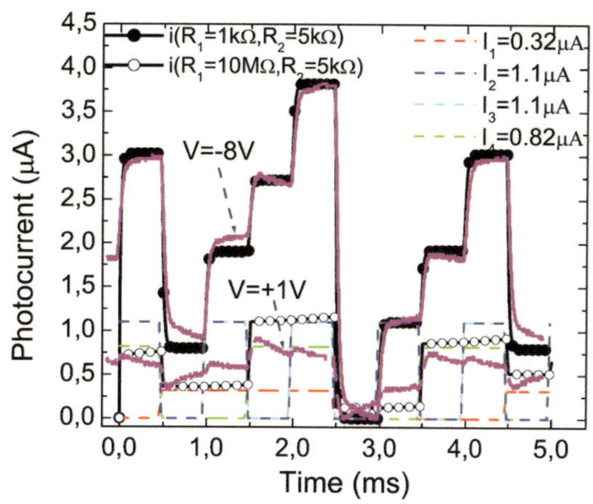

(a)

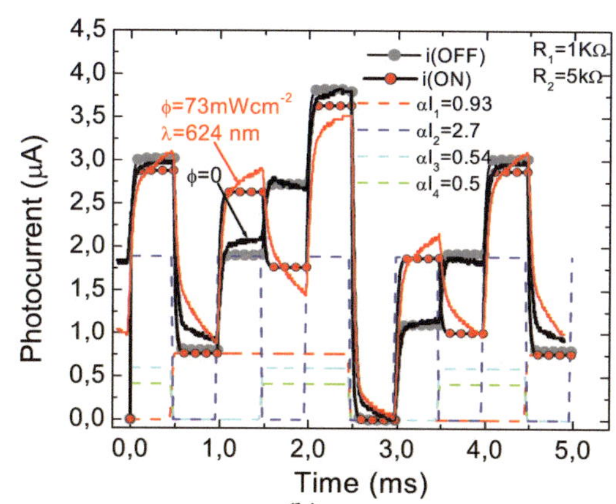

(b)

Fig. 21. Multiplexed simulated (symbols), current sources (dash lines) and experimental (solid lines): a) Positive (R_1=10MΩ; +1V) and negative dc bias. b) Negative dc bias (R_1=1kΩ; -8V) with and without red irradiation.

Good agreement between experimental and simulated data was observed. Under negative bias and if no optical bias is applied, the expected eight levels are detected, each one corresponding to the presence of three, two, one or no color channel ON. Under positive bias or steady state irradiation the levels are amplified or reduced depending on the external control. The expected optical amplification is observed due to the effect of the active multiple-feedback filter when the back diode is light triggered.

7.2 Sensing applications

The optical amplification under transient condition also explains the use of the same device configuration in the Laser Scanned Photodiode (LSP) image and color sensor. Here, if a low power monochromatic scanner is used to readout the generated carriers the transducer recognizes a color pattern projected on it acting as a color and image sensor. Scan speeds up to 10^4 lines per second are achieved without degradation in the resolution

For the color image sensor only the red channel is used. To simulate a color image at the XY position, using the multiplexing technique, a low intensity moving red pulse scanner (Φ_S, λ_S), impinges in the device in dark or under different red, green and blue optical bias (color pattern, Φ_L, $\lambda_{RGB,L}$, $\Phi_L > \Phi_S$). Fig. 22 displays the experimental acquired electrical signals. The image signal is defined as the difference between the photocurrent with (light pattern) and without (dark) optical bias.

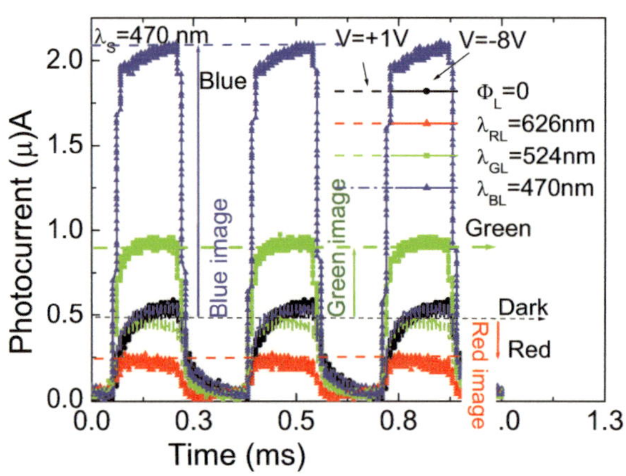

Fig. 22. Experimental color recognition using the WDM technique.

Without optical bias (Φ_L =0) and during the red pulse, only the minority carriers generated at the base of Q_2 by the scanner, flow across the circuit (I_2) either in reverse or forward bias. Under red irradiation (red pattern, $\Phi \neq 0$, λ_{RL}) the base-emitter junction of Q_2 is forward bias, the recombination increases reducing I_2 thus, a negative image is observed whatever the applied voltage. Under blue ($\Phi \neq 0$, λ_{BL}) or green ($\Phi \neq 0$, λ_{BL}) patterns irradiations the signal depends on the applied voltage and consequently, on R_1. Under negative bias, the charge transferred from C_1 to C_2, reaches the output terminal as a capacitive charging current. An optical enhancement is observed due to the amplifier action of adjacent collector junctions which are always polarized directly. Under positive bias the device remains in its non conducting state, unless the red pulse (I_2, dark level) is applied to the base of Q_2. Here, Q_2 acts as a photodiode for one polarity of the current. No amplification occurs and the red channel is strongly reduced when compared with its value under negative voltage. Under blue irradiation, the internal junction becomes reverse biased at +1 V (blue threshold) allowing the blue recognition. The behavior under a green pattern depends on the balance

between the green absorption into the front and back diodes that determines the amount of charges stored in both capacitors. Under negative bias both the green component absorbed in the front diode (blue-like) or at the back diode (red-like) reaches the output terminal while for voltages at which the internal junction n-p becomes reversed (green threshold), the blue-like component is blocked and the red-like reduced. So, by using a thin a-SiC:H front absorber optimized for blue collection and red transmittance and a back a-Si:H absorber to spatially decouple the green/red absorption, the model explains why a moving red scanner (probe beam) can be used to readout the full range of colors at each location without the use of a pixel architecture.
CCD digital image sensors are only able of recording one color at each point. Since, under steady state illumination (optical image) each phototransistor acts as a filter this multiplexing readout technique can also be used if the stack device is embedded in silicon forming a two layer image sensor that captures full color at every point. Here, a demosaicing algorithm is needed for color reconstruction.

8. Conclusions

Single and stacked pin heterojunctions based on a-SiC:H alloys were compared under different optical and electrical bias conditions using different readout techniques. Several applications is presented. A theoretical model gives insight into the physics of the device. Results show that the device can be considered as three integrated transducers in a single photodetector. If a light scan of fixed wavelength is used for readout it is possible to recognize a color pattern. On the other hand, if the photocurrent generated by different monochromatic pulsed channels or their combination is readout directly the information is multiplexed or demultiplexed. Finally, when triggered by light with appropriated wavelengths, it can amplify or suppress the generated photocurrent working as an optical amplifier. A self bias model was presented to support the sensing methodologies.

9. References

A. Fantoni, M. Vieira, R. Martins, (1999), *Mathematics and Computers in Simulation*, Vol. 49. pp. 381-401.

A. Zhu, S. Coors, B. Schneider, P. Rieve, M. Bohm, (1998) IEEE Trans. on Electron Devices, Vol. 45, No. 7, July, pp. 1393-1398.

D. E. Carlson and C. R. Wronski, (1976) *Applied Physics Letters* 33 935.

F. Palma (2000) in "Technology and Applications of Amorphous Silicon", Edited by R. A. Street, Spring-Verlag, Berlin Heidelberg New York.

G. de Cesare, F. Irrera, F. Lemmi, F. Palma, (1995) IEEE Trans. on Electron Devices, Vol. 42, No. 5, May, pp. 835-840.

H.K. Tsai, S.C. Lee, (1987) *IEEE electron device letters*, EDL-8, pp.365-367.

K.C. Chang, Chun-Yen Chang, Y.K. Fang and S.C. Jwo, (1987) "The amorphous Si/SiC heterojunctions colour-sensitive phototransistor" *IEEE electron device letters*, Vol. EDL-8,No. 2 pp.64-65.

M. A. Vieira, M. Vieira, M. Fernandes, A. Fantoni, P. Louro, M. Barata (2009), Amorphous and Polycrystalline Thin-Film Silicon Science and Technology 2009, MRS Proceedings Vol. 1153, A08-0.

M. Bas, Fiber Optics Handbook,(2002) Fiber, *Devices and Systems for Optical Comm.*, Chap, 13, Mc Graw-Hill.

M. J. Connelly (2002), *Semiconductor Optical Amplifiers*. Boston, MA: Springer-Verlag, 2002. ISBN 978-0-7923-7657-6.

M. Mulato, F. Lemmi, J. Ho, R. Lau, J. P. Lu, R. A. Street, (2001) *J. of Appl. Phys.*, Vol. 90, No. 3, pp. 1589-1599.

M. Topic, H. Stiebig, D. Knipp, F. Smole, J. Furlan, H. Wagner, (2000) *J. Non Cryst. Solids* 266-269 1178-1182.

M. Vieira, A. Fantoni, M. Fernandes, P. Louro, I. Rodrigues (2003) Mat. Res. Soc. Symp. Proc 762@2003 A.18.13.

M. Vieira, A. Fantoni, P. Louro, M. Fernandes, R. Schwarz, G. Lavareda, and C. N. Carvalho, (2008) *Vacuum*, Vol. 82, Issue 12, 8 August 2008, pp: 1512-1516.

M. Vieira, M. Fernandes, A. Fantoni, P. Louro, and R. Schwarz, (2002) "A new CLSP sensor for Image recognition and color separation", Mat. Res. Soc. Symp. Proc., Editors: Cohen, Abelson, Matsumura, Robertson, 715 (2002) pp.695-700.

M. Vieira, M. Fernandes, J. Martins, P. Louro, R. Schwarz, M. Schubert, (2001) *IEEE Sensor Journal*, 1, no.2 (August) pp. 158-167.

M. Vieira, M. Fernandes, P. Louro, R. Schwarz, M. Schubert (2002), *J. Non Cryst. Solids* 299-302 pp.1245-1249.

Michael Bas (2002), *Fiber Optics Handbook, Fiber, Devices and Systems for Optical Communication*, Chap, 13, Mc Graw-Hill, Inc. 2002.

P. Louro, M. Vieira, Yu. Vygranenko, A. Fantoni, M. Fernandes, G. Lavareda, N. Carvalho (2007) Mat. Res. Soc. Symp. Proc., 989, A12.04.

P. Louro, M. Vieira, Yu. Vygranenko, M. Fernandes, R. Schwarz, M. Schubert (2001), *Applied Surface Science* 184, 144-149 (2001).

R.C. Chittick, J.H. Alexander, H. F. Sterling, (1969) *J. Electrochem. Soc.* 116, 77 .

R.F. Wolffenbuttel, (1987) "Color filters integrated with he detector in silicon" *IEEE electron device letters*, Vol. EDL-8,No. 1 (1987) pp.13-15.

W. E. Spear and P. G. Le Comber (1975), *Solid State Commum.* 17, 1193 (1975)

Y.K. Fang, S.B. Hwang, Y.W. Chen, and L.C. Kuo, (1987) "vertical type a-Si:H back-to-back Schottky diode for high-seed colour image sensor" *IEEE electron device letters*, Vol. EDL-8,No. 2 pp.64-65.

20

InAs Infrared Photodiodes

Volodymyr Tetyorkin[1], Andriy Sukach[1] and Andriy Tkachuk[2]
[1]*V. Lashkaryov Institute of Semiconductor Physics NAS of Ukraine*
[2]*V.Vinnichenko Kirovohrad State Pedagogical University*
Ukraine

1. Introduction

The performance of infrared InAs homojunction and heterojunction photodiodes (PDs) and possibilities of its improvement are analyzed both theoretically and experimentally. The PDs are diffusion-limited at room temperature. An excess current at low temperatures is analyzed within a model of non-homogeneous p-n junction. The excess current is shown to be caused by tunneling of carriers via deep defect states in the gap. The carrier lifetime in InAs is calculated for radiative and Auger recombination mechanisms using three- and four-band Kane models. Theoretical limits of threshold parameters in homojunction and heterojunction PDs are calculated. Experimentally proved that p^+-InAsSbP/n-InAs heterojunction PDs can be more effective as sensitive element in gas sensors operated at room temperature in comparison with commercially available PDs. They have more broad spectral response and higher values of the resistance-area product at zero bias voltage.

2. Current status of InAs photodiodes

Mid-wavelength infrared (MWIR) photodetectors, light-emitting diodes (LEDs) and lasers are important components of modern infrared optoelectronic sensors. Despite the fact that MWIR region can be covered by different IV-VI and II-VI photodetectors (e.g., PbS and PbSe photoconductors, HgCdTe photodiodes and photoconductors, InAs PDs are expected to be the best choice from both practical and economic point of view due to mature technology of low-cost InAs substrates. The main advantage of InAs PDs is that they can operate at room and near room temperatures. The elimination of cryogenic cooling system results in significant reduction in cost and more reliable operation of infrared devices on their base. InAs PDs exhibit better noise characteristics at low frequencies and offer superior pulse response for detecting rapid processes in comparison with IV-VI photoconductors. Their characteristics (sensitivity, differential resistance, peak wavelength) can be optimized for specific purposes. The important fact is that InAs and ternary compounds on its base are also used for manufacture of LEDs (Zotova, 1991; Popov,1997; Matveev, 2002). The emission and photoresponse spectra of LEDs and PDs can be easily matched at different temperatures. In comparison with II-VI and IV-VI IR detectors, InAs PDs exhibit more stable characteristics. Commercially available InAs PDs are produced by Judson Technologies LLC and Hamamatsu Corporation as single-element structures. Apparently, the multielement structures exist only as laboratory samples. InAs infrared PDs have different applications such as conventional and laser spectroscopy, gas analysis, thermal imaging, remote sensing.

At the same time these PDs are characterized by rather low values of the differential resistance - area product R_0A at room temperature. This seems to be the main reason that InAs infrared photodetectors operate in sub-BLIP regime (Rogalsky, 1995). Further improvement in their performance is possible with a decrease in dark current and supression of Auger recombination.

3. Preparation of photodiodes and experimental techniques

Homojunction PDs were prepared by short-term (20-30 min) diffusion of Cd into n-InAs substrates at T=875 K. The substrates of n-type conductivity were cut from single-crystal ingots grown in joint-stock company "Pure metals plant" (Svetlovodsk, Ukraine). The damaged surface layers were removed by using dynamic chemical-mechanical polishing in 2% Br_2–CH_3OH solution. Their structural quality was controlled by X-ray diffraction method. In the chemically polished substrates the rocking curves half-width was 25-27″. Electrical parameters were controlled by van der Pauw technique at 295 K. The carrier concentration and mobility were found to be $n=(2\div3)\times10^{16}$ cm^{-3} and $\mu_n=(2\div2.5)\times10^4$ $cm^2/V\times s$, respectively. The density of dislocations was in the range $(2\div4)\times10^4$ cm^2. The substrates with the diffused p-type layer had mirror-like surfaces free of structural damages such as inclusions of impurity atoms. They were characterized by the rocking curve half-width of 32-35″. Typical profiles of Cd atoms in a substrate are shown in Fig.1. As seen, at low values of depth x they can be approximated by two exponential dependences (shown by solid lines in Fig. 1). Similar profiles of impurity atoms were previously observed in GaAs (Grigor'ev & Kudykina, 1997) and explained by generation of non-equilibrium vacancies at the substrate's surface. In such a case, an impyrity profile is determined by distribution of vacancies under the surface. The junction depth was determined from the probe thermo-emf measurements during careful chemical etching of diffused layers. Mesa structures were prepared on (111)A side of substrates by chemical etching in 2% Br_2-HBr solution. In order to eleiminate the surface leakage current mesas were passivated by etching in HNO_3-based solution followed by deposition of anode oxide doped with fluorine with thickness of 0.3 μm. After passivation they were covered by thin layer of ZnTe thermally deposited in a vacuum chamber at temperature 150 °C.

The heterojunction p^+-InAsSbP/n-InAs PDs were prepared by LPE technique in IOFFE Physico-Technical Institute, St.-Petersbur Russia (Zotova, 1991; Matveev, 1997; Matveev, 2002). For this purpose, lattice matched InAsSbP epitaxial layers of approximately 3 μm thickness were grown on (111)B surfaces of InAs substrates. The quaternary InAsSbP compound had energy gap of 0.43 eV at T = 297 K. The epilayers were doped to about 10^{18} cm^{-3} by addition of Zn to the melt. The substrates were n-type single crystals with electron concentration $n=(2\div3)\times10^{16}$ cm^{-3}. In homojunction and heterojunction PDs mesa structures have active area $A=4x10^{-2}$ cm^{-2} and 1.45×10^{-3} cm^2, respectively.

In order to characterize the prepared PDs, the dark current as well as the high-frequency (1 MHz) barrier capacitance were measured as a function of bias voltage and temperature. The carrier lifetime was determined from the photoconductivity signal decrease recorded by a memory oscilloscope in the temperature range 77-300 K. The investigated samples were illuminated by Nd:YAG pulsed laser with the pulse width around 2 ns. It has been observed that the signal decrease was not exactly exponential: a very fast decrease was observed immediately after the end of illumination followed by an exponential slow decrease. The excess carrier lifetime was associated with the slow decrease. The fast decrease, not typical for recombination processes, was associated with a high-level excitation of excess carriers. The photoresponse spectra were measured in the short-circuit current mode using grating monochromator and halogen lamp as a source of light. The photoresponse signal was recorded by a lock-in amplifier and processed by a personal computer.

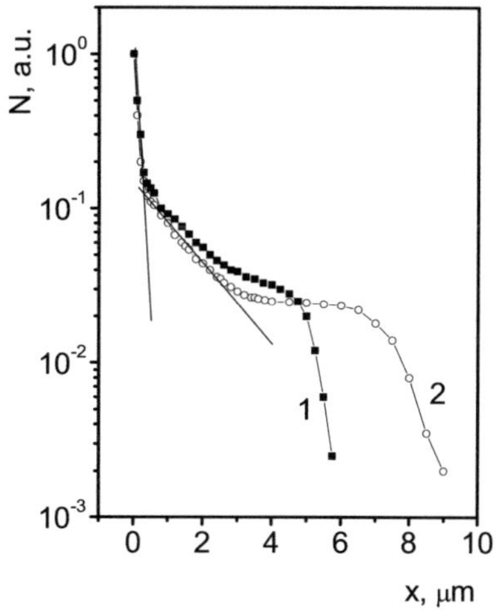

Fig. 1. SIMS profiles of Cd in InAs. The diffusion time is 20 min (1) and 30 min (2).

4. Recombination mechanisms in InAs

The carrier lifetime in narrow-gap semiconductors is determined by a number of recombination mechanisms, such as radiative, Auger and Shockley-Read-Hall recombination. The theory of radiative recombination has been developed by Van Roosbroeck and Shockley (Van Roosbroeck & Shockley, 1959; Blakemore, 1962). Ten types of possible band-to-band Auger recombination processes in n- and p-type semiconductors were determined by Beattie (Beattie, 1962). The Auger 1 recombination mechanism in n-type material with InSb-like band structure was firstly considered by Beattie and Landsberg (Beattie & Landsberg, 1959). The Auger 7 process is important in p-type material if the split-off band can be ignored (Beattie & Smith, 1967; Takeshima, 1972). The role of the split-off band in InAs was firstly recognized by Takeshima (Takeshima, 1972). Since the split-off energy Δ in InAs is close to the energy gap E_g, the Auger S process may be dominant in p-type material. Further investigations of recombination processes in InAs were based on these pioneering works (Blakemore, 1962; Rogalski, 1995).
Three types of Auger recombination mechanisms (Auger 1, Auger 7 and Auger S) have been shown to be important in InAs (Takeshima,1972; Takeshima, 1975; Rogalski. & Orman, 1985). According to Rogalski and Orman, contributions of Auger and radiative mechanism are comparable at room temperatures in samples with the carrier concentration of the order of 10^{16} cm^{-3}. The Auger lifetime was theretically analyzed by Barishev (Barishev, 1964). The well known problem connected with the Auger recombination processes is the uncertainty introduced by the overlap integrals F_1 and F_2. For instance, in n-type InAs the value of $|F_1F_2|^2$ is ranged within an order of magnitude. Appropriate analysis in p-type materials is

even more difficult because of experimental data are rather scarce. An improvement in the calculation of overlap integrals was done by Gelmont with co-authors. The Auger S and Auger 7 processes were reexamined for the three- and four-band Kane models of band structure (Gelmont, 1978; Gelmont, 1981; Gelmont, 1982; Gelmont & Sokolova, 1982). The SRH recombination process in InAs is investigated unsufficiently. On the other hand, frequently observed large scattering of experimental data in samples with close values of free cariers density clearly specifies an essential role of this process.

The lifetime in InAs was calculated as a fuction of temperature and density of carriers. The calculated lifetime is compared with experimental data available in the literature (Datal, 1979; Wider & Collins, 1974; Andrushko, 1986; Fomin, 1984; Blaut-Blachev, 1975; Bolgov, 1997). For small departure from equilibrium the radiative lifetime is given by

$$\tau_R = \frac{1}{B(n_o+p_o)} \tag{1}$$

where n_o and p_o are the equilibrium electron and hole concentrations. The capture probability B for parabolic conduction and valence bands is given by

$$B = 5.8\times 10^{-13}\varepsilon_\infty^{1/2}\left(\frac{m}{m_e^*+m_h^*}\right)^{3/2}\left(1+\frac{m_o}{m_e^*}+\frac{m_o}{m_h^*}\right)\left(\frac{300}{T}\right)^{3/2}E_g^2 \tag{2}$$

where ε_∞ is the high-frequency dielectric constant, m_e^* and m_h^* are the effective masses of electrons and holes, respectively, the energy gap E_g is expressed in eV (Rogalski & Orman, 1985). For n-type material the Auger 1 recombination lifetime is equal to (Rogalski, 1995; Rogalski & Orman, 1985)

$$\tau_{A1} = \frac{n_i^2}{n(n_o+p_o)g_{A1}} = \frac{2n_i^2\tau_A^i}{n_o(n_o+p_o)} \tag{3}$$

where τ_A^i is the carrier lifetime in the intrinsic material. The Auger 7 lifetime in the limit of small excitation is equal to

$$\tau_{A7} = \frac{2\tau_{A7}^i}{1+p_o/n_o} \tag{4}$$

where

$$\tau_{A7}^i = \frac{p_0}{2g_{A7}^G} \tag{5}$$

Analogous equations for the carrier lifetime can be written for the Auger S process. The effective carrier lifetime is given by

$$\frac{1}{\tau} = \sum_i \frac{1}{\tau_i} \tag{6}$$

where τ_i is the carrier lifetime determined by the *i*th recombination process. The generation rate for the Auger 1 process g_{A1} obtained by Beattie and Landsberg is given by

$$g_{A1} = \frac{8(2\pi)^{5/2} e^4 m_e^* |F_1 F_2|^2 n_o}{h^3 \varepsilon^2 (1+\mu)^{1/2} (1+2\mu)} \left(\frac{kT}{E_g}\right)^{3/2} \exp\left[-\left(\frac{1+2\mu}{1+\mu}\right)\frac{E_g}{kT}\right] \tag{7}$$

According to Gelmont, in the three-band Kane model the generation rate is given by

$$g_{A1}^G = 3\left(\frac{2}{\pi}\right)^{1/2} \frac{e^4 m_e^* n_o}{\hbar^3 \varepsilon^2} \left(\frac{kT}{E_g}\right)^{5/2} \exp\left[-(1+2\mu)\frac{E_g}{kT}\right] \tag{8}$$

For the Auger 7 lifetime the generation rate is

$$g_{A7}^G = \frac{18 m_o (m_{hh}^*/m_o) e^4}{\pi \hbar^3 \varepsilon^2} p_o \left(\frac{kT}{E_g}\right)^{7/2} \exp\left[-(1+\frac{m_{hl}^*}{m_{hh}^*})\frac{E_g}{kT}\right] g(\alpha) \tag{9}$$

In the last three expressions, ε is the static dielectric constant, m_{hl}^* and m_{hh}^* are the effective masses of light and havy holes, respectively, μ is the ratio of the electron to the heavy-hole effective mass. The generation rate in the Auger S process depends on difference between the split-off and band gap energies. The most frequently used expressions for these energies are as follows (Rogalski. & Orman, 1985, Madelung, 1964):

$$E_g = 0.44 - 2.8 \times 10^{-4} T \tag{10}$$

and

$$\Delta = 0.43 - 1 \times 10^{-4} T \tag{11}$$

Since in the temperature range 77 – 300 K the condition $\Delta > E_g$ is fulfilled, the generation rate is given by (Gelmont, et al., 1982)

$$g_{AS}^G = \frac{27}{5} \pi^4 n_i^2 p_o \frac{e^4 \hbar^3 (m_s^*)^{5/2} (\Delta - E_g) \exp[-(\Delta - E_g)/kT]}{\varepsilon^2 (m_{hh}^*)^3 (m_e^*)^{3/2} kT \Delta^2 (E_g + \Delta)} \tag{12}$$

where m_s^* is the effective mass of holes in the split-off band, n_i is the intrinsic concentration. The concentration dependences of the carrier lifetime in n-type InAs are shown in Fig.2 and 3. At low temperature T=77 K experimental data are well described by the radiative recombination mechanism. The contribution of the Auger 1 recombination increases with an increase in temperature. The pronounced scattering of experimental data at room temperature may be caused by contribution of any SRH recombination process at high temperatures to the measured lifetime.

Experimental and calculated dependences for *p*-InAs are shown in Fig.4 and 5.

Three recombination mechanisms radiative, Auger 7 and Auger S were taken into account. The main result of theoretical calculations is as follows: the Auger 7 process is dominant recombination process in p-InAs. The contribution of the radiative mechanism is essential at a concentration of holes around 10^{16} cm^{-3}. The Auger S recombination mechanism is weakly dependent on temperature and its contribution to the lifetime is less important. Since the

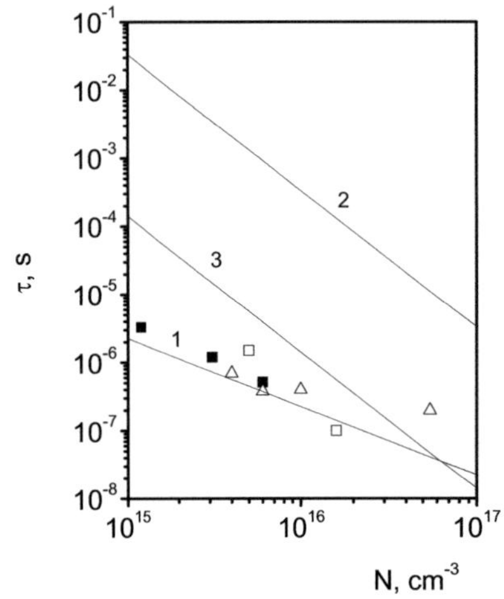

Fig. 2. Carrier lifetime vs doping level in *n*-InAs at 77 K (Tetyorkin, 2007). Solid curves are calculated for recombinations: 1 - radiative, 2 - Auger 1 (Gelmont), 3 – Auger 1 (Beattie and Landsberg, $|F_1F_2|^2$ equals 0.079).

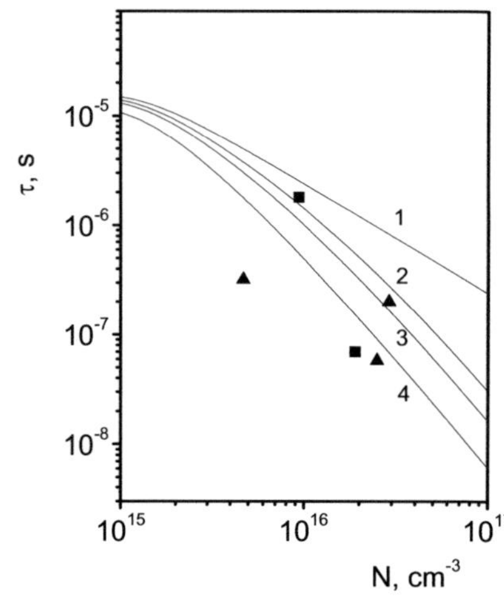

Fig. 3. Carrier lifetime vs doping level in *n*-InAs at 300 K (Tetyorkin, 2007). Calculated curves are obtained for: 1- radiative lifetime, 2 - radiative and Auger 1 (Gelmont), 3,4 -radiative and Auger 1 (Beattie and Landsberg, $|F_1F_2|^2$ equals 0.014 and 0.079, respectively).

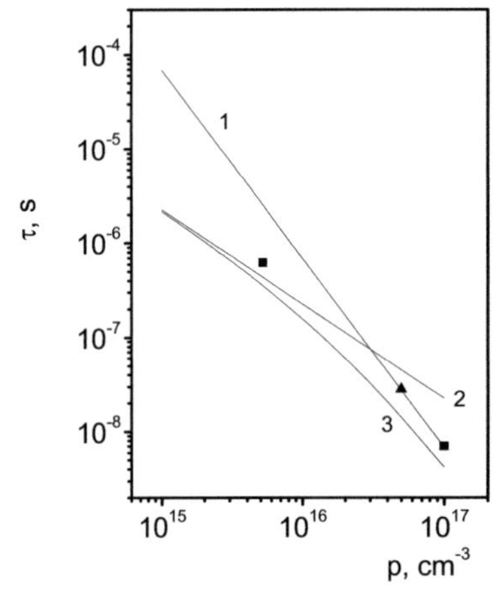

Fig. 4. Carrier lifetime vs. hole concentration at T=77 K (Tetyorkin, 2007). Calculated curves represent Auger 7 (1), radiative (2) and effective (3) lifetime which includes radiative, Auger 7 and Auger S recombinations.

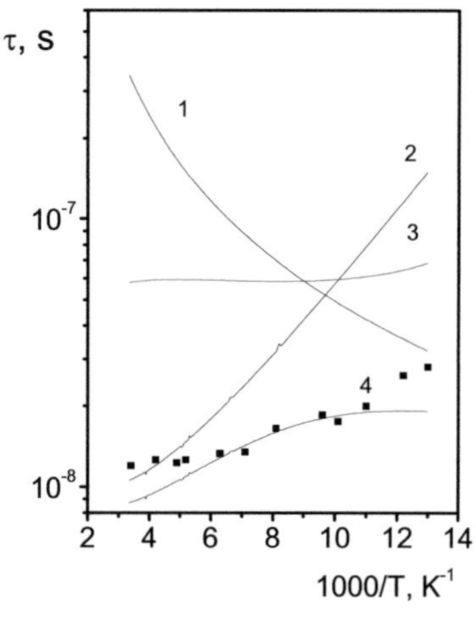

Fig. 5. Carrier lifetime vs temperature in *p*-InAs with p=5×10^{16} cm^{-3} (Tetyorkin, 2007). Calculated curves represent radiative (1), Auger 7 (2), Auger S (3) and effective (4) lifetime.

lifetime calculated for the Auger S mechanism is strongly dependent on the difference between Δ and E_g, this conclusion is based on values which were most often published in the literature.
Careful analysis of experimental values of the carrier lifetime obtained from the early 1960s to our days clearly indicate tendecy to their decrease. Gradually they have reaches theoretical limit which is determined by the radiative and Auger recombination processes. Obviously, this result is caused by improvements in technology of InAs single crystalls. On the other hand, the SRH recombination process still remains unsufficiently investigated. The role played by deep defects in the carrier transport mechanisms in PDs is analysed below.

5. Carrier transport mechanisms

The carrier transport mechanisms were analysed within one-dimensial analytic model. This model includes current from quasineutral and depletion regions of the junction (diffusion and generation-recombination currents), trap-assisted and band-to-band tunneling across the junction. In the investigated PDs the leakage current was treated as a bulk phenomenon. The diffuson and generation current density was calculated by using the formulas:

$$I_D = qn_i^2\left(\frac{1}{p}\sqrt{\frac{D_n}{\tau_n}} + \frac{1}{n}\sqrt{\frac{D_p}{\tau_p}}\right) \tag{13}$$

and

$$I_{GR} = \frac{qn_i W}{2\tau_o} \tag{14}$$

where W is the depletion region width, τ_0 is the effective lifetime in the depletion region, D_n, τ_n and D_p, τ_p are the diffusion coefficient and the carrier lifetime for electrons and holes, respectively. As a rule, the effective lifetime served as adjusting parameter in calculations.
The band-to-band tunneling current density was calculated for the triangular barrier model (Sze, 1981).

$$I_{BTB} = \frac{\sqrt{2m_e^*}q^3FU}{4\pi^2\hbar^2E_g^{1/2}}\exp\left[-\frac{4\sqrt{2m_e^*}E_g^{3/2}}{3q\hbar F}\right] \tag{15}$$

where F is the electric field in the junction.
The trap-assisted tunneling current in InAs photodiodes was firstly investigated by A. Sukach and V. Tetyorkin (Sukach, 2005; Tetyorkin, 2005). For this purpose, a model of the trap-assisted tunneling developed earlier for HgCdTe photodiodes (Nemirovsky & Unikovsky, 1992; Rosenfeld & Bahir, 1992) was used. This model is based on number of simplifying assumptions, namely: shallow impurities and traps are uniformly distributed in the depletion region; the electric field across the depletion region is constant; transitions of electrons and holes occure via single level in the gap; initial states are occupied and final states are empty; in degenerate diodes the Burstein-Moss shift results in higher potential barriers for thermal and tunnel transitions of cariers. Traps can exchange carriers with the valence and conduction bands by tunnel and thermal transitions. Under these assumptions, the rates of tunneling transitions between the allowed bands and a trap in the gap is given by

$$\omega_v N_v = \frac{\pi^2 q m_h^* F M^2}{h^3 (E_g - E_t)} \exp\left[-\frac{4\sqrt{2m_h^*}(E_g - E_t)^{3/2}}{3q\hbar F} \right] \tag{16}$$

and

$$\omega_c N_c = \frac{\pi^2 q m_e^* F M^2}{h^3 E_t} \exp\left[-\frac{4\sqrt{2m_e^*} E_t^{3/2}}{3q\hbar F} \right] \tag{17}$$

where the trap energy E_t is measured from the edge of the conduction band. The matrix element M between the trap wave function and the conduction band Bloch wave function in silicon is of the order of 10^{-29} $(eV)^2cm^3$ (Rosenfeld & Bahir, 1992). Its value can be also estimated by using the formula (Wenmu He & Zeynep Celik-Batler, 1996).

$$M = \frac{2\hbar^2 \sqrt{2\pi}}{m_o} \left(\frac{2m_o}{\hbar^2} \right)^{1/4} \frac{E_g}{E_t^{3/4}} \tag{18}$$

Then the trap-assisted tunneling current is given by

$$J_{tat} = qWN_t \left(\frac{1}{\omega_v N_v + c_p p_1} + \frac{1}{\omega_c N_c + c_n n_1} \right)^{-1} \tag{19}$$

where $c_p p_1$ and $c_n n_1$ are the rates of thermal transitions. The concentration of traps N_t, trap energy E_t and capture cross section for electrons and holes c_n and c_p are obtained from the fitting calculations.

Typical forward and reverse I-U characteristics of a representative diffused PD are shown in Fig.6 and 7. It is necessary to note presense of two exponential regions at temperatures $T \leq 180$ K, Fig. 6. The slope of the first region measured at lower biase voltages increases with temperature increasing. The second region exhibits an opposite dependence of the slope on temperature. At higher temperatures the only exponential region is obseved.

For analysis of carrier transport mechanisms in diffused PDs, the current-voltage characteristics should be corrected to the series resistance R_s. This correction is especially important due to possible formation of high-resistance compensated region. The presence of this region has been proved with capacitance-voltage measurements shown in Fig. 8.

The following peculiarities should be pointred out: experimental curves are linearized in coordinates typical for abrupt p-n junctions; the slope of C^{-2}-U dependences is weak function of temperature; the concentration of free carriers determined from their slope is ranged from 1.5×10^{15} at 77 K to 3.0×10^{15} cm^{-3} at 295 K. These values are approximately one order of magnitude lower than the carrier concentration in the starting material. The high-resistance compensated region results in the capacitance saturation at forward biase voltages. The width of the compensated region, estimated from this saturation, was in the range 2.5 - 3 μm. Since these values substantially exceed the depletion region width ($W\approx0.6$ μm at zero bias voltage), the p-n junction is presumably located inside the compensated region. It has been assumed that the diffused PDs have p^+-p_o-n_o–n junction structure. Experimental I-U characteristics were approximated by exponential dependence

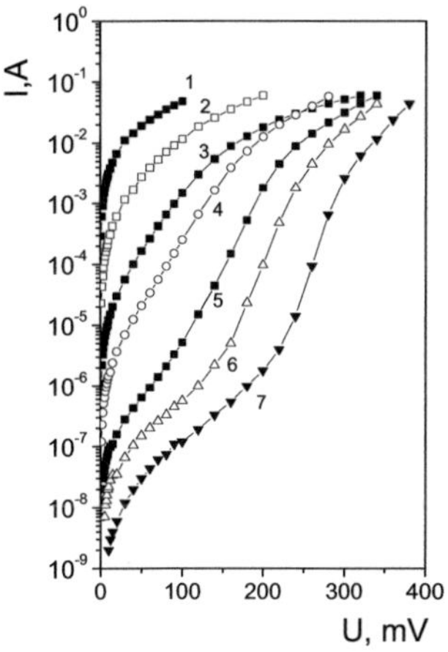

Fig. 6. Forward I-U characteristics in a diffused PD at temperatures, K: 1- 290, 2 - 255, 3 - 218, 4 - 197, 5 - 158, 6 - 135, 7 - 77. (Sukach, 2005).

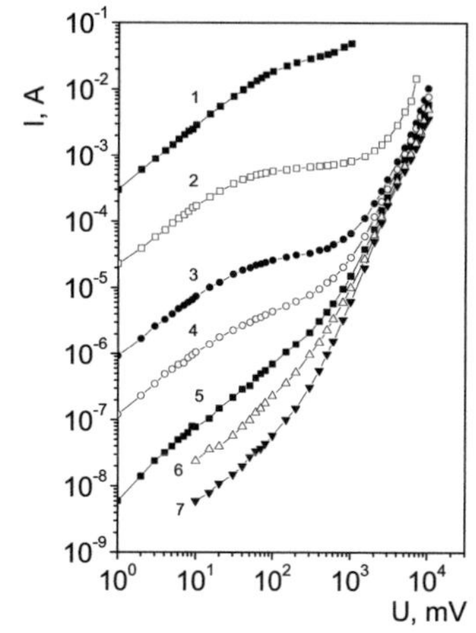

Fig. 7. Reverse current-voltage characteristics measured at the same temperatures as in Fig.6. (Sukach, 2005).

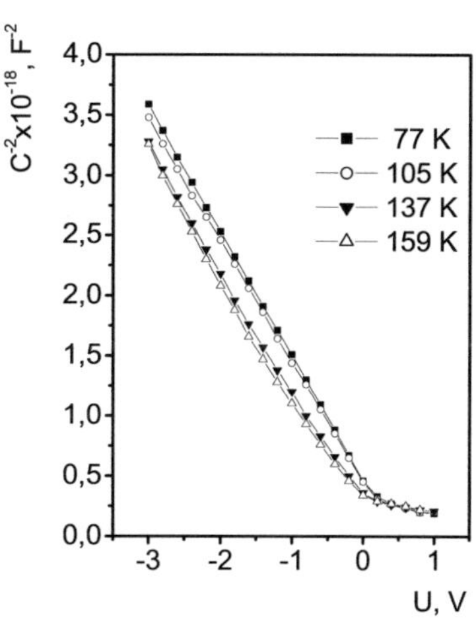

Fig. 8. Capacitance-voltage characteristics in a diffused photodiode at different temperatures (Sukach, 2005).

$$I = I_O \left\{ \exp[(qU - IR_S)/\beta kT] - 1 \right\} \tag{20}$$

The ideality coefficient extracted from the slope of R_s-corrected I-U curves is ranged from 5.9 to 1.1 and from 1.6 to 1.1 for the first and second exponential regions in Fig. 6, respectively. The reverse current saturation is observed only at high temperatures for narrow interval of biase voltages. At low temperatures the current-voltage characteristics have a form typical for the soft breakdown.

In the diffusion-limited diode the ideality coefficient $\beta = 1$. So, experimental data shown in Fig.6 and 7 clearly indicate presence of additional currents in the investigated PDs. Based on experimental values of the ideality coefficient, one can conclude that the diffusion current is dominant at room temperature. Predominance of diffusion and generation currents at reverse biase voltages was confirmed by theoretical calculations within the model of symmetrical p^+-p_o-n_o–n junction, Fig.9. As seen, the contribution of the generation current is increased with the reverse bias increasing, Fig.9.

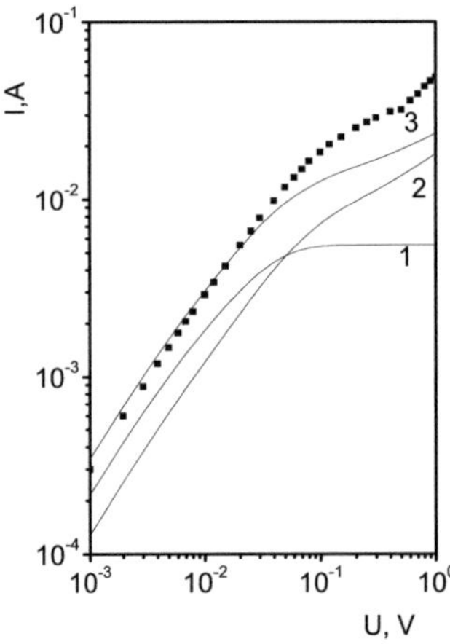

Fig. 9. Measured (dots) and calculated (solid lines) reverse current-voltage characteristics of a diffused PD at 290 K. Calculated curves represent diffusion (1), generation (2) and total (3) current. Parameters used for calculation: $\tau_o = 1.5\times10^{-9}$ s, $n_i = 5.6\times10^{14}$ cm^{-3}, $\tau_n = \tau_p = 1\times10^{-9}$ s, $n_o = p_o = 3\times10^{15}$ cm^{-3}, $\mu_n = 10^3$ cm^2/V×s, $\mu_p = 150$ cm^2/V×s, $U_D = 110$ mV (Sukach, 2005).

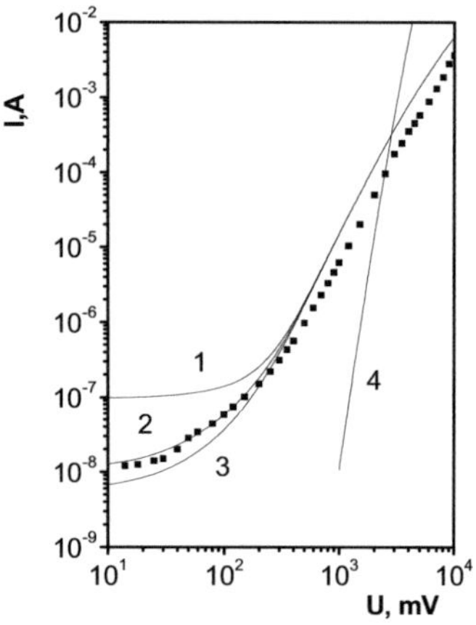

Fig. 10. Calculated (1-3) trap-assisted tunneling current (see text) in a model of nonhomogeneous p-n junction. Curve 4 represents the band-to-band tunneling current (Sukach, 2005). Experimental data are obtained for a diffused PD.

Large values of the ideality coefficient at low temperatures most probably indicate that the additional current is tunneling. To prove this assumption, band-to-band and trap-assisted tunneling currents were calculated using the above formulas. In order to reduce possible uncertanities in theoretical calculations caused by large amount of unknown adjusting parameters, the trap-assisted tunneling current was calculated for the midgap traps. Such traps with density 10^{13} - 10^{14} cm^{-3} were experimentally obseved in InAs (Fomin, 1984; Kornyushkin, 1996; Kurishev, 2001). The tunneling transitions of light holes were only considered because they have much greater tunneling probability compare to havy holes. It has been assumed that the electric field strength in the depletion region has maximum value. The correct choice of the electric field strength is important since it appears in the exponent of expressions (15-17). It must be pointed out that previously values of average and maximum electric field were used in theoretical calculations (Rosenfeld & Bahir, 1992; Kinch, 1981). The performed calculations showed that in homogeneous p^+-p_o-n_o–n junctions the band-to-band and trap-assisted tunneling currents are substantially less than the measured current at reverse bias voltages U<1 V. The same result was obtained for asymmetrical p-n junctions in the case of the carrier concentartin is less than 10^{16} cm^{-3}. The reason of this result is that the electric field in the depletion region is too low to enable effective band-to-band and trap-assisted tunneling. To explain experimental results, models of a non-homogeneous p-n junction with non-uniform distribution of shallow and deep defects in the depletion region should be considered.

In such a junction the electric field and depletion region width are fluctuated. In a model of the band-to-band tunneling current proposed by Raikh and Ruzin (Raikh & Ruzin,1985; Raikh & Ruzin, 1987), the non-homogeneous distribution of impurities results in fluctuations of the depletion region width and, therefore, the tunneling distance for carriers. The electric conductivity of such a junction is mainly determined by small regions with high local concentration of impurities. Unfortunately, Raikh and Ruzin's model is valid at large reverse biase voltages $qU > E_g$ and can't be used for analysis of tunnell current in the investigated PDs at actual low voltages. Large deviation of impurity concentration in local regions of the junction results in increase in the electric field strength over the value typical for uniform distribution of charged defects. There are some arguments that these regions are related with Cottrell's atmospheres around dislocations (see below). So, it has been assumed that each region has the effective area of the order of 1 μm^2. Experimentally, atmospheres of impurity atoms around dislocations with similar dimensions were observed in GaAs (Bruk, 1982). The density of regions was assumed to be equal to the density of dislocation in the starting material. The electric field strength F served as an adjusting parameter in calculations. The relation between F and effective concentration of charged defects N is given by

$$F = \frac{qNW}{\varepsilon_s \varepsilon_o} \tag{21}$$

Shown in Fig.10 the trap-assisted tunneling current was calculated for the following carrier transition paths: traps are exchanged with both bands by thermal and tunnel transitions (curve 1), tunnel transitions of carriers from the valence band to traps followed by thermal and tunnel transitions to the conduction band (curve 2), tunnel transitions of carriers from the valence band to the conduction band (curve 3). The concentration of traps was varied in the range of 10^{13} - 10^{15} cm^{-3}. The best fit was obtained for concentrations of charged defects in the local regions of the order of 4×10^{16} cm^{-3}. This value is more than one order of magnitude higher than the mean concentration of free carriers obtained from the capacitance-voltage measurements. Obviously, the capacitance is determined by the homogeneous part of the junction, whereas the current flows mainly through the non-homogeneous regions. As seen, discrepancies between the calculated characteristics for different transition paths of carriers are not so significant to make an unambiguous choice. However, conclusion about the dominant contribution of the trap-assisted tunneling current at the reverse bias voltages U < 1 V seems to be apparent. At higher voltages both band-to-band and trap-assisted tunneling currents have comparable magnitude. The convergence of I-U curves in Fig.7 < which were measured at different temperatures, clearly indicates tunneling nature of the reverse current at large voltages U > 1 V. Despite the fact that the best fit was obtained for the trap-assisted tunneling current, additional investigations of tunneling mechanismsm are needed.

In order to clarify the role of the trap-assisted current in diffused PDS effect of ultrasonic treatment on the current-voltage characteristics was investigated (Sukach & Tetyorkin,

2009). Representative PDs were subjected to ultrasonic vibration with frequency 5-7 MHz and intensity ~0.4 W/cm^2 during 1 hour at room temperature. The second two-hour ultrasonic treatment with the same intensity was repeated after 72 hours. At last, experimental measurements were repeated afte nine-month storage of PDs at laboratory conditions.

Experimental data are shown in Fig. 11. Drastic increase in the measured forward current was revealed as a result of the second ultrasonic treatment. After nine-month storage the forward current is practically decreased to the starting values. It must be pointed out, that only the part of the current-voltage characteristic associated with the tunneling current is changed under ultrasonic treatment. Because of the pre-threshold intensity regime was used for ultrasonic treatment, experimental results can be explained by transformation of existing complex defects rather than generation of new point defects. Most probably that this transformation is connected with Cottrell's atmospheres around dislocations which intersect the p-n junction. In accordance with the vibrating string model of Granato-Luecke (Granato & Luecke, 1966), the intensive sonic-dislocation interaction results in an effective transformation of the absorbed ultrasonic energy into the internal vibration states of a semiconductor stimulating different defect reactions. The driving force of relaxation may be deformation and electric fields around dislocations.

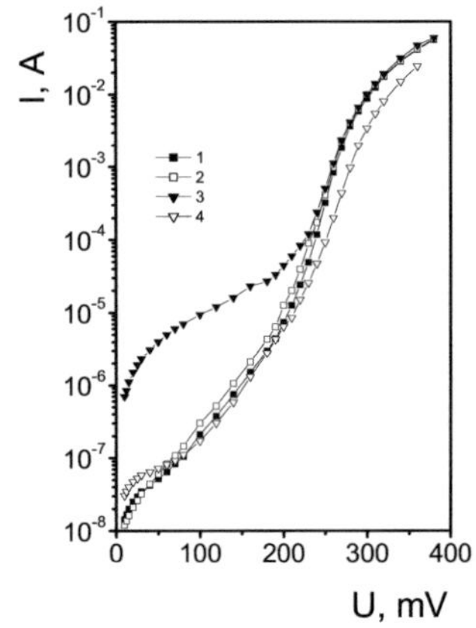

Fig. 11. Forward current vs voltage in a PD before (1) and after two sequential ultrasonic treatments (curves 2 and 3, respectively), and after nine-month storage (4), (Sukach & Tetyorkin, 2009).

The current-voltage characteristics of a representative heterojunction p^+-InAsSbP/n-InAs PD are shown in Fig. 12 and 13. The abrupt p-n junction formation in heterojunction PDs has been obtained from the barrier capacitance measurements. From the slope of C^{-2}-U characteristics the concentration of free carrirs in the range of $(2\div4)\times10^{16}$ cm^{-3} was determined. These values are in accordance with the electron concentration in the starting material. At two actual temperatures 290 K and 221 K the diffusion potential equals 0.2±0.01 V and 0.27±0.01 V, respectively.

Contrary to a diffused PD, the forward current in a heterojunction PD is characterized by one exponential dependence, Fig. 12. The ideality factor is changed from 1.1 at room temperature to 2.0 at 193 K. Thus, we can conclude that the bulk diffusion and recombination in the depletion region are dominant transport mechanisms at forward

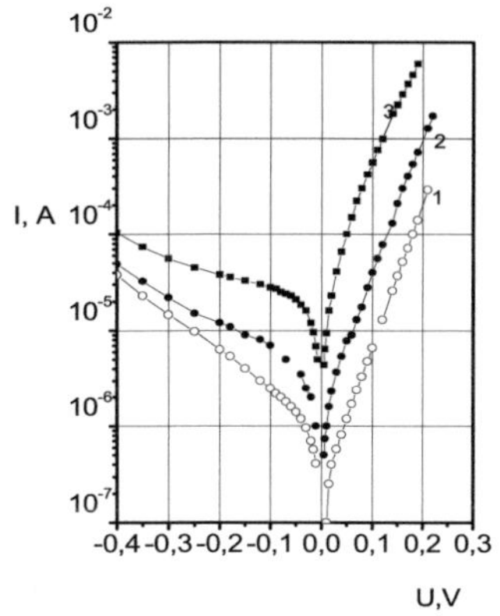

Fig. 12. Current-voltage characteristics of a representative heterojunction PD at temperatures, K: 193 (1), 228 (2) and 290 (3) (Tetyorkin, 2005).

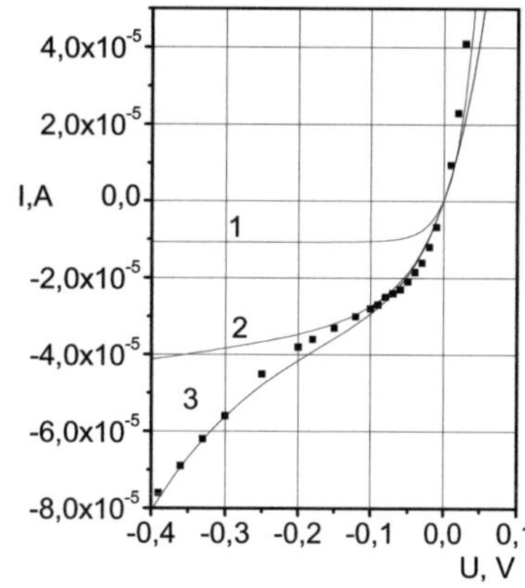

Fig. 13. Measured (dots) and calculated (lines) currents in a heterojunction PD at 295 K. Calculated curves represent diffusion (1), generation-recombination (2) and trap-assisted tunneling (3) mechanisms. The fit was obtained for $N_t = 6\times10^{13}$ cm^{-3}, n = 4×10^{16} cm^{-3}, $E_t = E_g/2$, $\tau_o = 6\times10^{-8}$ s, $\mu_p = 150$ cm^2/V×s. (Tetyorkin, 2005).

biases. At the same time, the reverse current was not saturated even at room temperature. As seen from Fig. 13, it has a form typical for the soft breakdown. The fitting calculation of the reverse current proved primary contribution of generation and trap-assisted tunneling currents at applied reverse bias voltages. The trap-assisted tunneling current was calculated for the following carrier transitions: traps are exchanged carriers with the valence band by thermal and tunnel transition, and with the conduction band by tunnel transitions only. Despite the fact that the fit was achieved for resonable values of trap concentration and energy, additional investigations are needed to clerify mechanisms of tunneling. In particular, the role played by the dislocation network at the InAs-InAsSbP heterojunction must be thoroughly investigated. At the reverse biase voltages U > 1.0 V the band-to-band tunneling seems to be dominant.

5. Performance of InAs photodiodes

5.1 Current sensitivity

The current sensitivity of PDs is given by

$$S_i = \frac{e}{hc}\eta\lambda = \frac{e}{hc}\lambda(1-R)\beta[1-\exp(-kd)]\alpha_{p-n} \tag{22}$$

where η is the quantum efficiency, β is the quantum yield of the internal photoeffect, d is the width of the photodiode's structure, and $\alpha_{p\text{-}n}$ is the collecting coefficient (G.S Oliynuk, 2004). It is known that three regions in the p-n junction can contribute to the photocurrent, namely: two quasineutral regions of p- and n-type conductivity and the depletion region. The excess carriers excited in these regions can be collected by the junction. In the diffused PDs the

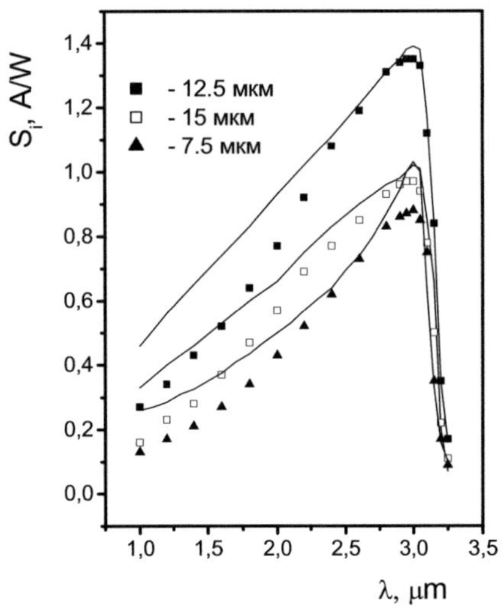

Fig. 14. Calculated (solid lines) and measured (dots) spectral dependences of current sensitivity in diffused PDs with different junction depth at 77 K and concentration of carriers in the compensated region, cm^{-3} : 8 10^{15} (▲), 5 10^{15} (□) and 2 10^{15} (■).

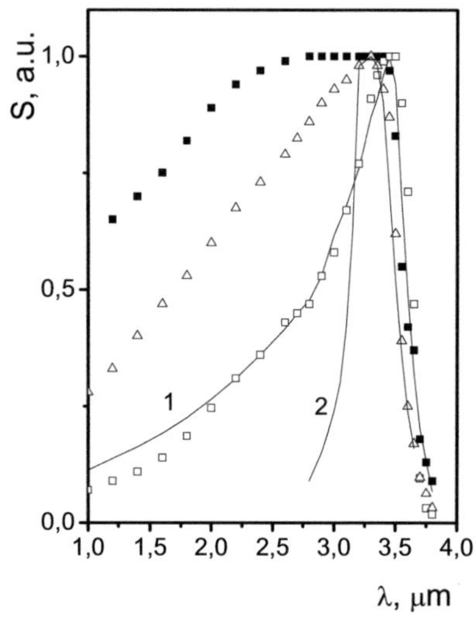

Fig. 15. Spectral dependences of sensitivity in homojunction (open dots and triangles) and heterojunction (close dots) PDs at 295 K (Tetyorkin, 2007). The junction depth in homojunction PDs equals 8 (□) and 4 (Δ) μm, respectively. Also shown is the emission spectrum of InAs LED (2).

current sensitivity $S_i(\lambda)$ is found to be basically determined by the quasineutral p-type region. The quasineutral n-type region contributes mainly to the long wavelength photosensitivity. The contribution of the depletion region is negligibly small in the whole spectral region (G.S Oliynuk, 2004). As seen from Fig. 14, the current sensitivity in the diffused PDs is not less than in commercially available InAs photodiodes.

The broadband spectrum shown in Fig. 15 is explained by contribution of both sides of the heterojunction PD, including heavily doped wide-gap InAsSbP constituent, to the photoresponse (Tetyorkin, 2007). The spectral dependence of photosensitivity in heterojunction PD is superior to homojunction one due to effect of "wide-gap window".

5.1 Resistance-area product

The differential resistance-area product at zero bias R_0A determines threshold parameters of infrared PDs. Theoretical limitations of threshold parameters in InAs PDs are related to the fundamental (radiative and Auger) recombination processes. The SRH recombination is considered as nonfundemental since it can be reduced by improvement in technology of PDs.

In the diffusion-limited asymmetrical p^+-n junction the product R_0A is given by

$$(R_0A)_D = \frac{(kT)^{1/2}}{q^{3/2}} \frac{n_o}{n_i^2} \left(\frac{\tau_p}{\mu_p}\right)^{1/2} \tag{23}$$

In the case of generation-recombination current it can be expressed as

$$(R_0A)_{GR} = \frac{4(kT/q)}{qn_iW/\tau_0} \tag{24}$$

The last formula is obtained by differentiating the well known expression $I=I_o[\exp(eU/2kT)1]$, where $I_o=qn_iW/2\tau_o$. Since experimental data were obtained at zero and small forward voltages (<10 mV) the depleted region width W was assumed to be independent on U.

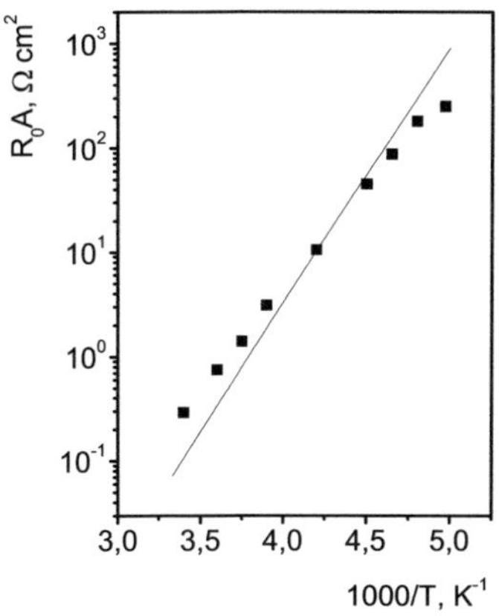

Fig. 16. Experimental (dots) and calculated (solid line) data dependences of R_oA vs. temperature in symmetrical p-n junction ($p=n=3\times10^{15}$ cm^{-3}) for diffusion mechanism of carrier transport. (Tetyorkin, 2007)

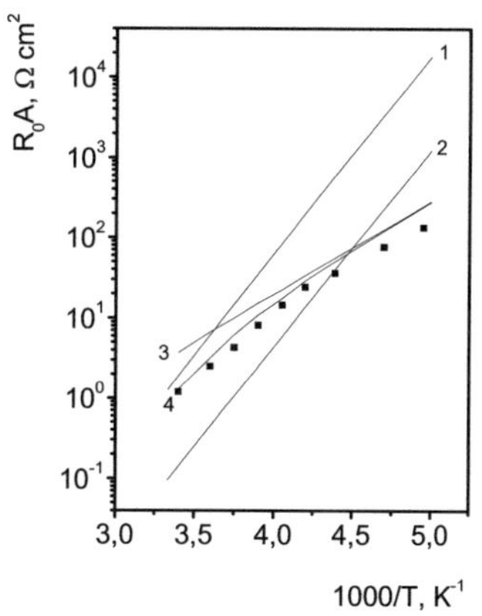

Fig. 17. Experimental (dots) and calculated (solid lines) dependences of R_oA vs. temperature in InAs p^+-n (1) and n^+-p (2) junctions. Calculated dependences were obtained for diffusion current (1,2) and recombination current in the depletion region (3). The doping level is equal to 3×10^{16} (1) and 5×10^{16} (2) cm^{-3}. Also shown is the R_0A product for diffusion and generation-recombination current in p^+-n junction with $n=3\times10^{16}$ cm^{-3} (4). Experimental data are measured in a p^+-n heterojunction PD with $n=3\times10^{16}$ cm^{-3} (Tetyorkin, 2007)

The measured and calculated values of R_oA in symmetrical homojunctin and asymmetrical heterojunction PDs are shown in Fig.16 and 17. The electron and hole mobility used in the calculation were approximated by the dependence $\mu(T)= \mu_0(T/300)^{-0.5}$, where μ_0 is the mobility at T=300 K. The effective lifetime was assumed to be determined by radiative and Auger 1 (Gelmont) recombination mechanisms. Since the electron mobility in InAs exceeds the hole mobility by approximately two orders of magnitude, the diffusion-limited PDs of p^+-n type can potentially have the highest values of R_oA, Fig.17. The calculated values of the current sensitivity S_i, differential-resistance product R_0A and specific detectivity are summarized in Table. The current sensitivity and detectivity was calculated for the peak wavelength λ_p. It should be pointed out that typical values of the specific detectivity in the investigated heterojunction PDs are of the order of $2\cdot10^9$ cm×$Hz^{1/2}$×W^{-1}. Approximately the same values of detectivity were obtained in commercially available PDs. However, at room temperature the resistance-area

Parameters of InAs photodiodes						
T_0, °C	A, cm^{-2}	λ_p, µm	$S_i(\lambda_p)$, A/W	R_0A, Ωcm^{-2}	D_λ^*, $cm{\cdot}Hz^{1/2}W^{-1}$	Manufacturer
22	$7.86{\cdot}10^{-3}$	3.35	1.0	0.12 - 0.20	$2.7{\cdot}10^9$	Judson
-85	$7.85{\cdot}10^{-3}$	3.20	1.5	196 – 393	$3.6{\cdot}10^{11}$	Judson
25	$7.85{\cdot}10^{-3}$	3.35	1.0	0.31 – 0.55	$(3.0 – 4.5){\cdot}10^9$	Hamamatsu
-196	$7.85{\cdot}10^{-3}$	3.00	1.3	$(0.8 – 8.0){\cdot}10^3$	$(3.5 – 6.0){\cdot}10^{11}$	Hamamatsu
20	$1.45\ 10^{-3}$	2.60-3.40	0.7-0.8	1.5-2.0	$(2.5\text{-}3.0)\ 10^9$	IOFFE PTI, St.-Petersburg
25	$1.0{\cdot}10^{-2}$	3.45 – 3,50	0.8	0.15 – 0.30	$2.0{\cdot}10^9$	ISP, Kiev
-196	$1.0{\cdot}10^{-2}$	3.00	1.2 – 1.3	$(0.5 – 2.0){\cdot}10^5$	$(5.0 – 6.0){\cdot}10^{11}$	ISP, Kiev
-196	$1.0\ 10^{-2}$	3.00	2.4	$1\ 10^{11}$	$2\ 10^{12}$ (BLIP)	ultimate parameters

product in the heterojunction PDs is five times higher. Taking into account their broadband spectral response, one can conclude that the heterojunction PDs can be more effective as sensitive element in gas sensors operated at room and near-room temperatures. The ultimate parameters shown in Table were calculated for the generation-recombination limited p^+-p_o-n_o-n junction with $n_o=p_o= 3\ 10^{15}$ cm^{-3}. The current sensitivity and specific detectivity were calculated using the formulas (22) and (24) for the experimentally measured parameters W=0.63 µm and $\tau_o = 8\ 10^{-8}$ s. It is assumed that the quantum efficiency was equal to 1.0. At 77 K the intrinsic concentration in InAs is 2.1 10^3 cm^{-3}. Parameters of PDs produced by Judson and Hamamatsu were taken from their web sites. As seen, in the generation-recombination limited PDs BLIP mode of operation can be realized.

6. New trends in development of InAs-based infrared detectors

InAs PDs are usually fabricated from bulk single crystall wafers. The p-n junctions are formed by ion (e.g. Be) implantation or Cd diffusion. Obviously, further progress in development of InAs infrared detectors including multielenment structures is closely connected with technology of epitaxial films. Diffetent epitaxial techniques including liquid-phase epitaxy (LPE), gas-phase epitaxy (GPE) and moleqular-beam epitaxy (MBE) were used in different laboratories for growth of InAs-based epitaxial films. Currently, their quality has not reached the level of maturity required for manufacture of electronically scanned multielement structures. As a rule, the as-grown LPE films has a high concentration of residual impurities which affect the lifetime and mobility of carriers. The low concentration of residual impurities in epitaxial layers is a crucial condition for improvement in performance of InAs-based infrared detectors. Effect of gadolinium doping on quality of InAsSbP epitaxial films was demostrated (Matveev, 2002). It is known that the rare earth impuruty doping results in a gettering effect in semiconductors. Epitaxial films grown by LPE technique from the melt doped with gadolinim exhibited better photoluminescence efficiency and higher mobility of carriers. As a result, the diffusion-limited InAs/InAsSbP heterosructure PDs with improved characteristics were manufactured (Matveev, 2002).

InAs PDs were also grown by molecular beam epitaxy (MBE) on alternative GaAs and GaAs-coated silicon substrates (Dobbelaere, 1992). The relatively high doping level (>10^{16} cm^{-3}) in the active region was used for the junction formation. The PDs were diffusion-limited at temeperatures as low as 160 K. At 77 K the dominant current is expected to be the defect-assisted tunneling current. Also, in these PDs rather high detectivity of the order of 7 10^{11} cm $Hz^{1/2}W^{-1}$ was achieved at the peak wavelength 2.95 μm. In opinion of the authors these results clearly demostrate the feasibility of the monolithic integration of InAs infrared detectors and GaAs or Si read-out electronics.

The cut-off wavelength in InAs PDs is 3-4 μm whicn is not enough to cover the atmospheric windows 3-5 entire μm. Therefore, ternary compounds InAsSb with more narrow bang gap were extensively investigated as a material for infrared detectors with longer cut-off wavelength. InAsSb epitaxial films were grown on GaAs substrates by MBE in IMEC, Belgium (Merken, 2000). Linear and two-dimensional focal-plane arrays with 256x256 pixels were realized. At room temperature the product R_oA was limited by the combined generation-recombination and diffusion currents.

Multielement InAs MOS capacitors were developed in A.V. Rzhanov Institute of Semiconductor Physics, Russia (Kuryshev, 2009). Auotoepitaxial films were grown on n-InAs substrates. The films were characterized by the electron concentration (1-5) $\cdot 10^{15}$ cm^{-3} and the carrier lifetime 0.3-1.8 μs at 77 K. The SiO_2 gate oxide with thickness of the order of 130 nm was deposited on a previously grown 15 nm thick anode oxide doped with fluorine. The surface states density of the order of to $2 \cdot 10^{10}$ cm^{-2} eV^{-1} was obtained compare to $3 \cdot 10^{11}$ cm^{-2} eV^{-1} in undoped films. Linear (1x384) and two-dimensional (128x128, 256x256) focal-plane arrays have been made. The specific detectivity in typical 128x128 assembly with pixel size 40x40 μm was $3 \cdot 10^{12}$ cm $Hz^{1/2}W^{-1}$ (λ=2.95 μm) at 80 K. Infrared devices (thermal imaging camera, microscope and spectrograf) with improved characteristics were designed.

A new type PDs based on InAs/GaSb superlattices have been recently developed in several laboratories (Rehm, 2006). They were grown by MBE on GaSb substrates. The PDs have p-i-n structure with the type-II short-period superlattice intrinsic region embedded between highly doped contact layers. The superlattice material has some advantages over bulk InAs. The band gap of the superlattice can be varied in a range between 0 and about 250 meV. The Auger recombination can be significantly suppressed, since electrons and holes are spatially separated in neighboring layers. In the single-element test diodes with the cut-off wavelength 5.4 μm at 77 K values of $R_oA= 4 \cdot 10^5$ $\Omega \cdot cm^2$ were measured. The diodes were limited by generation-recombination currents and show background limited performance. The quantum efficience as high as 60% and current responsivity of 1.5 A/W were achieved. High-performance 256x256 focal plane arrays on InAs/GaSb superlattice PDs were manufactured designed for 3-5 μm and 8-12 μm spectral regions (Rehm, 2006; Hill, 2008). Excellent thermal images with noise equivavlent temperature difference below 10 mK were realized. Despite these advantages, several problems such as the surface leakage current, band-to-band and trap-assisted tunneling currents should be solved for improving the superlattice PDs performance.

7. Conclusions

1. The carrier lifetime is investigated in *n*- and *p*-type InAs as a function of carrier concentration and temperature. It is proved that experimental data can be correctly explained by radiative recombination mechanism in both *n*- and *p*-type InAs at

temperatures close to 77 K. The lifetime in *p*-InAs is determined by three recombination mechanisms - radiative, Auger 7 and Auger S. The role of the Auger S recombination in p-InAs seems to be overestimated in the developed theoretical models. The contribution of the Shockley-Read-Hall recombination should be clarified. It is shown that the developed models of recombination can correctly predict the most important parameters of InAs-based infrared PDs.

2. The diffused homojunction PDs have threshold parameters comparable with commercially available ones. It is proved that p^+-InAsSbP/n-InAs heterojunction PDs may be more suitable for application in gas sensors which are operated at room temperature. The threshold parameters in conventional PDs may be improved by supression of the Auger recombination and reduction of the trap-assisted tunneling current.
3. Furter progress in manufacture of conventional single-element PDs is most likely associated with epitaxial films grown on InAs or alternative substrates. Linear and two-dimensional photodiode arrays based on InAs bulk technology which can be attributed to the second generation infrared detectors are in the early stage of development.
4. The results achived in InAs/GaSb type-II superlattice PDs confirm that InAs-based technology is now competitive for manufacture infrared devices with high performance.

8. References

Andrushko, A.I.; Selihov, A.I., Slobodchikov, S.V. (1986). On the recombination mechanism in indium arsenide crystals, *Fiz. Tekh. Poluprov.* (In Russia), Vol. 20, N3, 403-406, ISSN 0015-3222.

Barishev, N.S. (1964). Band-to-band recombination of electrons and holes in indium arsenide, *Fiz. Tverdogo Tela* (In Russia), Vol. 5, N10, 3027-3030, ISSN 0367-3294.

Beattie, A.R. (1962). Quantum Efficiency in InSb, *J.Phys.Chem.Solids*, Vol. 23, 1049-1056, ISSN 0022-3697.

Beattie, A. & Landsberg, P.T. (1959). Auger effect in semiconductors, *Proc. Roy. Soc. A.*, Vol. 249, 16-29, ISSN 0308-2105.

Beattie A.R & Smith G. (1967). *Phys. status solidi*, Vol.19, N3, 577-586, ISSN 0031-8965.

Blakemore J.S. (1962). *Semiconductor Statistics*, Pergamon Press, Oxford, ISBN 0486495027.

Blaut-Blachev, A.N.; Balagurov, L.A., Karatayev, V.V. et. al. (1975). Carrier recombination in n-InAs, *Fiz. Tekh. Poluprov.* (In Russia), Vol. 9, N4, 782-784, ISSN 0015-3222.

Bolgov, S.S., Malyutenko, V.K, Savchenko A.P. (1997). Exclusion of carriers in InAs, *Fiz. Tekh. Poluprov.* (In Russia), Vol. 31, N5, 526-527, ISSN 0015-3222.

Bruk, A.S.; Govorkov, L.I., Kolesnik , L.I. (1982). The role of interaction dislocation-point defect-dopant in the recombination processe in gallium arsenide, *Fiz. Tekh. Poluprov.* (In Russia), Vol. 16, N8, 1510-1512, ISSN 0015-3222.

Datal, V.L.; Hicinbothem, W.A. & Kressel, H. (1979). Carrier lifetimes in epitaxial InAs, *Appl. Phys. Lett.*, Vol. 24, 184-185, ISSN 0003-6951.

Dobbelaere W.; De Boeck G., Heremans P., Mertens R., and Borghs G. (1992). InAs p-n diodes grown on GaAs and GaAs-coated Si by molecular beam epitaxy, *Appl. Phys. Lett.*, Vol. 60, N7, 868-870, ISSN 0003-6951.

Fomin, I.A.; Lebedeva, L.V., Annenko, N.M. (1984). Investigation of deep defect levels in InAs by capacitance measurements of MIS structures, *Fiz. Tekh. Poluprov.* (In Russia), Vol. 18, N3, 734-736, ISSN 0015-3222.

Gelmont, B.L. (1978). Three-Band Kane Model of Auger Recombination, *Zh. Eksper. Teor. Fiz.* (In Russia), Vol. 75, N2, 536-544, ISSN 0044-4510.

Gelmont, B.L. (1981). Auger Recombination in Narrow-Gap p-Type Semiconductor, *Fiz. Tekh. Poluprov.* (In Russia), Vol. 15, N7, 1316-1319, ISSN 0015-3222.

Gelmont, B.L.; Sokolova, Z.N. & Yassievich, I.N. (1982). Auger Recombination in Direct-Gap p-Type Semiconductors, *Fiz. Tekh. Poluprov.* (In Russia), Vol. 16, N3, 592-600, ISSN 0015-3222.

Gelmont, B.L. & Sokolova, Z.N. (1982). Auger Recombination in Direct-Gap n-Type Semiconductors, *Fiz. Tekh. Poluprov.* (In Russia), Vol. 16, N9, 1670-1672, ISSN 0015-3222.

Granato A.V. & Luecke K. (1966), in *Physical Acoustic*, ed. W.P. Mason, Academic Press, New York, ISBN 0387984356.

Grigor'ev, N.N. & Kudykina, T.A. (1997). Recombination model of Zn diffusion in GaAs, *Fiz. Tekh. Poluprov.* (In Russia), Vol. 31, N6, 697-702, ISSN 0015-3222.

Kinch M.A. (1981). Metal-insulator semiconductor infrared detectors. In *Semiconductor and Semimetals*, vol. 18, R.K. Willardson and A.C. Beer, Eds., New York: Academic Press, 1981, ch.7.

Kornyushkin, N.A.; Valisheva N.A., Kovchavtsev A.P., Kuryshev G.L. (1996) Effect of interface and deep levels in the gap on capacitance-voltage characteristics of GaAs MIS structures, *Fiz. Tekh. Poluprov.* (In Russia), Vol. 30, N5, 914-917, ISSN 0015-3222.

Kuryshev G.L.; Kovchavtsev A.P., Valisheva N.A. (2001), Electronic properties of metal-insulator-semiconductor structures, *Fiz. Tekh. Poluprov.* (In Russia), Vol. 35, N9, 1111-1119, ISSN 0015-3222.

Kuryshev G.L.; Lee I.I., Bazovkin V.M. et al. (2009). Ultimate parameters of multielement hybrid MIS InAs IR FPA and devises based of them, *Prikladnaya Fizika* (In Russia), N2, 79-93, ISSN 1996-0948.

Madelung, O. (1964). *Physics of III-V semiconductors.* New York-London-Sydney.

Matveev, B.A.; Gavrilov, G.A., Evstropov, V.V. et. al. (1997). Mid-infrared (3-5 μm) LEDs as sources for gas and liquid sensors, *Sens. Actuators B*, Vol. 38-39, 339-343, ISSN 0925-4005.

Matveev, B.A.; Zotova, N.V., Il'inskaya, N.D. et al. (2002). Towards efficient mid-IR LED operation: optical pumping, extraction or injection of carriers?, *J. Mod. Opt.*, Vol. 49, N5/6, 743-756, ISSN 0950-0340.

Merken P.; Zimmermann L., John J., Nemeth S., Gastal M., Van Hoof C. (2000). InAsSb and InGaAs linear and focal-plane arrays, *Proc. SPIE*, Vol.4028, 246-251, ISSN 0277-786X.

Nemirovsky, Y. & Unikovsky, A. (1992). Tunneling and 1/f noise currents in HgCdTe photodiodes, *J. Vac. Sci. Technol. B*, Vol. 10, N4, 1602-1610, ISSN 0022-5355.

Oliynuk G.S, Sukach A.V., Tetyorkin V.V, Stariy S.V, Lukyanenko V.I, Voroschenko A.T. (2004), Uncooled InAs photodiodes for optoelectronic sensors, *Proc. SPIE*, Vol.5564, 143-148, ISSN 0277-786X.

Popov, F.A.; Stepanov, M.V., Sherstnev, V.V. & Yakovlev, Yu.P. (1997). 3.3 μm LEDs for measurements of methane, *Tech. Phys. Lett.*, Vol. 23, 24-31, ISSN 1063-7850.

Raikh, M.E. & Ruzin I.M. (1985). Fluctuation mechanism of excess tunneling currents in reverse biased p-n junctions", *Fiz. Tekh. Poluprov.* (In Russia), Vol. 19, N7, 1216-1225, ISSN 0015-3222.

Raikh, M.E. & Ruzin I.M. (1987). Temperature dependence of excess fluctuation currents through meta-semiconductor contact, *Fiz. Tekh. Poluprov.* (In Russia), Vol. 21, N3, 456-460, ISSN 0015-3222.

Rogalski, A. & Orman, Z. (1985). Band-to-band recombination in $InAs_{1-x}Sb_x$, *Infrared Phys.*, Vol. 25, N3, 551-560, ISSN 0020-0891.

Rogalski, A. (ed.) (1995). Infrared Photon Detectors, *SPIE Optical Engineering Press*, ISBN, 0 8194 1798 X, N.Y.

Rosenfeld D. & Bahir G. (1992). A model for the trap-assisted tunnelling mechanism in diffused n-p and implanted n^+-p HgCdTe photodiodes, *IEEE Trans. On Electron Devices*, Vol. 39, N7, 1638-1645, ISSN 0018-9383.

Sukach A., Tetyorkin V., Olijnyk G, Lukyanenko V., Voroschenko A. (2005), Cooled InAs photodiodes for IR applications, *Proc. SPIE.*, Vol.5957, 267-276, ISSN 0277-786X.

Sukach A.V.& Tetyorkin V.V. (2009) Ultrasonic treatment-induced modification of the electrical properties of InAs p-n junctions, *Tech. Phys. Lett.*, Vol.36, N6, 514-517, ISSN 1063-7850.

Sze, S.M. (1981). *Physics of Semiconductor Devices*, second edition, Wiley, N.Y. ISBN 0471056618.

Takeshima, M. (1972). Auger Recombination in InAs, GaSb, InP and GaAs, *J. Appl. Phys.*, Vol. 43, N10, 4114- 4119, ISSN 0021-8979.

Takeshima M. (1975). Effect of electron-hole interaction on the Auger recombination process in a semiconductor, *J. Appl. Phys.*, V. 46, N7, 3082-3088, ISSN 0021-8979.

Tetyorkin V. V., Sukach A. V., Stariy S. V., Zotova N. V., Karandashev S. A., Matveev B. A., Stus N. M. (2005). p+-InAsSbP/n-InAs photodiodes for IR optoelectronic sensors, *Proc. SPIE*, Vol.5957, 212-219 ISSN 0277-786X.

Tetyorkin V.V., Sukach A.V., Stary S.V., Zotova N.V., Karandashev S.A., Matveev B.A., Remennyi M.A.,Stus N.M. (2007), Performance of InAs-based Infrared Photodiodes, *Proc. SPIE.*, Vol. 6585, 185-194, ISSN 0277-786X.

Van Roosbroeck W. & Shockley W. (1954). Photon-radiative recombination of electrons and holes in germanium, *Phys. Rev.*, Vol. 94, 1558-1560, ISSN 1098-0121.

Wenmu He, Zeynep Celik-Batler. (1996). 1/f noise and dark current components in HgCdTe MIS infrared detectors, *Solid-State Electron,* Vol. 19, N1, 127-132, ISSN 0038-1101.

Wider, H.H. & Collins, D.A. (1974). Minority carrier lifetime in InAs epilayers, *Appl. Phys. Lett.*, Vol. 25, 742-743, ISSN 0003-6951.

Zotova, N.V.; Karandashev, S.A., Matveev, B.A. et. al. (1991). Optoelectronic sensors based on narrow band A^3B^5 alloys, *Proc. SPIE*, Vol. 1587, 334-345, ISSN 0277-786X.

Rehm, R; Walther M., Schmitz J. Et al. (2006). InAs/GaSb superlattice focal plane arrays for high-resolution thermal imaging, *Opto-Electronics Review*, Vol.14, N1, 19-24, ISSN 1230-3402.

Hill, C. J.; Soibel, A., Keo, S.A., et al (2008). Infrared imaging arrays based on superllatice photodiodes, *Proc. SPIE*, Vol. 6940, 69400C-1 – 69400C-10, ISSN 0277-786X.

21

The InAs Electron Avalanche Photodiode

Dr. Andrew R. J. Marshall
Lancaster University
England

1. Introduction

Avalanche photodiodes (APDs) exploit the process of impact ionisation to amplify the primary, or unity gain, photocurrent generated by the absorption of incident photons. In all APDs the signal enhancing avalanche multiplication is accompanied by an increase in the signal's noise current, in excess of shot noise. Hence APDs have found application in detection systems where the electrical noise introduced by following circuitry is greater than the noise introduced by a unity gain photodiode. These principally include detection systems which need to operate under low incident photon fluxes or with high bandwidths. In such systems an APD's multiplication can provide a desirable enhancement in the overall system sensitivity. Increasing an APD's operational gain only enhances a system's sensitivity whilst the APD's noise is less than the noise of the following circuitry. Hence the rate at which an APD's noise increases with increasing multiplication is a key performance parameter. The noise power (I_n^2) generated by an APD can be described by equation 1,

$$I_n^2 = 2\, q\, I_{pr}\, M^2\, F\, BW \tag{1}$$

where q is the electron charge, I_{pr} the primary photocurrent, M the avalanche multiplication factor, F the excess noise factor and BW the bandwidth.

An APD's excess noise results from the stochastic nature of the impact ionisation process, which leads to fluctuations in the instantaneous multiplication as individual injected carriers undergo different levels of multiplication. The impact ionisation of electrons and holes is described by the ionisation coefficients $a(\xi)$ and $\beta(\xi)$ respectively, representing the mean number of impact ionisation events per unit length travelled, as a function of electric field ξ. These ionisation coefficients vary from material to material and their accurate determination is essential to support the assessment of a material's suitability for use in APD applications, as well as the modelling of an APD's noise. Equations 2 and 3 (McIntyre, 1966) describe how, under the local model of impact ionisation, an APD's excess noise factor is related to its operational multiplication and the ratio of the ionisation coefficients, k.

$$F_e = kM_e + (1-k)\left(2 - \frac{1}{M_e}\right) \text{ where } k = \beta / a \tag{2}$$

$$F_h = kM_h + (1-k)\left(2 - \frac{1}{M_h}\right) \text{ where } k = \beta / a \tag{3}$$

Here M_e and F_e are the average multiplication and excess noise initiated by a primary photocurrent consisting of only electrons, injected from the p-type side of the depletion region. Similarly M_h and F_h are the average multiplication and excess noise initiated by a primary photocurrent consisting of only holes, injected from the n-type side. The relationship between F, M and k defined by equations 2 and 3 is plotted in figure 1.

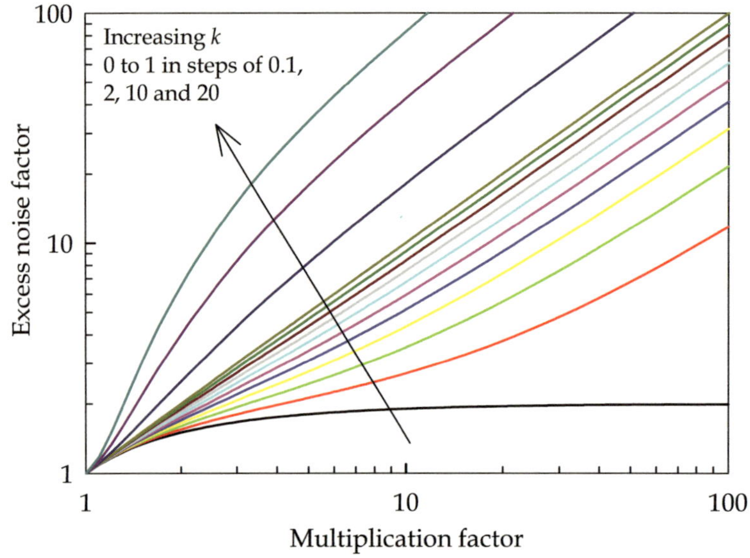

Fig. 1. The dependence of an APD's excess noise factor on its operational multiplication factor and the k of its multiplication medium, as defined by the local model (McIntyre, 1966).

Two important APD design principles can be taken from equations 2 and 3. Firstly, excess noise is always lower when only the carrier type with the highest ionisation coefficient is injected into the multiplication region, making $k \leq 1$. Secondly, in order to minimise the excess noise factor it is desirable to fabricate the multiplication region of an APD from a material with highly disparate ionisation coefficients, ideally one in which one of the ionisation coefficients is zero such that k also becomes zero.

The aggregate influence of an APD's multiplication and excess noise on the overall sensitivity of a light detecting system clearly varies depending on the system considered. To illustrate a typical case, figure 2 shows the sensitivity of a 10 Giga bit per second (Gbps) optical communications receiver, modelled as a function of its APD's multiplication and the k of the APD's gain medium. The APD's gain-bandwidth product limit is not considered in this illustrative case. From the results shown in figure 2 it can be seen that the lower the k of the APD's gain medium, the better the receiver sensitivity, and the higher the optimum APD gain in the absence of gain-bandwidth product limits. In the optimum case where k = 0, substantial improvements in receiver sensitivity are predicted as the APD's multiplication is increased. Furthermore it has been shown that both an APD's transit time limited bandwidth and its gain-bandwidth product limit increase as k reduces (Emmons, 1967).

The clear advantage afforded by employing materials with disparate ionisation coefficients in APDs, has led to a long term effort to characterise the ionisation coefficients in most common semiconductor materials (Stillman and Wolfe, 1977; Capasso, 1985; David and Tan,

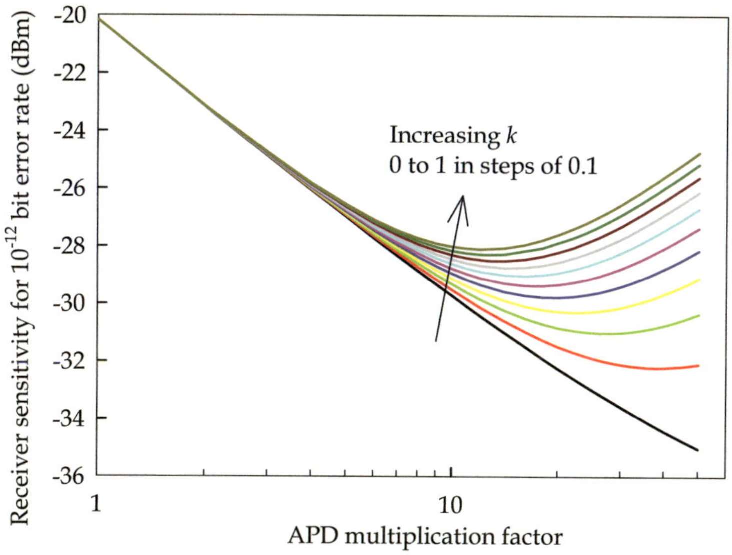

Fig. 2. The modelled receiver sensitivity of a 10 Gbps receiver as a function of its APD's operational multiplication factor and the k of its APD's gain medium.

2008). Until recently the optimum case, where $k = 0$, remained an unachievable theoretical ideal, with most materials exhibiting $0.1 < k < 1$. Indeed, unable to identify sufficiently capable materials, some researchers resorted to trying to engineer superlattice structures in which the ionisation coefficients were more disparate (Capasso *et al.*, 1982; Yuan *et al.*, 2000). Beck *et al.* were the first to report APD characteristics consistent with $k = 0$ in 2001, when they reported results from $Hg_{0.7}Cd_{0.3}Te$ APDs (Beck *et al.*, 2001). They have since shown that for a number of compositions β remains essentially zero in $Hg_{x-1}Cd_xTe$ APDs detecting in the short, mid and long wave infrared (SWIR, MWIR and LWIR) (Beck *et al.*, 2006). They coined the phrase electron-APD (e-APD) to describe such APDs where only electrons undergo impact ionisation. As desirable as some of the properties of $Hg_{x-1}Cd_xTe$ e-APDs undoubtedly are, $Hg_{x-1}Cd_xTe$ itself remains a challenging material to work with. It is not readily available through commercial foundries, unlike group IV and III-V materials, and is relatively expensive. It is also becomes unstable at lower temperatures than other established semiconductors. Furthermore it can suffer from compositional non-uniformity issues over imaging array sized areas and in some cases cannot be as highly doped as III-V materials. Hence it remains desirable to identify a more widely available III-V material which exhibits comparable e-APD properties. Recent characterisation and development work on InAs APDs has shown that they can meet this desire for the first time (Marshall *et al.*, 2008; 2009; 2010).

This chapter presents the emerging InAs e-APD, summarising its properties using both recently published data and new results. It is shown that multiplication and excess noise in InAs APDs match those expected for the emerging e-APD subclass. Furthermore the specific and at times unique characteristics of electron avalanche multiplication in InAs are discussed. The ability to characterise InAs e-APDs and demonstrate their desirable properties has been underpinned by the development of new fabrication procedures, the key aspects of which are also discussed here. Finally the potential for deploying InAs e-APDs in several significant applications is discussed. All results presented here were obtained from homojunction InAs *p-i-n* and *n-i-p* diode structures, grown by molecular beam epitaxy (MBE) or metal organic vapour phase epitaxy (MOVPE), in the EPSRC national centre for III-V technologies at The university of Sheffield, UK. The principle difference between the various structures characterised was the intrinsic region width. Hence whenever experimental results are presented here, the type of diode structure measured and its intrinsic region width are detailed. All device fabrication and characterisation work was undertaken within the Electronic and Electrical Engineering department at The University of Sheffield.

2. Avalanche multiplication and excess noise in InAs e-APDs

2.1 Avalanche multiplication

The magnitude of the impact ionisation coefficients α and β are usually determined through measurements of the photomultiplication factors M_e and M_h. It has been shown (Marshall *et al.*, 2010) that in InAs *p-i-n* diodes significant electron initiated multiplication can be achieved whilst hole initiated multiplication in InAs *n-i-p* diodes remains negligible across the same electric field range. The M_e measured on three *p-i-n* diodes with a range of intrinsic widths and the M_h measured on a *n-i-p* diode, are shown in figure 3a. The measurements were taken using a lock-in amplifier and phase sensitive detection of the photocurrent. This was generated by an appropriate laser wavelength such that all absorption took place within the doped *p-* and *n*-type cladding layers, allowing M_e to be measured on *p-i-n* diodes and M_h to be measured on *n-i-p* diodes. The results clearly show that $\beta \sim 0$ in InAs, within the electric field range exercised, also making $k \sim 0$. It should be noted that this finding is in contradiction to the only previously reported experimental study for avalanche multiplication in InAs. Mikhailova *et al.* reported that β was approximately 10 times greater than α in InAs, at 77K (Mikhailova *et al.*, 1976). This discrepancy is given more consideration in a number of journal papers (Marshall *et al.*, 2008; 2009; 2010) ; here it will simply be noted that during the new study of InAs e-APDs reviewed in this chapter, more than 20 different InAs diode structures have been characterised at room temperature, and all results are consistent with the finding that $\beta \sim 0$.

The most robust determination of the relative magnitude of α and β in any material comes from the measurement of M_e and M_h on a single diode structure, eliminating any uncertainty over variations in layer thickness and electric field profiles. In order to achieve this for InAs, the substrate was removed from a sample of fully fabricated *n-i-p* diodes. This was achieved through a combination of mechanical thinning and selective wet etching. This made it possible to measure M_h by illuminating the top side of the diodes and M_e by illuminating the substrate side of the same diodes. The photomultiplication results taken in this way are shown in figure 3b and confirm that $\alpha >> \beta$ in InAs at room temperature, with $\beta \sim 0$, making it possible to realise the first III-V based e-APDs from InAs.

The avalanche multiplication characteristics measured on e-APDs differ from those of all conventional APDs. In conventional APDs not only does the injected carrier type (e.g. electrons) undergo impact ionisation when transiting the depletion region (from *p-* to *n*-type claddings), but secondary carriers of the other type (holes) generated by impact ionisation, also undergo impact ionisation themselves when transiting the depletion region in the opposite direction (towards the *n*-type cladding), generating yet more carriers of the injected type. One possible sequence of impact ionisation events in a conventional APD is shown schematically in figure 4. If the electric field within such an APD is increased, in turn

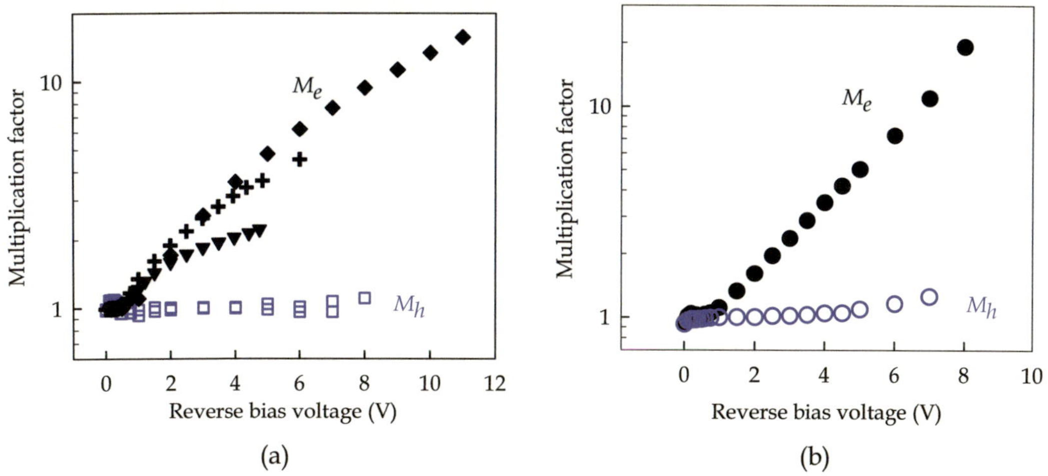

Fig. 3. Photomultiplication characteristics M_e and M_h for InAs diodes, measured on;
(a) three *p-i-n* diodes and one *n-i-p* InAs diode, with intrinsic region widths of 3.5µm (◆), 1.9µm (✚), 0.8µm (▼) and 1.8µm (□) respectively (Marshall *et al.*, 2010)
(b) one n-i-p diode with an intrinsic region width of 6µm doped at ~7×10^{14} cm^{-3}, with its substrate removed allowing both topside (●) and substrate side (○) illumination.

increasing α and β, the avalanche multiplication can rise very rapidly due to the feedback in this avalanche. Indeed if the magnitude of α and β are sufficient that each carrier ionises on average at least once before leaving the depletion region, the multiplication factor becomes infinite and avalanche breakdown occurs.

By contrast in e-APDs the feedback provided by hole impact ionisation is absent. As a result the avalanche of electron impact ionisation events, from which the multiplication is solely derived, builds up in a single transit of the depletion region. Again one possible sequence of impact ionisation events within an e-APD is shown schematically in figure 4. This avalanche is more analogous with naturally occurring avalanches, where the material involved in the avalanche builds up as it falls in a single trip down a hill. The maximum number of impact ionisation events in an avalanche without feedback is limited since in practice neither α or the depletion width can become infinite. Hence true e-APDs never undergo an avalanche breakdown, instead exhibiting a progressively increasing multiplication as the bias voltage and commensurate electric field are increased. This is evident from the expression for multiplication in e-APDs under a constant electric field, given by equation 4 where w is the depletion width.

$$M_e = \exp(\alpha w) \quad (4)$$

Figure 5 compares the multiplication characteristics of some e-APDs with that of an InAlAs APD (Goh *et al.*, 2007), representative of conventional APDs. The multiplication factor minus one scale is used because it allows both the low and high gain characteristics to be presented clearly. There is essentially no discernible multiplication in the InAlAs APD below 7V, however once multiplication starts it rises quickly with increasing bias voltage and the APD breaks down at approximately 15V. In contrast multiplication is discernable in the e-APDs from lower voltages, in some cases less than 1V and it rises much more progressively with increasing bias voltage. On the logarithmic scale the rise in multiplication is approximately

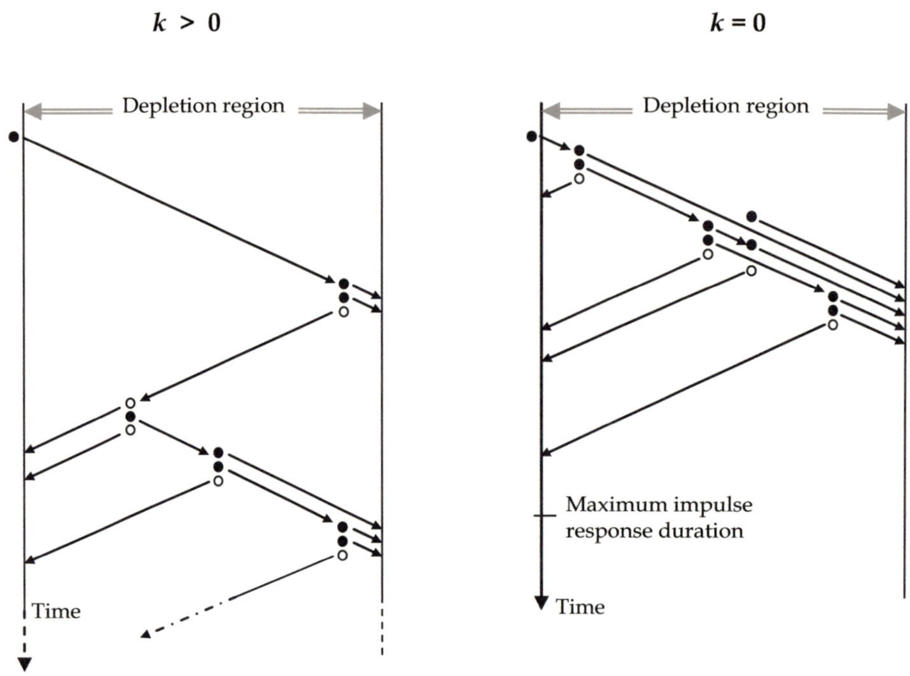

Fig. 4. Schematic representations of potential avalanches of impact ionisation events in multiplication regions where $k > 0$ and $k = 0$, showing the spatial and temporal distribution of impact ionisation by electrons (•) and holes (○).

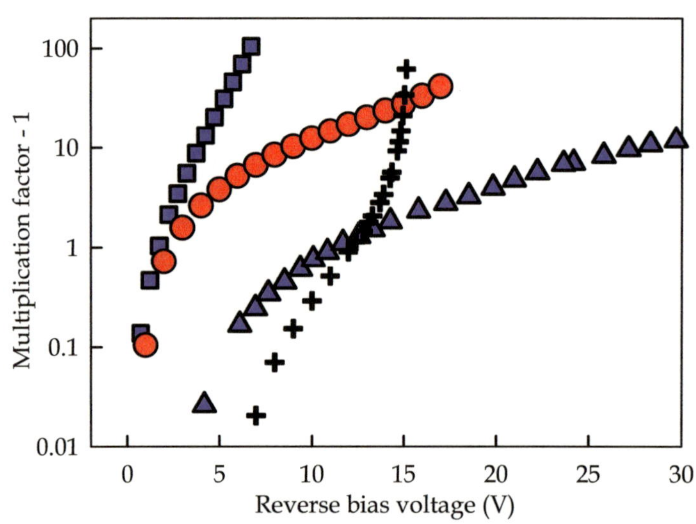

Fig. 5. A comparison between the M_e reported on APDs of different materials including, an InAs diode with a 3.5µm intrinsic width (●) (Marshall *et al.*, 2010), $Hg_{x-1}Cd_xTe$ diodes with cut-off wavelengths of 4.2µm (■) and 2.2µm (▲) (Beck *et al.*, 2006), and an InAlAs diode (✚) (Goh *et al.*, 2007).

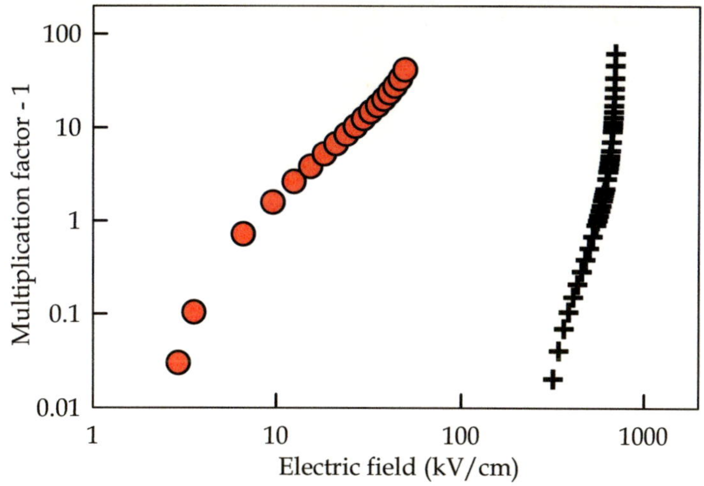

Fig. 6. A comparison between the M_e reported for an InAs diode with a 3.5µm intrinsic width (●) (Marshall *et al.*, 2010) and an InAlAs diode (✚) (Goh *et al.*, 2007), plotted against the electric field in the multiplication region.

linear after the initial turn on. As expected there is no breakdown observed in the e-APDs. A similar comparison against electric field is given in figure 6 and shows the same distinct characteristics, while also highlighting that avalanche multiplication in InAs e-APDs occurs at a much lower electric field than in conventional APDs.

It is worthwhile considering the bias dependence of the multiplication characteristics shown in figure 5, in relation to the concomitant biasing circuit requirements for each APD. Multiplication in all APDs is by nature dependent on the bias voltage applied to them; however it is undesirable for the multiplication to vary dramatically in response to small unintentional fluctuations in the nominal bias voltage. Conventional APDs need to be biased close to their breakdown voltage to provide significant multiplication, and hence are sensitive to bias voltage fluctuations. For example if the InAlAs APD reported in figure 5 were to be biased for a nominal multiplication of 10, a fluctuation of ± 0.25V about the nominal bias voltage would give rise to the actual multiplication varying between approximately 7 and 18. If a nominal gain of 100 were to be considered the variation in gain would be even more dramatic. By comparison if the InAs e-APD reported in figure 5 were to be biased for a nominal multiplication of 10, the same fluctuation of ± 0.25V about the nominal bias voltage would only result in the actual multiplication varying between approximately 9.5 and 10.6. Hence a further advantage of e-APDs, particularly when an APD needs to be operated at a high multiplication, is that their gain is less sensitive to fluctuations in their bias voltage.

2.2 Excess noise

The e-APD nature of InAs APDs has been further confirmed by excess noise measurements (Marshall *et al.*, 2009). The F_e measured on InAs *p-i-n* diodes, shown in figure 7, falls slightly below the local model prediction for $k = 0$ as given by equation 2. This excess noise is comparable to that reported for SWIR sensitive $Hg_{x-1}Cd_xTe$ e-ADPs, although it is somewhat higher than that reported for MWIR sensitive $Hg_{x-1}Cd_xTe$ e-APDs (Beck *et al.*, 2006). To allow comparison with the characteristic of a conventional APD, the excess noise measured on an InAlAs APD is also shown in figure 7. As with all conventional APDs in which both carriers under go impact ionisation, the excess noise in the InAlAs APD rises with increasing multiplication. In comparison away from the lowest gains, the excess noise in the e-APDs does not continue to rise. This is clearly a desirable characteristic for APDs, as it improves the overall system sensitivity as shown earlier.

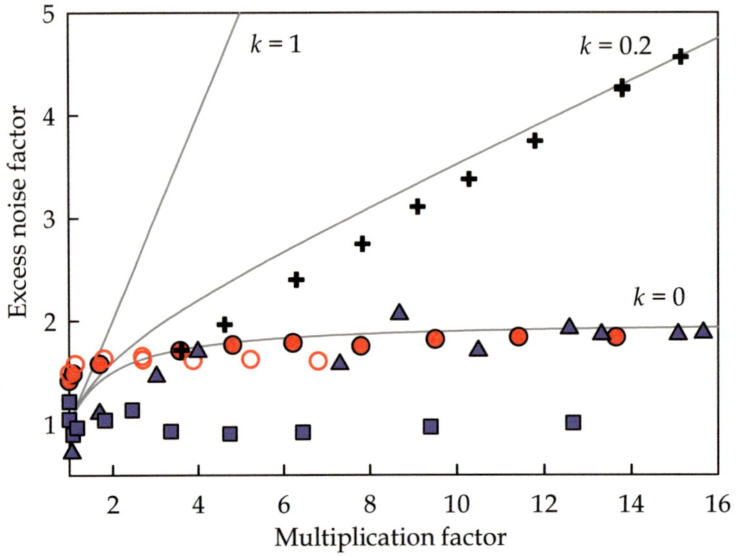

Fig. 7. A comparison between the F_e reported on APDs of different materials including, InAs diodes with a 3.5μm intrinsic width and radii of 50μm (●) and 100μm (○) (Marshall *et al.*, 2009), $Hg_{x-1}Cd_xTe$ diodes with cut-off wavelengths of 4.2μm (■) and 2.2μm (▲) (Beck *et al.*, 2006), and an InAlAs diode (✚) (Goh *et al.*, 2007).

In figure 7 it can be seen that the excess noise measured on the largest InAs APDs, for which the purest electron injection photocurrent was achieved, falls notably below the k = 0 local model prediction. To explain such excess noise below the lower limit case of the local model, it is necessary to consider the influence of deadspace, which is neglected from the local model. Deadspace has been described as the distance travelled by a carrier while it attains the ionisation threshold energy, or the distance travelled by a carrier while its energy rises into equilibrium with the electric field. Both descriptions attempt to address the reality that a carrier's ionisation probability does not become a non-zero function, described by its non-local ionisation coefficient, until it has travelled some distance. It is simplest to consider that it travels this distance with an ionisation probability of zero, leading to the first description. The effect of deadspace is to introduce determinism into both the spatial distribution of the impact ionisation events and the resulting multiplication experienced by individual carriers. This increased determinism leads to a reduction in the excess noise factor.

It is noted that deadspace is typically only considered to be of significance in conventional APDs with thin multiplication regions, less than a few hundred nanometres wide (Plimmer *et al.*, 2000). However it has been found that even in e-APDs with thick multiplication regions, such as the ones reported in figure 7, the deadspace can become significant with respect to the mean ionisation path length of electrons, α^{-1}, and hence noticeably affect the excess noise factor. When k = 0 the deadspace causes the excess noise factor to remain below

$F = 2$, reducing it towards the ultimate limit of $F = 1$. To achieve $F = 1$ it would be necessary for electrons to undergo impact ionisation immediately after transiting their deadspace, such that they transit the depletion region moving through a series of deadspaces between delta function ionisation probability density functions. The experimental results reported for $Hg_{0.3}Cd_{0.7}Te$ (Beck *et al.*, 2006) indicate that they are operating in approximately this ideal way.

2.3 Electron ionization coefficient

To allow modelling of the multiplication within arbitrary InAs e-APDs, an electron ionisation coefficient has been reported based on the multiplication results presented in figure 3 (Marshall *et al.*, 2010). This coefficient is parameterised as shown in equation 5.

$$\alpha = 4.62\times10^{4}\exp\left[-\left(\frac{1.39\times10^{5}}{|\xi|}\right)^{0.378}\right] \quad \text{cm}^{-1} \tag{5}$$

The new room temperature electron ionisation coefficient for InAs is shown in figure 8 together with selected other electron ionisation coefficients. Amongst the materials in which impact ionisation has been well characterised, $In_{0.53}Ga_{0.47}As$ is considered to have an atypically high electron ionisation coefficient at low electric fields (Ng *et al.*, 2003). In comparison significant electron ionisation occurs in InAs from much lower electric fields. Indeed the maximum electric field for which *a* has been calculates in InAs is lower than the minimum electric field for which *a* could be determined in $In_{0.53}Ga_{0.47}As$. This greatly enhanced electron impact ionisation at low electric fields is of pivotal importance for InAs APDs. It is considered to be both the reason that meaningful avalanche multiplication can be achieved at all in practical devices and the reason that they operate as e-APDs.

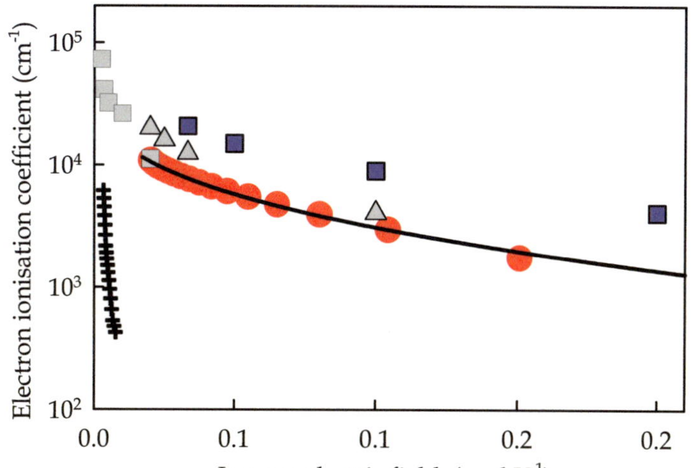

Fig. 8. A comparison between the electron ionization coefficients reported for different materials including, InAs as calculated from experimental results (●) (Marshall *et al.*, 2010), as parameterised (line) and as modelled by both Bude and Hess (1992) (■) and Brennan and Mansour (1991) (▲), $Hg_{0.3}Cd_{0.7}Te$ as modelled by Brennan and Mansour (1991) (■) and $In_{0.53}Ga_{0.47}As$ as calculated from experimental results by Ng et al. (2003) (✚).

$In_{0.53}Ga_{0.47}As$ has not been used as the gain medium of APDs because it generally suffers from excessive tunnelling current before the electric field becomes high enough for significant multiplication to be obtained. InAs has a significantly smaller bandgap and commensurately higher tunnelling current at a given electric field, however because α is high enough at low electric fields, practical APDs can still be realised without excessive tunnelling current. Furthermore InAs APDs operate as e-APDs, whereas other III-V based APDs do not, due the atypically enhanced α within their operational electric field range, rather than an atypically suppressed β. Indeed it is expected that could InAs be characterised at higher electric fields, hole impact ionisation would be found to commence somewhat below the electric field required for hole impact ionisation in $In_{0.53}Ga_{0.47}As$, ~ 150 kVcm^{-1}. However the tunnelling current at such electric fields is likely to make characterisation impossible.

As shown in figure 8, $Hg_{0.3}Cd_{0.7}Te$ is modelled to exhibit an even higher α than InAs, which is consistent with the higher gain and lower noise reported, as shown in the comparisons in figures 5 and 7 respectively. Brennan and Mansour (1991) and Bude and Hess (1992) both modelled α in InAs and reported results broadly in line with the new experimentally derived α. Brennan and Mansour suggested α should be slightly higher at 77K than the newly derived room temperature α. Working purely theoretically Bude and Hess modelled α for a higher electric field range than it has so far been possible to exercise in practice, however their lowest data point aligns with the new α well.

The combination of α being only weakly dependent on electric field and β being approximately zero, results in a final atypical characteristic of InAs e-APDs. This trend can be observed in the M_e results shown in figure 3 and should be explained since it has significant implications for the design of InAs e-APDs. Usually when the multiplication characteristics measured on *p-i-n* diodes with different intrinsic widths are compared as a function of the applied voltage, the multiplication at any given voltage is highest in the diode with the thinnest depletion width. However in InAs *p-i-n* diodes this trend is not seen, instead the highest multiplication at any given voltage is achieved in the diode with the widest intrinsic region and hence also the widest depletion region and lowest electric field. Uniquely in InAs e-APDs an increase in the depletion width over which the unidirectional electron avalanche can build up, has a greater influence on the APD's multiplication factor than the concomitant reduction in α due to the lower electric field. This unique trend can be exploited to improve the characteristics of InAs e-APDs, unimpeded by some of the classical APD design trade-offs. Increasing the depletion width in an InAs e-APD increases the multiplication achieved at low bias voltages, making it easier to integrate the APD into a system. Furthermore it also leads to a reduction in the electric field within the APD, improving its reliability and reducing tunnelling current. As a result of this it is desirable for almost all applications, that the intrinsic width in InAs e-APDs is increased as much as practical, since this results in demonstrably better device performance parameters.

3. The fabrication of practical InAs e-APDs

It was only possible to successfully undertake the characterisation of InAs e-APDs reported in the previous section, following the development of a growth and fabrication process which was capable of producing InAs diodes with reduced and controlled reverse leakage current. Minimising the reverse leakage current in InAs diodes is particularly challenging due to the low bandgap energy of InAs and its predisposition towards forming low impedance surfaces (Noguchi *et al.*, 1991).

Efforts to reduce the reverse leakage current in InAs APDs started with development of the epilayer growth conditions. The InAs used in this work was grown by MBE and MOVPE on *p*-type InAs substrates. Following a RHEED monitored clean-up at 500°C, MBE growth was performed at ~ 0.8 monolayers per a second with a substrate temperature of 470°C. MOVPE growth commenced with a 620°C substrate clean-up, followed by growth at ~10 Å/s with a substrate temperature of 600°C. During all growths the two key aims were to:

- Minimise the defect density so as to minimise the bulk leakage current and increase the maximum bias voltage which could be applied without the diodes failing.
- Minimise the background doping concentration in the intrinsic region, so that the depletion width and hence also the multiplication, was maximised.

Minimising the background doping was found to be easier using MBE, with background doping densities ≤ $1x10^{15}$ cm^{-3} routinely achievable and a minimum electrically active doping density of ~ $2x10^{14}$ cm^{-3} measured. However maintaining the crystal quality during the growth of diode structures > 5μm thick was challenging and ultimately MOVPE was found to be the preferred technique for growing the thickest InAs diode structures. The higher growth rate made it reasonably practical to grow total epitaxial thicknesses of 10μm. Furthermore the MOVPE grown InAs was also found to be electrically more robust than MBE grown InAs. MOVPE grown diodes were less prone to non ideal increases in bulk leakage under higher bias voltages and were able to withstand higher maximum voltages without failing. Keeping the background doping down to acceptable levels was more of an issue than with MBE, however following optimisation work it was possible to obtain doping densities slightly below $1x10^{15}$ cm^{-3}.

Fabricating InAs mesa diodes with low leakage currents is arguably more challenging that growing good quality InAs epilayers. The principle issue is the predisposition of etched InAs surfaces to become low impedence. Such surfaces link the *p*- and *n*-type regions of the mesa diode with a low resistance sidewall, down which significant surface leakage current can readily flow. Wet chemical etching typically produces mesa sidewalls with less damage than dry etch etching does, and for InAs e-APD fabrication wet etchants were again found to be preferable. A number of etchants were tested during this work (Marshall *et al.*, 2007) and the optimum etching routine developed was as detailed below.

1. Etch the mesa to approximately 0.5μm less than the desired total depth in a 1 : 1 : 1 mixture of $H_3PO_4 : H_2O_2 : H_2O$
2. Etch the mesa for 30 seconds in a 1 : 8 : 80 mixture of $H_2SO_4 : H_2O_2 : H_2O$.
3. Remove the resist from the sample using acetone only
4. Dip the unmasked sample in the 1 : 1 : 1 mixture of $H_3PO_4 : H_2O_2 : H_2O$ for 10 seconds, quench in deionised water and then immediately dip the sample in the 1 : 8 : 80 mixture of $H_2SO_4 : H_2O_2 : H_2O$ for 20 seconds
5. Avoid further immersion of the mesa sidewall

The sequential use of the two etchants consistently produces a better result that using either of them individually. It is postulated that the first etchant has a tendency to leave an indium rich surface whereas the second has a tendency to leave an arsenic rich surface. Using them sequential with the appropriate etch durations may produce a balanced InAs surface. Returning the sample to the etchants after the mask has been removed results in all exposed InAs surfaces being etched slightly. Whilst this is generally not desirable it is possible to design a fabrication process and diode structure which can tolerate it and the procedure does produce diodes with consistently lower surface leakage current.

As with many aspects of InAs diode fabrication, there is little information in the literature regarding the formation of ohmic contacts with InAs. However in this respect the surface properties of InAs are favourable and for the majority of this work Ti / Au contacts, 20nm / 200nm thick, were found to be adequate for both *n*- and *p*-type contacts. Indeed using this metallisation the typical contact resistance was in the order of 10Ω.

Using the optimised etching routine it was possible to fabricate InAs e-APDs with negligible surface leakage current across a wide range of bias voltages. Figure 9 shows the reverse leakage current characteristics measured on InAs mesa diodes with four different radii, along with the current densities calculated for the different diodes. The excellent consistency between the current densities calculated for the diodes with different areas, indicates that the surface leakage current was negligible in all diodes. If etched incorrectly the leakage current in diodes like these can reach 10mA at a reverse bias voltage as low as 1V or less.

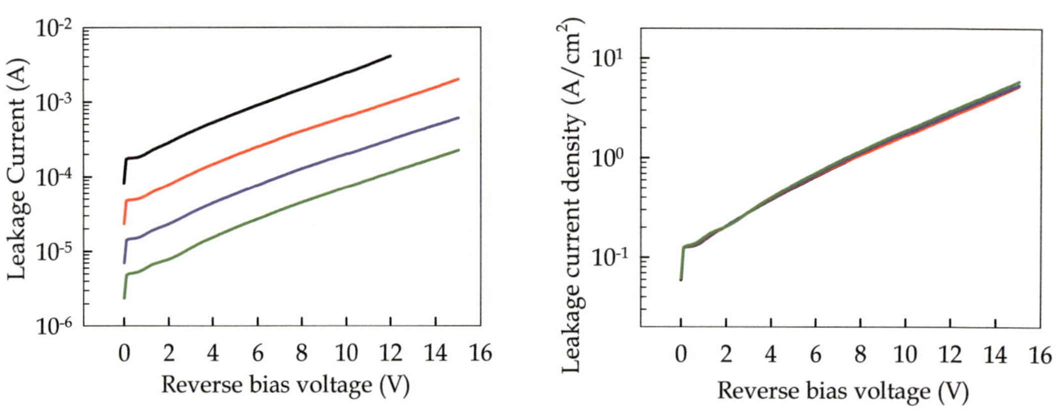

Fig. 9. The reverse leakage current measured on InAs *n-i-p* diodes with 200μm (black), 100μm (red), 50μm (blue) and 25μm (green) radii and intrinsic region widths of 6μm doped at ~1x10^{15} cm^{-3}, together with the commensurate leakage current density calculated.

The leakage characteristics of InAs photodiodes are rarely reported beyond a reverse bias of 0.5V, because to date they have been exclusively used as unity gain detectors. Not only does the leakage current in the InAs e-APDs developed during this work remain controlled under previously unreported high bias voltages, but it also compares favourably with reported unity gain InAs photodiodes at low bias voltages. Figure 10 shows a comparison of the leakage current densities in a number of InAs photodiodes. Under low bias voltages the lowest leakage current is observed in an MBE grown *p-i-n* diode. This diode structure includes a lattice matched AlAsSb layer immediately under the *p*-type contact, designed to block the diffusion of minority electrons from the surface or contact (Marshall *et al.*, 2007). Diodes with this blocking layer routinely yielded the lowest leakage currents at low reverse bias voltages. The MOVPE grown *n-i-p* diode exhibits only slightly higher leakage at low reverse bias, remaining below the level reported by others for InAs diodes. Under higher bias voltages this diode exhibits the lowest leakage current, typical of MOVPE grown diodes.

4. The potential for exploiting InAs e-APDs

4.1 Leakage current

The new InAs e-APD technology offers a III-V based alternative to the high performance but exotic $Hg_{x-1}Cd_xTe$ e-APD technology. The core multiplication and excess noise characteristics of these new APDs are undoubtedly desirable for a number of applications, however to asses their true suitability the most important parameter to consider is the leakage current. Due to their narrow bandgap, InAs APDs will inevitably exhibit higher leakage than similar APDs fabricated from wider bandgap materials. Whether or not this leakage can be tolerated, or suppressed through cooling, ultimately depends upon the specific application of interest. Based on the characterisation and development work carried out to date it is possible to make some observations and predictions regarding leakage current in InAs e-APDs, so as to support assessment of their potential. The leakage currents which affect InAs e-APDs under low and high bias voltages are considered separately. It has already been shown that surface leakage current can be adequately suppressed at room temperature, and with further development it is likely that the same can be achieved at lower temperatures. Hence surface leakage is not included in this consideration of the unavoidable mechanisms contributing to the leakage current in InAs e-APDs.

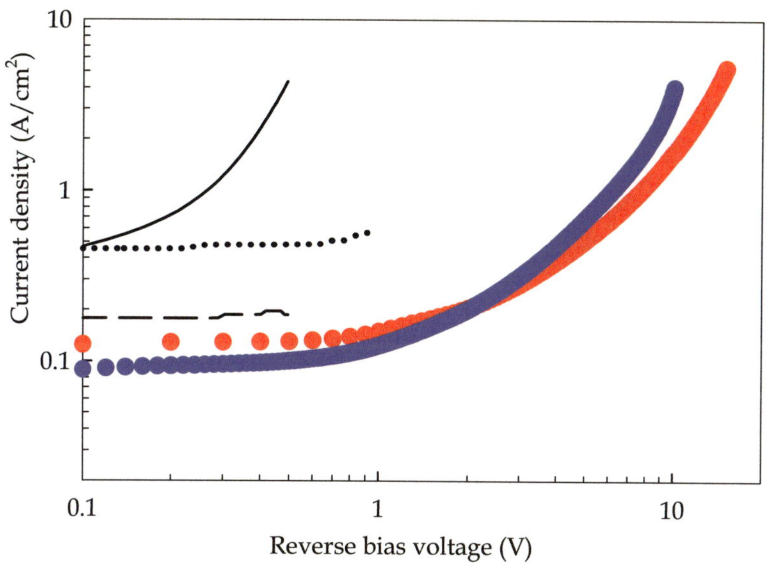

Fig. 10. A comparison between the reverse bias leakage characteristics measured on two of the new InAs e-APDs and those reported for or measured on other InAs diodes including, a commercial diode (solid line), a planar diode (Iwamura and Watanabe, 2000) (dotted line) and the best prior mesa diode (Lin *et al.*, 1997) (dashed line). The two InAs diodes were a *p-i-n* diode with a 3.5µm intrinsic width (●) and a *n-i-p* diode with an intrinsic region width of 6µm doped at ~$1x10^{15}$ cm^{-3} (●).

At room temperature and low reverse bias the leakage current in InAs e-APDs is dominated by bulk diffusion current. The two InAs e-APDs reported in figure 10 show a low bias leakage current density of ~ $100mAcm^2$. This level is typical of the InAs e-APD technology at present; however lower leakage current densities have been measured, down to ~

30mA/cm2. It is considered that there remains considerable scope for reducing the defect density in the epitaxial InAs through further development of the epitaxial growth conditions, and hence it is likely that this leakage current density can be reduced further. Beyond improving the crystal quality, the devices will need to be cooled to suppress the leakage current density even more. As the temperature is reduced it is typical for the leakage to change from being diffusion dominated to generation and recombination dominated (Krier *et al.*, 1998). At this time it is not known at which temperature this transition will occur for the new InAs e-APDs. Figure 11 provides an estimate of the upper and lower limits within which the leakage current density is likely to fall, as the temperature is reduced. The upper and lower limits were calculated based on the measured room temperature leakage current density falling in line with generation and recombination and diffusion current theory respectively, using the published temperature dependence of the intrinsic carrier concentration in InAs (Rogalski, 1989; Mikhailova, 1996). It is expected that initially as the temperature is reduced the leakage current will remain diffusion dominated and follow the lower of the lines, before changing to become generation and recombination dominated and fall further with the gradient of the higher line.

As the reverse bias voltage is increased to multiply the photocurrent, the leakage current also undergoes multiplication. In practice it has been found that the leakage current usually increases at approximately the rate of M_e. There is considered to be little scope for reducing this. Hence to obtain an estimate of the leakage current in an InAs e-APD of arbitrary size, operating at an arbitrary gain, the likely low bias leakage current density can be multiplied by the APD's area and the desired gain.

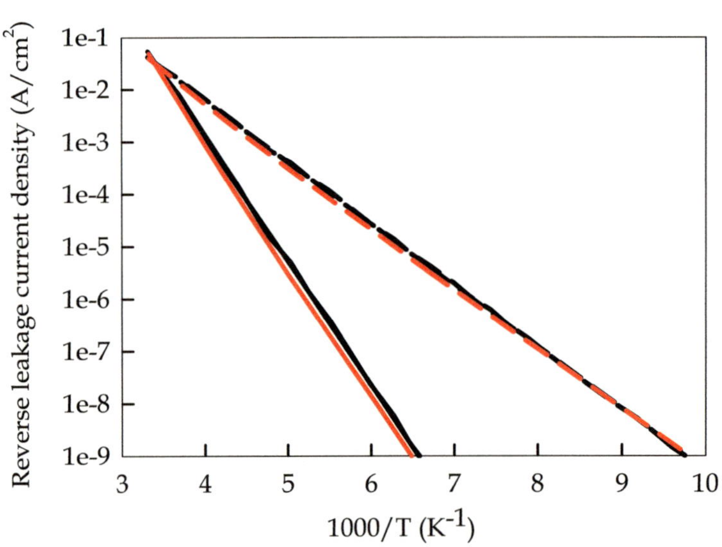

Fig. 11. Predicted boundary limits for the temperature dependant reverse leakage current density in InAs APDs under a low 0.25V reverse bias. Extrapolated from a room temperature result using published intrinsic carrier concentrations (Rogalski, 1989) (red lines) (Mikhailova, 1996) (black lines), considering diffusion limited leakage (solid lines) and G&R limited leakage (dashed lines) and excluding surface leakage.

The final leakage current related concern for APDs made from narrow bandgap materials is tunnelling current. InAs has both a low electron effective mass and a narrow bandgap, which combine to give significant band-to-band tunnelling current at much lower electric fields than in other materials used in established APDs. As identified earlier in this chapter, the deleterious effect of this is mitigated by α being significantly higher at low electric fields than in established APD materials. As a result it has been possible to fabricate many InAs e-APDs in which significant multiplication can be achieved while tunnelling current remains negligible. To illustrate this figure 12 shows the multiplication and leakage current density measured on one such e-APD, together with the band-to-band tunnelling current density expected from its structure. In practice it is considered that an intrinsic region width of > 3μm with a background doping concentration ≤ 1×10^{15} cm^{-3}, will be sufficient to avoid tunnelling current affecting most applications.

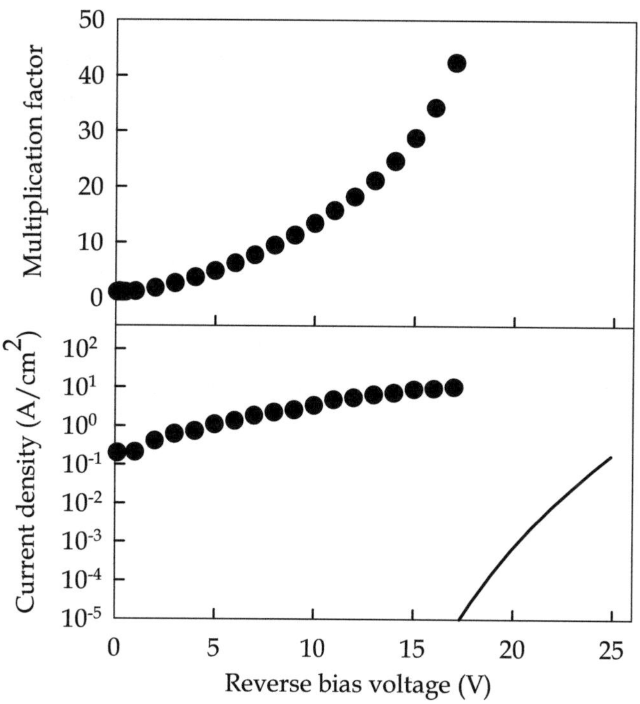

Fig. 12. The multiplication factor and reverse leakage current density measured on an InAs *p-i-n* diode with a 3.5μm intrinsic region width, together with the expected tunnelling current density for the structure, shown against reverse bias voltage.

4.2 Potential applications

InAs e-APDs will probably find application in systems where their extended spectral response, lower excess noise and increased bandwidth in the presence of gain, offer a clear advantage. Where this is not the case, the level of leakage current commensurate with the narrow bandgap of InAs and the increased cost per unit area are likely to make them unattractive, compared to existing detector options. Below are some of the applications which are likely to suit the unique characteristics of InAs e-APDs.

Imaging arrays and LIDAR

InAs e-APD arrays could be used for passive imaging across the full SWIR range, offering extended spectral sensitivity over InGaAs arrays. However since their response only reaches the bottom of the MWIR window, their advantage in such passive applications would be limited. More promising applications lie in the area of active imaging or LIDAR (light detection and ranging), where $Hg_{x-1}Cd_xTe$ e-APDs have already started to find applications (Baker *et al.*, 2004; Beck *et al.*, 2007). Currently such systems operate at a wavelength of 1.55μm due to the availability of cheap sources and InAs e-APDs can also offer high responsivity at this wavelength. Valuably InAs e-APDs could also support the use of longer SWIR wavelengths which would not be detectable with standard InGaAs detectors, affording a degree of covertness when desired. Thermoelectric cooling is likely to be required in such applications.

When APDs are used in array applications, gain uniformity across the array is an important consideration and in this respect InAs e-APDs can provide an advantage compared other APD options. Firstly, the InAs back-plane can be highly doped to reduce voltage drops across the array area. Secondly, InAs does not suffer from the compositional non-uniformity that $Hg_{x-1}Cd_xTe$ or other ternary alloys can exhibit, and hence all APDs on the array should have near identical voltage dependent gain characteristics. Furthermore should the bias voltage vary slightly across the array, the gain at individual pixels will vary much less significantly than it would for non e-APD technologies as highlighted in section 2.1.

Gas detection or monitoring

Many important gases have absorption lines in the spectral range between visible and 3.5μm wavelenghts, over which InAs e-APDs are sensitive, these include CO_2 at 2.05μm. Optimum applications for InAs e-APDs are likely to be those which require the profiling of gas concentrations across a significant distance. In such applications the ability of InAs e-APDs to greatly amplify weak signals will be advantageous, as will their ability to maintain a high bandwidths when operating at high gains, something which conventional APDs cannot do.

Optical communications

InAs e-APDs may appear to be an unlikely choice for high bit rate optical communication systems, due to their wide depletion regions. In conventional materials such wide depletion regions would result in the APD having an unacceptably low transit time limited bandwidth. However because there is no feedback within the avalanche multiplication in InAs e-APDs, their maximum impulse response duration (t_{max}) is the sum of the transit times for electrons and holes, irrespective of operational gain. In terms of the depletion width (w) and the average velocities of electrons (v_e) and holes (v_h), this is given by equation 6. Using the electron saturated drift velocity calculated by satyandah *et al.* (2002) and the hole saturated drift velocity for InGaAs, it is possible to estimate this maximum impulse response duration to be only ~60 ps, for an InAs e-APD with a 3 μm wide intrinsic region. The high speed potential of InAs e-APDs is further assisted by the low capacitance associated with the wide depletion region and the very low cladding and contact resistances which are achievable. These combine to make RC bandwidth limiting less of a concern than in established APD technologies.

$$t_{\max} = \frac{w}{v_e} + \frac{w}{v_h} \tag{6}$$

Because InAs e-APDs are likely to require some thermoelectric cooling to meet the leakage current targets for communications applications, it is unlikely that InAs e-APDs would be considered as an alternative for the established InAlAs/InGaAs APDs in high volume applications. However they may find selective application in systems where high gain and the maximum possible sensitivity are required, without a drop in the available bandwidth. Free space optical links are considered a potential application, since unimpeded by a classical gain-bandwidth product limit, InAs e-APDs could provide a greatly enhanced sensitivity dynamic range. This would allow the link to be maintained in bad weather by increasing the APD gain freely as required.

5. Conclusion

In this chapter the emerging InAs e-APD has been introduced. Experimental results have been presented, which confirm that it exhibits the fundamental characteristic of an e-APD, namely that only electrons undergo appreciable multiplication within it. The key advantage of e-APDs, their reduced excess noise, has been demonstrated and the potential benefit this affords a system has been introduced. Furthermore many of the unique characteristics of InAs e-APDs have been discussed in detail. Hence this work provides an up to date summary of the fundamental properties of InAs e-APDs. It is noted that further fundamental characterisation would be desirable, particularly assessing the temperature dependence of the multiplication and leakage characteristics. Detailed physical modelling of the impact ionisation in e-APDs would also be desirable, to improve understanding of the physical processes involved.

Beyond the fundamental characterisation results, some of the fabrication processes which have enabled the realisation of practical InAs e-APDs, have also been presented. The leakage current to be expected in such devices has been discussed. Importantly it has also been shown that high gain APDs can be designed and fabricated to operate with negligible tunnelling current, despite the narrow bandgap of InAs.

The realisation of APDs in InAs has brought the ideal avalanche multiplication and excess noise characteristics of e-APDs into the readily available III-V material system for the first time. This brings with it the potential for more wide spread application of e-APDs, previously only achievable in the less readily available $Hg_{x-1}Cd_xTe$ system. Some of the applications where InAs e-APDs may offer an advantage have been highlighted, as have the specific characteristics which make then ideally suited to such applications.

6. Acknowledgements

The work reported here was carried out in the Electronic and Electrical Engineering department at the at the University of Sheffield, UK, within the research group of Dr. Chee

Hing Tan and Prof. John David, whom the author thanks most sincerely for securing the necessary funding and helping to direct the work.
The author also thanks Dr. Mathew Steer and Dr. Andrey Krysa for their expert MBE and MOVPE growth.
This work was substantially funded by the Electro-Magnetic Remote Sensing (EMRS) Defence Technology Centre, established by the UK Ministry of Defence and run by a consortium of SELEX Sensors and Airborne Systems, Thales Defence, Roke Manor Research and Filtronic.

Patent no. GB0723858.7

7. References

Baker, I., Duncan, S., & Copley, J. (2004). A low noise laser-gated imaging system for long range target identification, *Proceedings of SPIE Vol. 5406 Infrared Technology and Applications*, pp. 133-144, Orlando, April 2004, SPIE, Bellingham

Beck, J.D., Wan, C-F., Kinch, M.A., Robinson, J.E. (2001). MWIR HgCdTe avalanche photodiodes, *Proceedings of SPIE Vol. 4454 Materials for Infrared Detectors*, pp. 188-197, Orlando, April 2001, SPIE, Bellingham

Beck, J., Wan, C., Kinch, M., Robinson, J., Mitra, P., Scritchfield, R., Ma, F. & Campbell, J. (2006). The HgCdTe electron avalanche photodiode, *J. Electronic Materials*, vol. 35, no. 6, (2006) 1166-1173

Beck, J., Woodall, M., Scritchfield, R., Ohlson, M., Wood, L., Mitra, P. Robinson, J. (2007). Gated IR imaging with 128x128 HgCdTe Electron avalanche photodiode FPA, *Proceedings of SPIE Vol. 6542 Infrared Technology and Applications*, pp. 654217, Orlando, April 2007, SPIE, Bellingham

Brennan, K.F. & Mansour, N.S. (1991). Monte Carlo calculation of electron impact ionization in bulk InAs and HgCdTe, *J. Appl. Phys.*, vol. 69, no. 11, (June 1991) 7844-7847

Bude, J. & Hess, K. (1992). Thresholds of impact ionization in semiconductors, *J. Appl. Phys.*, vol. 72, no. 8, (October 1992) 3554-3561

Capasso, F., Tsang, W.T., Hutchinson, A.L. & Williams, G. F. (1982). Enhancement of electron impact ionization in a superlattice: A new avalanche photodiode with a large ionization rate ratio, *Appl. Phys. Lett.*, vol. 40, no. 1, (January 1982) 38–40

Capasso, F. (1985). Physics of avalanche photodiodes, in *Semiconductors and Semimetals - Lightwave Communications Technology*, vol. 22-D, Ed. W. T. Tsang, 1-172, Academic Press Inc., Orlando

David, J.P.R. & Tan, C.H. (2008). Material considerations for avalanche photodiodes, *IEEE J. Sel. Top. Quantum Electron.*, vol. 14, no. 4, (July 2008) 998-1009

Emmons, R.B. (1967). Avalanche-photodiode frequency response, *J. App. Phys.*, vol. 38, no. 9, (August 1967) 3705-3714

Goh, Y.L., Marshall, A.R.J., Massey, D.J., Ng, J.S., Tan, C.H., Hopkinson, M., David, J.P.R., Jones, S.K., Button, C.C. & Pinches, S.M. (2007). Excess avalanche noise in $In_{0.52}Al_{0.48}As$, *IEEE J. Quantum Electron.*, vol. 43, no. 6, (June 2007) 503-507